D1554754

Pharmacokinetic Profiling in Drug Research

Biological, Physicochemical, and Computational Strategies

Pharmacokinetic Profiling in Drug Research

Biological, Physicochemical, and Computational Strategies

Bernard Testa, Stefanie D. Krämer,
Heidi Wunderli-Allenspach, Gerd Folkers (Eds.)

Prof. Bernard Testa
Pharmacy Department
University Hospital Centre
CHUV - BH 04
46 rue du Bugnon
CH-1011 Lausanne

Prof. Heidi Wunderli-Allenspach
Institute of Pharmaceutical Sciences
Dept. of Chemistry and Applied Biosciences
Swiss Federal Institute of Technology
ETH Hönggerberg, HCI H490
CH-8093 Zürich

Dr. Stefanie D. Krämer
Institute of Pharmaceutical Sciences
Dept. of Chemistry and Applied Biosciences
Swiss Federal Institute of Technology
ETH Hönggerberg, HCI G490.2
CH-8093 Zürich

Prof. Gerd Folkers
Institute of Pharmaceutical Sciences
Swiss Federal Institute of Technology
ETH Zentrum, STW/C17
Schmelzbergstr. 25
CH-8092 Zürich

Published jointly by
VHCA, Verlag Helvetica Chimica Acta, Zürich (Switzerland)
WILEY-VCH, Weinheim (Federal Republic of Germany)

Editorial Director: Thomas Kolitzus
Production Manager: Bernhard Rügemer

Cover Design: Jürg Riedweg

Library of Congress Card No. applied for.

A CIP catalogue record for this book is available from the British Library.

Die Deutsche Bibliothek – CIP-Cataloguing-in-Publication-Data

A catalogue record for this publication is available from Die Deutsche Bibliothek

ISBN 3-906390-35-7

Printed on acid-free paper.

Printing: Konrad Triltsch, Print und Digitale Medien, D-97199 Ochsenfurt-Hohestadt
Printed in Germany

Preface

The two Symposia organized at the University of Lausanne in March 1995 and 2000 met with considerable success, and the two resulting books [1][2] continue to receive frequent citation. In March 2004, *LogP2004 – The Third Lipophilicity Symposium* was organized at the Swiss Federal Institute of Technology (ETH) in Zurich as a logical sequel. Its theme (*Physicochemical and Biological Profiling in Drug Research*) is of the greatest current significance in drug research. A total of 26 invited lectures and 94 free communications were presented, most of the latter being also submitted for inclusion in the attached CD-ROM. The book with its 28 chapters and the CD-Rom form the Proceedings of the Symposium.

Informatics and robotics are the workhorses of a technological revolution in drug research. On them are based combinatorial chemistry which yields compounds by the many thousands, and high-throughput bioassays which screen them for bioactivity. The results are avalanches of hits which invade the databases like swarms of locusts. But far from being a plague, these innumerable hits become a blessing if properly screened for 'drug-ability', *i.e.*, for 'drug-like' properties such as good pharmacokinetic (PK) behavior. Pharmacokinetic profiling of bioactive compounds has thus become a *sine qua non* condition for cherry-picking the most-promising hits. Just as important, but less visible, are the structure–property and structure–ADME relationships which emerge from PK profiling and provide useful feedback when designing new synthetic series.

Absorption, distribution, metabolism, and excretion (ADME) are the focus of this book. Since the previous Symposium in 2000, many advances have been made both in methods and concepts, and many impressive successes have been reported. Schematically, they can be categorized as biological, physicochemical, or computational strategies. Recently, a synergistic use of these various strategies has emerged under the *in combo* roof, *i.e.*, combined and multidisciplinarity approaches whose potential goes well beyond that of the individual methodologies.

The book is structured according to these three strategies, themselves framed by two introductory and two concluding chapters. In the first introductory chapter, *David Triggle* takes a bird-eye view of scientific and societal issues in drug research. *Han van de Waterbeemd* then narrows the focus by taking a global look at property-based lead optimization.

Part II is dedicated to major biological strategies in ADME profiling. The role and significance of membranes receive an in-depth treatment in *Ole G. Mouritsen*'s chapter. The state of the art in cell cultures as absorption models is examined next by *Artursson* and *Matsson*. The two following chapters present an expert's view on the tools used in early metabolic screening (*Walther* and co-workers), and on the assessment of a candidate's capacity to induce drug-metabolizing enzymes (*Meyer et al.*). The significance of uptake and efflux transporters is gaining an ever increasing recognition in drug research and is expertly reviewed here by *Kusuhara* and *Sugiyama*. The newly recognized role of plasma protein binding in drug discovery is given convincing treatment in the next chapter by *Fessey et al. Part II* concludes with an insightful theoretical chapter on *in vivo* pharmacokinetic profiling.

Part III covers recent advances in the physicochemical strategies used to predict the ADME profile of bioactive compounds. Automated parallel synthesis is considered first, with a marked emphasis on drug-like features. The connected properties of ionization and lipophilicity continue to occupy a significant portion of the scene, as insightfully discussed by *Caron* and *Ermondi*. A novel HPLC technology to assess lipophilicity is presented by *Lombardo et al.*, while the role of lipid bilayers as permeation barriers receives an apt recognition in the next chapter. The PAMPA technology has progressed markedly since the last symposium, as explained by *Avdeef*. Interestingly, PAMPA permeation is correlated with some lipophilicity parameters, as demonstrated by *Comer* and co-workers. Predicting the intestinal solubility of poorly soluble drugs is a challenge faced head-on by *Dressman* and her co-workers in the next chapter. *Part III* then ends with two chapters covering properties, which until recently were in the province of late drug development and are now receiving earlier attention, namely chemical stability (*Kerns* and *Di*) and solid-state properties (*Giron*).

Part IV is devoted to many recent advances in the computational strategies that are having such a major impact on early predictions of physicochemical and pharmacokinetic profiles. This part opens with an overview by *Mannhold*, who offers a systematic overview of traditional and recent methods to calculate lipophilicity. In the next chapter, *Vistoli et al.* present and illustrate a concept, whose significance in structure–activity relationships remains to be better understood and recognized, namely the property space of molecules. Predicting the interactions between bioactive compounds and drug-metabolizing enzymes is a topic of the highest importance in drug discovery, and no less than four chapters are devoted to it. The computation of pharmacophores to predict biotransformation is exemplified by *Cruciani et al.*, and by *Clement* and *Güner*. *Barbosa* and her co-workers show how enzyme inhibition can be predicted based on large databases. *Judson* then presents an expert system created and developed to offer global predictions of drug metabolism. A higher order of biological complexity is covered in the chapter by *Lavé et al.*, who discuss physiologically based pharmacokinetic models. The last chapter in *Part IV* is a dense and demanding presentation of a very powerful in-house network of expert systems to process the flood of data generated during biopharmaceutical profiling.

The two concluding chapters open the reader's horizon by addressing two essential issues in PK profiling. First, *Borchardt* summarizes his views on the education of, and the communication among, scientists in drug discovery. *Cautreels et al.* then offer provocative ideas on the present and future significance of ADMET profiling in industrial drug research.

The mission of ADME profiling is to increase the clinical relevance of drug design, and to eliminate as soon as possible compounds with unfavorable physicochemical properties, pharmacokinetic profiles or toxicity. The objective of this book is to show how modern drug research achieves this mission. International authorities and practicing experts from academia and industry have been generous with their time and offer state-of-the-art presentations of concepts, methods, and technologies now in use or development in drug research.

The book would not exist were it not for help given by the members of our Scientific and Advisory Board, the technical organizers of the Symposium, the graduate students of the Pharmacy Department at the ETH, and various institutions and sponsors. Editing this volume has been a challenge and a memorable experience. We hope that our readers will find it a source of information, knowledge, and inspiration.

The Editors

REFERENCES

[1] 'Lipophilicity in Drug Action and Toxicology', Eds. V. Pliska, B. Testa, H. van de Waterbeemd, VCH Publishers, Weinheim, 1996.

[2] 'Pharmacokinetic Optimization in Drug Research: Biological, Physicochemical, and Computational Strategies', Eds. B. Testa, H. van de Waterbeemd, G. Folkers, R. Guy, Verlag Helvetica Chimica Acta, Zurich, 2001.

Contents

LogP2004 – The Third Lipophilicity Symposium

Sponsors

American Association of Pharmaceutical Scientists – AAPS, Arlington, USA

AstraZeneca, Macclesfield, Cheshire, UK

Bayer AG, Pharma Forschung, Wuppertal, Germany

BD Biosciences, Thrapston, UK

CombiSep, Ames, USA

The European Federation for Pharmaceutical Sciences EUFEPS, Stockholm, Sweden

GlaxoSmithKline, King of Prussia, USA

F. Hoffmann-La Roche AG, Basel, Switzerland

Krämer AG, Bassersdorf, Switzerland

Novartis Pharma AG, Basel, Switzerland

Pfizer Global Research and Development, Sandwich, Kent, UK

Schering AG, Berlin, Germany

Serono Pharmaceutical Research Institute, Plan-les-Ouates, Switzerland

Sirius Analytical Instruments Ltd., Forrest Row, UK

Swiss Federal Institute of Technology – ETH Zürich, Switzerland

Verlag Helvetica Chimica Acta AG, Zürich, Switzerland

Wyeth Research, Princeton, USA

Part I. Setting the Scene

Pharmaceutical Research: For What, for Whom? Science and Social Policies

by **David J. Triggle**

126 Cooke Hall, State University of New York, Buffalo, NY 14260, USA
(phone: 716 645 7315/716 983 6430; fax: 716 645 2941; e-mail: Triggle@buffalo.edu; davidtriggle@hotmail.com)

1. Introduction

'Now that the liability to, and danger of disease are to a large extent circumscribed – the effects of chemotherapeutics are directed as far as possible to fill up the gaps left in this ring.'
Paul Ehrlich, 1913

To deny that advances in health delivery and research, including therapeutic medicines during the past sixty years, have not been of significant benefit to mankind is to deny reality. Equally, few will be prepared to deny that the future will be one of at least equal promise. Children will be born with their genes profiled, 'personalized' medicines will be a reality, gene and stem cell therapies will be mature disciplines with major implications for the degenerative disorders of an aging world. This new world will be one of artificial cells and machines, many specifically created *de novo* with an expanded genetic code and that will execute unique tasks ranging from the site- and disease-specific delivery of drugs, genes, and gene repair instructions to neuronal- and DNA-based computers. These advances will have been made possible by a remarkable several generations of scientific research, culminating in the reading of multiple genomes, including the human genome.

The promise of *Ehrlich*, written ironically enough on the eve of World War 1, remains unfulfilled. Indeed, the world now faces challenges at least as large as those that existed at the turn of the 20th century. Two thirds of the world – the 'poor world' – still lives without adequate education, food, health care, sanitation and water, whilst the 'rich world' follows policies that largely ensure the continuation of this division, despite the spectacular advances in science and technology over the past one hundred years. Nowhere have these advances been as dramatic, spectacular or promising as in medicine and the pharmaceutical sciences, yet nowhere is there greater inequity of application, distribution, or benefits.

Indeed, in many important respects, the material gap between the rich and the poor worlds has increased rather than decreased. Some 11 million children die every year from starvation and other largely preventable diseases, almost 2 billion people live on less than one dollar a day, some 1.5 billion people routinely lack clean drinking water and sanitation, and malaria and other tropical diseases affect almost one billion people and account for some 5 million annual deaths. And this year, worldwide deaths from AIDS reach 3 million. The United Nations Human Development Report for 2003 notes that the 1990s, far from being a decade of progress, have actually seen remarkable reversals: 54 out of 175 countries are poorer in 2001 than in 1991; in 14 out of 175 countries, more children are dying before the age of five; in 21 out of 175 countries, more people are starving; and in 12 out of 175 countries, fewer children are being educated [1]. The gap between the rich and the poor worlds has actually increased in several areas of the world. Science has delivered for the rich world, but party and politics have blinded our eyes and have limited the participation of the poor world. Progress will not be possible until we break their cycle of poor health driving poverty: this is not likely to occur in the present Washington-driven 'free market Darwinism' model of economic development. Indeed, in the United States, where this policy is most slavishly advocated and followed, there has been a remarkable increase in income inequality between the richest and the poorest segments together with a considerable weakening of the social infrastructure of the country [2–6]. Such market-driven ideologies provide little or no incentive for the development of drugs for the diseases of the poor world, and alternative models must be adopted [7].

The challenges for the poor world in the 21st century are many. In particular, the absence of an adequate scientific and educational infrastructure confers an enormous disadvantage in an environment dominated increasingly by trade and intellectual property imbalance. The ongoing efforts to impose the existing standards of patent and copyright protection on the poor world are, in fact, likely to exacerbate the cycle of poor health and poverty. The selfish discussions over the past several years on making AIDS and other drugs available to the poor world provide adequate, and offensive, testimony to this point. Furthermore, the increasing enclosure of the scientific commons, to which aim universities that have always been the major contributor to this commons are now enthusiastic partners, will only exacerbate the problems of the poor world by diminishing their access to scientific and technological knowledge.

A recent issue of *The Economist* [8] observed, '*That the mental landscape today is almost unrecognizable from that of, say two centuries ago, is due almost entirely to the work of two groups of thinkers – scientists and economists. Add engineers to that and you have an explanation of why the physical, commercial and political landscapes have changed just as radically*'. This is true: science is mankind's greatest intellectual achievement, but its full realization will come only when it is placed fully in the service of man. We are a long way from that goal and in the absence of that achievement, particularly in the delivery of critical medicines and health services, our science will be naught for our comfort – physically or spiritually.

2. The Drug Discovery Process

'*They are ill discoverers that think there is no land, when they can see nothing but sea.*'
Francis Bacon, 1561–1626

2.1. *Overview*

The traditional process of drug discovery has been directed by target generation from observations of the biological activity of a natural product or synthetic entity on a physiological or pathological process [9]. Typically, the identification of a lead active structure was followed by iterative structural modification and biological testing. This process, essentially a 'one molecule at a time' approach relying heavily on trial and error, serendipity, scientific intuition, genius, and luck has achieved many notable therapeutic successes. Prominent examples include the development of antibiotics, β-adrenoceptor blockers, histamine H_2 receptor antagonists, ACE inhibitors, calcium blockers, and angiotensin II receptor blockers [10]. The essential mechanism of action or the structure of the underlying target not being a necessary prerequisite, the characteristics of this process are that the target is phenotypically defined and validated – blood pressure, acid secretion, smooth muscle relaxation or contraction *etc*. In contrast, the advent of genomics has led to the development of genotypically defined targets with defined structure, but frequently with undefined or only hypothetical phenotypical function.

It has been estimated that currently available drugs are directed towards some five hundred molecular targets with membrane receptors, notably G protein coupled entities, constituting almost 50% of the total. The heady promise of the genome project was that the human genome would be composed of perhaps as many as 150,000 genes generating on a 'one gene = one protein' rationale a *ca.* 300-fold increase in the number of possible drug targets. This number, together with the targets potentially realizable from bacterial and parasite genomes, was predicted to change dramatically the scale of the drug discovery enterprise. Simultaneously, the development of the new technologies of combinatorial chemistry, high-throughput screening, and informatics generated the *Viagra*-fueled 'bigger is better' model of drug development – the larger the company and the greater the throughput from chemistry and screening, the greater would be the output.

That the human genome expresses only some 30,000 genes (more than, but not dramatically so, our less complex fly, worm, and mouse relatives) means necessarily that the complexity of human organization is defined by multiple use of the same gene – splice variants, alleles in the population, post-translational modification *etc.* – and by the combinatorial diversification of regulatory and signaling pathways. From this relatively limited gene catalog, the human probably expresses in spatially and temporally limited manner as many as 100,000 proteins. A protein target may not be druggable because of its intrinsic properties or expression, but also because of a role it may play in regulatory networks other than the one of pathological interest. The elucidation of the cellular signaling network is therefore a critical component of the target validation problem [11]. Increasingly, a systems biology based approach is needed whereby an integrated approach, rather than a reductionist component analysis, is employed to understand the relationship between the overall function of a biological system and the effects of perturbations such as disease or small molecules [12]. Additionally, for a protein target to be druggable, there must be certain characteristics of the protein binding site: if one member of a gene family can bind a drug then it is assumed that other members will likely share this property. Using this approach, it has been estimated that *ca.* 10% of the proteins expressed by the human genome will fall into the druggable category. Three to four thousand targets is a far cry from the in excess of one hundred thousand targets originally claimed, although the former number may well increase as the roles of genes of previously unknown function are discovered [13].

The power of genomics to generate targets of well-defined proteins – receptors, channels, enzymes, *etc.* – to use in high-throughput screening has been extremely useful for generating 'hits' of appropriate affinity and, in a number of systems, to generate functional activity also. However, since these same systems usually, if not invariably, lack the complex signaling characteristics of 'real' cells and the integrated functional physiological properties of organ systems, they remain limited in their predictive properties and have probably served to consume very large amounts of research capital investment to the satisfaction of narrowly focused basic science without increasing the productivity of drug discovery.

Thus, critical to genomics-based target discovery is the issue of target validation – the determination of the actual role(s) of any potential gene target – the linkage between gene and phenotype. The technologies involved are several and include the analysis of gene and protein expression in normal and diseased tissues, knockout, conditional knockout, and knock-in animals, the creation of mutant (ethyl-nitrosourea-induced) mice, the use of model organisms including *Drosophila melanogaster* and *Caenorhabditis elegans*, and most recently the use of small interfering RNA. The elucidation of the mouse genome will place increased emphasis on this animal for target validation, the modeling of human diseases and drug discovery platforms. However, a too facile assumption of identity between animal and human models may be exceedingly counterproductive to therapy discovery and advancement. In any event, much human disease is almost certainly due to the influence of multiple genes, and, for those relatively few human diseases that are single gene failures, we do not need animal models with which to understand the problem [14][15].

The transition from phenotype-based to genotype-based drug discovery has brought with it the realization that biology is governed by a set of basic themes – *diversity*, *replication*, *evolution*, and *self-organization* – that are now recognized as generally applicable to disciplines from anthropology to zoology, including engineering and synthetic chemistry, and that are intimately linked through the process of biological recognition. These themes have had a major impact on chemistry, a discipline that remains fundamental to the drug-discovery process.

2.2. *Chemical Diversity*

A simplistic view of combinatorial chemistry suggests that by synthesizing all possible molecules and screening against all possible targets all possible drugs will be discovered. This view, expressed here in grossly exaggerated fashion, has yielded to a much more nuanced view of combinatorial chemistry, whereby the real issue is generating the maximum possible diversity within chemical libraries that encompass both the structural prerequisites for biological activity and for the appropriate pharmacokinetic and toxicological properties.

Nature is, of course, the ultimate combinatorial chemist. A limited repertoire of 20 amino acids and a rather larger number of protein folds has generated the several thousand catalytic, regulatory, immune, and structural proteins that constitute the existing cellular repertoire. Combinatorial chemistry in its various guises has proven to be extremely useful both in generating 'hits' and in exploiting molecular space around a 'lead' structure. In principle, outside of considerations of the amount of matter in the universe and of the database problems of tracking and compiling compounds made, there are few limitations to the number of molecules that can be made by combinatorial chemistry techniques. It has been estimated that the number of potential small drug molecules that could be made lies between 10^{62} and 10^{63} [16]. To attempt such a synthesis would be a mindless effort, and in practice, very careful consideration is required to ensure that an appropriate diversity of chemical space is explored and that this space is focused around 'drug-like' or pharmacophoric structures. The existence of such pharmacophoric or 'privileged' structures derives from the repeated presence in proteins of folds and domains that recognize generic structural skeletons. Intuitively, this has been recognized by medicinal chemists for decades as with the repeated presence of the diphenylmethyl and related hydrophobic double ring systems in many active drugs. Nature has, of course, linked combinatorial peptide and protein chemistry with biological selection to generate the most biologically fit molecules. The cone snails of the *Conus* genus with some 500 species generating as many as 50,000 toxins provide a potent example of this strategy played out in Nature: these venomous snails produce disulfide-bridged toxins of rigid three dimensional structure that exhibit both high affinity and selectivity for a variety of ion channels and neurotransmitter receptors. *Conus* appears to follow a combinatorial approach whereby the peptides are biosynthesized as larger precursors with a stable N-terminus and a hypervariable C-terminus region, the latter permitting amino acid change in discrete regions to tailor pharmacological specificity from ion channels to neurotransmitter receptors [17]. This strategy presumably permits *Conus* to match its venom production with its prey preference.

2.3. *Self-Organization*

The fundamental importance of template-guided reactions in biological systems is well known. Now, template-guided synthesis – '*click chemistry*' – is achieving significance in drug design [18]. The use of an enzyme active site to guide selectively molecular building blocks to a target structure and then permit them to link covalently has been described to yield an inhibitor of the enzyme acetylcholinesterase with femtomolar affinity – several hundred times more potent than existing inhibitors [19]. The use of a biological macromolecule as a template to both select and synthesize potent and specific ligands would appear to offer significant opportunity for the self-synthesis and targeting of new and active drug molecules.

2.4. *Evolution*

One of the major achievements of chemistry over the past decade has been the translation to the test tube of biological (Darwinian) evolution. Darwinian evolution exhibits three fundamental processes – selection, amplification, and mutation – regardless of whether it takes place in molecules or organisms. The *in vitro* evolution of DNA, RNA, and proteins to generate molecules of altered and desired properties has met with considerable success: the process is ideal for the optimization of protein therapeutic molecules where *de novo* design is difficult [20]. The process is now being applied to small molecules by the strategy termed '*Dynamic Combinatorial Chemistry*'. Dynamic combinatorial chemistry provides for the synthesis of molecules under reversible conditions – thermodynamic control *vs.* the kinetic control of conventional combinatorial chemistry – in the presence of a selection mechanism, a template for which some molecules will have enhanced affinity, thus shifting the equilibrium to favor production of this molecular species. In principle, an appropriate supply of elementary small molecule building blocks and the presence of the appropriate template will serve as a chemical factory for the production of '*lead*', '*candidate*', and '*drug*' molecules.

2.5. *Replication*

The genesis of molecular replication is a phenomenon of epochal significance and is classically embodied in nucleotide sequences [21]. Outside of DNA, an increasing number of self-replicating systems exist that permit the replication of both peptides and small molecules. Self-replication demands that a molecule be able to serve as a template to pre-organize molecular fragments for reaction – '*template-guided synthesis*'. When the molecule produced is identical to the template then an auto-catalytic cascade can be initiated.

Increasingly complex self-reproducing molecules are being described. A number of systems of small-molecule replication are known based on the self-complementarity of base-pairing mechanisms analogous to those that occur in nucleic acid replication. These systems can also show a behavior that incorporates 'evolution' and 'mutation' into the generation of enhanced replication processes. Self-replicating peptides are of particular importance because of their relationship to prebiotic conditions [22]. The description by *Lee et al.* [23] that a helically structured 32-residue peptide can autocatalyze its own synthesis provides proof of the concept that *in vitro* replicating systems are not limited to small molecules only.

3. The Shape of Things to Come

> '*Ruin is the destination toward which all men rush, each pursuing his own interest in a society that believes in the freedom of the commons*' *Garret Hardin*, 1960

Since the time of *Paul Ehrlich*, a principal goal of medicine has been the development of the '*magic bullet*' targeted only to those specific cells or pathways that are defective or are expressed only in disease states. Such a magic bullet would be without undesirable side effects since it would target only the component unique to the disease state. Although substantial selectivity of action has been obtained for a number of drugs, it is exceedingly rare that complete specificity is achieved. However, progress is being made through an increased understanding of the principles of biological recognition processes, processes that typically occur with uniquely defined specificity as revealed, for example, in the immune system and demonstrated with therapeutic antibodies [24]. Molecules such as *Gleevec*™ and *Herceptin*™ that target a tyrosine kinase overactive in chronic myelogenous leukemia and the overexpressed growth factor receptor Her2 in breast cancer, respectively [25][26], provide contemporary examples of such targeted molecular specificity.

Recent developments in '*viraceuticals*' exploit the tools of molecular biology to ensure that engineered viruses interact only with cells expressing a specific pathology. The tumor suppressor gene that encodes the protein p53, often described as a '*guardian of the genome*' is defective in over 50% of human cancers. Hence, approaches to restore its function are an attractive form of chemotherapy [27]. The E1B gene of the adenovirus encodes a protein that inactivates p53: a virus lacking this protein can replicate in and destroy p53-deficient cells present in tumors, but cannot do so in cells with functional p53 [28]. Similarly, an engineered vesicular stomatitis virus (VSV) that expresses CD4 and CXCR4 chemokine receptors, the coreceptors for HIV cell fusion and entry, will fuse and lyse only those cells – HIV-infected cells – that express the viral protein gp120 as the indicator of infection [29]. The clinical limitations to this approach are real, but they derive not from lack of specificity of action, but rather from the use of replication-competent viruses and the potential detrimental consequences of such replication.

Engineered viruses can be thought of as '*nano-factories*' capable of replicating in specific environments to produce specific therapeutic effects. There are clearly serious limitations to any consideration of their clinical use at the present time, but the concept of a nano-factory for drug synthesis and delivery remains an attractive one [30]. Already genetically engineered bacteria are employed as '*factories*' for the production of novel polyketide antibiotics and nonribosomal peptides [31], and yeast has been engineered to synthesize hydrocortisone [32]. It is not difficult to contemplate such '*bacteria*' or '*yeast*' cells being constructed *de novo* with the sole designed functions of synthesizing specific drugs and targeting diseased or infected cells and tissues.

The influence of the genome project and the paradigms of biology will be profound indeed on virtually all aspects of the human enterprise. Nowhere are they perhaps larger than in medicine and the pharmaceutical sciences. Not just in the prospects and promises of gene and stem-cell therapy, but in the application of new diagnostic procedures, new and more-selective and more-efficacious medicines, the generation of personalized medicine, and the actual elimination of diseases. The recent reporting of the genome of the malaria parasite and mosquito will ultimately be very bad news for the disease of malaria since we now have the genomes from all three participants in this most costly disease – man, the mosquito, and the parasite. From this knowledge should emerge a cure. Whether this is so will now depend critically on public policy: science counts for naught in the absence of public and political will and the integrity to use it to beneficial ends [33].

4. Balancing the Promises and the Problems

'Alas, how easily things go wrong.' *George MacDonald*, 1824–1905

The postgenomic era has brought with it the promise of both dramatically increased productivity of drug discovery, of increased creativity of exploitation of novel targets and mechanisms, and the introduction of personalized medicine that better matched disease, patient, and drug. It is difficult to believe that from our knowledge of the human, bacterial, and parasitic genomes, from the increasingly sophisticated technologies of structure-based design, combinatorial chemistry, and screening approaches plus the arrival of human-genome databases that can be mined for genetic links to individual variants of disease, that the predicted success will not ultimately arrive. However, the path will be longer and more expensive than was originally advocated: to date and for a number of reasons, this promise has not been fulfilled either quantitatively or qualitatively.

The original anticipation that the new technologies of combinatorial chemistry, high-throughput screening, and structure-based design would, together with the more than 100,000 new targets anticipated from the human-genome project, generate an arsenal of new and more-efficacious drugs has not been realized. Indeed, the recognition that the human genome perhaps codes as few as 30,000 genes means that the complexity of the human is *not* determined by numbers alone, but rather by multiple use of the same gene *and* by the signaling networks. Thus, the issue of target validation – the linkage between the gene and the phenotypic and disease states–assumes critical significance [7]. There is also no evidence that the series of mergers of companies has yielded either efficiencies of operation or enhanced creativity [34–36]. Indeed, with the introduction of new technologies and the increased search for new targets, global research and development costs have more than doubled over the past decade (it is claimed that the cost of introducing a new drug now exceeds $800 million) whilst the number of new molecular entities introduced has decreased by *ca.* 50%. Part of the problem is that each newly introduced technology, from structure-based design, to combinatorial chemistry, to high-throughput screening, and genomics, has been regarded as *the* savior of the discovery process. The reality is, of course, that all of these technologies are useful, but only to the extent of the creativity of the minds employing them. There are, in fact, several likely contributors to the increased cost and decreasing productivity currently seen. First, the time line for payoff by the new technologies is going to be significantly longer than was originally assumed; second, many of the 'easy' diseases have already been tackled with the consequence that there are many useful drugs available for such diseases. A case in point is hypertension with in excess of one hundred drugs in some ten mechanistic categories. In contrast, the neurodegenerative disorders, increasingly common in aging societies are far more difficult to study both clinically and preclinically. Third, the industry has been self-seduced with the goal of ever increasing returns on investment with the consequence that many potential areas with medical need have been neglected because they will not produce such levels of profit.

A combination of increasing costs, decreased productivity together with a decreased level of innovation in newly introduced drugs are all indications of an industry in major trouble. However, at 18%, the profits of the pharmaceutical industry remain amongst the highest, if not the highest, of any industry, whilst the costs of drugs to the public has escalated in the United States, the largest single market, at rates significantly in excess of the general rate of inflation. These issues have been subject to extensive discussion [7] [9] [37–46]. At the same time, the pharmaceutical industry has fallen from grace in the public eye and it is too frequently regarded by large segments of the public, because of its ill-conceived efforts to preserve its intellectual property rights at virtually any cost, to be just another greedy multinational concern. This was particularly clear to the world in 2001 when thirty nine multinational pharmaceutical companies sued the government of South Africa headed by President *Nelson Mandela* over the protection of their intellectual property rights for AIDS drugs [47].

Finally, the statements from *Alan Roses* of *GlaxoSmithKline* that '*The vast majority of drugs – more than 90% – only work in 30 or 50% of the people*', '*I wouldn't say most drugs don't work. I would say that most drugs work in 30 to 50% of people. Drugs out there on the market work, but they don't work on everybody*' [48]. This is an unsurprising clinical statement, but one that is significantly at odds with the general marketing messages of major pharmaceutical companies.

In the face of decreasing productivity and the desire to maintain the very high profit levels to which this business-oriented industry is now accustomed, the major pharmaceutical companies have adopted an approach that emphasizes the following general strategies:

1) Focus on so-called 'blockbuster' drugs – drugs that have market sales in excess of $1 billion per year.
2) Focus on chronically *vs.* acutely used drugs – HMGCoA inhibitors *vs.* antibiotics.
3) Focus on the United States market.
4) De-emphasize drugs that are primarily applicable in the 'poor world'.
5) Emphasize marketing – including direct-to-consumer advertising on a mass scale – to spur consumer-based demand with the message that a 'pill-a-day' is the route to the pursuit of happiness.
6) Emphasize drug use over acceptance of life changes.
7) Increase efforts to maintain and extent patent life.
8) Emphasize 'life-style' diseases – hair loss, erectile dysfunction, *etc.*
9) Enlarge the role of drugs in existing disorders and/or exaggerate the seriousness of existing disorders – attention deficit disorder, irritable bowel syndrome, mild depression, *etc.*
10) Invent new diseases – 'social phobia', 'female sexual dysfunction', *etc.*
11) Work closely with Congress to ensure favorable legislation is passed.
12) Ensure that intellectual property rights are maintained worldwide.

To be sure, the industry has received some bad publicity, most notably from its wrong-headed approach to the availability of drugs in the poor world. However, overall, this has been an extremely successful approach for the major pharmaceutical industry. The US market now constitutes *ca.* 50% of the world market and increasingly major pharmaceutical companies have shifted more of their research development and marketing to the United States where the drug prices are the highest in the world. To maintain this position and high profitability, the industry employs more than 650 Washington lobbyists and is a major and enthusiastic contributor of 'campaign funds' to largely Republican members: total expenditures on political activities since 1997 are in excess of $600 million. Recent successes of the industry in the USA include a provision in the new Medicare prescription bill that prohibits the Federal government from negotiating lower drug prices and the prevention of imports of cheaper drugs from other countries, notably Canada [49][50].

Nothing in the present political climate suggests that significant change in this approach is likely to occur or to be resisted by Congress. Indeed, it is probable that there will be increased pressure from the US government to ensure that drug prices in countries that are currently regulated as part of their comprehensive health care systems be allowed to rise to unregulated levels as part of 'free trade' agreements [51].

5. For Whom, for What?

'Am I my brother's keeper?' Genesis 4:9

The delivery of and access to health care and medicines is deficient and defective in both the rich and the poor worlds. For all of its scientific promise, the current model of pharmaceutical development is flawed, probably fatally, and needs major surgery. There are two principal issues. First, recognition that ill-health and disease is a driving force for the economic and physical inequality that characterizes at least 50% of the world population. Second, recognition that the benefits of science, including medicines development, must be more equitably shared and that the current Western trend towards the privatization of science makes this goal progressively less attainable.

5.1. *Health and Inequalities*

The former President of the United States, Jimmy Carter, observed on receiving the *Nobel* Peace Prize in November 2002, 'I was asked to discuss, here in Oslo, the greatest challenge that the world faces. I decided that the most serious and universal problem is the growing chasm between the richest and the poorest people on earth'. This chasm is, of course, not new but it is much to the shame of the rich world that the promises of economic coprosperity implicit in the globalization imperative have been too often hollow indeed [2][52–54]. The most-recent data from the *Food and Agricultural Organization of the United Nations* reveal that, even in a time of worldwide food availability, the number of undernourished actually increased from 1995–1997 to 1999–2001: '*Bluntly stated, the problem is not so much a lack of food as a lack of political will*' [55]. A significant contributor to this food insecurity is the increase in the AIDS population in the developing world [56]: in 2003 an estimated 40 million people are afflicted with the virus, some 5 million contracted the virus and a record 3 million died form AIDS. In the absence of far greater public health, scientific and financial resources, it is estimated that there will be 100 million cases of AIDS worldwide by the end of this decade. Efforts by the United States to ban the use of the phrase 'reproductive health' in reports from the *United Nations Population Fund* and to advocate 'abstinence only' policies in the developing world should, together with the most-recent statements from the Vatican arguing against condom use in AIDS-inflicted countries, be treated with the contempt that such primitive philosophies deserve [57–59].

In fact, the relationship between health, poverty, and economic development is well recognized with the following consequences [60]:

1) Poor health reduces healthy life expectancy and educational achievements.
2) Poor health reduces investments and returns on investment.
3) Poor health reduces parental investment in children.
4) Poor health reduces social and political stability.

Once established, the cycle of poor health–poverty–economic deprivation becomes difficult to break, in significant part because efforts to break the cycle have focused principally on providing economic aid (and this too frequently in 'tied' form), rather than the creation and provision of health services that could break the cycle at its inception.

The relationship between health and poverty is, of course, long established, it being generally well recognized that increased wealth brings, within limits, greater health. This relationship extends both between countries and regions of the world, but also between regions and populations of a given country. Two principles appear to operate. First, the *absolute level of wealth*, and second, the relative distribution of wealth. As expected, with increased societal wealth, expressed as gross domestic product *per capita* (GDPc), life expectancy increases: this relationship plateaus above GDPc levels of $10,000–20,000. This relationship is scarcely surprising since wealth generates at both societal and personal levels the infrastructure of education, sanitation, and public health, and transportation that form critical components of a contemporary society. Nonetheless, GDPc is a blunt instrument in assessing national wealth as *Paul Krugman* has observed in a trenchant comparison of health outcomes, poverty, and living standards between Sweden and the United States [61]. Thus, the second component of the relationship between wealth and health is the *distribution* of wealth within and between societies: societies that have significant relative differences in wealth have lower life expectancies than societies with a more-egalitarian wealth distribution. This appears to hold regardless of absolute wealth levels. The relationship between income inequality and health has been described by a number of workers [62–67] and likely has a number of origins. These include the underinvestment by society in health and physical infrastructure in discrete regions, and the fragmentation of psychosocial relationships with increasingly hierarchical social structure.

These observations are of considerable significance to considerations of the future of health care in both the rich and the poor worlds. Clearly, there is the need for a necessary investment to provide the necessary health infrastructure including medicines, hospitals, and public health. However, there is also a significant 'fine tuning' effect on health outcomes that appears to originate from relative wealth distribution within a society. How to ensure both the absolute increases in wealth necessary for health care in the poor world without increasing further the distribution gap is a major issue in an era of galloping globalization and the attendant maldistribution of intellectual property rights.

An additional contributing factor to the infrastructure gap between the rich and the poor worlds is the relative production and availability of scientists, engineers, and health personnel. The United States, Europe, and Japan have some 70 engineers and scientists per 100,000 population: sub-Saharan Africa has less than one. There are, for example, more African engineers and scientists working in the United States than in the entire continent of Africa [68]. Of even greater immediate significance is the equally large disparity in the distribution of health personnel across the rich and poor worlds. The United States has some 300 physicians per 100,000 population, and in excess of 900 nurses; in contrast, Botswana has fewer than 20 physicians and 200 nurses per 100,000 population and Chad is even worse off [69]. These numbers represent significant global problems that, when coupled with the increasing privatization of knowledge, will serve to exacerbate the already dangerous levels of worldwide inequality. Compounding this problem is the ongoing and increasing ability of the rich world to attract scientists, engineers, and health personnel from the poor world. Thus, *ca.* 50% of the graduate students in the United States and postdoctoral fellows are of nondomestic origin, principally from the so-called 'developing countries' [70][71]. Similarly, the aggressive recruitment of health personnel, notably nurses and physicians from the poor world, represents a brain drain that impacts immediately the already fragile health-care infrastructure of the poor world [72–74]. Absent a significant return by these individuals to their country of origin, the net result is a continued impoverishment of the country of origin, and, not coincidentally, a significant benefit to the recipient country which, thus, spared the cost of developing its own adequate scientific and health personnel.

These issues of inequality are not, of course new, and were noted admirably by *Anatole France* (1844–1924) in his famous irony, '*The law in its majestic equality, forbids the rich as well as the poor to sleep under bridges.*'

6. Science and the Social Order

> '*People of the same trade seldom meet together, even for merriment and diversion, but the conversation ends in a conspiracy against the public*'
>
> 'The Wealth of Nations', *Adam Smith*, 1776

Over half a century ago, *Robert Merton* defined an ethos of science based on the following values: the free and open exchange of knowledge, an unrestricted pursuit of this knowledge independent of self-interest, and an acceptance that science is a product of nature and not of politics, religion, or culture in general [75][76]. Accordingly, *Merton* defined four sets of norms that define science: 'universalism', 'communalism', 'disinterestedness', and 'organized skepticism' [76]. These norms contribute to what may be defined as an 'intellectual commons' of science – a freely generated and freely available pool of 'certified knowledge' [77]. It has been a particular role of universities and similar institutions to contribute to and maintain this commons. However, *Merton* also recognized that science as an organized social activity interacts with society itself: science and society are thus interdependent entities and the conduct of science is influenced by the imperatives of society.

These norms are, in fact, subject to continual societal challenge. The imposition of 'Aryan' and 'Marxist' dogma are major examples from the 20th century, and the United States remains prominent for efforts to impose fundamentalist religious values on the teaching of biology. Additionally, the norms are challenged from within as the traditional role of the universities is altered through the impact of commercial funding sources and priorities on communalism and disinterestedness [76][78].

6.1. *Intellectual Property and the Commons of Science*

Two principal actions in 1980 served to define new boundaries of patentable knowledge and have dramatically impacted both the pharmaceutical sciences and the intellectual commons and are simultaneously serving to redefine the nature of the university. First, the *United States Supreme Court* decision in *Diamond vs. Chakrabarty* has enabled the patenting of living things from bacteria to DNA sequences and genes and transgenic animals: in principle, the ruling permits the patenting of human clones, and, in the United States, legislative action will be necessary to ban this. Second, the passage of the *Bayh–Dole* act permitted universities (and their faculty members) to obtain title to inventions from research supported by Federal funds. This latter action, together with a variety of legislative actions designed to foster university–industrial cooperation, initiated a significant increase in industry-sponsored university research, particularly in the biological and biomedical (life) sciences [7][76][78][79]. In the year 2002, American universities collected *ca.* $1 billion in royalty income, filed 6500 patents, executed some 3700 licensing deals and created over 400 companies [80–83]. Universities now own in excess of 5% of U.S. patents, and the majority of this activity has occurred in the life sciences area [84]. The issues of intellectual property policy on the openness of science have been discussed in detail from a general perspective by the *Royal Society of London* [85] and by the *Nuffield Council on Bioethics* from the perspective of patents on genes [86]. A recent report from the *Federal Trade Commission of the United States* appears to recognize that the patent process has become too easy and that, in fact, too many patents are issued: once issued, a patent is extremely expensive to refute [87].

There are at least three important consequences to this transformation of university research activity in the life sciences:

1) Knowledge that would have entered the science commons is now being patented and available only through licensing mechanisms – exclusive or nonexclusive. We run the very real danger of creating a scientific 'anti-commons' [88] whereby information is not freely available or available only at an unaffordable price. Furthermore, in the life sciences arena, patenting activity has moved increasingly 'upstream' from the chemical composition of the actual drug molecule to the gene sequence of the drug target. Thus, patents have become increasingly enclosing of the scientific commons and can actually restrict rather than advance progress in a field – the very antithesis of the purpose of a patent. This has a dual impact: it prevents new therapies from being developed because they will infringe the gene patent and it prevents the use of genes and gene products as research and diagnostic tools [89] [90].

2) The poor world is doubly impacted by this. First, the knowledge is locked away and may require unaffordable access. Second, the tools and technologies necessary to develop such knowledge and obtain the patents are less-available (or unavailable) in the poor world, which thus falls further behind in the knowledge economy. This issue was well recognized by the *Commission on Intellectual Property Rights* [91], which advocated that the developing world should accept an international intellectual property system, but one that is crafted and nuanced to the needs of the developing countries and that is modifiable with increasing economic development.

3) Universities may be hoist with their own petard. It has been generally assumed that there is a legitimate 'research and educational' exclusion for the use of patented material. Recent court cases in the United States indicate that this is not so, a reasoning based at least partially on the grounds that universities are avid patent seekers and engaged in commercial activities [92–94]. If this ruling is upheld, university research will become progressively encumbered by the necessity to navigate financially and intellectually around patents that cover components of ongoing research.

Finally, for universities that are progressively embracing the industrial model of research, the traditional role of 'free and unencumbered inquiry' that has generally been thought to define academic work will be lost if faculty and students are unable to have such intercourse when the academic commons have been enclosed [95]. Furthermore, organizations that choose to impose on themselves restrictions on freedom and openness of inquiry should not be surprised if outside influences, including legislatures, act similarly: universities may, in fact, be 'hoist with their own petard'. We have forgotten the words of *Thomas Jefferson* who wrote in 1813, '*The exclusive right to invention is given not of natural right, but for the benefit of society.*'

6.2. *Ethical Issues*

Although the commercial support of university research is not new, the past decade has seen increasing concern over conflict-of-interest issues, bias in research, and loss of public confidence of the public-interest role of the university. These issues have been well reviewed by *Krimsky* [76]. Typical examples include industry sponsors refusing publication of unfavorable research results, research publications that report more-favorably on particular drugs or drug classes when support is provided by industry, conflicts when scientists and clinical investigators have financial ties to the companies whose drugs or protocols they are investigating, and data withholding by investigators to avoid information access (*inter alia* [96–104]). Most recently, a series of conflicts have been reported for senior clinical scientists from the *National Institutes of Health* some of whom consulted for the very firms over whose drugs they exerted regulatory approval [105].

Collectively, these ethical problems represent a major challenge to the integrity of the core values of the university and to the biomedical sciences in particular and they further diminish the role of the university as a venue of public intellectualism and public-interest science [76].

7. Conclusions

'Are you unaware that vast numbers of your fellow men suffer or perish from need of things that you have in excess?' *Jean-Jacques Rousseau*, 1712–1778

There are several obvious conclusions that may be drawn from this brief survey of current pharmaceutical research. First, the current business model of 'big' pharmaceutical research is broken. It is too expensive and not productive enough. Second, despite the promise of genomics, it is probably true that many of the 'easy' targets – hypertension, hyperacidity *etc.* – have largely been satisfied. No one doubts that genomics will generate new targets and that it will also generate 'personalized' medicine to make more-effective use of existing medicines and to facilitate the development of new ones. However, these approaches will take time, and a longer time than we had so confidently predicted a decade ago. Third, society needs to see medicines as only one component of total health care. Many of the disorders for which we use or seek drugs are essentially completely or largely man-made and could be better and more cheaply approached by public health and environmental approaches. Prominent examples here include lung cancer, obesity, and an increasing number of behavioral disorders. Finally, the pharmaceutical industry is an easy target to criticize – it is profitable, arrogant, and its products are indispensable [106]. Nonetheless, many of us, or our family members, may owe our lives to a particular drug: this should not be forgotten. The tragedy is that more than 50% of the world population does not have that choice: that is unacceptable.

For the latter population new methods must be found to deliver medicines and health care, since it is quite evident that, in its present form, the market-driven pharmaceutical industry intent on preserving its intellectual property rights has neither the motive nor the intention to do so. Charity is not the answer, since that does not guarantee sustainable relief and it breeds both resentment in the recipient and hubris in the donor. Ideally perhaps, individual countries would have the scientific ability and the infrastructure to generate their own medicines by cost-effective processes. This is not now possible save for a few countries in the poor world, including Brazil, China, and India. Clearly, the sub-Saharan African countries that are being devastated by AIDS fall completely outside of this possibility. Although the WTO has now agreed on the principle of compulsory licensing to permit import of needed drugs into countries that lack manufacturing infrastructure from countries that are making them under compulsory licensing conditions. Given the reluctance of the rich world to give up these intellectual property rights, it will be of interest to see how well this process will actually work in practice. An alternative for existing drugs, notably AIDS drugs, would be for an organization such as the *World Health Organization* to 'buy out' the patent holders thus providing the patent holder with a 'market return' and simplifying the issue of drug availability in the poor world. For new drugs – against tropical and parasitic diseases – several possibilities exist accepting that these diseases will not be a major priority of the existing pharmaceutical industry. Such possibilities include efforts by nongovernmental organizations such as the '*Drugs for Neglected Diseases*' sponsored by *Medecins sans Frontieres*, the *Global Alliance for TB Development*, the *Medicines for Malaria Venture*, and the *Institute for One World Health* [107–109]. These will not be simple ventures to run, and in the meantime millions of significantly avoidable deaths will occur. The rich world can take little comfort from this.

> '*It is easier for a camel to go through the eye of a needle, than for a rich man to enter the kingdom of God.*'
> Matthew 19:24

REFERENCES

[1] United Nations Development Program, Human Development Report 2003; www.undp.org.
[2] J. K. Galbraith, *Daedulus* **2002**, Winter, 11.
[3] P. Krugman, *New York Times Magazine* **2002**, October 20th.
[4] N. Klein, 'No Logo', St. Martins Press, New York, 2001.
[5] C. Jencks, *Daedulus* **2002**, Winter, 49.
[6] T. Skocpol, 'The Missing Middle', Norton, New York, 2000.
[7] D. J. Triggle, *Drug Dev. Res.* **2003**, *59*, 269.

[8] 'With all thy getting, get understanding', *The Economist* **2002**, January 5th; www.economist.com.
[9] J. Drews, *Science* **2000**, *287*, 1960.
[10] D. J. Triggle, in 'Encyclopedia of Molecular Cell Biology and Molecular Medicine', Ed. R. A. Meyers, Wiley-VCH, New York and Weinheim, 1996.
[11] S. Bunk, *The Scientist* **2003**, February 14th, 24.
[12] E. J. Davidov, J. M. Hollan, E. W. Marple, S. Naylor, *Drug Disc. Today* **2003**, *8*, 175.
[13] A. L. Hopkins, C. R. Groom, *Nature Rev. Drug Disc.* **2002**, *1*, 727.
[14] R. A. Coleman, *Drug Disc. Today* **2003**, *8*, 233.
[15] D. F. Horrobin, *Nature Rev. Drug Disc.* **2003**, *2*, 151.
[16] R. S. Bohacek, C. McMartin, W. C. Guida, *Med. Res. Rev.* **1996**, *16*, 3.
[17] D. J. Adams, P. E. Alewood, D. J. Clark, R. D. Drinkwater, R. J. Lewis, *Drug Dev. Res.* **1999**, *46*, 219.
[18] H. C. Kolb, M. G. Finn, K. B. Sharpless, *Angew. Chem., Int. Ed.* **2001**, *40*, 2005.
[19] W. G. Lewis, L. G. Green, F. Grynszpan, Z. Radic, P. R. Caeliec, P. Taylor, M. G. Finn, K. B. Sharpless, *Angew. Chem., Int. Ed.* **2002**, *41*, 1053.
[20] A. P. Vasserot, C. D. Dickinson, Y. Tang, W. D. Huse, K. S. Manchester, J. D. Watkins, *Drug Disc. Today* **2003**, *8*, 118.
[21] J. D. Watson, F. H. C. Crick, *Nature* **1953**, *171*, 737.
[22] R. Issac, Y.-H. Ham, J. Chmielewski, *Curr. Opin. Struct. Biol.* **2001**, *11*, 458.
[23] D. H. Lee, J. R. Granjc, J. A. Martinez, K. Severin, M. R. Ghadiri, *Nature* **1996**, *382*, 525.
[24] L. Takacs, M.-D. Vazquez-Abad, E. A. Elliott, *Ann. Rep. Med. Chem.* **2001**, *36*, 237.
[25] R. Capdevilla, E. Buchdinger, J. Zimmerman, A. Matter, *Nature Rev. Drug Disc.* **2002**, *1*, 493.
[26] D. S. Leonard, A. D. K. Hill, L. Kelly, B. Dijastra, E. McDermott, N. J. O'Higgins, *Br. J. Surg.* **2002**, *89*, 262.
[27] B. Vogelstein, K. W. Kinzler, *Nature* **2001**, *412*, 865.
[28] J. R. Bischoff, D. H. Kirn, A. Williams, C. Heise, S. Horn, M. Mina, *Science* **1996**, *274*, 373.
[29] M. J. Schnell, J. E. Johnson, L. Buonocore, J. K. Rose, *Cell* **1997**, *90*, 849.
[30] D. J. Triggle, *Ann. Pharmacother.* **1999**, *33*, 241.
[31] M. D. Burkhart, *Org. Biomol. Chem.* **2003**, *1*, 1.
[32] F. M. Szezabara, C. Chantelier, C. Villeret, A. Masurel, *Nature Biotech.* **2003**, *21*, 143.
[33] D. F. Wirth, *Ann. Rep. Med. Chem.* **1999**, *34*, 349.
[34] R. Mullin, *Chem. Eng. News* **2003**, January 27th, 27; http://pubs.ac.org/cen/.
[35] 'Drug-induced seizures. Drug mergers are all the rage, but few make much sense', *The Economist* **1999**, November 11th; www.economist.com.
[36] 'Big trouble for big pharma', *The Economist* **2003**, December 4th; www.economist.com.
[37] J. Drews, *Drug Disc. Today* **2003**, *8*, 411.
[38] J. A. DiMasi, R. W. Hansen, H. G. Grabowski, *J. Health Econ.* **2003**, *22*, 151.
[39] D. Horrobin, *J. R. Soc. Med.* **2000**, *93*, 341.
[40] D. Henry, J. Leschin, *Lancet* **2002**, *360*, 1590.
[41] R. Moynihan, I. Heath, D. Henry, *Br. Med. J.* **2002**, *324*, 886.
[42] 'Factors Affecting the Growth of Prescription Drug Expenditures', NIHCM Foundation, Washington, DC, 1999, July; www.nihcm.org.
[43] 'Changing Patterns of Pharmaceutical Innovation', NIHCM Foundation, Washington, DC, 2002, May; www.nihcm.org.
[44] '2002 Drug Industry Profits. Corporate Watch 2003, June', Public Citizen; www.citizen.org.
[45] A. Saer, D. Socolar, 'A Prescription Drug Peace Treaty', 2000, October 5th; www.nysenior.org.
[46] K. J. Watkins, *Chem. Eng. News* **2002**, June 28th, 27; http://pubs.ac.org/cen/.
[47] 'Drug-Induced Dilemma', *The Economist* **2001**, April 19th; www.economist.com.

[48] S. Connor, 'Glaxo Chief: our drugs do not work in most patients', *The Independent* (UK) **2003**, December 8th; http://news.independent.co.uk.
[49] B. Simon, 'Curtailing Medicines from Canada. Drug Makers Strike Back with Price Increases', *New York Times* **2003**, November 11th; www.nytimes.com.
[50] B. Simon, 'Canada resists U.S. pressure on drug sales', *New York Times* **2003**, November 19th; www.nytimes.com.
[51] E. Becker, 'Drug Industry Seeks to Sway Prices Overseas', *New York Times* **2003**, November 26th; www.nytimes.com.
[52] J. Stiglitz, *The American Prospect* **2002**, *13*, June 1–14th; www.prospect.org.
[53] J. Stiglitz, 'Globalization and Its Discontents', Norton, New York, 2002.
[54] C. E. Weller, A. Hersh, *The American Prospect* **2002**, *13*, June 1–14th; www.prospect.org
[55] 'The State of Food Insecurity in the World', United Nations Food and Agricultural Organization, 2003; www.fao.org.
[56] UNAIDS, 'AIDS Epidemic Update 2003'; www.unaids.org.
[57] 'Vatican in HIV Condom Row', BBC News, published on line 10/09/2003; http://newsvote.bbc.co.uk.
[58] N. Kristoff, 'Theology, Condoms and Deadly AIDS', *New York Times* **2003**, November 27th; www.nytimes.com.
[59] L. McClure, 'Bush's Drive for Global Abstinence', www.salon.com/news/feature/2002/12/19/populations/print/html.
[60] R. Feachem, 'Forum 5. Global Forum for Health Research 2001', Geneva, Switzerland.
[61] P. Krugman, 'For Richer', *New York Times Magazine* **2002**, October 20th; www.nytimes.com.
[62] C. Hertzman, *Am. Scient.* **2002**, *89*, 538.
[63] R. G. Wilkinson, *Br. Med. J.* **1992**, *304*, 165.
[64] R. G. Wilkinson, 'Mind the Gap', Yale University Press, New Haven, 2000.
[65] B. P. Kennedy, I. Kawachi, D. Prothrow-Stith, *Br. Med. J.* **1996**, *312*, 1004.
[66] G. A. Kaplan, E. R. Pamuk, J. W. Lynch, R. D. Cohen, J. L. Balfour, *Br. Med. J.* **1996**, *312*, 999.
[67] G. A. Kaplan, in 'Income, Socioeconomic Status and Health', Eds J. A. Auerbach, B. K. Krimgold, Academy for Health Services Research and Health Policy, Washington, DC, 2001, p. 137–148.
[68] N. E. Kimwell, 'Brain Drain in Sub-Saharan Africa (Kenya as an Example)', 2003; www.africabraingain.org.
[69] WHO Estimates of Health Personnel, 2003; www3.who.int/whosis/health_personnel/health_personnel.cfm.
[70] K. W. Miller, D. J. Triggle, *Am. J. Pharm. Ed.* **2002**, *60*, 287.
[71] W. Zumeta, J. S. Raveling, 'The Best and the Brightest for Science. Is There a Problem Here?', Commission on Professions in Science and Technology, Washington, DC, 2002; www.epst.org.
[72] A. S. Muula, J. M. Mfutso-Bango, I. Makoza, E. Chapitwa, *Nursing Ethics* **2003**, *10*, 433.
[73] A. Padarath, C. Chamberlain, D. McCoy, A. Ntuli, M. Rowson, R. Loewenson, 'Health Personnel in Southern Africa. Confronting Maldistribution of Brain Drain', Regional Network for Equity in Health in Southern Africa (Equinet), Health Systems Institute (South Africa) and MEDACT (UK); www.equinet.africa.org.
[74] V. Patel, *Br. Med. J.* **2003**, *327*, 926.
[75] R. K. Merton, *Phil. Sci.* **1938**, *5*, 321.
[76] S. Krimsky, 'Science in the Private Interest', Rowman and Littlefield, Lanhan, MD, 2003.
[77] R. K. Merton, *Am. Sociol. Rev.* **1957**, *22*, 635.
[78] D. Bok, 'Universities in the Market Place', Princeton University Press, Princeton NJ, 2003.

[79] A. K. Rai, R. S. Eisenberg, *Am. Scient.* **2003**, January–February, 52.
[80] G. Blumenstyk, *Chronicle of Higher Education* **2003**, December 9th.
[81] H. W. Bremer, 'The First Two Decades of the Bayh-Dole Act as Public Policy', www.nasulgc.org/COTT/Bayh-Dole/Bremer_speech.htm.
[82] K. Singer, *Matrix* **2001**, February, 32.
[83] T. Staedter, *Technol. Rev.* **2003**, December, 24.
[84] T. Agres, *The Scientist* **2002**, October 14th; www.the-scientist.com.
[85] 'Keeping Science Open: the Effects of Intellectual Property Policy on the Conduct of Science', Royal Society of London, 2003; www.pubs.royalsoc.ac.uk.
[86] 'The Ethics of Patenting DNA. A Discussion Paper', Nuffield Council on Bioethics, 2002; www.nufieldbioethics.org.
[87] 'To Promote Innovation: the Proper Balance of Competition and Patent Law and Policy', Federal Trade Commission, Washington, DC, 2003, October 20th; www.ftc.gov; see also D. Wessel, 'The Link Between Patents and Prosperity', *Wall Street Journal* **2003**, December 17th; www.wsj.com.
[88] M. A. Heller, R. S. Eisenberg, *Science* **1998**, *280*, 698.
[89] J. F. Merz, A. G. Kriss, D. G. B. Leonard, M. K. Cho, *Nature* **2002**, *415*, 577.
[90] B. Williams-Jones, *Health Law J.* **2002**, *10*, 121.
[91] The Commission on Intellectual Property Rights (CIPR), London, 2002, September; www.iprcommission.org.
[92] P. Brickley, *The Scientist* **2003**, March 10th, 42; www.the-scientist.com.
[93] S. R. Morrisey, *Chem. Eng. News* **2003**, May 26th, 21; http://pubs.acs.org/cen/.
[94] R. S. Eisenberg, *Science* **2003**, *299*, 1018.
[95] D. Bollier, 'The Enclosure of the Academic Commons', *Academe* **2002**; www.aaup.org/publications/academe.
[96] M. Angell, *N. Engl. J. Med.* **2000**, *342*, 1516.
[97] T. Bodenheimer, *N. Engl. J. Med.* **2000**, *342*, 1539.
[98] E. G. Campbell, B. R. Clarridge, M. Gokhale, L. Birnbaum, N. A. Holtzman, D. Blumenthal, *J. Am. Med. Assoc.* **2002**, *287*, 473.
[99] E. Ernst, P. H. Canter, *Trends Pharmacol. Sci.* **2003**, *24*, 219.
[100] J. Lexchia, L. A. Bero, B. Djulbegovic, O. Clark, *Br. Med. J.* **2002**, *326*, 1167.
[101] H. Melander, J. Ahlquist-Rastad, G. Meijer, B. Beerman, *Br. Med. J.* **2002**, *326*, 1171.
[102] D. C. Nathan, D. J. Weatherall, *N. Engl. J. Med.* **2002**, *347*, 1368.
[103] F. van Kolfschooten, *Nature* **2002**, *416*, 360.
[104] A. Wazana, *J. Am. Med. Assoc.* **2000**, *283*, 373.
[105] D. Willman, 'Stealth Merger: Drug Companies and Government Medical Research', *Los Angeles Times* **2003**, December 7th; www.Latimes.com.
[106] 'Follow the Money. The Pharmaceutical Industry: the Other Drug Cartel', Minnesota Attorney General, 2003, September 30th; www.ag.state.mn.us.
[107] 'Exotic Pursuits. Drug Firms Are Starting to Tackle Tropical Diseases', *The Economist* **2003**, January 30th; www.economist.com.
[108] S. Nwaka, R. G. Ridley, *Nature Rev. Drug Disc.* **2003**, *2*, 919.
[109] A. M. Rouhi, *Chem. Eng. News* **2002**, September 23rd, 67; http://pubs.acs.org/cen/.

Property-Based Lead Optimization

by **Han van de Waterbeemd**

AstraZeneca, DMPK, C-Lab Global Project Leader, Building/room 50S39, Mereside, Alderley Park, Macclesfield, Cheshire SK10 4TG, UK
(phone: +44-1625-518 472 (int. 28472) (UK), +46-31-776 1642 (int. 61642) (Sweden); e-mail: han.vandewaterbeemd@astrazeneca.com)

Abbreviations
ABC: ATP binding cassette; ADME: absorption, distribution, metabolism, excretion; ADME: 'automated decision-making engine'; AP_{SUV}: absorption potential parameter; BBB: blood–brain barrier; BBMEC: bovine brain microvessel endothelial cells; BCS: biopharmaceutics classification system; BMC: biopartitioning micellar chromatography; Caco-2: adenocarcinoma cell line derived from human colon; CNS: central nervous system; CSF: cerebrosal spinal fluid; CYP: cytochrome P450; DDI: drug–drug interactions; DMSO: dimethylsulfoxide; HLM: human liver microsomes; HTS: high-throughput screening; IAM: immobilized artificial membrane; ILC: immobilized liposome chromatography; MAD: maximum absorbable dose; MDCK: *Madin–Darby* canine kidney; MEKC: micellar electrokinetic chromatography; MLR: multiple linear regression; NMR: nuclear magnetic resonance; PAMPA: parallel artificial membrane permeation assay; PASS: prediction of activity spectra for substances; PBPK: physiologically based pharmacokinetic modeling; P-gp: P-glycoprotein; PK: pharmacokinetics; *PSA*: polar surface area; QSAR: quantitative structure–activity relationships; QSPR: quantitative structure–property relationships; R&D: Research and Development; Ro5: rule of five; *SITT*: small-intestinal transit time; *SIWV*: small-intestinal water volume; SPR: surface plasmon resonance; VS: virtual screening; WDI: World Drug Index.

1. Introduction

The design of new chemical entities intended as drugs is a challenging endeavor. Despite enormous investments in pharmaceutical R&D, the number of approved new drugs has fallen in recent years. Several reasons contribute to this state of affairs, *e.g.*, regulatory hurdles which have become higher and the number of potential targets which now appears lower as believed before the full human genome was unraveled. The attrition of compounds under development is dramatically high. Lack of efficacy, toxicology, and ADME problems are often cited as the responsible factors. Of course, such factors often contribute simultaneously to the misfortune of a compound.

Several decades ago, much of drug discovery was based on the tedious synthesis of new compounds followed by *in vivo* screening in animals. The gradual introduction of *in vitro* screening and later high-throughput screening stimulated new ways of doing chemistry. Combinatorial chemistry in various forms was born. It was quickly realized that, despite the shear power of large numbers, drug discovery did not become easier and attrition did not decrease. Good drugs must also have adequate pharmacokinetic, metabolic, biopharmaceutical, and physicochemical properties [1–5]. A great interest has developed in understanding what the ideal properties of a drug are and how these relate to molecular structure.

In the present chapter we will review technologies that may help to profile new compound libraries or singletons at early R&D stages with respect to their physicochemical and ADME properties in the hope to select only those compounds with the desired properties often seen in drugs. The intention of this approach is to contribute to a lower attrition by avoiding to bring inappropriate candidates into development.

Drug-like properties can be measured and evaluated using medium-to-high-throughput screening technologies. *In silico* methods are also being used increasingly. In the future, we are likely to see a full integration of *in silico* and *in vitro* screening [6] to become the ADME *in combo* approach [7]. Thus, the acronym ADME may in the future stand for 'Automated Decision-Making Engine'.

2. Lead-Like and Drug-Like Properties

Drugs have distinct properties, which differentiate them from other chemicals [8–10]. Limiting values of physicochemical properties are not historical artefacts but are under physiological control [9]. Using for example neural networks, decision trees, or a program such as PASS originally designed to predict potential biological activities [11], a compound can be predicted as being 'drug-like' with an error rate of *ca.* 20%. Similarly, in a study on drugs active as central nervous system (CNS) agents and using neural networks based on Bayesian methods, CNS-active drugs could be distinguished from CNS-inactive ones [12].

Further analyses led to the concept of 'lead-like' structures [13][14]. These tend to be smaller and less lipophilic, and to have less H-bonding groups than drugs. A summary of important properties of leads and drugs is given in *Table 1* [15–17]. From an analysis of the key properties of compounds in the World Drug Index (WDI), the now well-accepted rule of five (Ro5) has been derived [18][19]. Compounds are most likely to

Table 1. *Lead-like and Drug-like Properties*

Properties[a])	Lead-like	Drug-like	CNS-like
M_r [Da]	< 350 to 450	< 500	< 450
ClogP	< 3.5 to 4.5	< 5	
ClogD	– 1 to 4	– 1 to 3	1 to 4
PSA [Å^2]		< 120 to 140	< 60 to 90
HD		< 5	
HA		< 10	
Rot		< 10	

[a]) M_r: molecular weight; calc. log *P*: calculated log *P*; calc. log *D*: calculated log *D* at pH 7.4; PSA: polar surface area; HD: number of H-bond donors; HA: number of H-bond acceptors; Rot: number of rotatable bonds

have poor absorption when their molecular weight (M_r) is > 500 Da, their calculated octanol/water partition coefficient ClogP is > 5, the number of their H-bond donor groups is > 5, and the number of their H-bond acceptor groups is > 10. Computation of these properties is now available as an ADME (absorption, distribution, metabolism, excretion) filter in commercial softwares. The Ro5 should be seen as a qualitative absorption/ permeability discriminator [20], but not as a quantitative predictor [21].

3. Structure-Based and Property-Based Drug Design

The property distribution in drug-related chemical databases has been studied as another approach to understand 'drug likeness' [22] [23]. These afore-mentioned analyses all point to a critical combination of physicochemical and structural properties, which to a large extent can be manipulated by the medicinal chemist. This approach in medicinal chemistry has been called property-based design [24]. In our context, the generic term 'properties' includes physicochemical as well as pharmaco- and toxicokinetic properties.

For a long time, these properties were neglected by most medicinal chemists, who, in many cases, had but the quest for strongest receptor binding as ultimate goal. In many cases, they were assisted in this objective by structure-based design. The structural information could either be the crystal structure of the target or a target–ligand complex. However, this has changed dramatically, and the principles of drug-like compounds are now being used in computational approaches towards the rational design of combinatorial libraries [25]. Of course, a combination of structure- and property-based design is the most-powerful strategy.

4. Design Strategies for Virtual Screening and Real Libraries

Based on the concepts of drug and lead likeness, the design of combinatorial libraries has now changed to take into account not only patentability and diversity, but clearly also druggability [26–30]. Virtual screening (VS) of synthetically accessible compound proposals is today often the first step before the production of the real library. The first selection is often based on the Ro5 or similar rules, and includes a filter for toxicophores. Less evident is whether and which further ADME filters should be applied. Some properties are easier to correct in later stages than others, and weeding potentially promising compounds too early can be contraproductive.

5. *In vitro* Physicochemical and ADME Profiling

Early profiling is increasingly the preferred option to avoid attrition during development. A range of *in vitro* screens has been further automated and adapted to medium- and high-throughput technologies. *In silico* approaches are also increasingly being added. Some of the most-used screens are discussed below. We look first at physicochemical properties, then at nonbiological permeability screens and finally at biopermeability and metabolism screens.

5.1. *Ionization* (pK_a)

The ionization state of a molecule affects its solubility and lipophilicity, and indirectly other properties such as permeability, volume of distribution, and metabolism and excretion. The dogma based on the pH-partition theory that only neutral species cross a membrane has been challenged [20]. Using cyclic voltammetry, it was demonstrated that compounds in their ionized form pass into organic phases and may cross membranes in this ionized form [31]. The various ways by which a charged species can cross a membrane include transport as ion (trans- and/or paracellular), ion pairing, or protein-assisted transport (using the outer surface of a protein spanning a membrane) [32].

The importance of drug ionization in the *in vitro* prediction of *in vivo* absorption is under discussion. When the apical pH used in Caco-2 studies was lowered from 7.4 to 6.0, a better correlation was obtained with *in vivo* data, demonstrating that careful selection of experimental *in vitro* conditions is crucial for a reliable model [33]. Studies with Caco-2 monolayers also suggest that the ionic species may contribute considerably to the overall drug transport [34]. Thus, a continued interest exists in the role of pK_a in absorption, which is often related to its effect on lipophilicity and solubility. New methods to measure pK_a values are being explored [35], *e.g.*, using electrophoresis [36] [37], and an instrument for high-throughput pK_a measurement has been developed [38] [39].

5.2. *Dissolution and Solubility*

Each cellular membrane can be considered as a combined physicochemical and biological barrier to drug transport. Poor physicochemical properties may sometimes be overcome by an active transport mechanism. Before any absorption can take place at all, the first important properties to consider are dissolution and solubility. Many cases of solubility-limited absorption have been reported, and solubility is now seen as a property to be addressed early during the stages of drug discovery, since only dissolved compounds are available for permeation across the gastrointestinal membrane [18] [19].

Excessive lipophilicity is also a common cause of poor solubility and can lead to erratic and incomplete absorption following oral administration. Estimates of desired solubility for good oral absorption depend on the permeability of the compound and the required dose, as illustrated in *Table 2* [19].

The concept of the maximum absorbable dose (*MAD*) [40] [41] relates drug absorption to solubility *via* the relation:

$$MAD = S \cdot k_{abs} \cdot SIWV \cdot SITT \tag{1}$$

where S is the solubility [mg ml^{-1}] at pH 6.5, k_{abs} is the transintestinal absorption rate constant [min^{-1}], $SIWV$ is the small-intestinal water volume [ml] (assumed to be *ca.* 250 ml), and $SITT$ is the small-intestinal transit time [min] (assumed to be 4.5 h = 270 min).

Table 2. *Desired Solubility Depending on the Predicted Dose and the Level of Permeability* [19]

Dose [mg kg^{-1}]	Permeability level [µg ml^{-1}]		
	High	Medium	Low
0.1	1	5	21
1	10	52	210
10	100	520	2100

Dissolution testing has been used as a prognostic tool for oral drug absorption [42]. The biopharmaceutics classification scheme (BCS) has been proposed under which drugs can be categorized into four groups according to their solubility and permeability properties [43]. The BCS has been adopted as a regulatory guidance for bioequivalence studies. Because both permeability as well as solubility can be further dissected into more-fundamental properties, it has been argued that the principal properties are not solubility and permeability, but rather molecular size and H-bonding [44].

High-throughput solubility measurements have been developed that can be used in early discovery [18][45–47]. This has dramatically increased the availability of good data under the same protocol to be used in predictive modeling. As a key first step towards oral absorption, considerable effort has gone into the development of computational solubility prediction [48–54]. However, partly due to a lack of large sets of experimental data measured under identical conditions, today's methods are often not robust enough for reliable predictions. As mentioned earlier, further fine-tuning of the models can be expected now that high-throughput data become available to construct such models.

5.3. *Lipophilicity*

Octanol/water partition (log P) and distribution (log D) coefficients are widely used to make estimates for membrane penetration and permeability, including gastrointestinal absorption [55][56] and BBB crossing [57][58], and correlations to pharmacokinetic properties. The current Proceedings underline the importance of lipophilicity and its influence on various other ADME properties. Already in 1995 and 2000, specialized but very well attended meetings in this series were held to discuss the role of log P in drug research [59][60].

Several approaches for higher-throughput lipophilicity measurements have been developed in the pharmaceutical industry including automated shake-plate methods [61]. A convenient method to measure octanol/water partitioning is based on potentiometric titration, called the pH-metric method [62]. Traditional octanol/water distribution coefficients are still widely used in QSAR and in ADME/PK prediction. However, alternatives have been proposed, *e.g.*, the cyclohexane/water system has been used as a mimic for the blood–brain barrier [57][63].

The two major components of lipophilicity are molecular size and H-bonding [64], as discussed below. A number of rather comprehensive reviews on *in silico* lipophilicity estimation have been published and are recommended for further reading [65–67]. Due to its key importance, a continued interest exists to develop good log *P* estimation programs. Most log *P* approaches are limited due to a lack of parameterization of certain fragments. For the widely used ClogP program, a new version avoiding missing fragments has become available [68]. Most log *P* programs are based on the octanol/water system. An exception is based on *Rekker*'s fragmental constant approach and calculates a log *P* for aliphatic hydrocarbon/water partitioning [69]. These values may offer a better predictor for brain uptake. The AbSolv program based on the work of *Abraham* and his group also contains a number of other solvent scales. Log *D* calculation is so far based on a combination of calculated log *P* and calculated pK_a(s) and using the appropriate equations. Obviously, if both log *P* and pK_a contain an error the resulting log *D* can easily be off by 1 or 2 log units.

5.4. *Hydrogen Bonding*

Molecular size and H-bonding are the two major components of log *P* or log *D* [64] [70]. The H-bonding capacity of a solute correlates reasonably well to passive diffusion. Δlog *P*, the difference between octanol/water and alkane/water partitioning, was suggested as a good measure of solute H-bonding [57] [70] [71]. However, this involves tedious experimental work and it appears that calculated descriptors for H-bonding can be assessed by other, more-convenient methods.

Considerable interest is focussed on the calculation of H-bonding capability for use in QSAR studies, design of combinatorial libraries, and correlation with absorption and permeability data [72–75]. A number of different descriptors for H-bonding have been discussed [76], one of the simplest being the count of the number of H-bond-forming atoms [77]. A simple measure of H-bonding capacity is the polar surface area (*PSA*), summing the fractional contributions to surface area (*PSA*) of all N- and O-atoms [78]. This was used to predict the passage of the blood–brain barrier [58] [79] [80], flux across a Caco-2 monolayer [70], and human intestinal absorption [81] [82]. The physical explanation is that polar groups are involved in desolvation when they move from an aqueous extracellular environment to the more-lipophilic interior of membranes. *PSA* thus represents, at least in part, the energy involved in membrane transport. *PSA* depends on conformation, and the original method [78] is based on a single minimum-energy conformation. Other authors [81] have taken conformational flexibility into account and used a dynamic *PSA* in which a *Boltzmann*-weighted average *PSA* is computed. However, it was demonstrated that *PSA* calculated for a single minimum-energy conformation is in most cases sufficient to produce a sigmoidal relationship to intestinal absorption, differing very little from the dynamic *PSA* described above [82][83]. A fast calculation of *PSA* as a sum of fragment-based contributions has been published [84], using these calculations for large data sets such as combinatorial or virtual libraries. The sigmoidal relationship can be described with *Eqn. 2*:

$$A\% = 100/[1 + (PSA/PSA_{50})^{\gamma}] \quad (2)$$

where A% is the percentage of orally absorbed drug, PSA_{50} the *PSA* at 50% absorption level, and γ is a regression coefficient [83]. Other authors have used a *Boltzmann* sigmoidal curve given by [82]:

$$y = \text{bottom} + (\text{top-bottom})/(1 + \exp((x_{50} - x)/\text{slope})) \quad (3)$$

Poorly absorbed compounds have been identified as those with a $PSA > 140$ Å^2. For larger series, considerable more scatter was found around the sigmoidal curve than for a smaller set of compounds [82]. This is partly due to the fact that many compounds do not permeate by simple passive diffusion only, but are affected by active carriers, efflux mechanisms involving P-glycoprotein (P-gp) and other transporter proteins, and gut wall metabolism. A further refinement in the *PSA* approach is by taking into account the strength of the H-bonds, which in principle forms the basis of the HYBOT approach [73][74].

5.5. *Molecular Size, Shape, and Flexibility*

Molecular size can be a further limiting factor in oral absorption [85]. *Lipinski*'s 'rule of five' proposes an upper limit of M_r 500 Da as acceptable for orally absorbed compounds [18]. Size and shape parameters are generally not measured, but rather calculated. A measured property is the so-called cross-sectional area, which is obtained from surface activity measurements [86].

Molecular weight is often taken as the size descriptor of choice, while it is easy to calculate and is in the chemist's mind. However, other size and shape properties are equally simple to calculate and may offer a better guide to estimate permeability potential. Thus far, no systematic work has been reported investigating this in detail.

The cross-sectional area A_D obtained from surface activity measurements has been reported as a useful size descriptor to discriminate compounds, which can access the brain ($A_D < 80$ Å^2) from those that are too large to cross the blood–brain barrier [86]. Similar studies have been performed to define a cut-off for oral absorption [87].

The number of rotatable bonds can be taken as a measure of a compound's flexibility. A study using rat bioavailability data concluded that compounds with less than ten rotatable bonds and twelve or fewer H-bonds tend to have good bioavailability, and this independently of molecular weight [88]. Molecular weight was taken as a surrogate for more-fundamental properties.

5.6. *Amphiphilicity*

The balance of hydrophilic and hydrophobic moieties in a molecule defines its amphiphilicity. A program has been described to calculate this property and calibrated against experimental values obtained from surface activity measurements [89]. These values may be used to predict effect on membranes leading to cytotoxicity or phospholipidosis, but may also contain information, yet unraveled, on permeability.

5.7. *Permeability*

Since the oral route is often the preferred one for drug administration, an early estimate of the absorption potential is highly desirable. The methods (see *Table 3*) used to assess permeability are a compromise between high throughput and high predictability [90]. They are in fact predictive for both permeability and absorption.

Table 3. In vitro *Models of Membrane Permeability*

Octanol/water distribution
Cyclohexane/water distribution
Phospholipid vesicles
Liposome partitioning
Immobilized artificial membranes (IAM)
Immobilized liposome chromatography (ILC)
Micellar electrokinetic chromatography (MEKC)
Biopartitioning micellar chromatography (BMC)
Impregnated (or artificial) membranes
PAMPA
Filter IAM
Hexadecane-coated polycarbonate filters
Transil™ particles
SPR Biosensor

Table 4. *Opening the Black Box*[a])

Solubility =	A1 + A3 + A6
Lipophilicity =	A2 + A3 + A5
Permeability =	A1 + A4 + A5 + A6
Transporters =	A2 + A3 + A5
Metabolism =	A1 + A4 + A7
Absorption =	A1 + A3 + A4 + A5 + A6

[a]) A1 ... A7 are hypothetical molecular properties (descriptors) used to model each of the individual contributing factors of absorption, or absorption itself. Decomposing the contributing factors of absorption will provide more mechanistic insight.

The *in silico* prediction of oral (or intestinal) absorption has extensively been reviewed [91]. Oral absorption is a composite property. A better understanding and a reliable prediction of each of the contributing factors (such as solubility, permeability, transporters, metabolism) as outlined in *Table 4* will hopefully give us a better absorption model and afford more mechanistic insight.

5.7.1. *Artificial Membranes – PAMPA*

When screening for absorption by passive membrane permeability, artificial membranes have the advantage of offering a highly reproducible and high-throughput system. Artificial membranes have been compared to Caco-2 cells [92] and found to behave similarly for passive diffusion. This was the basis for the development of a parallel artificial membrane permeation assay (PAMPA) for rapid prediction of transcellular absorption potential [93] [94]. In this system, permeability is assessed through a membrane formed by a mixture of lecithin and an inert organic solvent on a hydrophobic filter support. Whilst not completely predictive for human oral absorption, PAMPA data show definite trends in the ability of molecules to permeate membranes by passive diffusion, which may be valuable in screening large compound libraries. This system is commercially available or can easily be implemented. Recently a PAMPA-BBB system has been developed [95]. Further optimizations of the experimental conditions have been investigated [96–98]. Predictability increases when a pH of 6.5 or 5.5 is used on the donor side [97]. It was also demonstrated that the effect of a cosolvent such as DMSO can have a marked effect depending on the basic or acidic nature of the compound [99]. PAMPA Results can be made more relevant by stirring, adjusted such that the unstirred-water-layer thickness matches the 30–100 μm range estimated in the human gut wall [99]. A further development is called

Double-Sink™ PAMPA, using a pH gradient and a chemical scavenger at the receiver site to mimic *in vivo* sink conditions and the presence of plasma proteins. The use of hydrophilic filters increases the rate of permeation and reduces transport time to 2 h compared to over 10 h using a hydrophobic membrane [100].

A similar system has been reported based on polycarbonate filters coated with hexadecane [101][102]. This system consists in a 9–10 µm hexadecane liquid layer immobilized between two aqueous compartments. It was observed that in this set up, diffusion through the unstirred water layer becomes the rate-limiting step for lipophilic compounds. To mimic the *in vivo* environment, permeability measurements were repeated at different pH values in the range of 4–8, and the highest transport value was used for correlation with the percentage absorbed in human. This gives a sigmoidal dependence, which is better than taking values measured at a single pH, *e.g.*, 6.8.

Models based on PAMPA data have been constructed [103] using QSAR and VolSurf tools. These studies show that PAMPA and log D_{oct} data differ only by a H-bonding descriptor. This highlights that PAMPA is just another lipophilicity scale (see also *Chapt. 15* on the linear relationship between PAMPA data and dodecane/water partitioning). Interesting is also the plot of Caco-2 against PAMPA values (*Fig. 1*). A distinction can

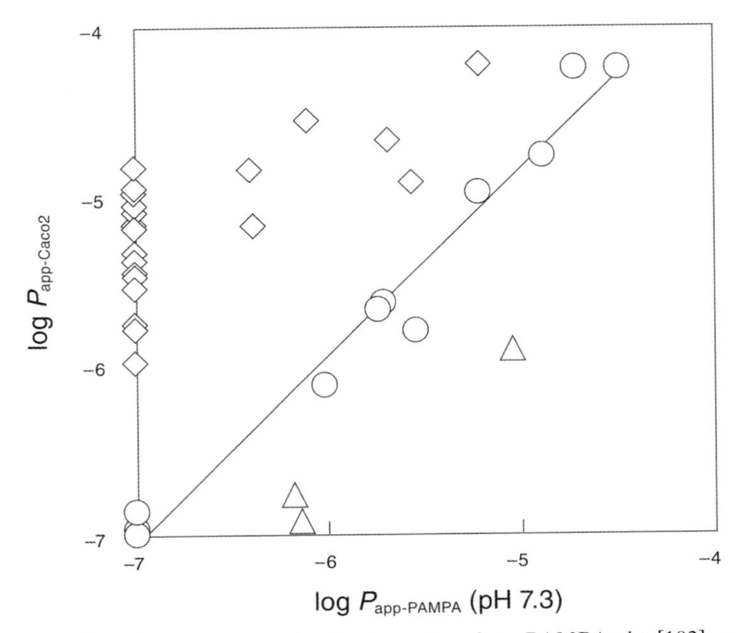

Fig. 1. *Unravelling the transport mechanism in a Caco-2* vs. *PAMPA plot* [103]. ○: passive diffusion, ◇: active transport (influx), △: efflux (P-gp and other transporters).

be made between compounds transported by simple passive diffusion, those influenced by P-gp (transporter) efflux, and those that are transported actively [103].

In summary, the PAMPA approach is believed to yield results comparable to Caco-2 studies [100]. However, this seems to contradict the fact that the PAMPA technology only measures passive diffusion and should be simpler in data interpretation than the Caco-2 assay in which various transporters are expressed. PAMPA is, however, easier to automate and cheaper than the Caco-2 assay. The question open to debate is whether we need another lipophilicity scale (PAMPA) to complement the octanol/water log *P*/*D* scale.

5.7.2. *Artificial Membranes – IAM, ILC, MEKC, and BMC*

Immobilized artificial membranes (IAM) are another means of measuring the lipophilic characteristics of drug candidates and other chemicals [104–109]. IAM Columns may better mimic membrane interactions than the isotropic octanol/water or other solvent/solvent partitioning system. These chromatographic indices appear to be a significant predictor of passive absorption through the rat intestine [110].

A related alternative is called immobilized liposome chromatography (ILC) [111][112]. Compounds with the same octanol/water log *P* were shown to have very different degrees of membrane partitioning on ILC depending on the charge of the compound [112].

Another relatively new lipophilicity scale proposed for use in ADME prediction is based on micellar electrokinetic chromatography (MEKC) [113]. A further variant is called biopartitioning micellar chromatography (BMC) and uses mobile phases of Brij35 (polyoxyethylene(23)lauryl ether) [114]. Similarly, the retention factors of 16 β-blockers obtained with micellar chromatography using sodium dodecyl sulfate as micelle-forming agent correlated well with permeability coefficients in Caco-2 monolayers and apparent permeability coefficients in rat intestinal segments [115]. Each of these scales affords a lipophilicity index related but not identical to octanol/water partitioning.

5.7.3. *Liposome Partitioning and Biosensors*

Liposomes, which are lipid bilayer vesicles prepared from mixtures of lipids, also provide a useful tool for studying passive permeability of molecules through lipid. For example, this system has been used to demonstrate the passive nature of the absorption mechanism of monocarboxylic acids [116]. Liposome partitioning of ionizable drugs can be determined by titration and has been correlated with human absorption [117–119]. A new absorption potential parameter AP_{SUV} has been suggested, as calculated from liposome distribution data and the solubility/dose ratio:

$$AP_{\mathrm{SUV}} = \log(\text{distribution} \cdot \text{solubility} \cdot V/\text{dose}) \tag{4}$$

which shows an excellent sigmoidal relationship with human passive intestinal absorption [117][118].

A further partition system based on the use of liposomes, and commercialized under the name *Transil*™, has been investigated [120][121]. It appears that such lipophilicity values are very useful in PBPK modeling, *e.g.*, in the program PK-Sim [122].

Liposomes have been attached to a biosensor surface, and the interactions between drugs and the liposomes can be monitored directly using surface plasmon resonance (SPR) technology. SPR measures changes in refractive index at the sensor surface caused by changes in mass. Drug–liposome interactions have been measured for 27 drugs and compared to fraction absorbed in humans [123]. A reasonable correlation was obtained, but most likely this method in its present form represents just another way of measuring 'lipophilicity'.

5.7.4. *Cell Lines – Caco-2, MDCK, and Beyond*

Several cell-based assays have been developed to screen the permeability/absorption potential. Most of these systems are intended to screen for oral absorption, others have been developed more specifically to mimic the transport through the blood–brain barrier (BBB). Such cell lines include Caco-2 [54], MDCK, and 2/4/A1 [124][125] cells. Some transporters are expressed in these cell lines albeit to different levels depending on the cell line. MDCK Cells grow faster than Caco-2, and 2/4/A1 have the advantage to have very low levels of expressed transporters and are thus suitable to study passive diffusion. MDCK Cells are also suitable to transfect one or more transporters to study their specific effect, such as MDCK-MDR1 cells to study the role of P-gp or MDCK-MRP2-OATP2 for more-complex transporter effects (*Fig. 2*).

These assays are quite costly and not always predictive for the *in vivo* situation. There is therefore an interest in establishing a 'pure' nonbiological permeability model, as discussed above.

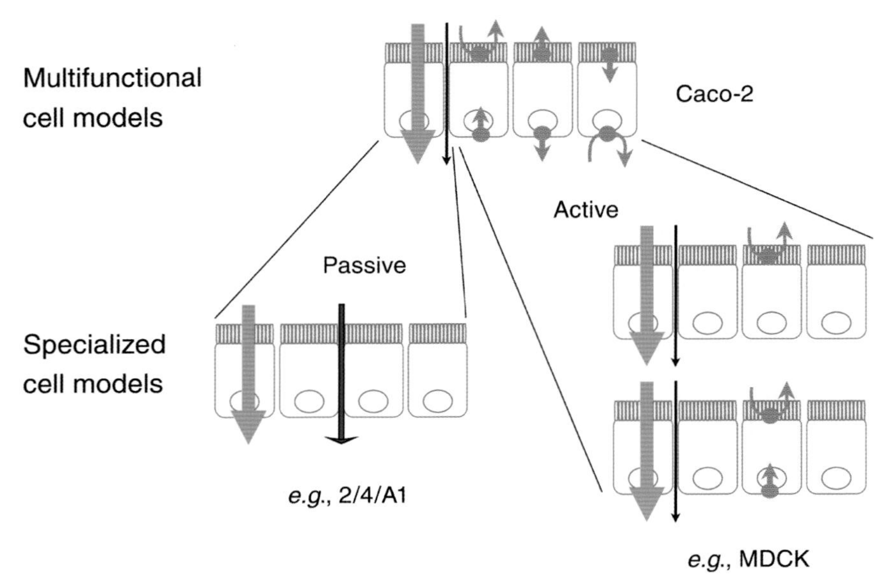

Fig. 2. *Studies of transport mechanism using different cell-based assays.* Caco-2 data may be too complex and can be decomposed in active and passive components [125].

5.8. *The Blood–Brain Barrier*

Similar to the Caco-2 model for gastrointestinal absorption, cell lines such as the bovine brain microvessel endothelial cells (BBMEC) have been proposed as a model for the BBB [126]. As discussed above, a PAMPA-BBB system has been developed [73], which is far easier to implement and run than the cell-based assay. More experience is needed, however, to decide whether the data are really useful.

Solvent systems such as cyclohexane/water have been suggested as a simple model for the blood–brain barrier partition [57][63]. In further studies, it appeared that the difference between log P in octanol/water and alkane/water (Δlog P) was even more meaningful for the prediction of BBB crossing [57]. As discussed above in more detail, a H-bonding descriptor such as *PSA* [78] is even easier to obtain. Despite the fact that many models have been published, the development of good predictive models for uptake in the human brain is hampered by the lack of good-quality data [127]. It is also still unclear what data would be best for such modeling efforts, and discussion is ongoing to compare whole brain *vs.* CSF/plasma ratios [126].

5.9. *Transporters*

The role of transporter proteins in drug disposition is still far from fully understood. These transporters are expressed in most organs involved in uptake and excretion such as gut wall, blood–brain barrier, hepatocytes, and kidney. Examples of clinical relevance of transporter-mediated effects have been reported [128]. It is also believed that some drug–drug interactions (DDI) are based on the competition for a particular transporter. Most studied so far is the ABC transporter P-glycoprotein (P-gp, product of the *MDR1* or *ABCB1* gene), although other transporters are increasingly investigated [129].

Several P-gp assays are in use. A monolayer efflux assay using the MDCK2-MDR1 cell line can identify substrates. The calcein-AM and ATPase assays identify P-gp inhibitors and modulators, which may or may not be substrates.

The P-gp efflux transporter is expressed in Caco-2 cell lines. By measuring the bidirectional transport ratio, this assay is therefore often used to flag potential P-gp-related absorption problems. However, such data should be handled with care. A recent study has shown that, although the BBB expresses P-gp, there is no relationship between the P-gp efflux ratio as measured with Caco-2 cells and limited brain penetration [130].

Assessing whether a given transporter is involved in drug–drug interactions (DDI) is not straightforward. In a study using Caco-2 cells, the pH-dependent bidirectional transport of some weakly basic drugs was measured [131]. A mixture of active and pH-dependent passive transport was seen for the basic P-gp substrates talinolol and quinidine, but not for the neutral drug digoxin. However, the clinically important quinidine–digoxin interaction depended on the presence of a pH gradient.

To date, efforts to predict whether a compound is a substrate or an inhibitor have focussed mainly on P-gp [7][132][133]. One of the incentives to work first on P-gp is that it has been shown to be involved in multidrug resistance (MDR), and to limit the oral uptake and brain access of drugs. It is sometimes observed that a good correlation is obtained only when subsets of closely related structures are considered [133]. This may be due to the existence of two or more binding sites on P-gp. The group of *Seelig* found that P-gp affinity increases with increasing lipophilicity and with increasing H-bonding capacity [87][134] (*Fig. 3*). Since these are rather unspecific criteria, the same group also suggested structural features typically associated with P-gp recognition.

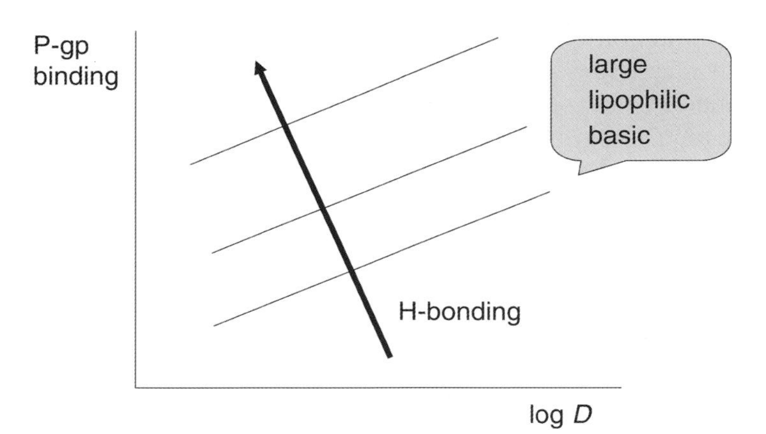

Fig. 3. *Affinity for P-gp increases with increasing lipophilicity* (log D) *and increasing H-bonding capability* [87] [134]

5.10. *Metabolic Stability*

Metabolic stability is a concern in most projects. Half-lives can be measured in human liver microsomes (HLM) or human hepatocytes, and these measurements can be automated and performed in 96-well format [135] [136]. Drawbacks of using liver microsomes include the lack of some soluble enzymes and variations in enzyme content dependent on preparation techniques and source [136]. Yet microsomes are easy and cheap to use [137]. Predictions of *in vivo* clearance can be made using *in vitro* intrinsic clearance values [138].

5.11. *Metabolism, Cytochrome P450 Inhibition, and Drug–Drug Interactions*

The cytochrome P450 (CYP) enzymes are involved in the metabolism of a wide range of drugs. CYP3A4 is the most abundant one, and is responsible for the metabolism of *ca.* 50% of known drugs. Inhibition of CYP3A4 by co-administered drugs can lead to adverse clinical drug–drug interactions. Early identification of CYP3A4 inhibition is therefore needed to minimize this risk. Other CYPs, in particular CYP2D6 and CYP2C19, are polymorphic. It is now recognized that DDI are often a combination of metabolism- and transporter-mediated effects affecting absorption and clearance [135].

Various *in vitro* inhibition assays have been developed such as liver microsomes, hepatocytes, and transfected microsomes [136]. High-throughput inhibition assays are now possible using various fluorescent probes available for individual CYPs. The percentage of inhibition can be determined at a single concentration or at a range of concentrations to estimate an IC_{50} value.

Predictive models have been developed, *e.g.*, for CYP3A4 inhibition [139]. The external validation ($n=9$) of this model based on a set of topological descriptors and using GA-PLS had a correlation coefficient of 0.74. Such models are encouraging, but clearly larger datasets are needed to make them more general. Such models are also likely to be generated in large or specialized companies using in-house screening data.

6. Conclusions: Towards the *in combo* Strategy – Combination of *in silico* and *in vitro* Screening Technologies

In this Chapter, I have described a number of *in vitro* screens currently used to filter out compounds with poor physicochemical and ADME properties. Profiling for drug-like properties has developed rapidly in recent years [140–142]. Increasingly, the accumulated data is being used to develop predictive *in silico* models for these endpoints [143]. In addition, human data such as bioavailability, oral absorption, and volume of distribution of marketed drugs is used to develop models [144]. In the future, both *in silico* and *in vitro* technologies will be combined in a strategy we like to call the *in combo* approach [7]. In this new paradigm, *in silico* descriptors, filters, models, and simulation tools [7][145] will be used to guide *in vitro* screening [6]. But before this agenda can be brought to full success, the predictability of computational approaches need to further improve, and sufficient validation in ongoing projects needs to be performed to prove the concept. It should not be forgotten that an *in silico* 'experiment', like *in vivo* and *in vitro* data, has some level of error. However, the rising costs of high-throughput *in vitro* screening as well as throughput and ethical limitations in *in vivo* studies will further stimulate the development and use of *in silico* ADME and similarly toxicology models [146]. Partly due to a lack of large sets of experimental data measured under identical conditions, today's methods are not robust enough for reliable predictions. This applies in particular to predictors of numerical values. One way out of this uncertainty in predictions is to use a classification model and binning of the compounds into 2–4 classes. Tools based on recursive partitioning or decision trees have become available for this purpose. Another approach is to use a consensus model by combining the output of several approaches into one predictor.

Thus, drug discovery projects will have access to a wide range of properties, *in silico* and *in vitro* in early stages, and *in vivo* in later stages, to allow truly property-based drug design.

REFERENCES

[1] D. A. Smith, H. van de Waterbeemd, D. K. Walker, 'Pharmacokinetics and Metabolism in Drug Design', Wiley-VCH, Weinheim, 2001.
[2] D. A. Smith, B. C. Jones, D. K. Walker, *Med. Res. Rev.* **1996**, *16*, 243.
[3] H. van de Waterbeemd, D. A. Smith, B. C. Jones, *J. Comput.-Aided Mol. Des.* **2001**, *15*, 273.
[4] H. van de Waterbeemd, in 'Drug Bioavailability – Estimation of Solubility, Permeability, Absorption and Bioavailability', Eds. H. van de Waterbeemd, H. Lennernäs, P. Artursson, Wiley-VCH, Weinheim, 2003, p. 3–20.
[5] D. Smith, E. Schmid, B. Jones, *Clin. Pharmacokinet.* **2002**, *41*, 1005.
[6] H. Yu, A. Adedoyin, *Drug Disc. Today* **2003**, *8*, 852.
[7] M. Dickins, H. van de Waterbeemd, *BioSilico* **2004**, *2*, 38.
[8] M. Feher, J. M. Schmidt, *J. Chem. Inf. Comput. Sci.* **2003**, *43*, 218.
[9] M. C. Wenlock, R. P. Austin, P, Barton, A. M. Davis, P. D. Leeson, *J. Med. Chem.* **2003**, *46*, 1250.
[10] M. Vieth, M. G. Siegel, R. E. Higgs, I. A. Watson, D. H. Robertson, K. A. Savin, G. L. Durst, P. A. Hipskind, *J. Med. Chem.* **2004**, *47*, 224.
[11] S. Anzali, G. Barnickel, B. Cezanne, M. Krug, D. Filimonov, V. Poroikov, *J. Med. Chem.* **2001**, *44*, 2432.
[12] Ajay, G. W. Bemis, M. A. Murcko, *J. Med. Chem.* **1999**, *42*, 4942.
[13] S. J. Teague, A. M. Davis, P. D. Leeson, T. I. Oprea, *Angew. Chem., Int. Ed.* **1999**, *38*, 3743.
[14] T. I. Oprea, A. M. Davis, S. J. Teague, P. D. Leeson, *J. Chem. Inf. Comput. Sci.* **2001**, *41*, 1308.
[15] I. Muegge, *Med. Res. Rev.* **2003**, *23*, 302.
[16] D. E. Clark, P. D. J. Grootenhuis, *Curr. Opin. Drug Disc. Dev.* **2002**, *5*, 382.
[17] M. M. Hann, A. R. Leach, G. Harper, *J. Chem. Inf. Comput. Sci.* **2001**, *41*, 856.
[18] C. A. Lipinski, F. Lombardo, B. W. Dominy, P. J. Feeney, *Adv. Drug Delivery Rev.* **1997**, *23*, 3.
[19] C. Lipinski, *J. Pharmacol. Toxicol. Methods* **2000**, *44*, 235.
[20] A. Pagliara, M. Reist, S. Geinoz, P.-A. Carrupt, B. Testa, *J. Pharm. Pharmacol.* **1999**, *51*, 1339.
[21] P. Stenberg, K. Luthman, H. Ellens, C. P. Lee, P. L. Smith, A. Lago, J. D. Elliott, P. Artursson, *Pharm. Res.* **1999**, *16*, 1520.
[22] A. K. Ghose, V. N. Viswanadhan, J. J. Wendoloski, *J. Combinat. Chem.* **1999**, *1*, 55.
[23] T. I. Oprea, *J. Comput.-Aided Mol. Des.* **2000**, *14*, 251.
[24] H. van de Waterbeemd, D. A. Smith, K. Beaumont, D. K. Walker, *J. Med. Chem.* **2001**, *44*, 1313.
[25] H. Matter, K. H. Baringhaus, T. Naumann, T. Klabunde, B. Pirard, *Combinat. Chem. High Throughput Screen.* **2001**, *4*, 453.
[26] D. Gorse, R. Lahana, *Curr. Opin. Chem. Biol.* **2000**, *4*, 287.
[27] S. Rose, A. Stevens, *Curr. Opin. Chem. Biol.* **2003**, *7*, 331.
[28] W. P. Walters, M. A. Murcko, *Adv. Drug Delivery Rev.* **2002**, *54*, 255.
[29] T. I. Oprea, *Molecules* **2002**, *7*, 51.
[30] R. D. Brown, M. Hassan, M. Waldman, *J. Mol. Graph. Mol.* **2000**, *18*, 427.
[31] G. Caron, P. Gaillard, P.-A. Carrupt, B. Testa, *Helv. Chim. Acta* **1997**, *80*, 449.
[32] G. Camenisch, H. van de Waterbeemd, G. Folkers, *Pharm. Acta Helv.* **1996**, *71*, 309.
[33] M. Boisset, R. P. Botham, K. D. Haegele, B. Lenfant, J. L. Pachot, *Eur. J. Pharm. Sci.* **2000**, *10*, 215.

[34] K. Palm, K. Luthman, J. Ros, J. Grasjo, P. Artursson, *J. Pharmacol. Exp. Ther.* **1999**, *291*, 435.
[35] J. Saurina, S. Hernandez-Cassou, R. Tauler, A. Izquierdo-Ridorsa, *Anal. Chim. Acta* **2000**, *408*, 135.
[36] Z. Jia, T. Ramstad, M. Zhong, *Electrophoresis* **2001**, *22*, 1112.
[37] C. E. Kibbey, S. K. Poole, B. Robinson, J. D. Jackson, D. Durham, *J. Pharm. Sci.* **2001**, *90*, 1164.
[38] J. Comer, K. Tam, in 'Pharmacokinetic Optimization in Drug Research. Biological, Physicochemical, and Computational Strategies', Eds. B. Testa, H. van de Waterbeemd, G. Folkers, R. Guy, Verlag Helvetica Chimica Acta, Zürich, 2001, p. 275–304.
[39] J. Comer, K. Box, *J. Assoc. Lab. Automat.* **2003**, *8*, 55.
[40] K. Johnson, A. Swindell, *Pharm. Res.* **1996**, *13*, 1795.
[41] W. Curatolo, Pharm. *Sci. Technol. Today* **1998**, *1*, 387.
[42] J. B. Dressman, G. L. Amidon, C. Reppas, V. P. Shah, *Pharm. Res.* **1998**, *15*, 11.
[43] G. L. Amidon, H. Lennernäs, V. P. Shah, J. R. A. Crison, *Pharm. Res.* **1995**, *12*, 413.
[44] H. van de Waterbeemd, *Eur. J. Pharm. Sci.* **1998**, *7*, 1.
[45] C. D. Bevan, R. S. Lloyd, *Anal. Chem.* **2000**, *72*, 1781.
[46] A. Avdeef, in 'Pharmacokinetic Optimization in Drug Research. Biological, Physicochemical, and Computational Strategies', Eds. B. Testa, H. van de Waterbeemd, G. Folkers, R. Guy, Verlag Helvetica Chimica Acta, Zürich, 2001, p. 305–325.
[47] A. Avdeef, C. M. Berger, *Eur. J. Pharm. Sci.* **2001**, *14*, 281.
[48] J. Huuskonen, *Combinat. Chem. High Throughput Screen.* **2001**, *4*, 311.
[49] J. W. McFarland, A. Avdeef, C. M. Berger, O. A. Raevsky, *J. Chem. Inf. Comput. Sci.* **2001**, *41*, 1355.
[50] D. J. Livingstone, M. G. Ford, J. J. Huuskonen, D. W. Salt, *J. Comput.-Aided Mol. Des.* **2001**, *15*, 741.
[51] P. Bruneau, *J. Chem. Inf. Comput. Sci.* **2001**, *41*, 1605.
[52] R. Liu, S.-S. So, *J. Chem. Inf. Comput. Sci.* **2001**, *41*, 1633.
[53] X.-Q. Chen, S. J. Cho, J. Sung, Y. Li, S. Venkatesh, *J. Pharm. Sci.* **2002**, *91*, 1838.
[54] A. Yan, J. Gasteiger, *QSAR Combinat. Sci.* **2003**, *22*, 821.
[55] S. Winiwarter, N. M. Bonham, F. Ax, A. Hallberg, H. Lennernäs, A. Karlen, *J. Med. Chem.* **1998**, *41*, 4939.
[56] P. Artursson, J. Karlsson, *Biochem. Biophys. Res. Commun.* **1991**, *175*, 880.
[57] Y. C. Young, R. C. Mitchell, T. H. Brown, C. R. Ganellin, R. Griffiths, M. Jones, K. K. Rana, D. Saunders, I. R. Smith, N. E. Sore, T. J. Wilks, *J. Med. Chem.* **1988**, *31*, 656.
[58] H. van de Waterbeemd, G. Camenisch, G. Folkers, J. R. Chrétien, O. A. Raevsky, *J. Drug Target.* **1998**, *2*, 151.
[59] 'Lipophilicity in Drug Action and Toxicology', Eds. V. Pliska, B. Testa, H. van de Waterbeemd, VCH, Weinheim, 1996.
[60] 'Pharmacokinetic Optimization in Drug Research. Biological, Physicochemical, and Computational Strategies', Eds. B. Testa, H. van de Waterbeemd, G. Folkers, R. Guy, Verlag Helvetica Chimica Acta, Zürich, 2001.
[61] L. Hitzel, A. P. Watt, K. L. Locker, *Pharm. Res.* **2000**, *17*, 1389.
[62] A. Avdeef, *J. Pharm. Sci.* **1993**, *82*, 183.
[63] P. Seiler, *Eur. J. Med. Chem.* **1974**, *9*, 473.
[64] H. van de Waterbeemd, B. Testa, *Adv. Drug Res.* **1987**, *16*, 85.
[65] P. Buchwald, N. Bodor, *Curr. Med. Chem.* **1998**, *5*, 353.
[66] P.-A. Carrupt, B. Testa, P. Gaillard, *Rev. Comput. Chem.* **1997**, *11*, 241.
[67] R. Mannhold, H. van de Waterbeemd, *J. Comput.-Aided Mol. Des.* **2001**, *15*, 337.
[68] A. J. Leo, D. Hoekman, *Perspect. Drug Disc. Des.* **2000**, *18*, 19.
[69] R. Mannhold, R. F. Rekker, *Perspect. Drug Disc. Des.* **2000**, *18*, 1.
[70] N. El Tayar, B. Testa, P. A. Carrupt, *J. Phys. Chem.* **1992**, *96*, 1455.
[71] T. W. Von Geldern, D. J. Hoffmann, J. A. Kester, H. N. Nellans, B. D. Dayton, S. V. Calzadilla, K. C. Marsch, L. Hernandez, W. Chiou, D. B. Dixon, J. R. Wu-Wong, T. J. Opgenorth, *J. Med. Chem.* **1996**, *39*, 982.

[72] J. C. Dearden, T. Ghafourian, *J. Chem. Inf. Comput. Sci.* **1999**, *39*, 231.
[73] O. A. Raevsky, K.-J. Schaper, *Eur. J. Med. Chem.* **1998**, *33*, 799.
[74] O. A. Raevsky, V. I. Fetisov, E. P. Trepalina, J. W. McFarland, K.-J. Schaper, *Quant. Struct.-Act. Relat.* **2000**, *19*, 366.
[75] H. van de Waterbeemd, G. Camenisch, G. Folkers, O. A. Raevsky, *Quant. Struct.-Act. Relat.* **1996**, *15*, 480.
[76] H. van de Waterbeemd, in 'Oral Drug Absorption', Eds. J. B. Dressman, H. Lennernäs, Dekker, New York, 2000, p. 31–49.
[77] T. Österberg, U. Norinder, *J. Chem. Inf. Comput. Sci.* **2000**, *40*, 1408.
[78] H. van de Waterbeemd, M. Kansy, *Chimia* **1992**, *46*, 299.
[79] J. Kelder, P. D. J. Grootenhuis, D. M. Bayada, L. P. C. Delbressine, J.-P. Ploemen, *Pharm. Res.* **1999**, *16*, 1514.
[80] D. E. Clark, *J. Pharm. Sci.* **1999**, *88*, 815.
[81] K. Palm, K. Luthman, A.-L. Ungell, G. Strandlund, F. Beigi, P. Lundahl, P. Artursson, *J. Med. Chem.* **1998**, *41*, 5382.
[82] D. E. Clark, *J. Pharm. Sci.* **1999**, *88*, 807.
[83] P. Stenberg, U. Norinder, K. Luthman, K. P. Artursson, *J. Med. Chem.* **2001**, *44*, 1927.
[84] P. Ertl, B. Rohde, P. Selzer, *J. Med. Chem.* **2000**, *43*, 3714.
[85] O. H. Chan, B. H. Stewart, *Drug Disc. Today* **1996**, *1*, 461.
[86] H. Fischer, R. Gottschlich, A. Seelig, *J. Membr. Biol.* **1998**, *165*, 201.
[87] H. Fischer, 'Passive Diffusion and Active Transport through Biological Membranes – Binding of Drugs to Transmembrane Receptors', Ph.D. Thesis, 1998, University of Basel, Switzerland.
[88] D. F. Veber, S. R. Johnson, H.-Y. Cheng, B. R. Smith, K. W. Ward, K. D. Kopple, *J. Med. Chem.* **2002**, *45*, 2615.
[89] H. Fischer, M. Kansy, D. Bur, *Chimia* **2000**, *54*, 640.
[90] K. A. Youdim, A. Avdeef, N. J. Abbott, *Drug Disc. Today* **2003**, *8*, 997.
[91] D. E. Clark, P. D. J. Grootenhuis, *Curr. Top. Med. Chem.* **2003**, *3*, 1193.
[92] G. Camenisch, G. Folkers, H. van de Waterbeemd, *Int. J. Pharm.* **1997**, *147*, 61.
[93] M. Kansy, F. Senner, K. Gubernator, *J. Med. Chem.* **1998**, *41*, 1007.
[94] M. Kansy, H. Fischer, K. Kratzat, F. Senner, B. Wagner, I. Parrilla, in 'Pharmacokinetic Optimization in Drug Research. Biological, Physicochemical, and Computational Strategies', Eds. B. Testa, H. van de Waterbeemd, G. Folkers, R. Guy, Verlag Helvetica Chimica Acta, Zürich, 2001, p. 447–464.
[95] L. Di, E. H. Kerns, K. Fan, O. J. McConnell, G. T. Carter, *Eur. J. Med. Chem.* **2003**, *38*, 223.
[96] K. Sugano, H. Hamada, M. Machida, H. Ushio, *J. Biomol. Screen.* **2001**, *6*, 189.
[97] K. Sugano, H. Hamada, M. Machida, H. Ushio, K. Saitoh, K. Terada, *Int. J. Pharm.* **2001**, *228*, 181.
[98] A. Avdeef, 'Absorption and Drug Development', Wiley, Hoboken, USA, 2003.
[99] M. Bermejo, A. Avdeef, A. Ruiz, R. Nalda, J. A. Ruell, O. Tsinman, I. Gonzalez, C. Fernandez, G. Sanchez, T. M. Garrigues, V. Merino, *Eur. J. Pharm. Sci.* **2004**, *21*, 429.
[100] C. Zhu, L. Jiang, T.-M. Chen, K.-K. Hwang, *Eur. J. Med. Chem.* **2002**, *37*, 399.
[101] B. Faller, F. Wohnsland, in 'Pharmacokinetic Optimization in Drug Research. Biological, Physicochemical, and Computational Strategies', Eds. B. Testa, H. van de Waterbeemd, G. Folkers, R. Guy, Verlag Helvetica Chimica Acta, Zürich, 2001, p. 257–274.
[102] F. Wohnsland, B. Faller, *J. Med. Chem.* **2001**, *44*, 923.
[103] R. Ano, Y. Kimura, M. Shima, R. Matsuno, M. Akamatsu, *Bioorg. Med. Chem.* **2004**, *12*, 257.
[104] C. Y. Yang, S. J. Cai, H. Liu, C. Pidgeon, *Adv. Drug Delivery Rev.* **1996**, *23*, 229.
[105] A. Taillardat-Bertschringer, P.-A. Carrupt, F. Barbato, B. Testa, *J. Med. Chem.* **2003**, *46*, 655.
[106] S. Ong, H. Liu, C. Pidgeon, *J. Chromatogr., A* **1996**, *728*, 113.
[107] B. H. Stewart, O. H. Chan, *J. Pharm. Sci.* **1998**, *87*, 1471.

[108] A. Ducarne, M. Neuwels, S. Goldstein, R. Massingham, *Eur. J. Med. Chem.* **1998**, *33*, 215.
[109] A. Reichel, D. J. Begley, *Pharm. Res.* **1998**, *15*, 1270.
[110] M. Genty, G. Gonzalez, C. Clere, V. Desangle-Gouty, J.-Y. Legendre, *Eur. J. Pharm. Sci.* **2001**, *12*, 223.
[111] P. Lundahl, F. Beigi, *Adv. Drug Delivery Rev.* **1997**, *23*, 221.
[112] U. Norinder, T. Österberg, *Perspect. Drug Disc. Des.* **2000**, *19*, 1.
[113] M. D. Trone, M. S. Leonard, M. G. Khaledi, *Anal. Chem.* **2000**, *72*, 1228.
[114] M. Molero-Monfort, L. Escuder-Gilabert, L. Villanueva-Camanoas, S. Sagrado, M. J. Medina-Hernandez, *J. Chromatogr., B* **2001**, *753*, 225.
[115] A. Detroyer, Y. VanderHeyden, S. Cardo-Broch, M. C. Garcia-Alvarez-Coque, D. L. Massart, *J. Chromatogr., A* **2001**, *912*, 211.
[116] M. Takagi, Y. Taki, T. Sakane, T. Nadai, H. Sezaki, N. Oku, S. Yamashita, *J. Pharmacol. Exp. Ther.* **1998**, *285*, 1175.
[117] K. Balon, B. U. Riebesehl, B. W. Muller, *Pharm. Res.* **1999**, *16*, 882.
[118] K. Balon, B. U. Riebesehl, B. W. Muller, *J. Pharm. Sci.* **1999**, *88*, 802.
[119] A. Avdeef, K. J. Box, J. E. A. Comer, C. Hibbert, K. Y. Tam, *Pharm. Res.* **1998**, *15*, 209.
[120] B. I. Escher, R. P. Schwarzenbach, J. C. Westall, *Environ. Sci. Technol.* **2000**, *34*, 3962.
[121] A. Loidl-Stahlhofen, A. Eckrt, T. Hartmann, M. Schottner, *J. Pharm. Sci.* **2001**, *90*, 599.
[122] S. Willmann, J. Lippert, M. Severstre, J. Solodenk, W. Schmitt, *BioSilico* **2003**, *1*, 121.
[123] E. Danelian, A. Karlén, R. Karlsson, S. Winiwarter, A. Hansson, S. Löfås, H. Lennernäs, D. Hämäläinen, *J. Med. Chem.* **2000**, *43*, 2083.
[124] S. Tavelin, J. Taipalensuu, I. Söderberg, R. Morrison, S. Chong, P. Artursson, *Pharm. Res.* **2003**, *20*, 397.
[125] P. Artursson, S. Tavelin, in 'Drug Bioavailability', Eds. H. van de Waterbeemd, H. Lennernäs, P. Artursson, Wiley-VCH, Weinheim, 2003, p. 72–89.
[126] F. Atkinson, S. Cole, C. Green, H. van de Waterbeemd, *Curr. Med. Chem.–CNS Agents* **2002**, *2*, 229.
[127] D. E. Clark, *Drug Disc. Today* **2003**, *8*, 927.
[128] A. Ayrton, P. Morgan, *Xenobiotica* **2001**, *31*, 469.
[129] H. Lennernäs, *J. Pharm. Pharmacol.* **2003**, *55*, 429.
[130] F. Faassen, G. Vogel, H. Spanings, H. Vromans, *Int. J. Pharm.* **2003**, *263*, 113.
[131] S. Neuhoff, A.-L. Ungell, I. Zamora, P. Artursson, *Pharm. Res.* **2003**, *20*, 1141.
[132] R. B. Wang, C. L. Kuo, L. L. Lien, E. J. Lien, *J. Clin. Pharmacol. Ther.* **2003**, *28*, 203.
[133] J. C. Dearden, A. Al-Noobi, A. C. Scott, S. A. Thomson, *SAR QSAR Environ. Res.* **2003**, *14*, 447.
[134] A. Seelig, E. Gatlik-Landwojtowicz, *Mini-Rev. Med. Chem.* **2005**, *5*, 135.
[135] R. J. Riley, I. J. Martin, A. E. Cooper, *Curr. Drug Metab.* **2002**, *3*, 527.
[136] S. A. Roberts, *Curr. Opin. Drug Disc. Dev.* **2003**, *6*, 66.
[137] E. H. Kerns, L. Di, *Drug Disc. Today* **2003**, *8*, 316.
[138] R. S. Obach, J. G. Baxter, T. E. Liston, B. M. Silber, B. C. Jones, F. MacIntyre, D. J. Rance, P. Wastall, *J. Pharmacol. Exp. Ther.* **1997**, *283*, 46.
[139] S. Wanchana, F. Yamashita, M. Hashida, *Pharm. Res.* **2003**, *20*, 1401.
[140] L. Di, E. H. Kerns, *Curr. Opin. Chem. Biol.* **2003**, *7*, 402.
[141] A. Avdeef, B. Testa, *Cell. Mol. Life Sci.* **2002**, *59*, 1681.
[142] A. Avdeef, *Curr. Top. Med. Chem.* **2001**, *1*, 277.
[143] H. van de Waterbeemd, *Curr. Opin. Drug Disc. Dev.* **2002**, *5*, 33.
[144] H. van de Waterbeemd, E. Gifford, *Nature Rev. Drug Disc.* **2003**, *2*, 192.
[145] F. Lombardo, E. Gifford, M. Y. Shalaeva, *Mini Rev. Med. Chem.* **2003**, *3*, 861.
[146] D. A. Smith, *Drug Disc. Today* **2002**, *7*, 1080.

Part II. Biological Strategies

Membranes – From Barriers to Magic Bullets

by **Ole G. Mouritsen**

MEMPHYS-Center for Biomembrane Physics, University of Southern Denmark,
Campusvej 55, DK-5230 Odense M
(e-mail: ogm@memphys.sdu.dk; http://www.memphys.sdu.dk)

Abbreviations
DAPC: diarachioyl phosphatidylcholine; DLPC: dilaureoyl phosphatidylcholine; DMPC: dimyristoyl phosphatidylcholine; DMPE: dimyristoyl phosphatidylcholine; DOPC: dioleoyl phosphatidylcholine; DPPC: dipalmitoyl phosphatidylcholine; DPPE: dipalmitoyl phosphatidylcholine; DSPC: distearoyl phosphatidylcholine; POPE: palmitoyl-oleoyl phosphatidylaminoethanol; SM: sphingomyelin.

---

1. Introduction: Membranes as Barriers, Carriers, and Targets

It is important to realize that a lipid bilayer membrane is not just a homogeneous thin slap of a dielectric medium immersed in water but that the bilayer is a highly stratified structure with a distinct trans-bilayer molecular profile and a particular heterogeneous lateral distribution of molecular species [1]. These structural properties determine the ability of the membrane to act as a barrier, a carrier, and a target. The transverse and lateral structures of lipid bilayers are of particular importance for understanding how proteins function in and at membranes and how amphiphilic drugs interact with membranes [2].

2. Transverse and Lateral Structure of Lipid Bilayers

2.1. *The Trans-Bilayer Profile*

The trans-bilayer profile is the best-characterized structural property of bilayers [3] since it most easily lends itself to be monitored by a number of techniques, *e.g.*, X-ray and neutron-scattering techniques, magnetic-resonance experiments, molecular-probe measurements, or computer-simulation calculations. Magnetic-resonance and molecular-probe techniques can give information about the structure and dynamics in various depths of the bilayer by using local reporter molecules or atoms.

The results of these studies have provided a picture of the lipid bilayer as a highly disordered liquid system with a distinct stratification [4]. Going from the outside, it can grossly be described in terms of four layers: *1*) a layer of perturbed water, *i.e.*, water that is structured and deprived of some of its H-bonds, *2*) a hydrophilic–hydrophobic region including the lipid polar head groups as well as both water and part of the upper segments of the fatty acid chains, *3*) a soft polymer-like region of ordered fatty acid chain segments, and *4*) a hydrophobic core with disordered fatty acid chain segments of a structure similar to that of a liquid oil like decane. Although the detailed nature of the trans-bilayer profile depends on the actual lipid species in question, the overall structural stratification is generic for aqueous lipid bilayers.

A striking observation to be made from this picture is that the region of space that makes up the hydrophobic–hydrophilic interface of the membrane occupies about one half of the entire lipid bilayer thickness. The presence of this spatially extended interface region, its chemical heterogeneity, as well as its dynamic nature is probably the single most-important quantitative piece of information on membrane structure and organization. The chemically heterogeneous nature of this extended interface region makes it prone for all sorts of noncovalent interactions with molecules, *e.g.*, peptides and drugs, that bind, penetrate, and permeate membranes [5]. This interface is thick enough to accommodate an α-helical peptide that lies parallel to the bilayer surface.

2.2. *The Lateral Pressure Profile*

Lipids in bilayers are kept in place because of the hydrophobic effect. This is a way to keep the oily fatty acid chains away from the water. It is not an entirely happy situation for the lipid molecules, however. They are subject to large stresses by being confined into a bilayer structure along with their neighbors [6].

In *Fig. 1,a*, a schematic illustration is given of a cross-section through a lipid bilayer indicating the forces that act to stabilize the layer. When the bilayer is in equilibrium, these forces have to sum up to zero. Since the forces, due to the finite thickness of the bilayer, operate in different planes, the pressures are distributed nonevenly across the bilayer as shown schematically by the profile in *Fig. 1,b*. This profile is called the lateral pressure or lateral stress profile of the bilayer.

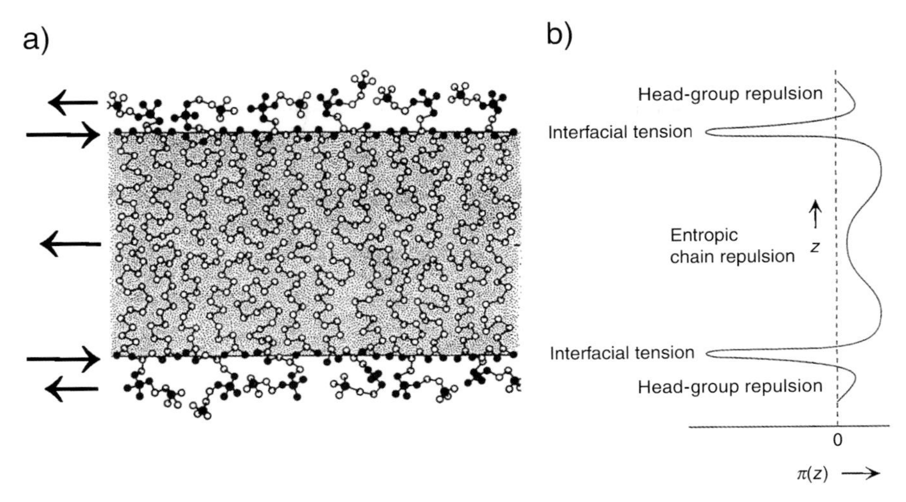

Fig. 1. *Trans-bilayer profile of a lipid bilayer. a*) Schematic illustration of a cross-section through a lipid bilayer with indication of the forces that act within the layer. A repulsive force acts between the lipid head groups, an attractive (tensile) force acts at the hydrophobic–hydrophilic interface, and a repulsive force acts between the lipid chains. *b*) The resulting lateral pressure profile, $\pi(z)$, denoting the local lateral pressure at depth z in the bilayer. Adapted from [7].

The lateral pressure profile is built up from three contributions. A positive lateral pressure resulting from the repulsive forces that act between the head groups, a negative lateral pressure (the interfacial tension) that acts in the hydrophobic–hydrophilic interface as a result of the hydrophobic effect, and a positive lateral pressure arising from the entropic repulsion between the flexible fatty acid chains (chain pressure). The detailed form of the pressure profile depends on the type of lipids under consideration. Due to the small thickness of the lipid bilayer, the rather large interfacial tension from the two interfaces of the bilayer has to be distributed over a very short range. This implies that the counteracting pressure from the fatty acid chains has to have an enormous density, typically around several hundreds of atmospheres per unit area. This is easily seen by noting that the interfacial tension, γ, at each of the two hydrophobic–hydrophilic interfaces of a lipid bilayer is *ca.* 50 mN/m. The lateral pressure of the interior of the lipid bilayer has to counterbalance this tension over a thickness, d, of only *ca.* 2.5–3 nm. The lateral pressure density then becomes $2\gamma/d$, which amounts to *ca.* 350 atm. Pressure densities of this magnitude are capable of influencing the molecular conformation of proteins imbedded in the membrane and hence provide a possible nonspecific coupling between the lipid membrane and the function of proteins [6].

It is clear from the description of the lateral pressure profile and *Fig. 1* that it is not possible to assign a well-defined shape to a lipid molecule imbedded in a bilayer. The stressed and frustrated situation that a lipid molecule experiences in a bilayer is better described by the pressure profile, although there is no simple relation between the molecular structure and the actual distribution of stresses in the bilayer. Therefore, it is the lateral pressure profile that is the more-fundamental physical property and which underlies the curvature stress field described in *Sect. 4* below.

2.3. *Lateral Membrane Structure*

Being many molecules together in a bilayer membrane, the lipids act cooperatively and organize laterally in the plane of the bilayer in a nonrandom and nonuniform fashion [8]. In contrast to the trans-bilayer structure described above, the lateral bilayer structure and molecular organization are less well characterized and its importance also generally less appreciated. This is particularly the case when it comes to the small-scale structure and micro-heterogeneity in the range from nanometers to micrometers. One reason for this is that this regime is experimentally difficult to access by direct methods. Another reason is that the small-scale structures are often dynamic and change in time.

Several physical mechanisms can lead to the formation of a highly nonrandom and nontrivial lateral organization of membranes. First, proteins anchored to the cytoskeleton can provide effective fences or corrals that lead to transient or permanent membrane domains. Second, phase separation can occur leading to large areas of different molecular composition. Finally, the molecular interactions between the membrane constituents, in particular the lipids, lead to cooperative behavior and phase transitions that can be associated with significant fluctuation effects. These fluctuations are the source for the formation of lipid membrane domains on different time and length scales.

Lipid bilayer fluctuations can be perceived as either local density variations or local variations in molecular composition. The range over which these variations occur is described by a coherence length. The coherence length is a measure of the size of the lipid domains. Obviously, these domains need not be sharply defined, and a certain gradual variation in the lipid bilayer properties is expected upon crossing a domain boundary. Lipid domains caused by fluctuations should be considered dynamic entities that come and go and which have life times that depend on their size and the thermodynamic conditions. We refer to this type of domain formation as dynamic heterogeneity.

Direct imaging of the lateral structure and possible domain formation can be performed on individual bilayers by the same techniques as often applied to lipid monolayers, specifically fluorescence microscopy [9] and atomic force microscopy [10]. It requires, however, that the membranes are fixated in some way. Imaging by fluorescence microscopy exploits the possibility that different fluorescent probes can localize differently in different membrane phases and domains. The contours of the domains then appear as contrasts of regions of different color. These techniques have been widely used to image the surface structure of whole cells and fragments of real biological membranes.

Some very significant and definite evidence for the presence of lipid domains in well-defined model membranes has been obtained from fluorescence microscopy on giant unilamellar vesicles of diameter 50–100 μm [9]. A range of different membranes have been investigated, including simple binary mixtures as well as lipid and protein extracts from real cell membranes. In *Fig. 2*, a gallery of images obtained for different giant vesicle membranes is shown. The images show that on the length scales accessible by microscopy based on light, lipid domains occur in these membrane systems. The observed domains in the mixtures can be related to the different lipid phases. Fluorescence microscopy can also be applied to lipid bilayers supported on a solid hydrophilic surface.

The presence of small-scale lateral structure in real biological membranes and its importance for biological activity have received an increasing attention in recent years [11]. The interest is fuelled by two types of information. First, it was discovered by single-particle tracking techniques that labeled lipid or protein molecules performed a certain type of lateral diffusive motion, which suggested that they were temporarily confined to a small region of the membrane surface. The other line of evidence derives from biochemical treatment of cold membrane samples treated with a harsh detergent (*Triton X-100*). It was discovered that a certain fraction of the membranes was resistant to the detergent and it was suggested that this fraction corresponds to supramolecular entities floating around in fluid membranes as a kind of 'rafts'. The rafts were surmised to behave as functional units supporting various functions.

The accumulated evidence in favor of rafts in a variety of membranes is gaining momentum although the whole topic is still somewhat controversial. A common characteristic of the rafts is that they contain high levels of cholesterol and sphingolipids, as well as unsaturated phospholipids. The presence of sphingolipids, such as sphingomyelin or glycosphingolipids, which often have high phase-transition temperatures, and

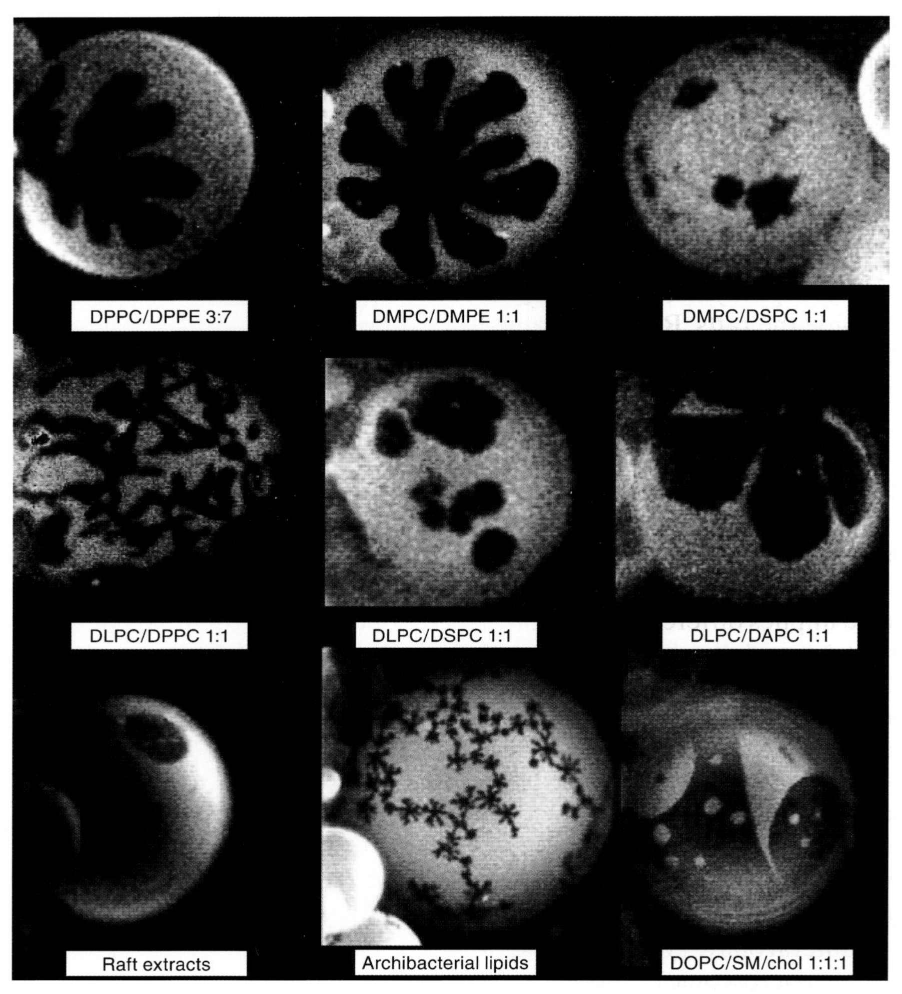

Fig. 2. *Gallery of fluorescence microscopy images of the lateral structure of different lipid bilayers forming giant unilamellar vesicles.* The typical size of the liposomes shown is 50 μm. The raft extracts refer to membrane compositions of lipids and proteins extracted form biological membranes with putative rafts. Courtesy of Dr. *Luis Bagatolli*.

cholesterol, which promotes ordering of the lipid chain, led to the suggestion that the rafts had a structure similar to the liquid-ordered phase in the lipid-cholesterol phase diagram [12]. Rafts are believed to be associated with mostly peripheral proteins that stabilize the rafts and function in connection with the rafts. Raft-like entities are however also found in simple lipid mixtures containing sphingomyelin, cholesterol, and unsaturated phospholipids.

There is now accumulating evidence that domains and rafts also support aspects of membrane function [13]. Certain proteins seem to prefer association with rafts. Many of these proteins carry a hydrocarbon chain anchor, which fits snugly into the tight packing of the raft. Recruitment of proteins to the rafts or detachment of proteins from the rafts can conveniently be facilitated by enzymatic cleavage or attachment of appropriate hydrocarbon chains. For example, long saturated acyl chain anchors have affinity for the ordered raft structure, whereas the more-bulky isopranyl chain anchors prefer to be in the liquid-disordered phase outside the rafts. Rafts have been shown to facilitate the communication between the two monolayer leaflets of the bilayer and to be involved in cell-surface adhesion and motility. Furthermore, there are indications that rafts are involved in cell-surface signaling and the intracellular trafficking and sorting of lipids and proteins. It is interesting to note that some of these functions become impaired when cholesterol, which appears to be a necessary molecular requirement for raft formation, is extracted from the membranes.

The formation of lipid domains of a particular composition and structure implies differentiation and compartmentalization of the lipid bilayer that control the association and binding of peripheral (*e.g.*, charged) macromolecules and enzymes [14][15]. For example, water-soluble, positively charged proteins (such as cytochrome c) exhibit enhanced binding to lipid membranes where a small fraction of negatively charged lipids form domains which have a local charge density large enough to bind the proteins. Conversely, the charged protein helps stabilize the charged micro-domain. Enzymes like protein kinase C and phospholipases display variations in activity that correlate with the occurrence of small lipid domains.

Lateral bilayer heterogeneity in terms of lipid domains implies changes in the macroscopic bilayer properties, *e.g.*, lateral compressibility, bending rigidity, permeability, binding affinity for various solutes, as well as the way the bilayer mediates the interaction and organization of membrane proteins and peptides.

3. Permeability and Adsorption Properties of Lipid Bilayers

The foremost mission of the lipid bilayer component of biological membranes is to act as a permeability barrier. Nonspecific passive permeation has to be avoided. The permeation of molecular species across lipid bilayers depends on both the diffusion rate and the solubility of the permeant in the membrane. The permeability, therefore, intimately reflects the inhomogeneous nature of the membrane, both transversely and laterally.

Whereas lipid bilayers are moderately permeable to water, gaseous substances like CO_2 and O_2, small hydrophobic molecules like benzene, ions, and larger molecular species such as glucose, amino acids, as well as peptides only pass very slowly across the bilayer. The passage of hydrophilic and charged compounds like ions is strongly inhibited by the hydrophobic bilayer core. For example, for an ion to passively cross a lipid bilayer it has to leave a medium with a high dielectric constant ε of *ca.* 80 and venture into a hydrocarbon medium with a low dielectric constant ε of *ca.* 1–3. This amounts to an enormous electrostatic barrier in the order of 100 $k_B T$.

Nevertheless, ions can pass through a lipid bilayer, and the lipid bilayer structure and organization are determining factors for the degree of permeability. This is where the lipid phase transitions and phase equilibria come in. In *Fig. 3,a* are shown data for the passive permeability of a small negative ion, $S_2O_2^{2-}$, through a lipid bilayer of DMPC. The remarkable observation is that the lipid phase equilibria have a strong effect on the leakiness of the bilayer. At the phase transition of the pure lipid bilayers and at the temperatures corresponding to the phase lines in the phase diagram of lipid mixtures, the permeability is anomalously large. In addition, in the phase-separation region of the mixture, the mixed bilayer is quite leaky. These observations are fairly generic and have also been found for other ions like Na^+. The leakiness at the transitions is directly

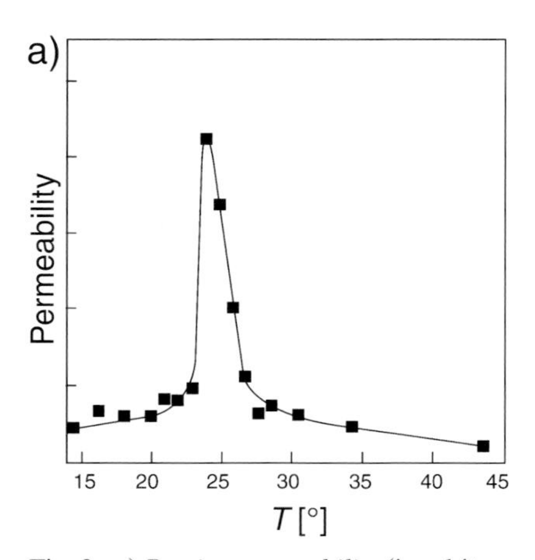

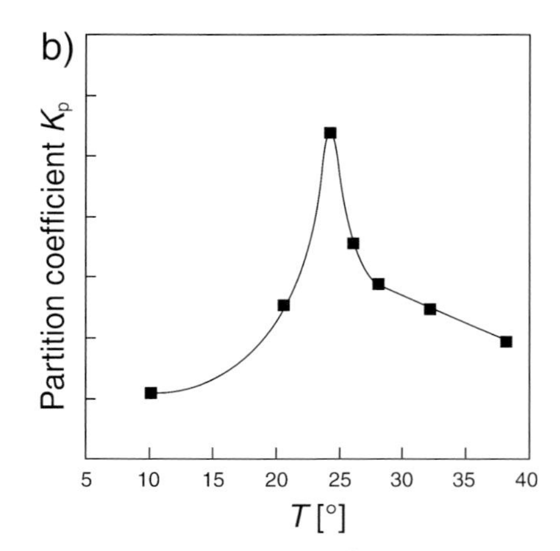

Fig. 3. a) *Passive permeability* (in arbitrary units) *of a small negative ion, $S_2O_2^{2-}$, through lipid bilayers of DMPC* [16a]. b) *Binding of ethanol to DMPC lipid bilayers.* The binding is given in terms of the partition coefficient, K_p, (in arbitrary units) which is a measure of the concentration of ethanol in the bilayer in relation to that in water [16b].

related to the small-scale structure and the lipid domains that develop as a consequence of the lipid phase transitions. The small-scale structure implies that the bilayer has a significant amount of defect and lines of defects through which the permeants can leak through the bilayer.

The second example demonstrates that foreign compounds that interact with membranes can sense the lipid phase transition. In *Fig. 3,b*, the binding of a simple alcohol, ethanol, to DMPC lipid bilayers is shown [16]. As is well known, ethanol has a strong effect on biological membranes, in particular those of nerve cells. The figure shows that the partitioning of ethanol into lipid bilayers is strongly enhanced in the transition region. These two examples illustrate that a lipid bilayer becomes vulnerable in its transition region. It gets leaky and it can be invaded by foreign compounds. There are many more examples known of dramatic events that become facilitated in the lipid phase transition region: membrane proteins are more easily inserted, cholesterol can more readily be exchanged between membranes, and the probability of membrane fusion and fusion of vesicles with lipid monolayers becomes enhanced.

Obviously, it is not very desirable for biological membranes to be as vulnerable to nonspecific invasion of foreign compounds as illustrated in *Fig. 3*. Eukaryotes have found a way of dealing with this by incorporating cholesterol into their plasma membranes. As shown in *Fig. 4*, large amounts of cholesterol both serve as to suppress the anomalous permeability behavior as well as act to inhibit the binding of ethanol. It is a peculiar observation, that small amounts of cholesterol have the opposite

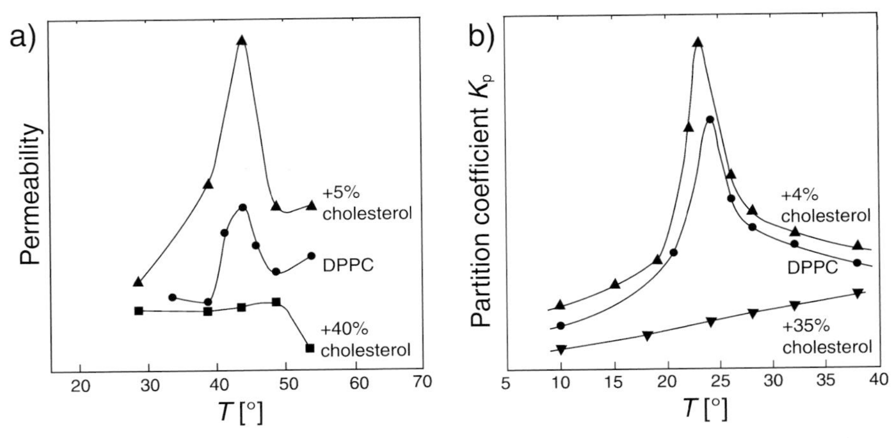

Fig. 4. a) *The effect of cholesterol on the passive permeability* (in arbitrary units) *of Na^+ ions through DPPC lipid bilayers* [17]. b) *The effect of cholesterol on the binding of ethanol to DMPC lipid bilayers* [16]. The partition coefficient is given in arbitrary units.

effect. They tend to soften the bilayer, leading, *e.g.*, to enhanced permeability and ethanol binding. Cholesterol in large amounts is often used to seal and tighten liposomes used for drug delivery.

Certain lysolipids and free fatty acids can act as drugs or permeability enhancers as we shall discuss in *Sect. 6* below. Due to their limited water solubility, lysolipids and free fatty acids will partition into lipid bilayers with a partition coefficient that depends sensitively on the length of the alkyl chains as well as on the phase state of the lipid membrane [18]. Whereas the partitioning of saturated fatty acids does not depend on the lipid phase state, the solubility of saturated lysolipids is typically an order of magnitude larger in the fluid phase than in the solid phase. The partitioning for both saturated fatty acids and lysolipids increases about an order of magnitude when the alkyl chain length is increased by two C-atoms. As an example, the partition coefficients of palmitic acid and 1-palmitoyl phosphatidylcholine in fluid DPPC lipid bilayers (at 50°) are 93 000 and 52 500, respectively, whereas the corresponding ones for capryl alkyl chains (ten C-atoms) are 52 and 1700.

The permeability-enhancing effect on liposomes due to lysolipids and free acids is demonstrated in *Fig. 5*. The figure clearly shows that each of

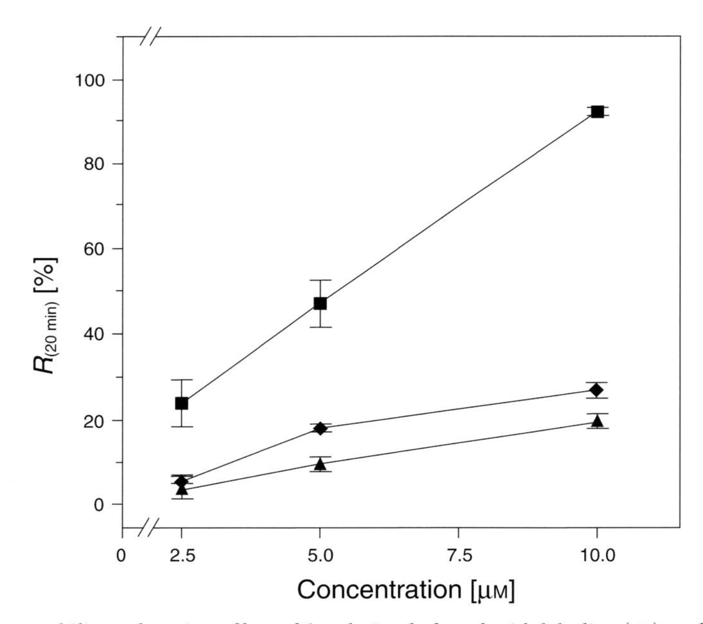

Fig. 5. *Permeability-enhancing effect of 1-palmitoyl phosphatidylcholine* (◆), *palmitic acid* (▲), *and 1:1 mixtures of the two* (■) *of fluid-phase liposomes monitored by the release of calcein after 20 min*. Results are shown as a function of the concentration of the enhancer. Adapted from [19].

the compounds lowers the permeability barrier which is likely to be related to the fact that both types of molecules increase the curvature stress in the bilayer. In addition, the figure demonstrates that there is a distinct synergistic effect when lysolipids and fatty acids are present at the same time [19]. This synergistic effect can be exploited in liposome-based drug delivery, which uses phospholipase A_2 as a trigger as discussed in *Sect. 6.2* below. This lipase produces equimolar mixtures of lysolipids and free fatty acids upon hydrolysis of a liposomal carrier.

4. Lipids' Sense for Curvature

The dimensions of a lipid molecule are determined by several factors. Obviously, there are geometric factors like the size of the polar head, the length of the fatty acid tail, and the degree of unsaturation of the fatty acid chains. Only lipids with a limited range of shapes will fit into a bilayer structure. In general, the average molecular shape has to be close to that of a cylindrical rod. The effective molecular shape of a lipid molecule is important for the ability of a lipid to form and participate in a bilayer structure. The effective shape is a property that is influenced by the geometrical constraints imposed by the lipid bilayer. In recent years, it has become increasingly clear that lipid shape is important for the control of the barrier properties of lipid membranes.

The effective shape of a lipid molecule is determined by the degree of compatibility between the size of the head group and the size of the hydrophobic tail. There are various ways of changing the effective shape of a lipid molecule by varying the relative sizes of the head and the tail. A small head and a bulky tail and a large head and a skinny tail will produce conical shapes of different sense as illustrated in *Fig. 6*.

The effective shape of lipid molecules determines their ability to form a stable bilayer and hence its barrier properties as well as its affinity for exogenous compounds. The more noncylindrical their shapes are, the less stable a bilayer they will form. This is illustrated in *Fig. 7*, where the two monolayers separately possess an intrinsic tendency to elastically relax towards a state of finite curvature. The monolayers display spontaneous curvature. When a bilayer is made of monolayers with nonzero spontaneous curvature it becomes subject to a built-in frustration, *i.e.*, a curvature stress field. If the spontaneous curvatures of the two monolayers are different, the bilayer becomes asymmetric and assumes itself a nonzero spontaneous curvature.

If, however, the cohesion of the bilayer cannot sustain the curvature stress, the stress will force nonlamellar structures to form such as emul-

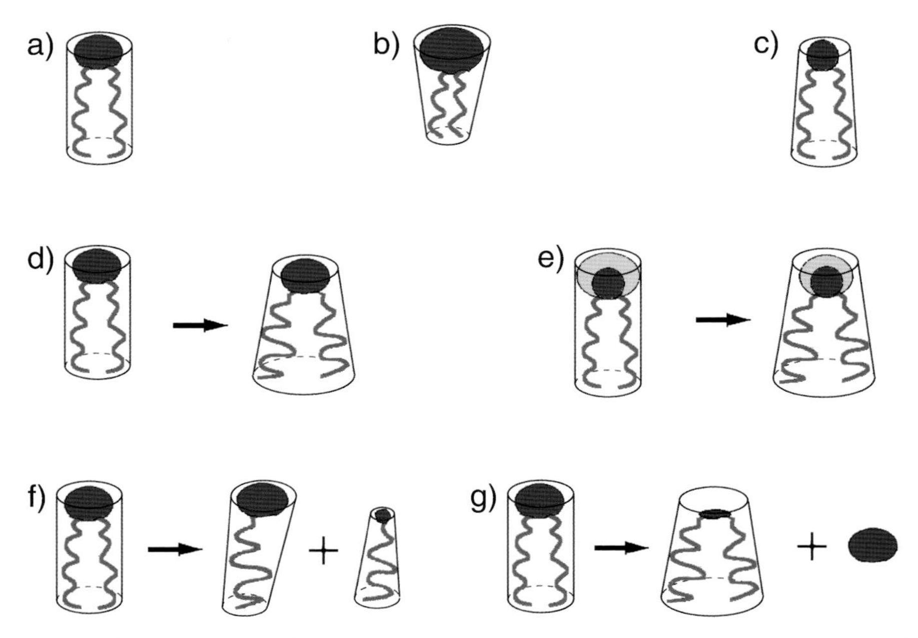

Fig. 6. *Effective shapes of lipid molecules. a*) Cylindrical: similar cross-sectional areas of head and tail. *b*) Cone: big head and skinny tail. *c*) Inverse cone: small head and bulky tail (*e.g.*, with unsaturated fatty acid chains). *d*) Becoming conical by increasing temperature. *e*) Becoming conical by changing the effective size of the head group, *e.g.*, by changing the degree of hydration or by changing the effective charge of an ionic head group. *f*) Becoming conical by chopping off one fatty acid chain, *e.g.*, by the action of phospholipase A_2, which forms a lysolipid molecule and a free fatty acid. *g*) Becoming conical by chopping off the polar head group, *e.g.*, by the action of phospholipase C.

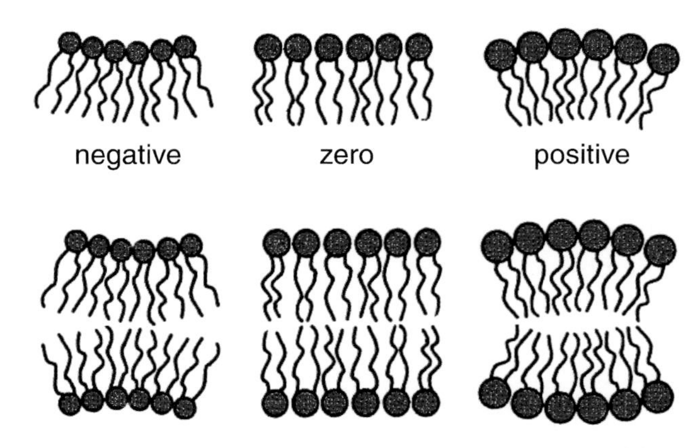

Fig. 7. *Illustration of the destabilization of a lipid bilayer composed of lipids with conical shapes producing a curvature stress field that leads to a tendency for the two monolayers to curve*

sions, hexagonal phases, cubic phases, and sponge structures. A particularly interesting structure is the inverted hexagonal structure H_{II} [20]. This structure is characterized by long cylindrical rods of lipids arranged as water-filled tubes. The diameter of the tubes can be varied by changing the type of lipid and by varying environmental conditions such as temperature, degree of hydration, as well as pH. The propensity of the lipids for forming and stabilizing the H_{II} structure is increased by shifting the balance between the size of the alkyl chain region and the size of the polar head as illustrated in *Fig. 6*.

Since the self-assembly process of lipid molecules into aggregates of different morphology implies a subtle competition between forces of different nature, and since many of the forces are of a colloidal and entropic nature, the relative stability of the resulting structures is intimately dependent on temperature, composition, and environmental conditions. The incorporation of various hydrophobic and amphiphilic solutes, such as hydrocarbons, alcohols, detergents, as well as a variety of drugs can shift the equilibrium from one structure to another. For example, monoacylglycerols, diacylglycerols, triacylglycerols, alkanes, and fatty acids promote the H_{II} structure, whereas lyso-phosphatidylcholine, digalactosyl diglyceride, certain antiviral peptides, as well as detergents inhibit the formation of the H_{II} structure [20].

The shape of the cholesterol molecule in relation to membrane curvature deserves a special remark. Compared to the small head group, which is just a OH group, the steroid ring structure, although hydrophobically rather smooth, is bulky, thus imparting cholesterol with an inverse conical shape. Cholesterol therefore displays propensity for promoting H_{II} structures.

5. Acylated Peptides at Membranes

Several molecular mechanisms can be imagined when water-soluble peptides and proteins associate with the surface of membranes, and a host of forces of different origin can be involved. Some of these forces are weak and nonspecific physical forces, others are strong and long-ranged electrostatic forces, while still others involve formation of H-bonds or even strong chemical bonds. In addition to these forces comes the possibility of a hydrophobic force that may drive an amphiphilic protein with a hydrophobic domain onto the membrane surface, caused by the hydrophobic effect, as is the case with many antimicrobial peptides.

A particular way of enhancing the affinity of a peptide to a membrane surface is to associate it with a hydrophobic anchor. This strategy of using a hydrocarbon chain as a membrane anchoring device is used by a large number of natural membrane proteins, *e.g.*, protein myristoylated alanine-rich C-kinase substrate (MARCKS) and those proteins called lamins which provide the scaffolding at the inner side of the membrane that bounds the cellular nucleus in eukaryotes. We consider here an example of a small synthetic and artificial polypeptide with only ten amino acids [21]. The polypeptide is acylated by a saturated hydrocarbon chain with 14 C-atoms. This chain can be used by the polypeptide to anchor to the membrane surface as illustrated in *Fig. 8*. The polypeptide is folded like a hairpin because it contains in the middle a particular amino acid (proline) that induces a turn. Furthermore it has another particular amino acid (tryptophan) near the acyl chain anchor. Tryptophans are known to be abundant in natural membrane proteins at the protein domains that locate in the hydrophilic–hydrophobic interface of the membrane.

The effects of drugs and hormones that act at membranes can be dramatically enhanced by the increased binding affinity that results from attaching a hydrophobic anchor to the substance. Examples include acylated insulin and the prodrug dipalmitoylated desmopressin. Desmo-

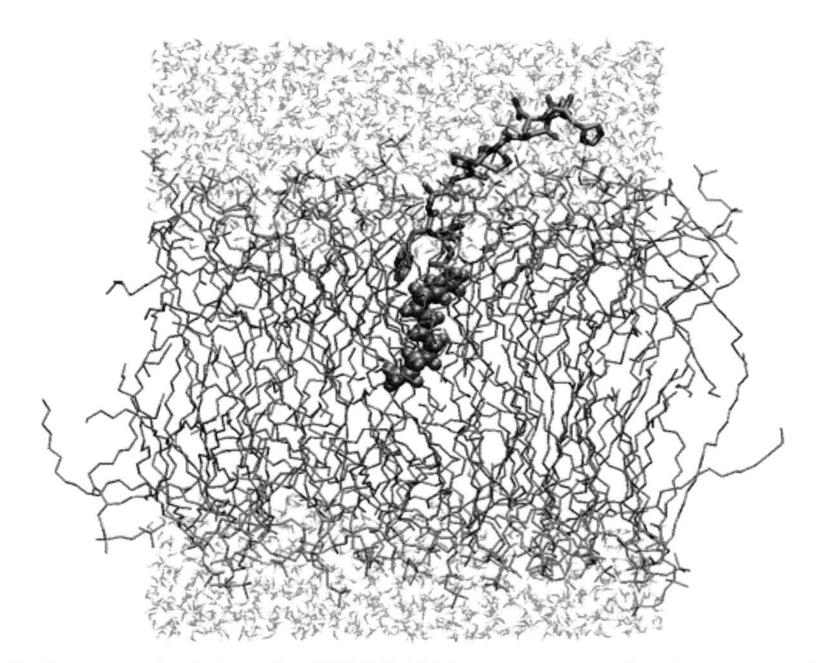

Fig. 8. *Computer simulation of a DPPC lipid bilayer incorporated with an myristoylated polypeptide* (C_{14}-His-Trp-Ala-His-Pro-Gly-His-His-Ala-amide). The polypeptide exhibits a hairpin conformation outside the membrane and the acyl anchor buried in the membrane core. Courtesy of Dr. *Morten Østergaard Jensen*.

pressin is an antidiuretic hormone peptide whose potency is increased 250-fold upon acylation [22].

6. The Magic Bullet Revisited

A successful administration of drugs requires a number of problems to be solved, many of which involve lipids and lipid membranes. Firstly, the free drug may need to be formulated which often requires lipid systems, such as micelles, emulsions, or liposomes. Secondly, the drug needs to be transported, targeted, and delivered to the target. Finally, the drug has to become adsorbed or adsorbed at this site and possibly be transported across a biological membranes. Also these steps often involve lipids. Below, we shall describe a situation where lipids are involved at every step on this route, and where the drug itself may even be a lipid molecule.

6.1. *Liposomes as Magic Bullets*

One of the key problems in the treatment of serious diseases is that many potent drugs have severe side effects. In the beginning of the twentieth century, the father of modern medicinal chemistry, *Paul Erlich*, envisioned the perfect drug as a 'bullet' that automatically targets and selectively kills the diseased cells without damaging healthy tissue. The term 'magic bullet' refers to this perfect drug. Dr. *Erlich*'s magic bullet has since been the holy grail in medicinal chemistry.

In a modern version, a magic bullet could be represented as shown in *Fig. 9*. The magic bullet contains a number of features [23]. First of all, it contains the drug or another related compound, *e.g.*, a prodrug that can be turned into a drug by an appropriate mechanism. The drug is attached to a carrier, which can transport the drug to the target. Secondly, the carrier may contain some kind of homing device that can search for and target the

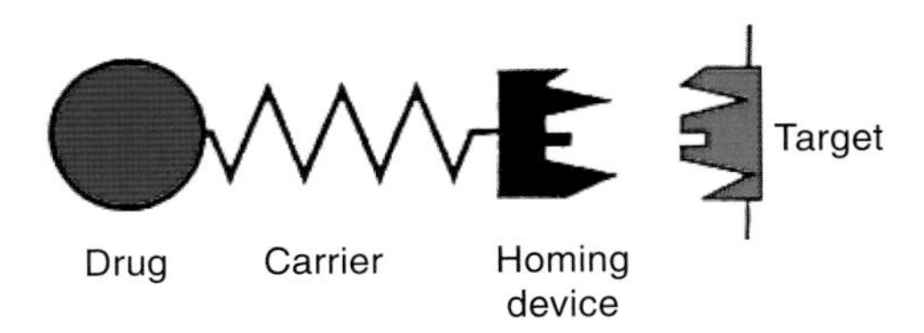

Fig. 9. *Schematic illustration of Dr.* Erlich*'s 'magic bullet' that can target and deliver a drug to a specific site.* The magic bullet consists of a drug, a carrier, and a homing device that can identify the target for the drug. Adapted from [23].

site where the drug is supposed to act. And finally, the carrier could also contain substances that act as enhancers or permeabilizers for the drug.

Ever since the British haematologist *Sir Alec Bangham* in the early 1960s identified liposomes as small water-containing lipid capsules, it has been a dream to use liposomes as magic bullets for drug delivery. Liposomes appear to be ideal for this purpose for several reasons. They are made of biocompatible, nontoxic, and biodegradable materials; they have an aqueous lumen that can contain hydrophilic substances; they are composed of a lipid bilayer that can accommodate hydrophobic or amphiphilic drugs; they can be manufactured in different sizes, some of them small enough to travel into the finest capillaries; and they can be associated with specific chemical groups at the liposome surface which can act as homing devices and thereby target specific cells.

Despite tremendous efforts made by a large number of researchers to devise liposome-based drug delivery systems, it is only in recent years that some success seems to be within reach. One of the major problems has been that conventional liposomes injected into the blood stream quickly become captured and degraded by macrophages of the immune system. When this happens, the drug is released into the blood where it becomes degraded or, even worse, may lyse red blood cells. Conventional liposomes therefore seldom make it to other sites in the body than the liver and the spleen.

A major step forward was made with the invention of the second-generation liposomes, the so-called '*Stealth* liposomes', that are screened from the macrophages by a polymer coat [24]. This coat is constructed by incorporating a certain fraction of lipopolymers into the liposome. A lipopolymer is a lipid molecule, to the head group of which is chemically linked a polymer molecule that is water soluble. The aqueous polymer coat exerts several physical effects. One is to provide an entropic repulsion between different liposomes and between liposomes and the special proteins that usually adsorb foreign particles in the blood as part of the immune system's defense strategy. Another effect is that the water-soluble polymers make the surface of the liposome look like water. The *Stealth* liposomes exhibit a circulation time in the blood that is far longer than that of bare conventional liposomes.

The increased stability of the *Stealth* liposomes implies that they can retain their poisonous load from the blood and get time to reach diseased sites before they eventually are cleared by the macrophages. Surprisingly and very fortunately, the *Stealth* liposomes are found to passively target to sites of trauma. The reason for this fortunate mechanism is that the liposomes, due to their small size and their long circulation times, can venture into the leaky capillary that are characteristic of tissues infested with tumors, inflammation, and infections. The diseased tissue, so to say, sucks up the circulating liposomes in their porous structure. This enhances the efficacy of the drug and limits severe side effects.

One of the first drugs that was successfully used in a liposomal formulation was amphotericin B. Amphotericin B is a very potent antibiotic which is used in the treatment of systemic infections that are very serious for immunodepressed patients such as AIDS patients and patients undergoing chemotherapy. Amphotericin B is extremely toxic but water soluble and can, therefore, be readily encapsulated in the aqueous lumen of the liposome. Another example is liposomal formulations of doxorubicin which is a potent anticancer drug that is used, *e.g.*, in the treatment of breast cancer in women. Doxorubicin is hydrophilic and it is incorporated in the carrier liposomes as a small solid crystalline particle. A few other liposomal formulations with anticancer drugs and vaccines have been approved for use in patients, and there are currently a number under development and clinical testing.

The properties of a liposomal carrier system can be optimized to the actual case by modulating the lipid composition of the liposomes. The lesson so far has been that it is necessary to go through an elaborate optimization procedure that takes a large number of details into account with respect to the actual drug, the actual disease, and the molecular composition of the liposomal formulation. With respect to active targeting of liposomes, only limited progress has been made so far. Hence, the full realization of Dr. *Erlich*'s magic bullet in *Fig. 9* by means of liposomes remains a visionary idea.

Paradoxically, one of the outstanding problems is not as much how to stabilize liposomes with encapsulated drugs as it is to destabilize them and arrange for the liposomes to release and deliver a sufficiently large part of their load exactly where it is needed. Moreover, the release should take place over a time span that is tuned to the mode of action of the drug. Below, we shall describe a couple of cases where the insight into the physics and physical chemistry of lipid bilayers and liposomes, in particular with respect to thermal phase transitions and enzymatic degradation of lipids, has provided a key to solve the problem of site-specific drug release.

6.2. *Liposomes in Cancer Therapy*

One of the problems using liposomes for cancer therapy is that, although it is possible to encapsulate the drug, *e.g.*, doxorubicin, and thereby significantly reduce the toxic side effects of the chemotherapy, the liposomes do not necessarily deliver more drug to the tumors than by applying the free drug. The reason is that the drug cannot get out of the capsule sufficiently rapidly and in sufficiently large local doses.

A team of scientists and medical doctors at Duke University may have solved this problem using hyperthermia, that is heating the tumor a few degrees above body temperature [25]. It turns out that heating has several beneficial effects. The heat opens the tiny blood vessels in the tumor making it possible for the liposomes to sneak in. Moreover, the heat enhances the uptake of the drug into the cancer cells and increases the damage which the drug does to the DNA of the cancer cell. The crucial point, however, is that the liposomes used by the Duke researchers are poised to become leaky at temperatures a few degrees above the body temperature. The mechanism to do so is a lipid phase transition. At the transition, the lipid bilayer becomes leaky and the encapsulated material flows out.

The American material scientist *David Needham* has used this phase transition phenomenon to construct liposomes whose drug-release mechanism is precisely the lipid phase transition [25]. By composing the liposomes of lipids that have a phase transition and become leaky slightly above body temperature, but otherwise are fairly tight at lower temperatures, *Needham* has succeeded in making a formulation for chemotherapy that can deliver as much as thirty times more drug at the tumor site than a conventional liposome. In *Fig. 10* is shown how sensitive the release in this system can be tuned to temperatures above body temperature in a range which can be clinically achieved by local heating using microwave, ultrasound, or radio-frequency radiation. The release of the drug from the heated liposomes is very fast, within 20 s after heating, which is a crucial factor for the therapeutic effect. The release is much faster than from ordinary liposomes.

There is an additional benefit of using soft matter like lipid aggregates for drug encapsulation and delivery of this type. When a leaking liposome leaves the tumor area that is heated, it seals again when the temperature drops, because the liposome is a self-assembled object. The remaining drug is retained in the liposome and therefore does not get out into possible healthy tissue. If this liposome later diffuses back into the heated tumor area, more drug can get released. The system developed at Duke University has shown some very promising results in the treatment of breast cancer and may eventually also be used for other cancers.

To use hyperthermia as a mechanism for drug release in chemotherapy, it is necessary to know which area to heat. The position and the size of the diseased tissue therefore have to be known beforehand. These conditions may not be fulfilled for many cancers, in particular in their early stages of

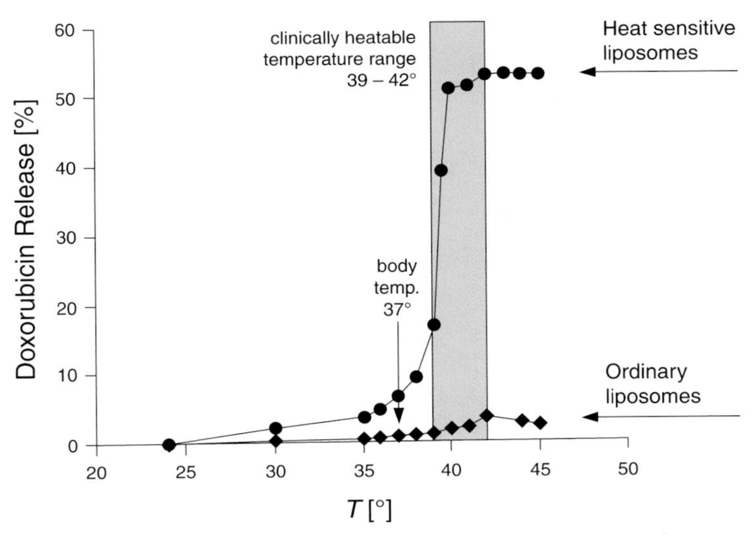

Fig. 10. *Release of the anticancer drug doxorubicin from liposomes triggered by a lipid phase transition in the liposome.* The liposomes sensitive to hyperthermia (●) are seen to release a very large part of the encapsulated drug over a narrow range of temperatures slightly above body temperature. In contrast, conventional liposomes (◆) release almost no drug in the same temperature range. Courtesy of Dr. *David Needham*.

development. To come closer to Dr. *Erlich*'s vision of a magic bullet, it would be desirable to have a liposome which could identify the sites of disease itself and by some appropriate automatic mechanism be triggered to unload the drug precisely at those sites.

For this purpose, it may be possible to use specific phospholipases to automatically trigger the opening of liposomes at diseased sites [26]. It is known that certain variants of secretory phospholipase A_2 are overexpressed in malignant tumors and sometimes occur in a concentration that is maybe ten times larger than in healthy tissue. Phospholipase A_2 catalyzes the hydrolysis of phospholipids into lysolipids and free fatty acids, leading to a leakage of bilayers and eventually to a breakdown of liposomes. Moreover, the activity of these enzymes is tightly regulated by the physical properties of the lipid bilayer. Hence, by tailoring liposomes to be sensitive to enzymatic breakdown under circumstances prevailing in the tumor on the one side and by taking advantage of the elevated levels of phospholipase A_2 in the tumor on the other site, a smart principle of automatic triggered drug release arises.

These examples show that lipids in the form of liposomes may be of great help in fighting cancer. But the role of lipids does not stop with that. In 1999, the Danish pharmacist *Kent Jørgensen* realized that it should be possible to use the phospholipase-induced triggering of drug release by the

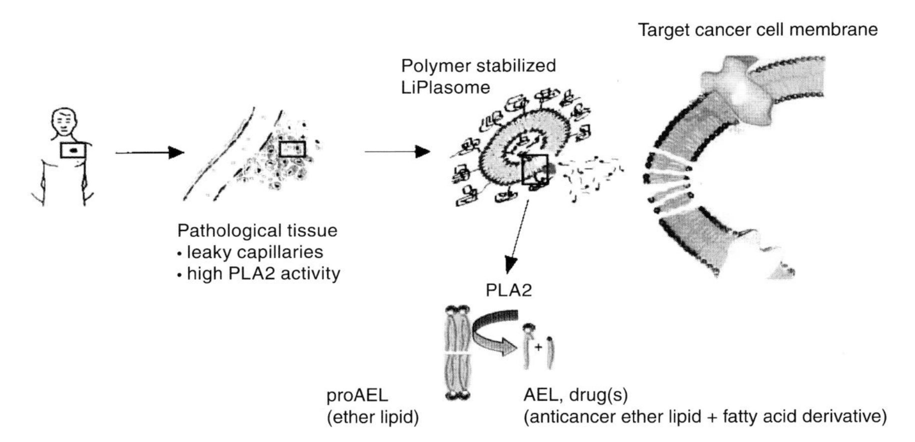

Fig. 11. *Schematic illustration of lipids fighting cancer.* LiPlasomes which are liposomes with ether lipids circulate in the blood stream for a long time because they have a polymer coat. The LiPlasomes accumulate in cancerous tissue by a passive mechanism caused by the leaky capillaries characteristic of solid tumors. The LiPlasomes are broken down in the tumor by the enzyme phospholipase A_2 (PLA2), which is upregulated in cancer. The products, anticancer lyso-ether lipid and free fatty acid, are released and transported into the cancer cell. Courtesy of Dr. *Jesper Davidsen*.

mechanism described above not only to release drugs but also produce a potent drug at the site of disease [27]. The idea is amazingly simple and illustrated schematically in *Fig. 11*. The trick is to use liposomes made of lipids that upon hydrolysis with the phospholipase lead to products that are drugs themselves. Compounds that can be turned into drugs but are not drugs themselves are so-called prodrugs. The prodrug in this case is lipid where the alkyl chain in *sn*-1 position is bound to the glycerol backbone by an ether bond and in the *sn*-2 position by an ester bond [28]. After hydrolysis catalyzed by phospholipase A_2, the products are a lyso-ether lipid and a free fatty acid. The lyso-ether lipid is an extremely potent anticancer drug which so far has found limited use in conventional chemotherapy because it kills red blood cells. However, in its masked prodrug form as part of a lipid it turns out to be completely harmless. Hence the prodrug can be incorporated into long-circulating *Stealth* liposomes which, upon accumulation in the capillaries of porous cancerous tissue, are broken down by phospholipase A_2. The drug is, therefore, produced exactly where it is needed, in fact without any prior knowledge of the localization of the tumor. In this system, the drug carrier and the drug are two sides of the same thing. Obviously, liposomes of this type, which are called LiPlasomes, can also be made to include conventional anticancer agents, like doxorubicin. This may prove useful in combination therapies. One could also imagine that the second hydrolysis product, the fatty acid, is chosen to have some additional therapeutic effect.

But this is not the full story of ether lipids fighting cancer. First, cancer cells are very vulnerable to ether lipids because they do not contain enzymes that can break down ether lipids. Except for red blood cells, other healthy cells do have such enzymes. Second, there are a number of added benefits from the LiPlasome concept. As an example, the lysolipids and the free acids produced by the phospholipase-triggered hydrolysis near the cancer cells have some beneficial effects from their propensity to form nonlamellar lipid phases. They facilitate the transport of the drug into the cancer cell by lowering the permeability barrier of the target cell. Compounds with this capacity are called drug enhancers. In the case of lysolipids and fatty acids, the mode of action of the enhancers is based on a purely physical mechanism caused by the effective conical shapes of the molecules. In this way, the LiPlasome not only carries a prodrug, it also carries pro-enhancers and pro-permeabilizers that are turned into enhancers at the target.

7. Conclusions

In this review, we discussed the biophysics of lipid bilayer membranes from the point of view that the membrane can act both as a barrier, as a carrier, and as a target for drugs. After a general description of the transverse and lateral structure of lipid bilayers composed of different lipids and cholesterol, we considered the permeability and absorption properties of lipid bilayers in relation to solutes such as alcohols, fatty acids, and lysolipids, with a focus on amphiphilicity and the propensity of the solutes to form nonlamellar phases. The binding of acylated peptides were briefly mentioned. Finally, it was shown how the insight into the biophysical properties of lipid membranes can be used for the design of novel systems for liposome-based, targeted drug delivery involving prodrugs, pro-enhancers, and pro-permeabilizers.

This work was supported by the *Danish National Research Foundation* with a grant to *MEMPHYS-Center for Biomembrane Physics*.

REFERENCES

[1] O. G. Mouritsen, 'Life – As a Matter of Fat. The Emerging Science of Lipidomics', Springer Verlag, Berlin, 2005.
[2] O. G. Mouritsen, K. Jørgensen, *Pharm. Res.* **1998**, *15*, 1507.

[3] S. H. White, M. C. Wiener, in 'Biological Membranes. A Molecular Perspective for Computation and Experiment', Eds. K. M. Merz and B. Roux, Birkhäuser Publ. Co., New York, 1996, p. 127–144.
[4] D. P. Tieleman, S. J. Marrink, H. J. C. Berendsen, *Biochim. Biophys. Acta* **1997**, *1331*, 235.
[5] R. S. Cantor, *Biochemistry* **2003**, *42*, 11891.
[6] R. S. Cantor, *Chem. Phys. Lipids* **1999**, *101*, 45.
[7] I. Israelachvili, 'Intermolecular and Surface Forces', Academic Press, London, 1991.
[8] O. G. Mouritsen, O. S. Andersen, 'In Search of a New Biomembrane Model', *Biol. Skr. Dan. Vid. Selsk.* **1998**, *49*, 1–214.
[9] L. A. Bagatolli, *Chem. Phys. Lipids* **2003**, *122*, 137.
[10] H. A. Rinia, B. de Kruijff, *FEBS Lett.* **2001**, *504*, 194; T. Kaasgaard, O. G. Mouritsen, K. Jørgensen, *FEBS Lett.* **2002**, *515*, 29.
[11] M. Edidin, *Annu. Rev. Biomol. Struct.* **2003**, *32*, 257.
[12] L. Miao, M. Nielsen, J. Thewalt, J. H. Ipsen, M. Bloom, M. J. Zuckermann, O. G. Mouritsen, *Biophys. J.* **2002**, *82*, 1429.
[13] R. G. W. Anderson, K. Jacobson, *Science* **2002**, *296*, 1821.
[14] P. K. J. Kinnunen, *Chem. Phys. Lipids* **1991**, *57*, 375.
[15] E. K. J. Tuominen, C. J. A. Wallace, P. K. J. Kinnunen, *J. Biol. Chem.* **2002**, *277*, 8822.
[16] a) J. S. Andersen, K. Jørgensen, O. G. Mouritsen, unpublished; b) C. Trandum, P. Westh, K. Jørgensen, O. G. Mouritsen, *Biophys. J.* **2000**, *78*, 2486.
[17] E. Corvera, O. G. Mouritsen, M. A. Singer, M. J. Zuckermann, *Biochim. Biophys. Acta* **1992**, *1107*, 261.
[18] P. Høyrup, J. Davidsen, K. Jørgensen, *J. Phys. Chem., B* **2001**, *105*, 2649.
[19] J. Davidsen, O. G. Mouritsen, K. Jørgensen, *Biochim. Biophys. Acta* **2002**, *1564*, 256.
[20] P. K. J. Kinnunen, 'Nonmedical Applications of Liposomes', Eds. D. D. Lasic and Y. Barenholz, CRC Press, Inc., Boca Raton, Florida, 1995, p. 153–171.
[21] M. Ø. Jensen, O. G. Mouritsen, G. H. Peters, *Biophys. J.* **2004**, *86*, 3556.
[22] T. B. Pedersen, S. Frokjaer, O. G. Mouritsen, K. Jørgensen, *Int. J. Pharm.* **2002**, *233*, 199.
[23] D. J. A. Crommelin, G. Storm, in 'Innovations in Drug Delivery. Impact on Pharmacotherapy', Eds. T. Sam and J. Fokkens, Houten: Stichting Orhanisatie Anselmus Colloquium, 1995, p. 122–133.
[24] A. S. Janoff, 'Liposomes–Rational Design', Marcel Dekker Inc., New York, 1999.
[25] D. Needham, G. Anyarambhatla, G. Kong, M. W. Dewhirst, *Cancer Res.* **2000**, *60*, 1197.
[26] J. Davidsen, K. Jørgensen, T. L. Andresen, O. G. Mouritsen, *Biochim. Biophys. Acta* **2003**, *1609*, 95.
[27] K. Jørgensen, J. Davidsen, O. G. Mouritsen, *FEBS Lett.* **2002**, *531*, 23.
[28] T. L. Andresen, J. Davidsen, M. Begtrup, O. G. Mouritsen, K. Jørgensen, *J. Med. Chem.* **2004**, *47*, 1694.

Cell Culture Absorption Models – State of the Art

by **Per Artursson*** and **Pär Matsson**

Department of Pharmacy, Center of Pharmaceutical Informatics, Uppsala University, Box 580, SE-751 23 Uppsala
(e-mail: per.artursson@farmaci.uu.se; Par.Matsson@farmaci.uu.se)

Abbreviations
2/4/A1: Conditionally immortalized rat intestinal epithelial cell line; Caco-2: adenocarcinoma cell line derived from human colon; *FA*: fraction of a dose that is absorbed after oral administration; f_i: fraction of total drug amount present in the ionized form; f_u: fraction of total drug amount present in the unionized form; HDM: filter-immobilized artificial hexadecane membrane; LC/MS/MS: high-performance liquid chromatography with mass spectrometric detection; log P_{oct}: octanol/water partition coefficient; P-gp: P-glycoprotein; P_m: membrane permeability coefficient; P_{mi}: permeability coefficient for the ionized species; P_{mu}: permeability coefficient for the unionized species; *PSA*: polar surface area; TEER: trans-epithelial electrical resistance.

---

1. Introduction

Intestinal drug permeability is considered to be one of the two major barriers to intestinal drug absorption, solubility being the other [1]. For the assessment of intestinal drug permeability, epithelial cell culture models such as Caco-2 are routinely used [2] [3]. One of the reasons for the widespread use of Caco-2 assays is the versatility of the cell line, which allows studies not only of passive diffusion processes, but after modifications of the experimental setup also of active drug transport and efflux systems as well as presystemic drug metabolism [4] [5]. However, this versatility is considered as a potential weakness of the Caco-2 cell culture model in the screening setting. For instance, the Caco-2 cell line forms very tight monolayers compared to human small intestine, which has been explained with the colonic origin of the cell line [6]. Also, in some combinatorial libraries, an unexpectedly large percentage of the compounds are found to be substrates for efflux proteins in Caco-2 cells, whereas the *in vivo* relevance of these findings remains unclear [7].

2. Alternative Models

A preferred solution to these issues is obtained *via* a reductionist approach by which different mechanisms affecting intestinal drug absorption are studied in separate experimental systems. Different alternative models have been used to study the dominating passive transport pathways in isolation from active transport, such as the intestinal epithelial cell line 2/4/A1 [8] and the filter-immobilized hexadecane membrane model (HDM) [9], while for active transport, studies in expression systems that overexpress the specific active drug transporter are generally favored [10] [11]. In this review, we will focus on cell culture models of the passive routes.

2/4/A1 Cells lack functional expression of several important active drug transporters [12] and form monolayers with a more leaky, small-intestinal-like paracellular pathway than Caco-2 monolayers (*Fig. 1*). As a result, the 2/4/A1 cell line is thought to better mimic the epithelial barrier to passive drug transport *in vivo* [13]. The different artificial membrane models also completely lack active transport pathways, but in addition, the paracellular pathway is absent in these models (*Fig. 1*) [9] [14–16].

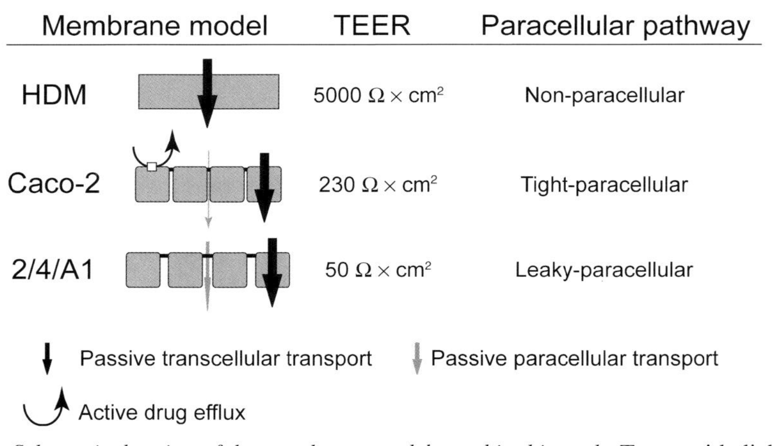

Fig. 1. *Schematic drawing of the membrane models used in this study.* Trans-epithelial electrical resistance (TEER) is a measure of the resistance to ion flux across the membrane. Since the paracellular pathway greatly affects TEER, a high value will reflect a tighter membrane.

3. Transcellular and Paracellular Drug Transport Pathways

The popularity of the artificial membrane models relies on the assumption that the transcellular barrier is the dominating barrier to drug absorption. This assumption is supported by numerous studies indicating that intestinal drug absorption of soluble drugs is correlated to passive drug permeability, *e.g.*, [17–19]. Passive membrane permeability is strongly related to relatively simple molecular descriptors, and this has made it possible to develop *in silico* models that predict permeability of passively transported drugs. Examples include rule-based models such as *Lipinski*'s rule-of-five [1] and correlations with H-bond descriptors [20][21] or molecular surface area descriptors such as the polar surface area (*PSA*) [18][22–24]. Influential descriptors in such relationships can give information about rate-limiting steps in the permeability process. For instance, molecular descriptors relating to H-bond interactions often significantly decrease permeability, whereas lipophilicity measures such as log P_{oct} are positively correlated with permeability [18][22][25]. Thus, in principle, it is possible to develop *in silico* models that predict drug permeability from molecular descriptors, but whether such models can compete with experimental permeability models in a screening setting is not clear yet [26].

If the only important drug transport route were the passive transcellular route, then the pH-dependent permeability of a charged drug would be in excellent agreement with the pH-partitioning theory [27], at least under well-controlled *in vitro* conditions. However, significant deviations from the pH-partitioning theory have been observed for passively transported compounds in epithelial cell culture models such as Caco-2 [28]. These deviations may be explained by permeation of charged drug through the paracellular route in these monolayers. The paracellular pathway can be considered to consist of dynamic water-filled channels and is believed to contribute significantly to the permeability of hydrophilic drugs such as H_2 receptor antagonists, some β-receptor antagonists, and hydrophilic peptides in the human small intestine. Since the paracellular route is permeable to hydrophilic drugs, it is likely that this route plays an important role for the permeability of the ionized fraction of drug, and causes deviations from the pH-partitioning theory of transport across epithelial cell layers. It should also be emphasized that deviations from the pH-partitioning theory need to be considered in the experimental design of permeability screening experiments to interpret passive as well as active transport data correctly [29]. Ignoring these basic concepts may result in erroneous conclusions with regard to transport mechanisms. For instance, when transport experiments are performed with different pH values on

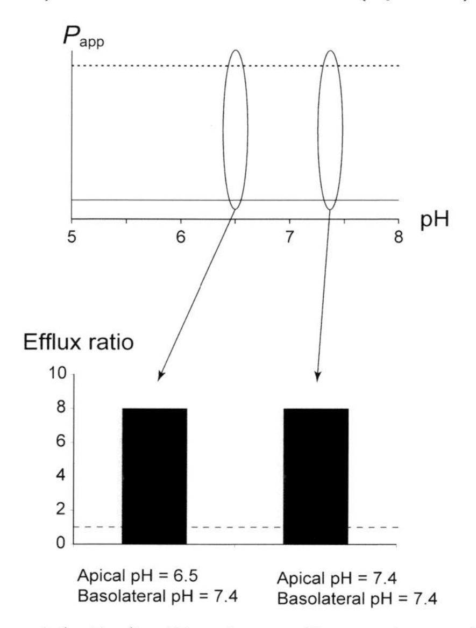

Fig. 2. *Schematic drawing of 'passive' and real* (active) *efflux in a cell monolayer. a*) Asymmetric pH-dependent transport of a basic drug (which is not an efflux substrate) caused by uneven distribution of the uncharged species, rather than by active transport. The observed efflux ratio in the left panel can erroneously be interpreted as active, P-gp-mediated efflux. *b*) Active pH-independent transport of an uncharged drug that is an efflux substrate for P-gp results in an efflux ratio that is independent of pH. An aprotic P-gp substrate (*e.g.*, digoxin) was chosen to avoid bias from the 'passive' efflux observed in *a*. Solid line: apical to basolateral permeability. Dashed line: basolateral to apical permeability. Efflux ratio = (basolateral to apical permeability)/(apical to basolateral permeability). P_{app} = apparent permeability coefficient.

the two sides of the cell monolayer, the observed drug efflux can be caused by pH-partitioning phenomena rather than by active efflux by for instance P-glycoprotein (P-gp) (*Fig. 2*) [29].

4. Comparative Studies

We reasoned that comparative studies of drug permeability in immobilized artificial membranes (which lack tight junctions), Caco-2 cell monolayers (which have low permeable tight junctions), and 2/4/A1 cell monolayers (which have tight junctions with higher, small-intestinal-like permeability) would give further insight into the contribution of the paracellular route to drug permeability.

4.1. *Examination of pH-Dependent Membrane Permeability*

In the first study, we used artificial membranes based on hexadecane (HDMs) together with Caco-2 and 2/4/A1 models to examine the pH-dependent membrane permeability (P_m) of the highly permeable drug alfentanil and the low-permeable drug cimetidine [30]. These models represent transcellular, transcellular + tight (colon-like) paracellular, and transcellular + leaky (small-intestinal-like) paracellular routes, respectively. Both alfentanil and cimetidine have pK_a values in the physiological pH range of the gastrointestinal tract (pH 5.0–8.0). The paracellular permeability was calculated using two approaches; one based on the assumption that the permeability of the ionized form (P_{mi}) permeates the cell monolayer exclusively by the paracellular route (*Fig. 3*), and another method based on pore restricted diffusion [31][32]. For both drugs, sigmoidal relationships between membrane permeability and pH were observed in all models. No significant permeability to the ionized species of the drugs could be observed in HDM, while the permeability to the ionized drugs was small and large in Caco-2 and 2/4/A1, respectively. The permeability of the ionized species was in excellent agreement with the paracellular permeability of cimetidine in the two cell models. Moreover, the permeability coefficient of cimetidine in 2/4/A1 cells was in excellent agreement with that obtained after perfusion of the human jejunum. We concluded that the paracellular route plays a significant role in the permeability of small hydrophilic drugs, such as cimetidine, in leaky, small-intestinal-like epithelial such as 2/4/A1 [30]. By contrast, in tighter epithelia such as Caco-2 and in artificial membranes such as HDM, the permeability of the ionized forms of the drugs and the paracellular permeability are lower and insignificant, respectively.

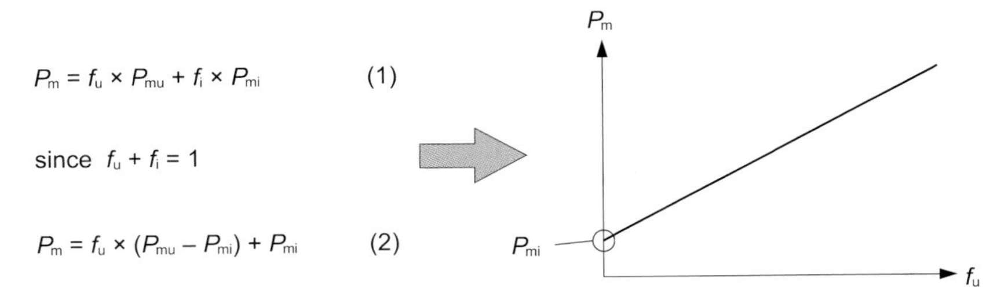

Fig. 3. *A simple approach to the calculation of the permeability of the unionized drug species from* in vitro *cell culture experiments.* The cellular permeability (P_m) is equal to the sum of the fractional permeabilites of the unionized (P_{mu}) and ionized (P_{mi}) drug species (*Eqn. 1*). Since the sum of f_u and f_i equals one, the linear relationship in *Eqn. 2* can be obtained. P_{mi} is obtained from the intercept of the P_m axis in the graph.

4.2. *Examination of the Influence of Different Drug Transport Routes in Intestinal Drug Permeability*

The aim of the second comparative study was to examine the influence of different drug transport routes in intestinal drug permeability screening assays, using a data set of drugs that was evenly distributed with regard to the absorbed fraction in humans (*FA* = 0–100%). For this purpose, the three experimental permeability models used to investigate the pH-dependent permeability were used once more: the small-intestinal-like 2/4/A1-cell model, the widely used Caco-2 epithelial cell line, which has more colon-like permeability, and the artificial hexadecane membranes (HDM), which exclusively models the passive transcellular pathway. The three models were investigated regarding their ability to: *1*) divide compounds into two permeability classes, and *2*) rank compounds in order of intestinal absorption. The experimental data were also used to develop *in silico* permeability models. All three studied experimental models performed equally well in classifying drugs into permeability classes above or below an *FA* cutoff of 80–90%.

A potential bias seen in studies of *in vitro* and *in silico* models aiming at predicting drug absorption is that the data sets are generally enriched in completely absorbed compounds [9][16][17][21][33–35]. This is probably due to that the data sets are based on orally administered marketed drugs that have been developed partly based on their favorable absorption properties. Models based on such skewed data sets are often able to distinguish fairly well between completely (≥80%) and poorly absorbed compounds, and can rapidly give useful indications with regard to the expected average permeability of compound libraries early in the drug discovery process. However, compounds with moderate absorption properties (*FA* ≥ 30%) may also be of interest in early drug discovery phases, and in our opinion it would be advantageous to include such compounds in the evaluation of permeability models. The need to include a significant number of incompletely absorbed compounds in the data set is further emphasized during lead optimization, when ranking of compounds according to absorption properties is desirable.

When we used a lower *FA* cut off of 30% to divide the compounds into two permeability classes, the cell-based experimental models performed better than the HDM model. The best results with regard to compound ranking were obtained with 2/4/A1 ($r = 0.74$; after removing actively transported outliers $r = 0.95$). Moreover, the *in silico* model based on 2/4/A1 permeability gave results of similar quality as when using experimental P_{app}, and was better than the HDM experimental model in ranking compounds in order of *FA* ($r = 0.85$ and 0.47 for the 2/4/A1 *in silico* model and the HDM experimental model, resp.). From these results, we concluded that the paracellular transport pathway present in the cell models plays a significant role in models used for permeability screening, and that the new 2/4/A1 model is a promising alternative for drug discovery permeability screening.

5. Conclusions

Previous studies have shown that the reductionist approach, by which different mechanisms of drug absorption are studied in separate experimental systems, can be advantageous for studies of active transport mechanisms [36]. Our ongoing investigations indicate that a reductionist approach, which allows the study of the dominating passive transport pathways in isolation from active transport, is advantageous and may be used to improve the predictivity of permeability screening in cell culture models, especially with regard to incompletely absorbed drugs.

Our findings also demonstrate a need to include the paracellular pathway in models used for drug permeability screening, especially in drug discovery programs where significant numbers of low-permeability compounds are expected. We propose that the small-intestinal-like 2/4/A1 epithelial cell line, which has the largest influence of the paracellular pathway of the experimental models we have studied, is a suitable experimental model for studies of passive permeability in drug discovery. Further, since the 2/4/A1 model is at least 100-fold more permeable to low-permeability drugs than Caco-2 and HDM, the need for time-consuming and sophisticated analytical equipment such as LC/MS/MS is probably eliminated. However, the results in our comparison of permeability screening models also show that none of the studied models were able to predict *FA* for all compounds with significant active transport mechanisms *in vivo*. Separate models for detecting active transporter affinities of discovery compounds are therefore warranted.

This work was supported by grant no. 9478 from the *Swedish Research Council*, the *Knut and Alice Wallenberg Foundation*, the *Swedish Fund for Research without Animal Experiments*, and the *Swedish Animal Welfare Agency*.

REFERENCES

[1] C. A. Lipinski, F. Lombardo, B. W. Dominy, P. J. Feeney, *Adv. Drug Delivery Rev.* **1997**, *23*, 3.
[2] P. Artursson, R. T. Borchardt, *Pharm. Res.* **1997**, *14*, 1655.
[3] P. Artursson, K. Palm, K. Luthman, *Adv. Drug Delivery Rev.* **2001**, *46*, 27.

[4] H. A. Engman, H. Lennernäs, J. Taipalensuu, C. Otter, B. Leidvik, P. Artursson, *J. Pharm. Sci.* **2001**, *90*, 1736.
[5] P. Schmiedlin-Ren, K. E. Thummel, J. M. Fisher, M. F. Paine, K. S. Lown, P. B. Watkins, *Mol. Pharmacol.* **1997**, *51*, 741.
[6] E. Grasset, M. Pinto, E. Dussaulx, A. Zweibaum, J. F. Desjeux, *Am. J. Physiol.* **1984**, *247*, C260.
[7] S. Tavelin, J. Gråsjö, J. Taipalensuu, G. Ocklind, P. Artursson, *Methods Mol. Biol.* **2002**, *188*, 233.
[8] S. Tavelin, V. Milovic, G. Ocklind, S. Olsson, P. Artursson, *J. Pharmacol. Exp. Ther.* **1999**, *290*, 1212.
[9] F. Wohnsland, B. Faller, *J. Med. Chem.* **2001**, *44*, 923.
[10] J. W. Polli, S. A. Wring, J. E. Humphreys, L. Huang, J. B. Morgan, L. O. Webster, C. S. Serabjit-Singh, *J. Pharmacol. Exp. Ther.* **2001**, *299*, 620.
[11] T. Nakanishi, L. A. Doyle, B. Hassel, Y. Wei, K. S. Bauer, S. Wu, D. W. Pumplin, H. B. Fang, D. D. Ross, *Mol. Pharmacol.* **2003**, *64*, 1452.
[12] S. Tavelin, J. Taipalensuu, F. Hallböök, K. S. Vellonen, V. Moore, P. Artursson, *Pharm. Res.* **2003**, *20*, 373.
[13] S. Tavelin, J. Taipalensuu, L. Söderberg, R. Morrison, S. Chong, P. Artursson, *Pharm. Res.* **2003**, *20*, 397.
[14] J. A. Ruell, A. Avdeef, C. Du, K. Tsinman, *Chem. Abstr.* **2002**, *224*, U35.
[15] K. Sugano, H. Hamada, M. Machida, H. Ushio, K. Saitoh, K. Terada, *Int. J. Pharmac.* **2001**, *228*, 181.
[16] M. Kansy, F. Senner, K. Gubernator, *J. Med. Chem.* **1998**, *41*, 1007.
[17] M. Yazdanian, S. L. Glynn, J. L. Wright, A. Hawi, *Pharm. Res.* **1998**, *15*, 1490.
[18] P. Stenberg, U. Norinder, K. Luthman, P. Artursson, *J. Med. Chem.* **2001**, *44*, 1927.
[19] P. Artursson, J. Karlsson, *Biochem. Biophys. Res. Commun.* **1991**, *175*, 880.
[20] J. T. Goodwin, R. A. Conradi, N. F. Ho, P. S. Burton, *J. Med. Chem.* **2001**, *44*, 3721.
[21] Y. H. Zhao, J. Le, M. H. Abraham, A. Hersey, P. J. Eddershaw, C. N. Luscombe, D. Butina, G. Beck, B. Sherborne, I. Cooper, J. A. Platts, D. Boutina, *J. Pharm. Sci.* **2001**, *90*, 749.
[22] C. A. Bergström, M. Strafford, L. Lazorova, A. Avdeef, K. Luthman, P. Artursson, *J. Med. Chem.* **2003**, *46*, 558.
[23] J. Kelder, P. D. Grootenhuis, D. M. Bayada, L. P. Delbressine, J. P. Ploemen, *Pharm. Res.* **1999**, *16*, 1514.
[24] K. Palm, K. Luthman, A. L. Ungell, G. Strandlund, F. Beigi, P. Lundahl, P. Artursson, *J. Med. Chem.* **1998**, *41*, 5382.
[25] H. van de Waterbeemd, G. Camenisch, G. Folkers, O. A. Raevsky, *Quant. Struct.–Act. Rel.* **1996**, *15*, 480.
[26] C. Zhu, L. Jiang, T. M. Chen, K. K. Hwang, *Eur. J. Med. Chem.* **2002**, *37*, 399.
[27] P. Shore, B. Brodie, C. Hogben, *J. Pharmacol. Exp. Ther.* **1957**, *119*, 361.
[28] K. Palm, K. Luthman, J. Ros, J. Gråsjö, P. Artursson, *J. Pharmacol. Exp. Ther.* **1999**, *291*, 435.
[29] S. Neuhoff, A. L. Ungell, I. Zamora, P. Artursson, *Pharm. Res.* **2003**, *20*, 1141.
[30] N. Nagahara, S. Tavelin, P. Artursson, *J. Pharm. Sci.* **2004**, *93*, 2972.
[31] F. Curry, in 'Handbook of Physiology', Eds. E. Renkin, C. Michel, American Physiology Society, Bethesda, 1984, p. 309–374.
[32] G. T. Knipp, N. F. Ho, C. L. Barsuhn, R. T. Borchardt, *J. Pharm. Sci.* **1997**, *86*, 1105.
[33] D. Sun, H. Lennernäs, L. S. Welage, J. L. Barnett, C. P. Landowski, D. Foster, D. Fleisher, K. D. Lee, G. L. Amidon, *Pharm. Res.* **2002**, *19*, 1400.
[34] J. D. Irvine, L. Takahashi, K. Lockhart, J. Cheong, J. W. Tolan, H. E. Selick, J. R. Grove, *J. Pharm. Sci.* **1999**, *88*, 28.
[35] P. Mattson, C. A. S. Bergström, N. Nagahara, S. Tavelin, U. Norinder, P. Artursson, *J. Med. Chem.* **2005**, *48*, 604.
[36] N. Mizuno, T. Niwa, Y. Yotsumoto, Y. Sugiyama, *Pharmacol. Rev.* **2003**, *55*, 425.

Metabolic Studies in Drug Research and Development

by **Benjamin Neugnot**[a]), **Marie-Jeanne Bossant**[b]), **Fabrice Caradec**[b]), and **Bernard Walther***[b])

[a]) Département de Biologie Cellulaire et Moléculaire, Commissariat à l'Energie Atomique, Service de marquage moléculaire et de chimie bioorganique, Département de Biologie Joliot-Curie, F-91191 Gif-sur-Yvette

[b]) *Technologie Servier*, 25–27 rue Eugène Vignat, F-45000 Orléans
(e-mail: bernard.walther@fr.netgrs.com)

Abbreviations
AMS: Accelerator mass spectrometry; APCI: atmospheric pressure chemical ionization; API: atmospheric pressure ionization; ARC: accurate radioisotope counting; BNMI: *Bruker* NMR/MS interface; CYP: cytochrome P450; ESI: electrospray ionization; IMS: ion mobility spectrometry; LC: liquid chromatography; LSC: liquid scintillation counting; MALDI: matrix-assisted laser desorption/ionization; MRM: multiple reaction monitoring; MS: mass spectrometry; NMR: nuclear magnetic resonance; Q-TOF: quadrupole time-of-flight; Q-TRAP: quadrupole ion trap; SELDI: surface-enhanced laser desorption/ionization.

1. Introduction

The role and timing of metabolic studies in pharmaceutical Research and Development (R&D) have constantly changed over the last decade. But whatever the balance between R&D metabolic studies will be in the future, metabolism will remain a continuum of expertise, building up knowledge, first on the metabolism of chemical series tested in parallel to pharmacology, then on a few chemical entities tested in humans. The ultimate objective of this domain will remain the understanding of the biotransformation of a drug in its target population, namely patients.

Today, this objective can only be partly met for specific drugs like cytotoxic agents tested only in patients. Most drugs are evaluated in specific metabolic studies and always in a limited number of healthy volunteers. Overall, very little information is available on the impact of the patient's status on metabolic pathways, and one could imagine making a better use of the large number of clinical studies.

Xenobiotic metabolism can be defined as the chemistry of enzymatic and nonenzymatic processes [1] [2]. It covers therefore the identification of the chemical entities formed *in vitro* during early research programs, and in the organism of animals and humans during classical *in vivo* metabolism studies [2]. The chemical nature of these metabolites is of great importance because they have to be tested for their pharmacological profile. Today, in the absence of guidelines, some authors have proposed that the metabolites participating to the overall pharmacological activity (at least 25% of the total activity) and representing more than 25% of the exposure of the circulating unchanged drug should be monitored in future animal and clinical studies [3]. In parallel, the safety profile of major metabolites will also be assessed, comparing exposures of individual metabolites or global routes of metabolism to those obtained in mutagenecity tests and animal species of the toxicological safety program, and some authors have highlighted the potential safety concerns of human-specific metabolites [4]. To ensure adequate coverage and safety ratios, a '25 times rule' difference in exposure between animals and man is been discussed for either the parent drug or the metabolites. In a number of cases, the net exposure, defined as the sum of the exposure of the drug and its metabolites, will have to be used.

Metabolic studies also include the determination of the main characteristics of the enzymes involved in these metabolic reactions, enabling to predict, understand, and better control potential interactions with co-administered drugs. In addition, the knowledge of these enzymatic properties can help explain the impact of metabolic variability on a drug's activity [5].

In the past decade, emphasis has been on the introduction of validated simplified metabolism tools in early discovery research projects to predict sooner some of the potential limiting factors for drug development [6]. This process has been very successful to integrate metabolism scientists in the drug discovery process, although the balance between the systematic and selective use of these tools is still a matter of debate. In parallel, new technological advances have considerably changed how to approach classical *in vivo* metabolism studies in man with the objective of collecting information during the various clinical studies.

The common cause of these technological changes is the development of analytical techniques such as mass spectrometry, which is able to detect and identify chemical entities in biological fluids at extremely low concentrations, together with laboratory automation and miniaturization. The evolution and integration of mass spectrometry in the metabolism field, rather than reaching a plateau, is still evolving. However, whilst mass spectrometry is an extremely powerful technique in the detection and identification of unknown chemical entities, it remains a qualitative technique, in the absence of synthesized reference metabolites. The quantitative aspects are covered mostly by the use of radiolabeled drugs and LC/NMR. The use of radioisotopes has been historically the technique of choice for metabolism scientists and they remain, thanks to new technological advances in detection techniques, particularly important in the determination of time–concentration profiles of circulating metabolites.

Using specific examples, this chapter focuses on some of the important technical changes that have a major impact on the metabolic studies carried out in both research and development programs of new chemical entities.

2. Integration of Mass Spectrometry in Metabolic Studies

Mass spectrometry (MS) has helped increase the knowledge of metabolism scientists on the drugs they evaluate. The role of this technique is best known in structural identification studies. But because of its versatility and sensitivity, mass spectrometry is used nowadays in all major steps of metabolism studies.

One example is the use of generic methods enabling to follow the entire range of cytochrome P450 substrates and their metabolites used in the determination of the inhibition potential of new candidates. Eleven metabolic reactions, each specific for one human CYP, are classically followed in these evaluations. This implies to separate and quantify eleven different substrates and their respective metabolites within the same analytical run (*Figure*).

Before the use of mass spectrometry, every single end point was measured separately with LC/fluorescence, LC/UV or LC/radioactivity detections. With mass spectrometry, the pooling of samples in combination with the single analytical method enables to rapidly evaluate the inhibition potential of drug candidates in the first steps of their evaluation.

As mentioned earlier, mass spectrometry is an important method which has contributed significantly to the monitoring of new drug entities and to the structural characterization of drug metabolites. LC/MS coupling began in the early 1970s, and only two interfaces remain from the wide variety developed over the years, namely electrospray ionization (ESI) and atmospheric-pressure chemical ionization (APCI), which are both atmospheric-pressure-ionization (API) techniques. The advents of matrix-assisted laser desorption/ionization (MALDI) and surface-enhanced laser desorption/ionization (SELDI) have extended the appli-

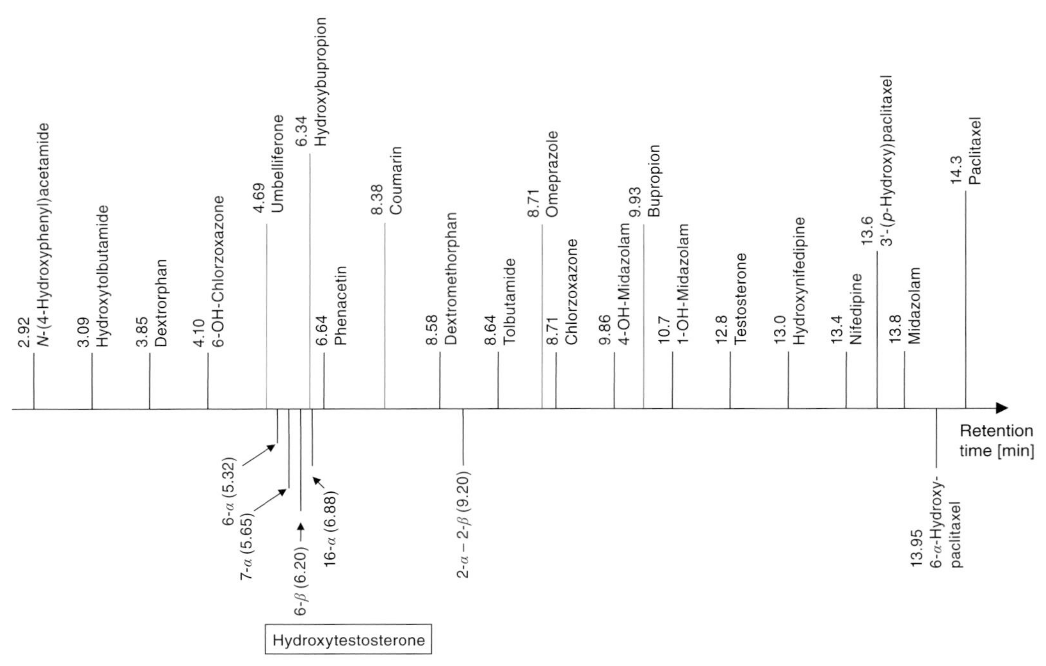

Figure. *LC/MS/MS Generic method profile for the* in vitro *evaluation of the inhibition potential of drug candidates using phenacetin* (CYP1A1/2)*, coumarin* (CYP2A6)*, bupropion* (CYP2B6)*, paclitaxel* (CYP2C8)*, tolbutamide* (CYP2C9)*, omeprazole* (CYP2C19)*, dextromethorphan* (CYP2D6)*, chlorzoxazone* (CYP2E1)*, and testosterone, midazolam and nifedipine* (CYP3A4) *specific substrates.* Sample analysis was performed using a *Waters Xterra MS C18* 100×4.6 mm $\times 3.5$ μm column (without any split), flow rate 0.5 ml min^{-1} for a total run of 15 min using a *Micromass Triple* quadrupole spectrometer (MRM transitions detection).

cation of MS to the study of proteins in complex biological systems [7]. Moreover, direct tissue analysis was performed by MALDI-MS on intact tissues to determine the spatial distribution of compounds and their relative expression levels without the need for specific exogenous compounds such as radiolabeled or immunochemical reagents [8].

2.1. *Atmospheric Pressure Ionization Mass Spectrometry*

LC/API-MS/MS, using a triple quadrupole instrument, is the standard technique in metabolite identification. The power of this approach is its ability to perform both MS/MS scans (*e.g.*, neutral-loss and precursor-ion scans) and fullscans that search for predicted molecular weight changes. For example, constant-neutral-loss scans can be used to identify the presence of glucuronide and sulfate conjugates by scanning for losses of 176 and 80 Da, respectively. The ion-trap mass spectrometer offers another means to acquire MS^n spectra (where $n \geq 2$). The advantages of MS^n are the ability to expedite the interpretation of collision-induced dissociation spectra and its low limit of detection (5 pg of metabolite), which is attainable because a specific population of ions can be accumulated in the trap.

Using the multiple reaction monitoring experiment (MRM), the quantitation limits of these methods in many common matrices (plasma, serum, or cellular media) are generally excellent (less than a few nanograms per milliliter of sample). One shortcoming is a scarcity of qualitative information needed to support the recognition and structural elucidation of metabolites.

2.2. *Tandem Quadrupole Time-of-Flight* (Q-TOF) *Mass Spectrometry*

High-resolution and qualitative data acquisition can be achieved using tandem quadrupole time-of-flight (Q-TOF) mass spectrometry. The Q-TOF-MS is capable of routinely providing a mass accuracy of <5 ppm for product ions, which allows the determination of the product ion's elemental composition and affords greater confidence in the interpretation of MS/MS spectra. The Q-TOF-MS has advantages for metabolite identification over ion trap or triple quadrupole mass spectrometers, including fast mass-spectral acquisition speed with high full-scan sensitivity, enhanced mass resolution, and accurate mass measurement capabilities which allow for the determination of elemental composition. Exact mass measurement is very useful for the confirmation of elemental composition and is a valuable tool to solve structure elucidation problems.

Recently, linear ion traps coupled with TOF-MS have been introduced, which combine the advantages of an ion-trap MS such as MS^n capabilities, and a TOF-MS instrument such as high resolution and fast duty cycle. In addition, a hybrid triple quadrupole linear ion trap mass spectrometer (Q-TRAP™) has been introduced and applied in metabolite identification and quantification of drugs in biological matrices [9].

Several analytical techniques can offer very good alternatives or be complementary to the currently used LC/API-MS techniques. Ion mobility spectrometry combined with mass-spectrometric detection (IMS/ESI-MS) has already been used for the analysis of opiates and their metabolites [10]. In IMS, a sample is introduced in the instrument and vaporized by flash heating, the vapor is ionized, and the resulting ions are introduced into a drift tube. Ion mobilities are determined from ion velocities measured in the drift tube at ambient pressure and are characteristic for the analyte. This method has achieved separation efficiencies comparable to common chromatographic techniques and is being evaluated as a high-resolution separation technique for, *e.g.*, the separation of metabolic isomers.

The on-line combination of mass-spectrometric characterization and biological screening based on ligand–receptor or antigen–antibody interactions can be attractive for some applications. An earlier example of such an approach is on-line affinity capillary electrophoresis/MS. The receptor is present in the electrophoresis buffer and the metabolite mixture is injected as the sample. Metabolites that show strong binding to the receptor are retained and thus separated from compounds that do not interact. On-line MS detection allows direct characterization of the interesting ligands [11].

One can easily imagine that, with all these enhanced technologies, mass spectrometry has become a key technique in every major step of the daily work of metabolism departments. However, the race towards greater sensitivity will probably reach some limits. This is already illustrated in reports on matrix effects observed with sensitive and fast turnover analytical methods. It is explained by the influence of endogenous factors inducing a suppression of the signal of the analyte.

In the mass spectrometry field, major advances can come from improvements in the treatment of data. One illustration is found in the metabolomic field where powerful statistical tools such as principal component analysis are used to monitor the changes in urinary and plasmatic patterns of endogenous markers after treatment with a drug. The same treatment of data can help understand matrix effects, separating drug metabolites from endogenous material, and helping in the early identification of metabolites.

3. Metabolic Studies: Major Advances in Development

3.1. *Progress in the Detection and Use of Radioisotopes*

For biotransformation investigations *in vivo* and *in vitro*, radioactively labeled drugs provide one of the most-important tools to completely track the metabolites in complex biological matrices such as blood, urine, faeces, bile, and *in vitro* samples. The most-frequently used approach is to administer the drug containing a radioactive isotope such as ^{14}C or ^{3}H and conduct liquid scintillation counting of collected fractions or directly monitor LC effluents. In combination with separation methods, *e.g.*, LC, radioactive labeling allows the highly selective, sensitive, and quantitative detection of unknown metabolites. Coupled with mass spectrometric and/ or other spectroscopic data, it allows to elucidate the complete biotransformation pathways of a given compound [12].

Flow-through radioactivity detectors have been used since many years in radio-LC for radioactive peak detection. But radio-LC can only detect down to 200–500 dpm with a liquid cell or 500–1000 dpm with a solid cell, depending on the hardware and software used. This technology was a limiting factor when assessing the exposure to circulating metabolites in man, an information systematically required for registration purposes. To analyze samples containing insufficient radioactivity for on-line radio-LC detection, it was necessary to extract large volumes of blood (up to 20 ml), or to collect LC elutes into a series of vials for subsequent liquid scintillation counting (LSC). Both techniques afforded limited information on circulating metabolites.

3.1.1. *Accurate Radioisotope Counting Technology*

A novel on-line radio-LC detection system, LC/ARC (LC accurate radioisotope counting), uses advanced stop-flow counting technologies to accurately detect/quantitate any portion of the radio chromatogram. LC/ ARC can also detect any volatile radioactive metabolite because no drying phase is necessary during analysis. The ARC flow cell being designed specifically for accurate radioisotope counting in radio-LC, background and counting efficiency are improved. The stop-flow AQ cocktail is compatible with LC solvents, produces neither gel nor luminescence and provides better mixing to homogenous mixtures. The typical limit of detection of LC/ARC is 15–20 dpm for ^{14}C and 10–40 dpm for ^{3}H (ARC after *D. Y. Lee* [13]) and background levels of 2–5 dpm. The LC/ ARC data system allows its user to control the desired limit of detection by choosing different stop-flow modes and counting times. The LC/ARC system dramatically improves the sensitivity, reduces the radioactive wastes, and eliminates the need to collect fractions for off-line counting of low-level radioactivity. Another advantage of this system is its easy interface with the mass spectrometer, which allows on-line acquisition of mass-spectrometric data [14].

3.1.2. *Micro Plate Counting Technology*

Alternative methods for the generation of radio chromatograms, which retain the sensitivity benefits of LC/LSC with reduced manual input and instrument time, is to use *TopCount*® microplate scintillation counting. When combined with μLC, the *TopCount* microplate scintillation and luminescence counter can dramatically improve the daily work. LC performs the chemical separation, and the samples are automatically dispensed into 96 deep-well microplates that contain a tritium silicate based solid scintillator. After sample drying, the *TopCount* system counts the microplates. This method eliminates the need for the scintillation cocktail, and it greatly reduces the labor required to prepare a large number of samples, the work being almost completely performed by the LC instrument, the fraction collector, and the *TopCount*. The combination of LC with *TopCount* offers high sensitivity (25–700 cpm) and high-resolution power simultaneously [15]. This combination is superior to the classical on-line radioactivity detection and at least equivalent to the classical thin-layer radio-chromatography regarding performance and sensitivity. A few limitations should be mentioned as well. The direct counting of plasma, urine, and faeces samples can lead to incorrect results, because the color and small particles of the sample can produce strong color quenching. The sample material shouldn't be volatile because the samples in a microplate must be completely dry before counting and volatile metabolites will be lost in the drying process. It should also be noted that some compounds may have a high affinity for the material of the microplates.

3.1.3. *Accelerator Mass Spectrometry*

A further step in the detection of radioisotopes has been reached with the use of accelerator mass spectrometry (AMS) to detect extremely low levels of substances such as drugs and their metabolites in blood, urine, and faeces. AMS is a mass-spectrometric method to quantify isotopes. AMS separates atoms on the basis of their mass, charge, and energy differences, and can individually quantify isotopes such as ^{12}C, ^{13}C, ^{14}C, ^{3}H, and ^{36}Cl.

Before biological samples can be analyzed for ^{14}C, which is the main application of AMS in drug discovery and development, the carbon in the sample has to be converted to graphite. This is achieved in two stages: oxidation to CO_2, then reduction to carbon (graphite). AMS is a type of tandem isotope ratio mass spectrometry in which negative ions undergo a high-energy collision process that removes electrons, converting them to purely atomic positive ions. These collisions generally require ion energies of mega-electron-volts (MeV) rather than the kilo-electron-volt (keV) energies used in most mass spectrometers. A tandem electrostatic accelerator with a high positive potential (0.5–10 MeV) at its midpoint attracts the negative ions through an evacuated beam tube. After electron removal in the collision cell at this high voltage, the now positive ions are further accelerated to even higher energies and separated by magnetic and electrostatic spectrometer elements before arriving at a detector that determines the isotopic identity of each ion. Radioactivity levels as low as 0.0001 dpm can be detected using AMS.

AMS is several orders of magnitude more sensitive than liquid scintillation counting, being able to detect radiolabeled drugs at amol (10^{-18} mol) levels, and with great precision. This reduces the total radiation exposure of volunteers and minimizes any chemical hazards. With the application of sample separation techniques such as LC, AMS can quantify levels of drugs and their metabolites, assess bioavailability, and measure plasma clearance at doses considerably below the pharmacological level. However, the challenges to be overcome before AMS can be used routinely are significant. Advances are needed to facilitate sample preparation, which at present is performed off-line in a labor-intensive manner. In addition, the large size of the AMS instruments precludes their general use and location.

The use of radioisotopes has several disadvantages. The synthesis and purification of radioactive compounds are expensive. Time-consuming synthesis stages and the requirements for handling radioactive material and wastes introduces additional costs. In addition, mass-balance and metabolic studies using radioisotopes cannot be conducted in women and children. Because synthesized references of metabolites are available only in late development stages, radiolabeled xenobiotics remain widely used for the detection and quantification of metabolites *in vitro* and *in vivo* samples. The average dose of radioactivity given in human metabolic studies is between 30 and 50 micro-Curies (1110–1500 MBq). This seems quite high in regards to the recent detection limits reached with LC/ARC and LC/*TopCount*.

With AMS detection, the dose of radioactivity proposed for metabolic studies is at nano-Curie levels. This radiation dose is exempt from regulatory approval and comparable to daily radiation exposure [16]. This low dose of radiotracer can theoretically be used in any clinical study allowing to get a first insight in the metabolism of a drug in most clinical studies.

3.1.4. *Direct Labeling of Metabolites in Biological Samples*

To avoid the direct administration of radiolabeled xenobiotics to patients, a metabolite can also be labeled *in situ* in the biological samples taken after administration of the nonlabeled drug. Isotopic exchange of ^{1}H- with a ^{2}H- and a ^{3}H-atom represents the easiest way to label compounds. In addition, these techniques can now be miniaturized and applied to label mixtures of compounds at microgram levels.

A successful labeling was obtained for indapamide and its metabolites *in situ* in both *in vitro* and *in vivo* biological samples. The mixtures of metabolites were labeled to different degrees by isotopic exchange (*Table*). The metabolites were separated by liquid chromatography and analyzed with a high-resolution mass spectrometer (Q-TOF spectrometer) enabling to evaluate the isotopic pattern of each metabolite. The isotopic enrichment of these metabolites (mean number of ^{3}H-atoms per molecule of metabolite) was determined and the specific activity of each metabolite calculated. In parallel to the mass analysis, radioactive counting allows the quantification of labeled metabolite present. Knowing both the response factor in mass spectrometry (triple quadrupole spectrometer) and the quantity of metabolite present in the synthetic mixture, one can quantify these metabolites in any other sample by LC/MS/MS.

This is an example of an approach that needs neither the synthesis of metabolites nor the administration of radiolabeled drug for metabolite identification. However, the isotopic exchange protocols will have to be adapted, on a case by case, to the drug of interest. The major advantage is that it can be applied to any clinical samples, especially phase-1 single- and repeated-escalating-dose studies in which the amount of metabolites in urinary samples can be quite high.

3.2. *Increasing Importance of Nuclear Magnetic Resonance*

Nuclear magnetic resonance (NMR) has always been a reference technique for metabolism scientists. Whilst mass spectrometry can provide structural information, the determination of the correct sites of biotransformation on a drug molecule is best done by NMR. This method measures the presence of atoms such as ^{1}H, ^{13}C, ^{19}F, which are nonradioactive stable isotopes, and ^{31}P. It is a nonselective, nondestructive detector of low molecular weight molecules in solution. Coupled with LC or LC/MS, NMR spectroscopy can determine drug metabolite profiles and the pharmacokinetics of multiple metabolites in one study. The detection limits of LC/NMR are continually being revised downwards as new technical advances are made. Recently, these have included the use of high magnetic field strengths (operating at 900 MHz for ^{1}H-NMR spectroscopy [17]), the incorporation of digital filtering and oversampling into NMR data acquisition and the introduction of microbore LC methods. It appears

Table. *Microsynthesis of Labeled Compounds* (indapamide and metabolites) *Using Two Different Catalysts and 5 μg of Starting Drug Material, and Respective Isotopic Yields Obtained in a Biological Matrix* (rat microsomes and/or human urine) *after Extraction of the Drug Material on* OASIS HLB *Columns*

Indapamide (S 1520) and metabolites	Isotopic enrichment of the metabolites (mean number of ^{3}H-atoms per molecule of metabolite)	
	Catalyst used	
	Crabtree	(^{2}H)TFA
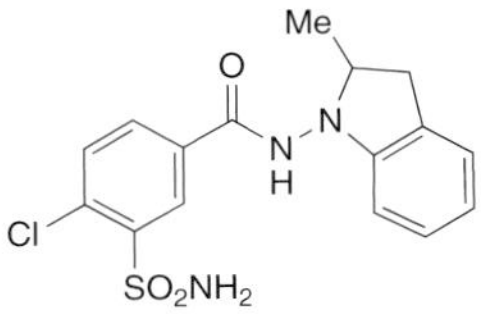 S 1520	0.6	1.4
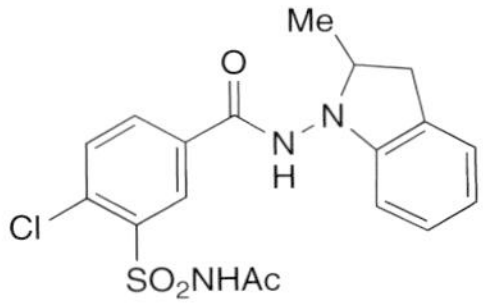 Y 1438	0.7	1.5
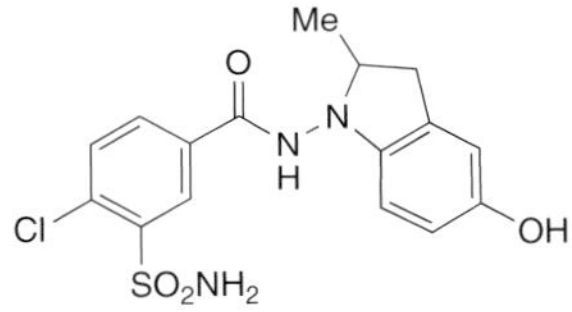 S 17610	0.7	1.5
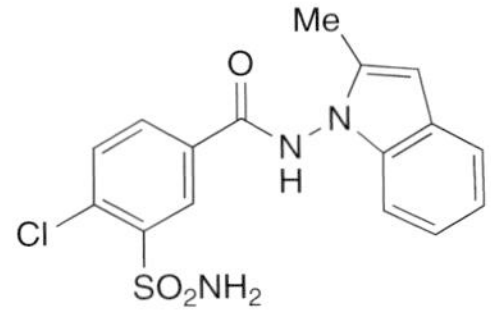 Y 38	0.4	0.8
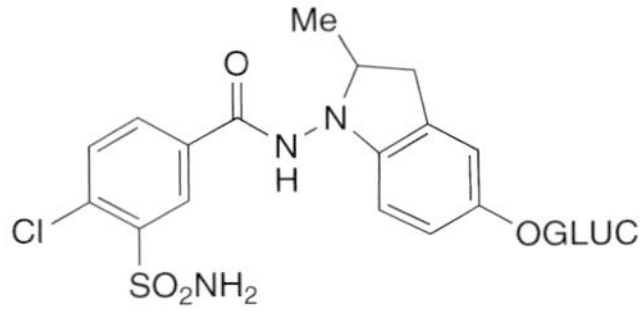 *O*-Glucuronide	0	0.8
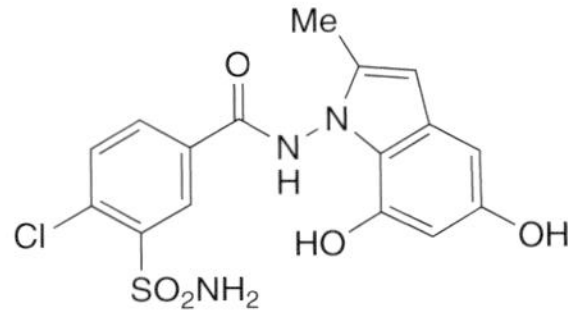 Dihydroxydehydro-indapamide	0.2	1.2

that detection limits for structural characterization can be in the region of 5 ng even at lower ^{1}H-NMR observation frequencies of 500 and 600 MHz. Moreover, recent innovations to improve NMR detection limits include cryogenic cooling of the probes [18]. Cryoprobes need very cold He gas to reduce thermal noise in the pickup coil and associated electronics, thereby obtaining a significant improvement in signal-to-noise ratios. Cryoprobe technology remains relatively expensive, and the associated hardware has both sitting and maintenance requirements which are not trivial. However, there is a great interest in the technology, due to the promise of obtaining data from very dilute samples or samples with low solubility.

Another potentially important advance is the design of microcoil probes; solenoid NMR transceiver coils can detect NMR signals in microliter samples. These detector coils have been coupled with LC and capillary electrophoresis separations.

As NMR is nondestructive, the sample can be recovered for further analysis and LC/NMR/MS (or LC/NMR/MS^{n}) using ion-trap mass spectrometer [19] combining two powerful and complementary analytical methods. A simple hyphenation of LC to NMR and MS is achieved using a post-column splitter. This directs 90–95% of the flow to the NMR and the remainder to the MS. A powerful alternative is the valve switching interface termed the BNMI (*Bruker* NMR/MS interface). This is a computer controlled splitter and a double dilutor for providing an appropriate make-up flow for optimal ionization in the MS. It also permits ^{1}H–^{2}H exchange to simplify MS spectra otherwise obtained in LC/NMR/MS. The limits of detection at the ^{1}H observation frequency of 600 MHz for 500-Da analytes are *ca.* 100 ng in stopped-flow and loop-storage-flow mode.

The highest magnetic field strengths with cryogenically cooled NMR probes and preamplifiers provides unsurpassed NMR sensitivity. Currently, the highest magnetic field used for LC/NMR/MS is 800 MHz [20].

In the absence of reference metabolites and without the use of radiolabeled drug, the sum of urinary metabolites identified by LC/NMR/MS affords a good evaluation of the excretion balance of the drug.

4. Metabolic Studies: Impact on Research Programs

Current drug research programs, encompassing early screening studies on large series of compounds and more-informative studies on selected candidates, allow a rapid understanding of the *in vivo* fate of a drug candidate. These studies are able to define systematically, at early stages, the metabolic stability, permeability, and solubility of the compounds tested, with the aim of rapidly selecting lead candidates [4]. These programs are then completed with *in vivo* studies of a few chemicals. Overall, the data available at the 'research stages' represent a combination of animal and human *in vitro* information as well as *in vivo* studies in animals. But most of this early information is oriented towards predicting the *in vivo* pharmacokinetic behavior of a candidate or the risk of drug–drug interaction. This latter step includes the identification of the nature and number of enzymes involved and the inhibition and induction potential of the drug candidate. These different parameters have to be known before the first administration to humans.

The weak point in these early evaluations remains the absence of a complete identification and quantification of the major metabolites needed to quantify species differences. This information is essential for the safety evaluations, in particular to select the large animal species used in the toxicological studies. But, depending on the type of metabolic reactions, *in vitro* studies are not always sufficient to identify the major metabolites for further pharmacological and toxicological evaluations.

Ideally, metabolic studies at this stage should allow the definitive identification of the major circulating metabolites and the comparison of the *in vitro* and *in vivo* biotransformation pathways in animals with those observed *in vitro* with human preparations. Attempts are made in research programs to identify the major metabolites. However, in the absence of references and without the use of radioactive substrates, the information on interspecies metabolic differences is usually limited due to the lack of definitive structural evidence and solid quantitative data.

Some of the technological advances described earlier, such as LC/NMR/MS and *in situ* metabolite labeling, could help in the identification, structural characterization, and quantification of the major metabolites in research programs, and provide a better support for toxicological assessment.

5. Conclusions

Progress in optimizing the way to carry out metabolic studies in research and development has been constant over the last decade. In a way, metabolism represents a quite unique domain within Research and Development. It is effectively one of the sole disciplines present at every major step in R&D, and provides pivotal information to pharmacologists, chemists, toxicologist, and clinical scientists. It has therefore evolved a large range of metabolic tools, allowing the rapid understanding of a drug's fate. These tests are now well established and have been completely integrated in R&D programs.

Technology advances rapidly, and the advent of analytical techniques such as LC/MS/MS, LC/NMR/MS and LC/*TopCount*, LC/ARC and AMS, is changing again the role and timing of metabolic studies in R&D programs. On the Research front, more information on the identity and concentration of metabolites will potentially be available helping to better interpret safety programs. In clinical studies, much is done with the same techniques to understand the behavior of the drug in patients and to make better use of the large number of studies realized during Development.

REFERENCES

[1] J. M. Mayer, H. van de Waterbeemd, *Environ. Health Perspect.* **1985**, *61*, 295.

[2] B. Testa, 'The Metabolism of Drugs and other Xenobiotics – Biochemistry of Redox Reactions', Academic Press, London, 1995.

[3] T. A. Baillie, M. N. Cayen, H. Fouda, R. J. Gerson, J. D. Green, S. J. Grossman, L. J. Klunk, B. LeBlanc, D. G. Perkins, L. A. Shipley, *Toxicol. Appl. Pharmacol.* **2002**, *182*, 188.

[4] K. L. Hastings, J. El-Hage, A. Jacobs, J. Leighton, D. Morse, R. E. Osterberg, *Toxicol. Appl. Pharmacol.* **2003**, *190*, 91.

[5] 'Guidance for Industry. Drug Metabolism/Drug Interaction Studies in the Drug Development Process: Studies *in vitro*', U.S. Food and Drug Administration, Center for Drug Evaluation and Research, April 1997.

[6] M. Bertrand, P. Jackson, B. Walther, *Eur. J. Pharm. Sci.* **2000**, *2*, S61.

[7] M. Merchant, S. R. Weinberger, *Electrophoresis* **2000**, *21*, 1164.

[8] S. A. Schwartz, M. L. Reyser, R. M. Caprioli, *J. Mass Spectrom.* **2003**, *38*, 6.

[9] Y. Xia, J. D. Miller, R. Bakthiar, R. B. Franklin, D. Q. Liu, *Rapid Commun. Mass Spectrom.* **2003**, *17*, 1137.

[10] L. M. Matz, H. H. Hill, *Anal. Chem.* **2001**, *73*, 1664.

[11] J. G. Krabbe, H. Lingeman, W. M. Niessen, H. Hirth, *Anal. Chem.* **2003**, *75*, 6853.

[12] D. I. Papac, Z. Shahrokh, *Pharm. Res.* **2001**, *18*, 131.

[13] A.-E. F. Nassar, Y. Parmentier, M. Martinet, D. Y. Lee, *J. Chromatogr. Sci.* **2004**, *42*, 341.

[14] A. E. Nassar, S. M. Bjorge, *Anal. Chem.* **2003**, *75*, 785.

[15] K. O. Boernsen, *Anal. Chem.* **2000**, *72*, 3956.

[16] R. C. Garner, I. Goris, A. A. E. Laenen, E. Vanhoutte, W. Meuldermans, S. Gregory, J. V. Garner, D. Leong, M. Whattam, A. Calam, C. A. W. Snel, *Drug Metab. Dispos.* **2002**, *30*, 823.

[17] O. Corcoran, M. Spraul, *Drug Disc. Today* **2003**, *8*, 624.

[18] M. Spraul, A. S. Freund, R. E. Nast, R. S. Withers, W. E. Maas, O. Corcoran, *Anal. Chem.* **2003**, *75*, 1536.

[19] G. J. Dear, J. Ayrton, B. C. Sweatman, I. M. Ismail, I. J. Fraser, P. J. Mutch, *Rapid Commun. Mass Spectrom.* **1998**, *12*, 2023.

[20] U. G. Sidelmann, E. Christiansen, L. Krogh, C. Cornett, J. Tjornelund, S. H. Hansen, *Drug Metab. Dispos.* **1997**, *25*, 725.

In vitro, *in vivo*, and *in silico* Approaches to Predict Induction of Drug Metabolism

by **Urs A. Meyer***, **Sharon Blättler**, **Carmela Gnerre**, **Mikael Oscarson**, **Anne-Kathrin Peyer**, **Franck Rencurel**, **Oktay Rifki**, and **Adrian Roth**

Division of Pharmacology/Neurobiology, Biozentrum of the University of Basel, Klingelbergstr. 50–70, CH-4056 Basel
(phone: +41 61 267 22 20; fax: +41 61 267 22 08; e-mail: Urs-A.Meyer@unibas.ch)

Abbreviations
BRET: Bioluminescent resonance energy transfer; CAR: constitutive androstane receptor; CARLA: coactivator-dependent receptor-ligand assay; CYP: cytochrome P450; FRET: fluorescent resonance energy transfer; GFP: green-fluorescent protein; PCN: pregnenolone 16α-carbonitrile; PXR: pregnane X receptor; RXR: retinoid X receptor; TCPOBOP: 1,4-bis[(3,5-dichloropyrid-2-yl)oxy]benzene (=2,2′-[1,4-phenylenebis(oxy)]bis[3,5-dichloropyridine]).

1. Introduction

Many drugs and other xenobiotics can increase their own metabolism and clearance, a phenomenon called induction. The major drug-metabolizing enzymes in the liver are members of the gene superfamily of cytochromes P450 (CYP), which in conjunction with other enzymes and transporter proteins cause the clearance of lipophilic substances [1]. Induction of these proteins by xenobiotics is an important part of inter-individual variation in drug response. Induction influences the pharmacokinetic and pharmacodynamic properties of the inducer itself but also of other compounds metabolized or transported by the induced proteins. Drug-mediated increase of these enzymes or transporters can cause drug–drug interactions, inefficacy of drug treatment, or adverse drug reactions [2–4].

The basic mechanism of induction is activation of gene transcription. This suggested that drug-responsive transcription factors may bind to regulatory regions of the genes encoding these enzymes. In 1995, the first drug-responsive enhancer sequence was identified in the 5′-flanking region of the rat *CYP2B2* gene [5]. This observation was rapidly followed by the detection of similar sequences in other rodent *CYP*s, and later in the genes of other drug-metabolizing enzymes and drug transporters (for a review, see [4]). These drug-responsive sequences had the general structure of hexamer repeats similar to those of nuclear hormone receptors.

In 1998, two members of the gene superfamily of nuclear receptors, namely the pregnane X receptor (PXR) and the constitutive androstane receptor (CAR), were identified in rodents and man as key transcription factors in hepatic drug induction (for a review, see [6]). More recently, orthologs of these receptors have also been isolated in chicken, pig, dog, monkey, and fish suggesting a mechanism of induction that is evolutionary conserved in all of these species (for a review, see [4]). PXR and CAR along with a number of other members of the nuclear receptor family bind to DNA as heterodimers with the retinoid X receptor (RXR). PXR and CAR heterodimers bind to repeats of core hexamers in different arrangements and localized in the 5′-flanking regions of the induced genes. Whereas PXR is predominantly activated by dexamethasone/rifampicin-type inducers and preferentially affects CYPs of the CYP3A subfamily, CAR activity is influenced more specifically by the phenobarbital (PB)-type class of compounds and increases transcription of *CYP2B* and *CYP2C* genes. However, there is a considerable overlap of these two receptor systems both in terms of their activator spectrum as well as in their affinity to DNA-response elements [7]. The importance of PXR and CAR in mediating hepatic drug induction was further underlined by the generation of the respective knock-out mouse lines that are characterized by severely impaired CYP induction [8–10]. Interestingly, PXR and CAR have different modes of activation. PXR is located in the nucleus and is directly activated by binding of a large number of compounds to its ligand-binding domain. The activation mechanism of CAR is more complex. Apparently, CAR is a constitutively active transcription factor, *i.e.*, it triggers gene transcription in the absence of ligands. In unchallenged mouse hepatocytes, CAR resides in the cytoplasm in a multiprotein complex until drugs trigger a cytoplasm–nucleus translocation of this receptor (for a review, see [11]). Phosphorylation events in the cytoplasm as well as the nucleus modulate CAR transfer and activity both positively and negatively [12]. Finally, androstanes have been found to act as inverse agonists on mouse CAR. Androstane repression of CAR can be reversed by a number of inducer compounds, including the potent mouse *Cyp2b* activator 1,4-bis[(3,5-dichloropyrid-2-yl)oxy]benzene (2,2′-[1,4-phenylenebis(oxy)]bis[3,5-dichloropyridine]; TCPOBOP). It is not known, however, if this reversal of inhibition is a direct or an indirect effect of the inducers. TCPOBOP can also directly activate CAR, but drugs such as the

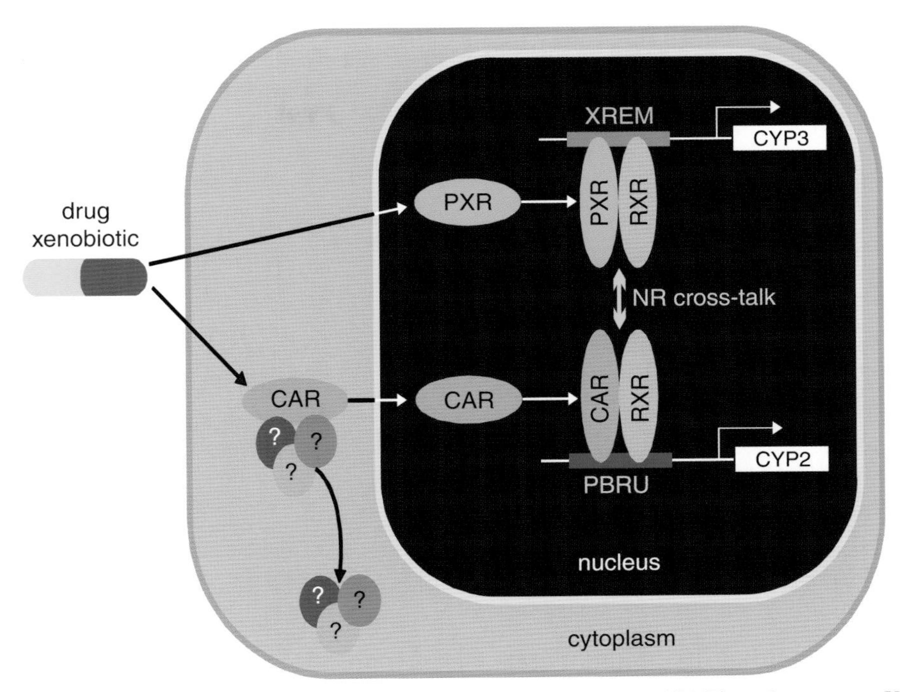

Figure. *Role of the xenosensors constitutive androstane receptor* (CAR) *and pregnane X receptor* (PXR) *in the induction of cytochromes P450 of the CYP2 and CYP3 subfamilies.* Activation of CAR occurs by an unknown indirect mechanism by which drugs cause its transfer into the nucleus, where it combines with the retinoid X receptor (RXR). The CAR/RXR heterodimer then binds to drug-response elements (PBRUs) in the flanking regions of inducible genes such as those of the *CYP2* subfamily. PXR is localized in the nucleus and activated directly by its ligands. PXR also combines with RXR and the PXR/RXR heterodimer binds preferentially to response elements (XREMs) on the flanking regions of genes of the *CYP3A* subfamily. The ↔ denotes cross-talk between the two xensosensors. CAR/RXR can also bind to DNA elements on *CYP3A* genes and PXR/RXR to DNA elements on *CYP2* genes.

classical inducer phenobarbital activate CAR transfer into the nucleus by a yet unknown mechanism (*Figure*).

A large variety of drugs used for the treatment of different clinical conditions have been found to affect PXR and CAR signaling and some examples are listed in the *Table*. There are no obvious chemical similarities between these compounds. 'Good' inducers are lipid-soluble, have molecular weights between 150 and 1000 Da, and are relatively slowly metabolized. A more-comprehensive listing of the most important compounds involved in evoking drug–drug interactions by induction is provided in the 'Cytochrome P450 Drug Interaction' website (http://medicine.iupui.edu/flockhart/).

Table. *Inducers of Cytochromes P450 in Man Which Act* via *CAR and/or PXR*

CYP Inducers	Examples
Antibiotics	Rifabutin
	Rifampicin
Anticonvulsants	Carbamazepin
	Oxcarbazepine
	Phenobarbital
	Phenytoin
	Primidone
	Telbamate
	Topiramate
	Vigabatrin
Antidepressants	Hyperforin (in *St. John*'s wort)
HIV Protease and reverse transcriptase inhibitors	Efavirenz
	Ritonavir
	Saquinavir
Others	Bosentan
	Dexamethasone
	Tamoxifen
	Troglitazone

Prediction of the human induction response to candidate therapeutic compounds has been difficult because of important species differences in regard to inducers and responses. This has limited the extrapolation of animal studies to man. The discovery of the xenobiotic-sensing nuclear receptors PXR and CAR and their DNA-response elements has changed this dramatically. It has provided tools that can be used to better predict drug induction and the associated drug–drug interactions and adverse drug reactions.

2. *In vitro* Assays

A range of *in vitro* assays have been developed to test compounds for their potential to interact with these receptors. In the case of PXR, the correlation between ligand binding, receptor activation, and target gene induction has been well established [13]. Thus, high-throughput screening of large compound libraries using different *in vitro* approaches has helped to identify PXR-activating compounds at a very early stage [14]. This can be achieved by ligand-binding assays using purified PXR ligand-binding domain and measuring displacement of a radiolabeled ligand such as tritiated 'SR12813', a very potent ligand of human PXR [15]. Alternatively, fluorescent and bioluminescent resonance energy transfer assays (FRET and BRET, resp.) or co-activator-dependent receptor-ligand assays (CARLA) measure the ligand-mediated association or dissociation of receptor proteins with co-factors [16][17]. In order to measure functional activation, expression plasmids for PXR and for a reporter gene under the control of a PXR-responsive element are cotransfected into a suitable cell line such as LS174T (human colon carcinoma cell), HepG2 (human hepatoma cells) or CV-1 (monkey kidney epithelial cells). All of these cell lines express a considerable amount of RXR, the PXR and CAR heterodimerization partner. The level of reporter gene expression after treatment with different compounds is then determined. A variant of this assay is the expression of the PXR ligand binding domain fused to the yeast GAL4-DNA binding domain and a reporter gene plasmid under the control of the GAL4 upstream activation sequence [18]. The latter is the method of choice if no or only a weak response element for a certain receptor is known. A combination of these approaches, ligand-binding assays as well as PXR activation assays, are widely used in early stages of drug development to identify potential inducers.

The mechanism underlying activation of the CAR system upon drug treatment is less amenable to *in vitro* studies. This is due to the drug-induced cytoplasm–nucleus translocation, the constitutive activity and different modes of receptor modulation in the cytoplasm and the nucleus (for a review, see [11]). Moreover, the CAR translocation into the nucleus has clearly been demonstrated only in rodent liver. Cell lines that are either transiently or stably transfected with CAR show an aberrant distribution of CAR in both the cytoplasm and the nucleus and are unlikely to reflect CAR-mediated induction *in vivo*. One way to overcome these problems and study the nuclear translocation of CAR is to express CAR tagged with green-fluorescent protein (GFP) in mouse or possibly human hepatocytes in primary culture or in mouse liver and to quantitate the cytoplasmic/nuclear ratios of CAR with time. Obviously, this approach is not adaptable to high-throughput screening of a large number of compounds. Other methods for determining the enrichment of CAR protein in the nucleus involve cellular fractionation with subsequent use of the nuclear fraction in Western blots, DNA-affinity columns or electromobility shift assays [19][20]. Unfortunately, whereas potent inverse agonists for mouse CAR have been discovered, androstanes inhibit only the mouse but not the human ortholog. The reversal of inhibition by inducer drugs has been correlated with inducer potential only with mouse CAR. One recently proposed strategy to develop an assay for human CAR activation is to permanently overexpress (by cDNA transfection) a tetratricopeptide repeat protein that was shown to participate in the retention of CAR in the cytosol [12]. Again, this assay is far from a physiological human liver situation and has not been validated. In summary, due to the peculiarities of the signaling mechanism of this xenosensor, no high-throughput *in vitro* assay for human CAR is available.

The current consensus recommendations state that reporter gene and ligand-binding assays are only appropriate for initial screens because their predictive value has not firmly been established. We believe that further improvement of these assays will serve well in the future to identify human inducers at an early stage of drug development [21].

3. Cell Culture Systems and Liver Slices

Induction of enzymes by drugs and other xenobiotics is a typical feature of differentiated hepatocytes. Therefore, immortalized liver-derived cell lines do not exhibit CYP induction or the pattern of expressed CYPs is aberrant [22]. The only cell line that has been found so far to be fully responsive to phenobarbital-type inducers comparable to primary cultures of hepatocytes is the LMH cell line [23]. These cells originate from a chicken hepatoma and thus respond in a chicken-specific way to xenobiotics. They have provided important insights into the molecular mechanisms of induction, but their use for prediction of drug induction in man is limited.

The current cell culture system of choice are primary cultures of human hepatocytes, which retain a human-specific drug-induction response. In the future, the potential of primary hepatocytes for large-scale compound screening might be enhanced by improvement in cryopreservation, long-term cultures, and re-use of cultures. Recent attempts to select 'more-differentiated' cells from virus-transformed hepatocytes or from hepatoma cells have not yet been very successful. They may provide cell systems that are easier to procure and handle than primary cells. Primary cultures have of course several major drawbacks including restricted availability. They are technically demanding and normally suffer from low cell homogeneity and stability. Furthermore, primary hepatocytes from different donors reflect the interindividual variability in drug response and might therefore lead to divergent results. In addition, primary cultures are not easily transfectable, although recent advances in the use of adenoviral vectors have improved transfection efficiency [24].

Some of the disadvantages of primary hepatocytes can be overcome by using precision-cut liver slices. The hepatic tissue architecture is preserved in liver slices including cell–cell contacts. Their preparation and culture is less demanding than that of primary hepatocytes and they do not require collagenase treatment of cells. Liver slices can be more easily cryopreserved without altering the induction potential of different CYPs [25]. However, liver slices have the same drawbacks as primary hepatocytes concerning availability and interindividual variability. In addition, the function of liver slices declines rapidly with time and allows only short exposures for a few hours to drugs.

In summary, an ideal *in vitro* system for prediction of the human drug-induction response is not available. In the future, genetically engineered cell culture systems or immortalized and selected cells might replace these rather cumbersome systems and eliminate the disadvantages associated with them.

4. Animal Models and Other *in vivo* Techniques

In drug development, testing of compounds in different animal models is one of the key steps. However, only limited information regarding drug pharmacokinetics and pharmacodynamics in humans can be derived from these tests because of the species differences in drug metabolism and drug targets. The discovery of PXR and CAR and the realization that the divergent ligand-binding domains or activation mechanisms of the receptor orthologs account for a large part of the species differences [15] led to the generation of so-called 'humanized' mice. In these animals, the endogenous gene encoding mouse PXR or mouse CAR were knocked out and the resulting mice were then 'rescued' by a transgenic expression of human PXR and CAR, respectively, under the control of a liver-specific promoter [9] [26]. When treated with species-specific inducer compounds, these mice exhibit regulation of *Cyp3a11* and *Cyp2b10* that resembles the regulation of *CYP3A*s and *CYP2B*s in man. For example, the rodent-specific CYP3A inducer pregnenolone 16α-carbonitrile (PCN) is no longer active in the humanized PXR mouse whereas these animals now strongly react to rifampicin treatment. Thus, humanized animal models might prove to be predictive tools for testing candidate drugs for their induction potential in man. So far, humanized mice which express either human PXR or human CAR have been published, but crossing of these two lines might result in an even better model concerning drug induction in man. Recently improved methods of *in vivo* DNA delivery systems combined with powerful noninvasive, extracorporal monitoring of bioluminescent reporter gene assays have further improved *in vivo* induction models, leading to reduction in the number of animals used [27]. In spite of these technical innovations, it should be cautioned that animal models still reflect our present limited knowledge regarding the drug induction signaling pathways. For instance, induction is a dose-dependent phenomenon and it has not been demonstrated that a dose that induces in mice will reliably predict the corresponding dose that will induce in man.

Probe drugs are the current method of choice to assess the potential of a compound to modify CYP activity in human subjects *in vivo*. A single drug or a combination of drugs is given in small (subtherapeutic) doses to a volunteer or patient and the parent drug and its metabolites are measured in saliva, blood, or urine. Some of these drugs are quite specific substrates for a single CYP or an other drug-metabolizing enzyme, others might target a variable battery of different enzymes and thus only vaguely reflect the respective CYP activity (for a review, see [28]). In the case of drug-mediated CYP induction, it would, of course, be useful to be able to measure an endogenous marker compound that closely correlates with inducible CYP levels. An increase of hydroxylated steroid metabolites is observed in patients after prolonged treatment with strong inducers such as rifampicin. However, the recent discovery that the CYP3A4 generated metabolite of cholesterol, 4β-hydroxycholesterol, is increased in plasma after drug treatment of patients treated with anticonvulsants could be a biomarker of induction of CYP3A4 [29][30]. Other endogenous metabolites generated by CYP3A4 have been recently described and include 25- or 6α-hydroxylated bile acids that are excreted in urine [31][32]. Similar markers that may reflect the activity of other reported drug-metabolizing enzymes, however, and the usefulness of endogenous metabolites generated by CYP3A4 has not been evaluated.

5. *In silico* Approaches

Recently, an algorithm called NUBIScan (accessible at http://www.nubiscan.unibas.ch) has been developed in this laboratory. It provides assistance in the identification of likely nuclear receptor binding sites in large sequences, including those activated by the xenobiotic-sensing nuclear receptors PXR and CAR [33]. Recognition of functional drug-responsive elements in regulatory regions of genes is based on a statistical algorithm that searches genomic DNA sequences in the flanking regions of genes for hexamer repeats with a number of nucleotides in between and determines the resemblance to a known functional site that binds CAR or PXR heterodimers. DNA sequence analysis with NUBIScan may thus provide insight into some questions in drug induction. Thus, it can highlight the probable location of a nuclear receptor–DNA interaction in a drug-responsive enhancer sequence. To understand the functionality of a complex drug-responsive enhancer element (which may be up to a few kilobases in length), it is vital to know where in the sequence xenobiotic-sensing receptors can bind, and *in silico* analysis followed by experimental verification is more efficient compared to classical methods. Application of genome-wide NUBIScan analysis can predict new target genes of particular receptors. Such an endeavor, however, must be combined with a previous selection of a subset of the genome, *i.e.*, focusing on stretches of DNA immediately around a gene's coding region. Otherwise, the vast amount of noncoding DNA confronts the algorithm with too much nonspecific data, impairing the production of useful predictions. Lastly, judicious NUBIScan analysis combined with expression array data from control and drug-treated samples is a promising two-pronged approach. Analyzing the regulatory regions of genes induced by drugs can immediately reveal candidate nuclear receptor response elements and also stipulate a distinction between genes directly affected and those likely to be regulated by downstream mechanisms.

6. Conclusions

Major breakthroughs in elucidating the molecular signaling pathways underlying drug induction of cytochromes P450, other drug-metabolizing enzymes and drug transporters in liver and intestine have been made in recent years. However, many questions regarding the details of activation of transcription remain. The xenosensors PXR and CAR explain the divergence in drug-responsive DNA enhancer elements in diverse *CYP*s and other genes and also explain the species differences in the induction response. Moreover, the structure of the PXR ligand binding domain provides explanations for the promiscuity of this receptor and for the structural variety of inducer compounds [34]. Therefore, several *in vitro*, *in vivo*, and *in silico* methods have been developed on the basis of PXR and CAR that now allow rapid screening of a large number of compounds. These assays increasingly permit early elimination of inducer compounds in the drug discovery process and a more-rational choice of the species for preclinical toxicology tests.

The application of present knowledge to noninvasively and without probe drugs assess induction in clinical settings in patients is not yet possible. Few of the currently used assays take into account the inter-individual variability of drug response. So far, this variability has been largely associated with polymorphisms in the *CYP* genes [35] but recent data suggest that polymorphisms might also occur in xenobiotic-sensing nuclear receptors and in drug-responsive enhancer elements [36–38]. In the future, it might be possible to genotype patients for polymorphisms in CYPs and other drug-metabolizing enzymes, variability in nuclear receptors and enhancer elements, and based on these data provide a treatment that has less risk for drug–drug interactions or adverse drug effects. It is obvious from the present data that genotyping (*e.g.*, for polymorphisms of *CYP*s) should always be accompanied by 'phenotyping' because induction, repression, or inhibition of CYPs and other enzymes can dramatically alter the pharmacokinetic behavior and pharmacodynamic properties of drugs [39]. For this purpose, the development of 'surrogate' biomarkers of induction and repression which can easily be measured in blood samples of patients will be a major challenge for the future.

Research in the laboratory of *Urs A. Meyer* is supported by the *Swiss National Science Foundation*.

REFERENCES

[1] D. W. Nebert, D. W. Russeld, *Lancet* **2002**, *360*, 1155.
[2] D. J. Waxman, L. Azaroff, *Biochem. J.* **1992**, *281*, 577.
[3] Y. Park, H. Li, B. Kemper, *J. Biol. Chem.* **1996**, *271*, 23725.
[4] C. Handschin, U. A. Meyer, *Pharmacol. Rev.* **2003**, *55*, 649.
[5] E. Trottier, A. Belzil, C. Stoltz, A. Anderson, *Gene* **1995**, *158*, 263.
[6] D. J. Waxman, *Arch. Biochem. Biophys.* **1999**, *369*, 11.
[7] P. Wei, J. Zhang, D. H. Dowhan, Y, Han, D. D. Moore, *Pharmacogenomics J.* **2002**, *2*, 117.
[8] P. Wei, J. Zhang, M. Egan-Hafley, S. Liang, D. D. Moore, *Nature* **2000**, *407*, 920.
[9] W. Xie, J. L. Barwick, M. Downes, B. Blumberg, C. M. Simon, M. C. Nelson, B. A. Neuschwander-Tetri, E. M. Brunt, P. S. Guzelian, R. M. Evans, *Nature* **2000**, *406*, 435.
[10] J. L. Staudinger, B. Goodwin, S. A. Jones, D. Hawkins-Brown, K. I. MacKenzie, A. LaTour, Y. Liu, C. D. Klaassen, K. K. Brown, J. Reinhard, T. M. Willson, B. H. Koller, S. A. Kliewer, *Proc. Natl. Acad. Sci. U.S.A.* **2001**, *98*, 3369.
[11] K. Swales, M. Negishi, *Mol. Endocrinol.* **2004**, *18*, 1589.
[12] K. Yoshinari, K. Kobayashi, R. Moore, T. Kawamoto, M. Negishi, *FEBS Lett.* **2003**, *548*, 17.
[13] S. A. Kliewer, T. M. Willson, *J. Lipid Res.* **2002**, *43*, 359.
[14] J. L. Raucy, *Drug Metab. Dispos.* **2003**, *31*, 533.
[15] S. A. Jones, L. B. Moore, J. L. Shenk, G. B. Wisely, G. A, Hamilton, D. D. McKee, N. C. Tomkinson, E. L. LeCluyse, M. H. Lambert, T. M. Willson, S. A. Kliewer, J. T. Moore, *Mol. Endocrinol.* **2000**, *14*, 27.
[16] G. Zhou, R. Cummings, Y. Li, S. Mitra, H. A. Wilkinson, A. Elbrecht, J. D. Hermes, J. M. Schaeffer, R. G. Smith, D. E. Moller, *Mol. Endocrinol.* **1998**, *12*, 1594.
[17] G. Krey, O. Braissant, F. L'Horset, E. Kalkhoven, M. Perroud, M. G. Parker, W. Wahli, *Mol. Endocrinol.* **1997**, *11*, 779.
[18] I. Dussault, H. D. Yoo, M. Lin, E. Wang, M. Fan, A. K. Batta, G. Salen, S. K. Erickson, B. M. Forman, *Proc. Natl. Acad. Sci. U.S.A.* **2003**, *100*, 833.
[19] P. Honkakoski, I. Zelko, T. Sueyoshi, M. Negishi, *Mol. Cell. Biol.* **1998**, *18*, 5652.
[20] T. Kawamoto, S. Kakizaki, K. Yoshinari, M. Negishi, *Mol. Endocrinol.* **2000**, *14*, 1897.
[21] G. T. Tucker, J. B. Houston, S. M. Huang, *Eur. J. Pharm. Sci.* **2001**, *13*, 417.
[22] P. Honkakoski, R. Moore, J. Gynther, M. Negishi, *J. Biol. Chem.* **1996**, *271*, 9746.
[23] C. Handschin, M. Podvinec, J. Stockli, K. Hoffmann, U. A. Meyer, *Mol. Endocrinol.* **2001**, *15*, 1571.

[24] C. Rodriguez-Antona, R. Bort, R. Jover, N. Tindberg, M. Ingelman-Sundberg, M. J. Gomez-Lechon, J. V. Castell, *Mol. Pharmacol.* **2003**, *63*, 1180.
[25] M. Martignoni, M. Monshouwer, R. de Kanter, D. Pezzetta, A. Moscone, P. Grossi, *Toxicol. in Vitro* **2004**, *18*, 121.
[26] J. Zhang, W. Huang, S. S. Chua, P. Wei, D. D. Moore, *Science* **2002**, *298*, 422.
[27] E. Schuetz, L. Lan, K. Yasuda, R. Kim, T. A. Kocarek, J. Schuetz, S. Strom, *Mol. Pharmacol.* **2002**, *62*, 439.
[28] S. Chainuvati, A. N. Nafziger, J. S. Leeder, A. Gaedigk, G. L. Kearns, E. Sellers, Y. Zhang, A. D. Kashuba, E. Rowland, J. S. Bertino Jr., *Clin. Pharmacol. Ther.* **2003**, *74*, 437.
[29] K. Bodin, L. Bretillon, Y. Aden, L. Bertilsson, U. Broome, C. Einarsson, U. Diczfalusy, *J. Biol. Chem.* **2001**, *276*, 38685.
[30] K. Bodin, U. Andersson, E. Rystedt, E. Ellis, M. Norlin, I. Pikuleva, G. Eggertsen, I. Bjorkhem, U. Diczfalusy, *J. Biol. Chem.* **2002**, *277*, 31534.
[31] C. Furster, K. Wikvall, *Biochim. Biophys. Acta* **1999**, *1437*, 46.
[32] Z. Araya, K. Wikvall, *Biochim. Biophys. Acta* **1999**, *1438*, 47.
[33] M. Podvinec, M. R. Kaufmann, C. Handschin, U. A. Meyer, *Mol. Endocrinol.* **2002**, *16*, 1269.
[34] J. T. Moore, L. B. Moore, J. M. Maglich, S. A. Kliewer, *Biochim. Biophys. Acta* **2003**, *1619*, 235.
[35] U. A. Meyer, *Lancet* **2000**, *356*, 1667.
[36] J. Zhang, P. Kuehl, E. D. Green, J. W. Touchman, P. B. Watkins, A. Daly, S. D. Hall, P. Maurel, M. Relling, C. Brimer, K. Yasuda, S. A. Wrighton, M. Hancock, R. B. Kim, S. Strom, K. Thummel, C. G. Russell, J. R. Hudson Jr., E. G. Schuetz, M. S. Boguski, *Pharmacogenetics* **2001**, *11*, 555.
[37] E. Hustert, A. Zibat, E. Presecan-Siedel, R. Eiselt, R. Mueller, C. Fuss, I. Brehm, U. Brinkmann, M. Eichelbaum, L. Wojnowski, O. Burk, *Drug Metab. Dispos.* **2001**, *29*, 1454.
[38] K. A. Arnold, M. Eichelbaum, O. Burk, *Nucl. Recept.* **2004**, *2*, 1.
[39] A. Sapone, M. Paolini, G. L. Biagi, G. Cantelli-Forti, F. J. Gonzalez, *Trends Pharmacol. Sci.* **2002**, *23*, 260.

Coordination of Uptake and Efflux Transporters in Drug Disposition

by **Hiroyuki Kusuhara*** and **Yuichi Sugiyama**

Department of Molecular Pharmacokinetics, Graduate School of Pharmaceutical Sciences, University of Tokyo, 7-3-1 Hongo, Bunkyo-ku, Tokyo, 113-033, Japan
(fax: +81-3-5841-4766; e-mail: kusuhara@mol.f.u-tokyo.ac.jp)

Abbreviations
ABC: ATP Binding cassette; ACE: angiotensin converting enzyme; BCRP: breast cancer resistance protein; BBM: brush border membrane; BSEP: bile salt exporting polypeptide; cMOAT: canalicular multispecific organic anion transporter; DHEAS: dehydroepiandrosterone sulfate ; E3040S: 6-hydroxy-5,7-dimethyl-2-(methylamino)-4-(3-pyridylmethyl)benzothiazole sulfate; $E_217\beta G$: estradiol 17-glucuronide; HMG-CoA: 3-hydroxy-3-methylglutaryl-CoA; MDCK-II cells: *Madin–Darby* canine kidney II cells; MDR: multidrug resistance; MRP: multidrug resistance associated protein; NTCP: sodium-taurocholate cotransporting polypeptide; OAT: (human) organic anion transporter; Oat: (rodent) organic anion transporter; OATP: organic anion transporting polypeptide; *PS*: permeability surface area product; rOat: rat organic anion transporter.

---

1. Introduction

Cumulative *in vivo* and *in vitro* studies have revealed that transporters play important roles in drug disposition. They have been recognized as target molecules for achieving optimum pharmacokinetic profiles in drug discovery and development and, from a clinical point of view, they can be sites of drug–drug interactions, and one of the factors responsible for interindividual differences in drug response. During the last decade, many human and animal transporter cDNAs have been isolated [1], which have enabled us to investigate their substrate specificity in expression systems. Overlapping substrate specificity between uptake and efflux transporters gives epithelial cells the ability to carry out efficient vectorial transport. This transport plays an essential role in the biliary and urinary excretion of organic compounds as well as the efflux transport across the barriers of central nervous system. As far as organic anions are concerned, organic anion transporting polypeptide (OATPs) and organic anion transporter (OATs) have been cloned, and they play a central role in the uptake process in the liver and kidney, respectively. ATP Binding cassette (ABC) transporters, such as multidrug resistance associated protein (MRPs), breast cancer resistance protein (BCRP), and bile salt export pump (BSEP) are involved in the subsequent excretion processes.

In this manuscript, recent progress made in our own laboratory and by others will be summarized, focusing particularly on the transporters involved in vectorial transport in the liver and kidney; in addition, the use of recently established double transfectants, which express both uptake and efflux transporters in polarized cells, is described.

2. Transport of Organic Anions in the Liver and Kidney

The transporters involved in the vectorial transport of organic anions in the liver are illustrated in *Fig. 1*. The hepatobiliary transport of amphipathic organic anions is achieved by a coordination of uptake and efflux transporters. OATPs play an important role in the uptake process across the sinusoidal membrane, while ABC transporters such as MRP2 mediate the subsequent excretion process into the bile. The characteristics of the uptake and efflux transporters for organic anions in the liver and their coordination are described in this section.

2.1. *Hepatobiliary Transport of Organic Anions*

2.1.1. *Organic Anion Transporting Polypeptide* (OATP)

OATPs were classified as members of the SLC21 family but, recently, their gene symbols have been incorporated in the SLCO family and they have been renamed based on their sequence similarity (http://www.pharmaconference.org/SLC21.htm; and for a review [2]). The members of the OATP family contain 12 putative transmembrane domains. In the human liver, three OATP isoforms have been identified on the sinusoidal membrane: OATP-B (*SLCO2B1*/OATP2B1), OATP-C (*SLCO1B1*/OATP1B1) and OATP8 (*SLCO1B3*/OATP1B3) (*Fig. 1*) [3–8]. OATP-B is ubiquitously expressed in the body [8], while OATP-C and OATP8 are expressed predominantly in the liver where OATP-C is expressed homogenously, and OATP8 is highly expressed around the central vein [3–6][8].

OATP-C exhibits broad substrate specificity including bilirubin and its mono- and bisglucuronides, bile acids, conjugated steroids, eicosanoids and other organic anions, such as pravastatin, bromosulfophthalein and

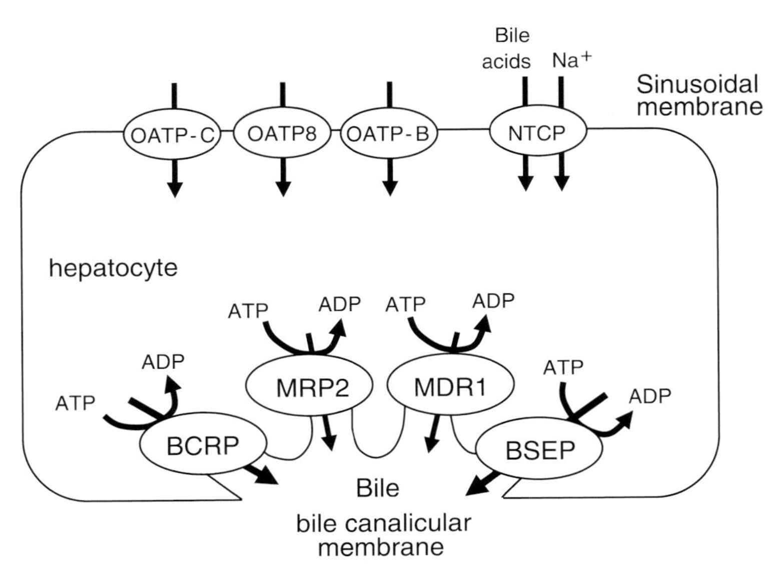

Fig. 1. *Schematic diagram of the organic anion transporters expressed in the liver.* OATP-B, OATP-C, and OATP8 have been identified on the sinusoidal membrane. MRP2 plays a major role in the excretion of amphipathic organic anions, such as glucuronides, glutathione conjugates, and nonconjugated amphipathic organic anions, across the bile canalicular membrane. In addition to MRP2, other ABC transporters, such as P-glycoprotein and BCRP, also accept certain kinds of organic anions as substrates. For bile acids, NTCP accounts for the sodium-dependent uptake, while OATP-C and OATP8 account for the sodium-independent uptake. Subsequently, bile acids undergo excretion across the bile canalicular membrane *via* an ABC transporter, BSEP. *Abbreviations:* BCRP, breast cancer resistance protein; BSEP, bile salt exporting polypeptide; MDR, multidrug resistance; MRP2, multidrug resistance associated protein 2; NTCP, sodium-taurocholate cotransporting polypeptide; OATP, organic anion transporting polypeptide.

rifampicin [2]. OATP8 shares a similar substrate specificity with OATP-C [3–7], but cholecystokinin and digoxin are specifically transported by OATP8 [7][9]. The contribution of OATP8 to the hepatic uptake of common substrates with OATP-C remains unknown. Due to the apparent narrow substrate specificity, the role of OATP-B in the liver remains unclear [7].

OATP-C has been suggested to be a site of drug–drug interaction of the immunosuppressant cyclosporin A. Cerivastatin, an HMG-CoA reductase inhibitor, was withdrawn from the market due to a drug–drug interaction with gemfibrozil. Previously, the pharmacokinetic profile of cerivastatin had been compared in kidney transplant recipients receiving individual immunosuppressive treatment and healthy subjects, and it was suggested that coadministration of cyclosporin A caused an increase in the blood concentration of cerivastatin in patients [10]. Cyclosporin A is an inhibitor of OATP-C, and its plasma unbound concentration is sufficient to inhibit OATP-C mediated uptake under clinical conditions. Currently, it is hypothesized that the OATP-C mediated uptake process is the site of the drug–drug interaction between cerivastatin and cyclosporin A [11]. OATP-C may account for the interaction between bosentan (an endothelin receptor antagonist) and cyclosporin A [12].

OATP-C gives us some insight into the interindividual differences in the plasma concentrations of its substrates. Recently, a genetic polymorphism of OATP-C was reported to influence the hepatic clearance of pravastatin [13]. Haplotypes of OATP-C nonsynonymous mutations are classified into sixteen types [13–15]. Of these haplotypes, OATP-C*15 (N130D and V174A) is a relatively common haplotype in the Japanese population (10%, $n=267$ in Ref [15], and 15%, $n=120$ in Ref [13]). The plasma concentration of pravastatin (an HMG-CoA reductase inhibitor) was higher in healthy Japanese volunteers with OATP-C*15/*15 or OATP-C*15/*1b than that in volunteers with OATP-C*1b/*1b (*Fig. 2*) [13]. The effect of the double mutations on the intrinsic transport activity and expression on the plasma membrane remains to be elucidated.

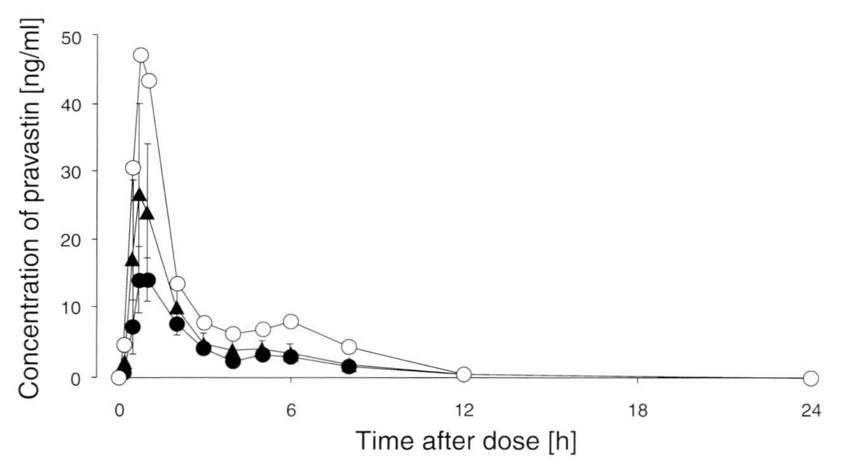

Fig. 2. *Plasma concentrations of pravastatin in Japanese healthy subjects.* Pravastatin was administered orally to Japanese healthy subjects (OATP-C*1b/*1b (●; $n=4$), OATP-C*15/*1b (▲; $n=15$), OATP-C*15/*15 (○; $n=1$)). The plasma concentration was greater in the subjects with OATP-C*15/*1b and OATP-C*15/*15 than in the subjects with OATP-C*1b/*1b. Nonrenal clearance was 50% in the subjects with OATP-C*15/*1b, and 10% in the subjects with OATP-C*15/*15 of that in the subjects with OATP-C*1b/*1b [13].

2.1.2. *Multidrug Resistance Associated Protein 2* (MRP2)

MRP2 (*ABCC2*) is a 190 kDa protein with 17 putative transmembrane domains and two cytoplasmically located ATP-binding cassettes, and a deficiency in MRP2 function is associated with the *Dubin–Johnson* syndrome (OMIM 237500). Mutant rats, such as TR^{-}/Groningen yellow and *Eisai* hyperbilirubinemic (EHBR) rats, are hereditarily Mrp2 deficient [16–18]. This deficit causes an inability to excrete many kinds of organic anions into the bile, suggesting that it has a key role in the biliary excretion of organic anions. Most of the Mrp2 substrates have been elucidated by comparing the biliary excretion in normal and mutant rats: glutathione and glucuronide conjugates, sulfo-conjugated bile acids, and nonconjugated amphipathic organic anions [19–21]. The physicochemical properties of rat Mrp2 substrates that correlate with recognition specificity were investigated using methotrexate derivatives [22]. The affinity constants closely correlated with the calculated octanol/water partition coefficient (ClogP), and a linear combination of polar and nonpolar surface areas. The affinity for MRP2 (*i.e.*, the human form) also closely correlated with the molecular weight, which also showed a significant correlation with the nonpolar surface area and ClogP. Accordingly, recognition by MRP2 depends on a balance of dynamic surface properties between the polar and nonpolar regions of methotrexate analogs. The so-called 'molecular weight threshold' for the MRP2 affinity of these compounds can be explained by their physicochemical parameters, especially their nonpolar surface areas.

2.1.3. *Co-Expression of Uptake and Efflux Transporters in Polarized Cells*

The substrate specificity of OATP-C/OATP8 and MRP2 overlaps, indicating a coordination in the efficient vectorial transport from blood to bile. When OATP-C and MRP2 are expressed in MDCK-II cells, they are localized on the basal and apical membrane, respectively [23]. Co-expression of OATP-C and MRP2 in MDCK-II cells increased the basal-to-apical transport of typical ligands such as $E_217\beta G$ and pravastatin, but did not affect the transcellular transport in the opposite direction (*Fig. 3*). PS_{apical}, an intrinsic parameter for the efflux across the apical membrane, was determined by dividing the rate of the transcellular transport of $E_217\beta G$ by its cellular concentration. The PS_{apical} was greater in the OATP2/MRP2 double-transfectant than in the control OATP-C and MRP2 MDCK-II cells (*Fig. 4*), which is in good agreement with the localization of MRP2 in MDCK-II cells. By correcting PS_{apical} by the factor

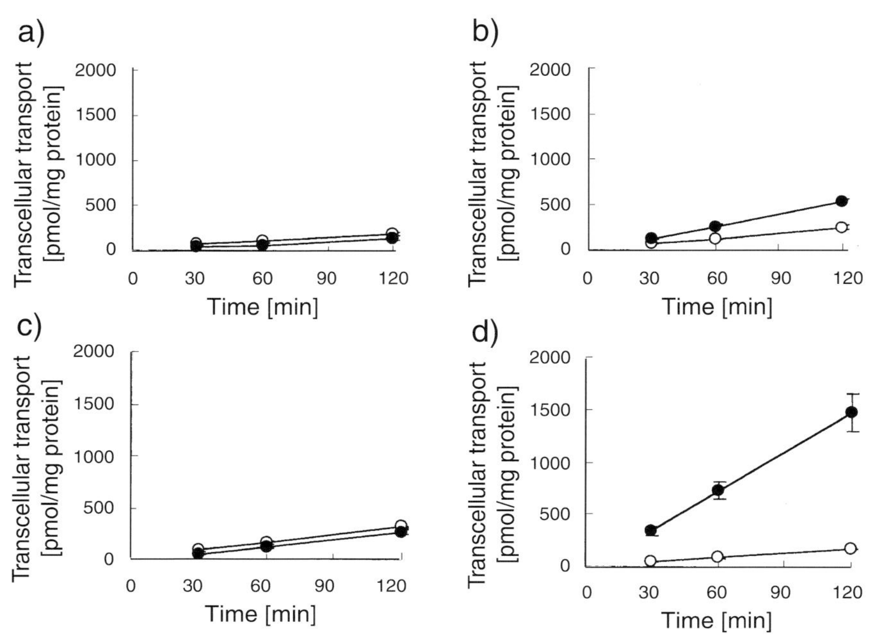

Fig. 3. *Transcellular transport of $E_217\beta G$ in OATP-C/MRP2 double transfectants in MDCK-II cells.* The transcellular transport of $[^3H]$-$E_217\beta G$ (1 μM) across MDCK-II monolayers expressing OATP (*b*), MRP2 (*c*), and both OATP2 and MRP2 (double transfectant; *d*) was compared with that across the control MDCK-II monolayer (*a*). *Open* and *closed circles* represent the transcellular transport in the apical-to-basal and basal-to-apical directions, resp. Each *point* and *vertical bar* represents the mean ± S.E. of three determinations. Where *vertical bars* are not shown, the S.E. was contained within the limits of the symbol [23].

of protein-expression ratio in the double transfectant and the human liver, we may be able to estimate the intrinsic parameter for the biliary excretion of drugs.

The basal-to-apical transport of $E_217\beta G$ was saturable, but *PS* showed a slight saturation at the concentrations examined (*Fig. 4*). Thus, the uptake process at the basal side is likely to be rate-limiting for all the transcellular transport systems of $E_217\beta G$. Compounds which are not substrates of MRP2, such as estrone-3-sulfate and DHEAS, exhibit no vectorial transport across the monolayer. *Cui et al.* established the double transfectant in a combination of OATP8 and MRP2 in MDCK-II cells, where there is basal-to-apical transport of organic anions such as bromosulfophthalein, $E_217\beta G$, DHEAS, taurocholate, and Fluo-3 [24]. The reason for the discrepancy in the vectorial transport of DHEAS in OATP-C/MRP2 and OATP8/MRP2 double transfectants remains unknown [23] [24].

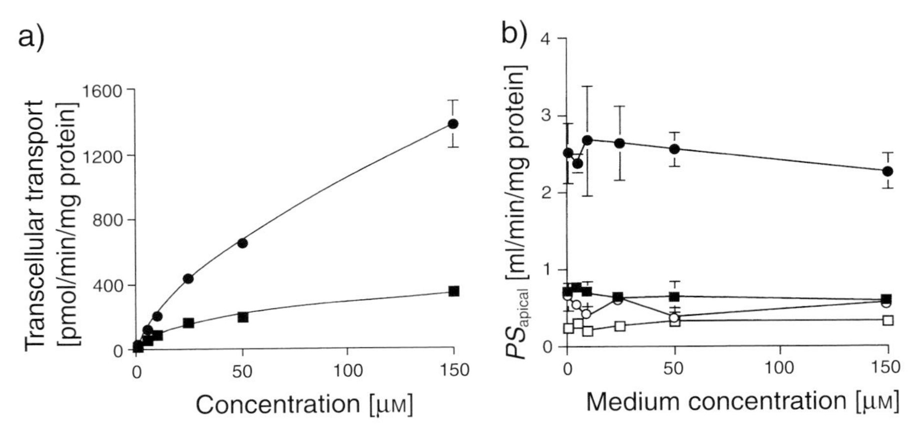

Fig. 4. *Concentration dependence of transcellular transport and the PS product of $E_2 17\beta G$ in OATP-C and OATP-C/MRP2 MDCK-II cells. a)* The transcellular transport of [^{3}H]-$E_2 17\beta G$ (1 µM) across MDCK-II monolayers expressing OATP2 (■) and both OATP2 and MRP2 (double transfectant; ●) was studied for 2 h in the presence and absence of unlabeled $E_2 17\beta G$ at 37°. Each symbol and bar represents the mean ± S.E. of three determinations. The solid lines represent the fitted line. *b)* The PS product for the transport of $E_2 17\beta G$ across the apical membrane of an MDCK-II monolayer (PS_{apical}) was determined by dividing the rate of transcellular transport of [^{3}H]-$E_2 17\beta G$ [dpm/min/mg of protein] for 2 h by the cellular concentration of $E_2 17\beta G$ determined at the end of the experiments (2 h). The PS_{apical} across MDCK-II expressing OATP2 (■), MRP2 (○), and both OATP2 and MRP2 (double transfectant; ●) was compared with that across control MDCK-II monolayers (□). The horizontal axis represents the medium concentration (basal compartment) of $E_2 17\beta G$. Each point and vertical bar represents the mean ± S.E. of three determinations. Where vertical bars are not shown, the S.E. was contained within the limits of the symbol [23].

Transcellular transport from the basal-to-apical side of the monolayer corresponds to biliary transfer in this system, and the double transfectants will be applied for high-throughput screening in identifying compounds, such as temocapril, an ACE inhibitor which undergoes efficient hepatobiliary transport. Temocaprilat, the active metabolite of the prodrug temocapril, is a typical case for multiple elimination pathways *via* both the liver and kidneys, thereby avoiding to a large extent the interindividual variability in its pharmacokinetic profile compared with other ACE inhibitors which are eliminated from the body *via* a single pathway [25]. The main elimination organ for the other ACEs is the kidney and, therefore, the plasma concentration–time profile of other ACE inhibitors in patients depends on their renal function [25]. In contrast, 85–90% of the administered dose of temocaprilat is excreted in the feces of animals [26], while 36–40% are excreted in the feces of humans [27]. Therefore, the plasma concentration–time profile of temocaprilat is hardly affected by interindividual differences in renal function [25]. The biliary excretion of temocaprilat is mediated by MRP2, while other ACE inhibitors have no effect on the ATP-dependent uptake of temocaprilat by canalicular membrane vesicles, suggesting that they are not substrates of MRP2 [26]. In addition, the system will allow us to investigate the drug–drug interaction in the hepatobiliary transport.

2.1.4. *Other ABC Transporters*

In addition to MRP2, three ABC transporters have been identified on the canalicular membrane: P-glycoprotein (MDR1; *ABCB1*) and bile salt exporting polypeptide (BSEP; *ABCB11*) and, in the mouse, breast cancer resistant protein (Bcrp1) (*Fig. 1*) [28]. Whether BCRP, the human homolog of Bcrp1, is expressed in human liver remains unknown. Although the substrates of MDR1 are generally hydrophobic or neutral compounds, there are some organic anions that are substrates of MDR1, such as $E_217\beta G$ and fexofenadine [29] [30].

BSEP is highly selective for bile salts [31], and lack of BSEP function is associated with type II progressive familial intrahepatic cholestasis (OMIM 601847). Bile acids undergo enterohepatic circulation [19] [32], and BSEP is responsible for the biliary excretion of bile acids [31]. The hepatic uptake of bile acids exhibits sodium-dependent and sodium-independent mechanisms [19] [32]. The sodium-dependent mechanism is accounted for by sodium-taurocholate cotransporting polypeptide (NTCP; *SLC10A1*), while OATP-C and OATP8 account for the sodium-independent mechanism.

BCRP is a half-size ABC transporter with one ATP-binding cassette, and it forms a homo-dimer to function as an efflux transporter. BCRP shows broad substrate specificity, including anticancer drugs such as topotecan, daunomycin, and mitoxantrone [33] [34]. Recently, we found that BCRP accepts as substrates endogenous and exogenous sulfate conjugates such as estrone 3-sulfate, dehydroepiandrosterone sulfate (DHEAS), E3040S, and 4-methylumbelliferone sulfate [35]. The hepatobiliary excretion of intravenously administered topotecan was reduced two-fold by oral GF120918, an inhibitor of P-glycoprotein and BCRP, which is consistent with an excretory role for bile canalicular Bcrp1 [36]. Recently, it was demonstrated that the hepatobiliary excretion of the dietary carcinogen, 2-amino-1-methyl-6-phenyl-1*H*-imidazo[4,5-*b*]pyridine, was greatly reduced in Bcrp1 (–/–) mice [37]. These ABC transporters are presumably candidate transporters involved in vectorial transport in concert with basolateral organic anion transporters.

2.2. *Renal Transport of Organic Anions*

The transporters involved in the vectorial transport of organic anions in the kidney are illustrated in *Fig. 5*. The renal transport of amphipathic organic anions across the proximal tubules is achieved by a coordination of uptake and efflux transporters. OAT1 and OAT3 play an important role in the uptake process at the basolateral membrane of the proximal tubules, while the transporters involved in the subsequent excretion process have not fully elucidated yet. The characteristics of the uptake and efflux transporters for organic anions in the kidney are described in this section.

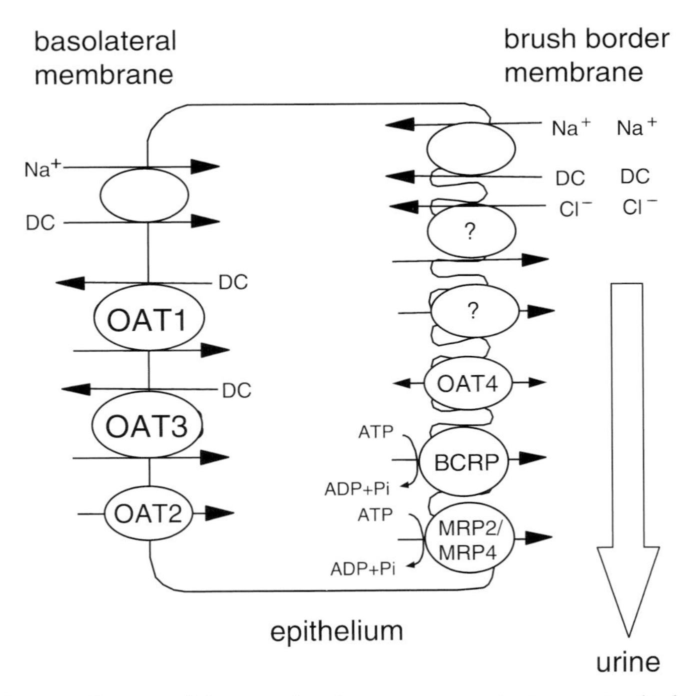

Fig. 5. *Schematic diagram of the organic anion transporters in the proximal tubules.* OAT1 and OAT3 have been identified on the basolateral membrane. According to the investigations in rats, OAT1 plays a major role in the renal uptake of small and hydrophilic organic anions, while OAT3 plays a major role in the renal uptake of more-bulky organic anions. The transporters involved in the excretion across the brush border membrane of the proximal tubules have not been fully described yet. Currently, three kinds of transporters have been proposed: *1*) an exchanger of Cl^- and organic anions, *2*) a facilitative transporter, and *3*) an ABC transporter (MRP2, MRP4, and BCRP). *Abbreviations:* BCRP, breast cancer resistance protein; MRP2 and -4, multidrug resistance associated protein 2 and 4; OAT, organic anion transporter.

2.2.1. *Organic Anion Transporters* (OAT)

OATs have been classified as members of the *SLC22A* family (http://www.pharmaconference.org/SLC22.htm). Members of this OAT family contain 12 putative transmembrane domains. In the human kidney, OAT3 mRNA is the most-abundant form followed by OAT1, while there is little OAT2 and OAT4 [38]. OAT1 and OAT3 have been identified on the basolateral membrane of the proximal tubules [38][39], while OAT4 is expressed on the BBM [40]. The substrate specificity of OAT1 and OAT3 has been investigated in *Xenopus laevis* oocytes and mammalian cells. OAT1 transports *p*-aminohippurate, antiviral nucleotides, nonsteroidal anti-inflammatory drugs, and uremic toxins [41–44], while OAT3 has a broad substrate specificity including hydrophilic and amphipathic organic anions, and the weakly basic cimetidine [39][42].

The contribution of Oat1 and Oat3 to the renal uptake of organic anions has been evaluated in rat kidney slices, and it has been proposed that rOat1 (rat organic anion transporter 1) is mainly responsible for the renal uptake of hydrophilic and small molecules, such as *p*-aminohippurate and 2,4-(dichlorophenoxy)acetate, while rOat3 is responsible for the renal uptake of amphipathic and bulky organic anions, such as pravastatin, benzylpenicillin, methotrexate, and DHEAS [42][45–47].

2.2.2. *Apical Transporters*

A series of studies using brush border membrane (BBM) vesicles suggests the presence of two transport mechanisms: *1*) facilitative transporter(s) driven by a luminal positive membrane voltage, or *2*) an exchanger which exports organic anions in exchange for luminal Cl^- [48]. The molecular characteristics of these transporters remain to be elucidated. Although OAT4 has been identified on the BBM of the proximal tubules, its role in tubular secretion and/or reabsorption remains unknown [40]. In addition, ABC transporters such as MRP2 and MRP4 are expressed on the BBM [49][50], and the expression of BCRP has also been detected on the BBM of mouse kidney [28].

Schaub et al. have demonstrated that MRP2 is localized on the BBM of the proximal tubules [49]. Comparison of the renal clearance of MRP2/cMOAT substrates between normal and mutant rats such as EHBR/TR^-/Groningen yellow will allow us to estimate the contribution of Mrp2 to renal excretion. *Masereeuw et al.* have shown that the renal clearance of calcein and fluo-3 is decreased in Mrp2-deficient mutant rats (TR^-), suggesting that it is involved in renal excretion [51].

MRP4 is a 170 kDa protein with 12 transmembrane domains and 2 cytosolic ATP-binding cassettes. MRP4 was initially reported as a homolog of MRP1 by screening databases of human expressed sequence tags [52]. It is expressed only at very low levels in a few tissues [52]. In the kidney, it has been shown to be expressed on the apical membrane of the proximal tubules [50]. The substrates of MRP4 include cyclic nucleotides (cAMP and cGMP), folate and its analog, methotrexate, steroid conjugates ($E_2 17\beta G$ and DHEAS), and prostaglandins (PGE_1 and PGE_2) [50][53–55], and it has been suggested that MRP4 is involved in the tubular secretion of these compounds. The renal clearance of lucifer yellow is not affected by a deficiency of Mrp2 [51]. Since lucifer yellow is a substrate of MRP4, the involvement of Mrp4 in its tubular secretion has been proposed [51].

These ABC transporters, as well as unknown transporters such as exchanger or facilitative transporters, may be involved in tubular secretion together with basolateral transporters (OATs). Co-expression of these transporters with OATs in polarized cells will help us to examine the contribution of these transporters to the tubular secretion process.

3. Conclusions

Coordination of uptake and efflux transporters allows the efficient vectorial transport of organic anions in the liver and kidney. We can reconstitute such systems in polarized cells by co-expression of uptake and efflux transporters. The double transfectants can be applied for high-throughput screening to obtain candidate compounds with optimal pharmacokinetic properties, for examining the possibility of drug–drug interactions, and for investigating the effect of genetic polymorphisms on the efficacy of vectorial transport across epithelial cells.

REFERENCES

[1] N. Mizuno, T. Niwa, Y. Yotsumoto, Y. Sugiyama, *Pharmacol. Rev.* **2003**, *55*, 425.
[2] B. Hagenbuch, P. J. Meier, *Biochim. Biophys. Acta* **2003**, *1609*, 1.
[3] T. Abe, M. Kakyo, T. Tokui, R. Nakagomi, T. Nishio, D. Nakai, H. Nomura, M. Unno, M. Suzuki, T. Naitoh, S. Matsuno, H. Yawo, *J. Biol. Chem.* **1999**, *274*, 17159.
[4] T. Abe, M. Unno, T. Onogawa, T. Tokui, T. N. Kondo, R. Nakagomi, H. Adachi, K. Fujiwara, M. Okabe, T. Suzuki, K. Nunoki, E. Sato, M. Kakyo, T. Nishio, J. Sugita, N. Asano, M. Tanemoto, M. Seki, F. Date, K. Ono, Y. Kondo, K. Shiiba, M. Suzuki, H. Ohtani, T. Shimosegawa, K. Iinuma, H. Nagura, S. Ito, S. Matsuno, *Gastroenterology* **2001**, *120*, 1689.
[5] J. Konig, Y. Cui, A. T. Nies, D. Keppler, *Am. J. Physiol. Gastrointest. Liver Physiol.* **2000**, *278*, G156.

[6] J. Konig, Y. Cui, A. T. Nies, D. Keppler, *J. Biol. Chem.* **2000**, *275*, 23161.

[7] G. A. Kullak-Ublick, M. G. Ismair, B. Stieger, L. Landmann, R. Huber, F. Pizzagalli, K. Fattinger, P. J. Meier, B. Hagenbuch, *Gastroenterology* **2001**, *120*, 525.

[8] I. Tamai, J. Nezu, H. Uchino, Y. Sai, A. Oku, M. Shimane, A. Tsuji, *Biochem. Biophys. Res. Commun.* **2000**, *273*, 251.

[9] M. G. Ismair, B. Stieger, V. Cattori, B. Hagenbuch, M. Fried, P. J. Meier, G. A. Kullak-Ublick, *Gastroenterology* **2001**, *121*, 1185.

[10] W. Muck, I. Mai, L. Fritsche, K. Ochmann, G. Rohde, S. Unger, A. Johne, S. Bauer, K. Budde, I. Roots, H. H. Neumayer, J. Kuhlmann, *Clin. Pharmacol. Ther.* **1999**, *65*, 251.

[11] Y. Shitara, T. Itoh, H. Sato, A. P. Li, Y. Sugiyama, *J. Pharmacol. Exp. Ther.* **2003**, *304*, 610.

[12] A. Treiber, R. Schneiter, S. Delahaye, M. Clozel, *J. Pharmacol. Exp. Ther.* **2003**, *308*, 1121.

[13] Y. Nishizato, I. Ieiri, H. Suzuki, M. Kimura, K. Kawabata, T. Hirota, H. Takane, S. Irie, H. Kusuhara, Y. Urasaki, A. Urae, S. Higuchi, K. Otsubo, Y. Sugiyama, *Clin. Pharmacol. Ther.* **2003**, *73*, 554.

[14] R. G. Tirona, B. F. Leake, G. Merino, R. B. Kim, *J. Biol. Chem.* **2001**, *276*, 35669.

[15] T. Nozawa, M. Nakajima, I. Tamai, K. Noda, J. Nezu, Y. Sai, A. Tsuji, T. Yokoi, *J. Pharmacol. Exp. Ther.* **2002**, *302*, 804.

[16] K. Ito, H. Suzuki, T. Hirohashi, K. Kume, T. Shimizu, Y. Sugiyama, *Am. J. Physiol.* **1997**, *272*, G16.

[17] C. C. Paulusma, P. J. Bosma, G. J. Zaman, C. T. Bakker, M. Otter, G. L. Scheffer, R. J. Scheper, P. Borst, R. P. Oude Elferink, *Science* **1996**, *271*, 1126.

[18] M. Buchler, J. Konig, M. Brom, J. Kartenbeck, H. Spring, T. Horie, D. Keppler, *J. Biol. Chem.* **1996**, *271*, 15091.

[19] H. Suzuki, Y. Sugiyama, in 'Membrane Transporters as Drug Targets', Eds. W. Sadée, G. Amidon, Kluwer Academic/Plenum Publishing, New York, 1998, p. 387–439.

[20] K. N. Faber, M. Muller, P. L. Jansen, *Adv. Drug Delivery Rev.* **2003**, *55*, 107.

[21] D. Keppler, J. Konig, *Semin. Liver Dis.* **2000**, *20*, 265.

[22] Y. H. Han, Y. Kato, M. Haramura, M. Ohta, H. Matsuoka, Y. Sugiyama, *Pharm. Res.* **2001**, *18*, 579.

[23] M. Sasaki, H. Suzuki, K. Ito, T. Abe, Y. Sugiyama, *J. Biol. Chem.* **2002**, *277*, 6497.

[24] Y. Cui, J. Konig, D. Keppler, *Mol. Pharmacol.* **2001**, *60*, 934.

[25] M. Nakashima, J. Yamamoto, M. Shibata, T. Uematsu, H. Shinjo, T. Akahori, H. Shioya, K. Sugiyama, Y. Kawahara, *Eur. J. Clin. Pharmacol.* **1992**, *43*, 657.

[26] H. Ishizuka, K. Konno, H. Naganuma, K. Sasahara, Y. Kawahara, K. Niinuma, H. Suzuki, Y. Sugiyama, *J. Pharmacol. Exp. Ther.* **1997**, *280*, 1304.

[27] H. Suzuki, T. Kawaratani, H. Shioya, Y. Uji, T. Saruta, *Biopharm. Drug Dispos.* **1993**, *14*, 41.

[28] J. W. Jonker, M. Buitelaar, E. Wagenaar, M. A. Van Der Valk, G. L. Scheffer, R. J. Scheper, T. Plosch, F. Kuipers, R. P. Elferink, H. Rosing, J. H. Beijnen, A. H. Schinkel, *Proc. Natl. Acad. Sci. U.S.A.* **2002**, *99*, 15649.

[29] M. Cvetkovic, B. Leake, M. F. Fromm, G. R. Wilkinson, R. B. Kim, *Drug Metab. Dispos.* **1999**, *27*, 866.

[30] L. Huang, T. Hoffman, M. Vore, *Hepatology* **1998**, *28*, 1371.

[31] T. Gerloff, B. Stieger, B. Hagenbuch, J. Madon, L. Landmann, J. Roth, A. F. Hofmann, P. J. Meier, *J. Biol. Chem.* **1998**, *273*, 10046.

[32] G. A. Kullak-Ublick, B. Stieger, P. J. Meier, *Gastroenterology* **2004**, *126*, 322.

[33] L. A. Doyle, D. D. Ross, *Oncogene* **2003**, *22*, 7340.

[34] J. D. Allen, A. H. Schinkel, *Mol. Cancer Ther.* **2002**, *1*, 427.

[35] M. Suzuki, H. Suzuki, Y. Sugimoto, Y. Sugiyama, *J. Biol. Chem.* **2003**, *278*, 22644.

[36] J. W. Jonker, J. W. Smit, R. F. Brinkhuis, M. Maliepaard, J. H. Beijnen, J. H. Schellens, A. H. Schinkel, *J. Natl. Cancer Inst.* **2000**, *92*, 1651.

[37] A. E. van Herwaarden, J. W. Jonker, E. Wagenaar, R. F. Brinkhuis, J. H. Schellens, J. H. Beijnen, A. H. Schinkel, *Cancer Res.* **2003**, *63*, 6447.

[38] H. Motohashi, Y. Sakurai, H. Saito, S. Masuda, Y. Urakami, M. Goto, A. Fukatsu, O. Ogawa, K. Inui, *J. Am. Soc. Nephrol.* **2002**, *13*, 866.
[39] S. H. Cha, T. Sekine, J. I. Fukushima, Y. Kanai, Y. Kobayashi, T. Goya, H. Endou, *Mol. Pharmacol.* **2001**, *59*, 1277.
[40] S. H. Cha, T. Sekine, H. Kusuhara, E. Yu, J. Y. Kim, D. K. Kim, Y. Sugiyama, Y. Kanai, H. Endou, *J. Biol. Chem.* **2000**, *275*, 4507.
[41] M. Hosoyamada, T. Sekine, Y. Kanai, H. Endou, *Am. J. Physiol.* **1999**, *276*, F122.
[42] T. Deguchi, H. Kusuhara, A. Takadate, H. Endou, M. Otagiri, Y. Sugiyama, *Kidney Int.* **2004**, *65*, 162.
[43] A. S. Mulato, E. S. Ho, T. Cihlar, *J. Pharmacol. Exp. Ther.* **2000**, *295*, 10.
[44] T. Cihlar, D. C. Lin, J. B. Pritchard, M. D. Fuller, D. B. Mendel, D. H. Sweet, *Mol. Pharmacol.* **1999**, *56*, 570.
[45] Y. Nozaki, H. Kusuhara, H. Endou, Y. Sugiyama, *J. Pharmacol. Exp. Ther.* **2004**, *309*, 226.
[46] M. Hasegawa, H. Kusuhara, H. Endou, Y. Sugiyama, *J. Pharmacol. Exp. Ther.* **2003**, *305*, 1087.
[47] M. Hasegawa, H. Kusuhara, D. Sugiyama, K. Ito, S. Ueda, H. Endou, Y. Sugiyama, *J. Pharmacol. Exp. Ther.* **2002**, *300*, 746.
[48] J. B. Pritchard, D. S. Miller, *Physiol. Rev.* **1993**, *73*, 765.
[49] T. P. Schaub, J. Kartenbeck, J. Konig, H. Spring, J. Dorsam, G. Staehler, S. Storkel, W. F. Thon, D. Keppler, *J. Am. Soc. Nephrol.* **1999**, *10*, 1159.
[50] R. A. van Aubel, P. H. Smeets, J. G. Peters, R. J. Bindels, F. G. Russel, *J. Am. Soc. Nephrol.* **2002**, *13*, 595.
[51] R. Masereeuw, S. Notenboom, P. H. Smeets, A. C. Wouterse, F. G. Russel, *J. Am. Soc. Nephrol.* **2003**, *14*, 2741.
[52] M. Kool, M. de Haas, G. L. Scheffer, R. J. Scheper, M. J. van Eijk, J. A. Juijn, F. Baas, P. Borst, *Cancer Res.* **1997**, *57*, 3537.
[53] Z. S. Chen, K. Lee, G. D. Kruh, *J. Biol. Chem.* **2001**, *276*, 33747.
[54] G. Reid, P. Wielinga, N. Zelcer, M. De Haas, L. Van Deemter, J. Wijnholds, J. Balzarini, P. Borst, *Mol. Pharmacol.* **2003**, *63*, 1094.
[55] Z. S. Chen, K. Lee, S. Walther, R. B. Raftogianis, M. Kuwano, H. Zeng, G. D. Kruh, *Cancer Res.* **2002**, *62*, 3144.

The Role of Plasma Protein Binding in Drug Discovery

by **Roger E. Fessey***, **Rupert P. Austin**, **Patrick Barton**, **Andrew M. Davis**, and **Mark C. Wenlock**

Department of Physical and Metabolic Science, *AstraZeneca* R&D Charnwood, Bakewell Road, Loughborough, Leics LE11 5RH, UK
(e-mail: Roger.Fessey@astrazeneca.com)

Abbreviations
AUC: Area under the curve (exposure); AUC_u: exposure to the unbound drug concentrations; c_b: total concentration of drug in whole blood; *CL*: clearance; CL_{int}: intrinsic clearance; c_p: total concentration of drug present in the plasma compartment of the blood; *F*: oral bioavailability; F_{abs}: fraction of the oral dose absorbed; fb_p: fraction of drug bound; fu_b: whole blood free fraction of a drug; fu_p: free fraction of a drug in plasma; fu_t: free fraction of drug in tissues; HSA: human serum albumin; $K^{B/F}$: pseudo-binding constant for plasma protein binding; PPB: plasma protein binding; Q_h: liver blood flow; QSPR: quantitative structure–property relationship; $t_{1/2}$: pharmacokinetic half-life; VHDL: very high density lipoproteins; VLDL: very low density lipoproteins; V_p: volume of the plasma compartment; V_{SS}: steady-state volume of distribution; V_t: volume of the extravascular tissue compartments.

---

1. Introduction

Measurement of the extent of plasma protein binding (PPB) and an understanding of the molecular properties that control binding is of fundamental importance in the drug discovery process. The extent of plasma protein binding of a candidate drug molecule has an influence on a number of critical areas listed below, and many of these areas will be reviewed in this article, *i.e.*:

- determination of margins in safety assessment/toxicology studies
- the efficacy of a drug
- drug metabolism and pharmacokinetics
- drug–drug interactions
- blood–brain barrier penetration

The term protein binding normally refers to the reversible association of a drug with the proteins of the plasma compartment of blood, and this binding is due to electrostatic and hydrophobic forces between drug and protein [1]. A drug which is bound reversibly will be in equilibrium with the free (unbound) drug, with the amount bound being dependent on both the affinity of the drug for the various proteins and the binding capacity of each protein.

Upon entering plasma, most drugs bind rapidly to the plasma constituents, which principally include albumin and α_1-acid glycoprotein [2][3]. Measurement of the extent of this binding requires either a physical separation of free from bound drug, or a technique that can distinguish some property of bound from unbound drug. The free fraction of a drug in plasma, fu_p, is defined by *Eqn. 1*:

$$fu_p = [\text{free drug in plasma}]/[\text{total drug in plasma}] \tag{1}$$

The fraction of drug bound, fb_p, is given by *Eqn. 2*:

$$fb_p = 1 - fu_p. \tag{2}$$

While plasma binding refers to the binding of a drug to the plasma compartment of blood, studies can also be carried out on binding to serum or to whole blood. The major difference between plasma and serum is the removal of fibrinogen from plasma, and since most drugs do not bind to fibrinogen, no differences in binding to plasma or serum are expected [4]. Plasma is the fluid that remains after blood cells have been removed by centrifugation and is usually obtained using an anticoagulant so that all clotting factors (*e.g.*, fibrinogen) are retained. Serum is the fluid that remains after clotting factors have been removed by first allowing clotting to take place. Proteins not involved in clotting, *e.g.*, albumin, are still present in serum. In whole blood, drugs can bind to plasma proteins and to blood cells, and, hence, there can be significant differences between blood binding and plasma binding [2]. The whole blood free fraction of a compound, fu_b, is related to the plasma free fraction by *Eqn. 3*. Here c_b/c_p is the experimentally accessible blood-to-plasma ratio of the drug, *i.e.*, the ratio of total concentration of drug in whole blood to the total concentration of drug present in the plasma compartment of the blood.

$$fu_b = \frac{fu_p}{c_b/c_p} \tag{3}$$

When $c_b/c_p > 1$, the free fraction of compound in whole blood will be less than in plasma, *i.e.*, higher binding in whole blood. When $c_b/c_p < 1$, the free fraction of drug in whole blood will be greater than in plasma. When $c_b/c_p = 1$, the free fractions of drug are equal in both the whole blood and plasma compartments.

It is generally assumed that only the unbound drug is able to passively transport across membranes and become subject to distribution, metabolism, and excretion processes, and subsequently bind to the target receptor or receptors to elicit a pharmacological or toxicological effect. This is known as the free drug hypothesis [5] [6]. Free levels of a drug drive efficacy and influence the steady-state volume of distribution (V_{SS}) and clearance (CL). An array of physicochemical properties control the extent of binding to plasma proteins. Amongst these are lipophilicity, ionization state, H-bonding potential and molecular size [2] [7–13].

2. Experimental Aspects

2.1. *Scales of Measurement of Plasma Protein Binding* (PPB)

Highly detailed studies of PPB allow derivation of several binding constants from the experimental data. Two different formulations may be used for this purpose, one being stoichiometric and the other being site oriented [14–16]. However, most plasma-binding data is simply quoted on a percentage scale (*Eqn. 4*):

$$\%\mathrm{PPB} = 100(1 - fu_p) \tag{4}$$

Because the percentage scale is bounded by 0 and 100, a pseudo-binding constant ($K^{B/F}$) for PPB is frequently used, and this is defined by *Eqn. 5*:

$$K^{B/F} = \frac{(1 - fu_p)}{fu_p} = \frac{\%\mathrm{Bound}}{\%\mathrm{Free}} \tag{5}$$

When $K^{B/F} = 1$, the extent of plasma binding is 50%. When $K^{B/F} = 10$, the extent of plasma binding is 90.9%, *etc.* In quantitative structure–property relationship (QSPR) analyses of PPB data, the logarithm of $K^{B/F}$ is often used since the logarithm of a binding (or pseudo-binding) constant is directly proportional to a *Gibbs* free energy change, ΔG. In addition, the errors in the %PPB data are not normally distributed and are heteroscedastic, and these problems are avoided through the use of log $K^{B/F}$ [7] [10].

2.2. *Experimental Methods for PPB Measurement*

There are many experimental methods which have been used to determine the extent of plasma protein binding, and these include equilibrium dialysis [17–19], ultrafiltration [20][21], microdialysis [19][22], dynamic dialysis [20][23], ultracentrifugation [20][24][25], charcoal-binding kinetic methods [26], various chromatographic techniques such as immobilized-albumin support coupled with high-performance liquid chromatography [7][27], high-performance frontal analysis [22], a multitude of spectroscopic techniques such as fluorescence spectroscopy [28], UV spectroscopy [29], circular dichroism [29], and resonance spectroscopies, *e.g.*, NMR [29].

Equilibrium dialysis is still regarded as the gold standard method by most researchers. In this technique, two cells are separated by a semipermeable membrane which precludes high-molecular-weight compounds from crossing from one cell to the other, but does allow transfer of low-molecular-weight compounds to occur. In one cell, an appropriate volume of plasma (containing compound spiked at the chosen concentration) is placed. In the other cell, an equivalent volume of an appropriate aqueous buffer solution is placed. The dialysis cell is then equilibrated at the desired temperature for the chosen dialysis time. The plasma side of the dialysis cell contains compound which is bound to plasma proteins and compound which is unbound and able to cross the semipermeable membrane into the buffer side, and *vice versa*. When the dialysis time is sufficiently long, an equilibrium is reached, where the free concentration of compound is the same on both the plasma side and buffer side of the dialysis cell. An aliquot of the buffer side is analyzed, which gives a measure of the free concentration of compound. An aliquot of the plasma side is also analyzed, and this gives a measure of the total concentration of compound (bound and free). The extent of plasma protein binding as measured by fu_p is then given by *Eqn. 6*:

$$fu_p = \left(\frac{c_{buffer}}{c_{plasma}}\right) \tag{6}$$

where c_{buffer} is the concentration of compound in the buffer side of the dialysis cell and c_{plasma} is the total concentration of compound in the plasma side of the dialysis cell.

2.3. *In-House Automated Equilibrium Dialysis PPB Assay*

We have developed and validated an automated assay, based on equilibrium dialysis. Up to 60 compounds per experiment can have plasma protein binding measured in duplicate, using a multitude of plasma species and strains. Compounds are dialyzed for 18 h at 37° in a purpose-built plate which permits facile access for robot dispensing and sampling. Good throughput can be achieved through dialysis of mixtures of compounds, with a maximum of five compounds per dialysis cell, each at a concentration of 10 μM. *Fig. 1* shows the experimental apparatus. In the top left photograph in *Fig. 1*, tubes containing the different species and strains of plasma are to the left, the purpose built 24-well plates are in the center of the photograph, whilst HPLC vials into which dialysis samples are dispensed for subsequent MS/MS analysis are shown to the far right. The other photographs in *Fig. 1* show in more detail how the dialysis plates are constructed and attached to a rotator unit. The filling of the plates with plasma and compounds, and subsequent preparation of samples for HPLC/MS/MS analysis is fully automated using robotic liquid handling. The custom-designed plate has been validated against the commercially available and widely used *Dianorm* apparatus using a diverse set of compounds which cover a wide range of fu_p. The validation data is shown in *Fig. 2,a*, where it can be seen that the methods have very good correspondence over the whole dynamic range of PPB measurement. From a statistical viewpoint, a two-tailed paired *t*-test indicates that the two methods are not different at the 95% confidence level.

The principal binding protein in plasma is albumin (plasma concentration *ca.* 600 μM), and if it is assumed that the majority of binding occurs at a single site on this protein then a theoretical binding curve for the drug can be generated as shown in *Fig. 3*. The curve in *Fig. 3* has been simulated for a highly bound compound and shows that significant saturation of albumin leading to a decrease in %Bound will not occur until compound concentrations greater than 300 μM are reached. This forms the basis for the use of mixtures of compounds in our automated PPB assay, where the total compound concentration of 50 μM is much less than the concentration of albumin. Hence the saturation of albumin is not possible and competitive displacement of compounds is not expected. However, interactions between compounds which bind to lower concentration proteins (*e.g.*, α_1-acid glycoprotein), giving rise to a change in the degree of plasma binding are still theoretically possible, although such interactions at lower concentration proteins have been shown to be dampened by the often rather dominant binding to albumin [30]. Detailed validation studies indicate that in our experimental method, displacement interactions do

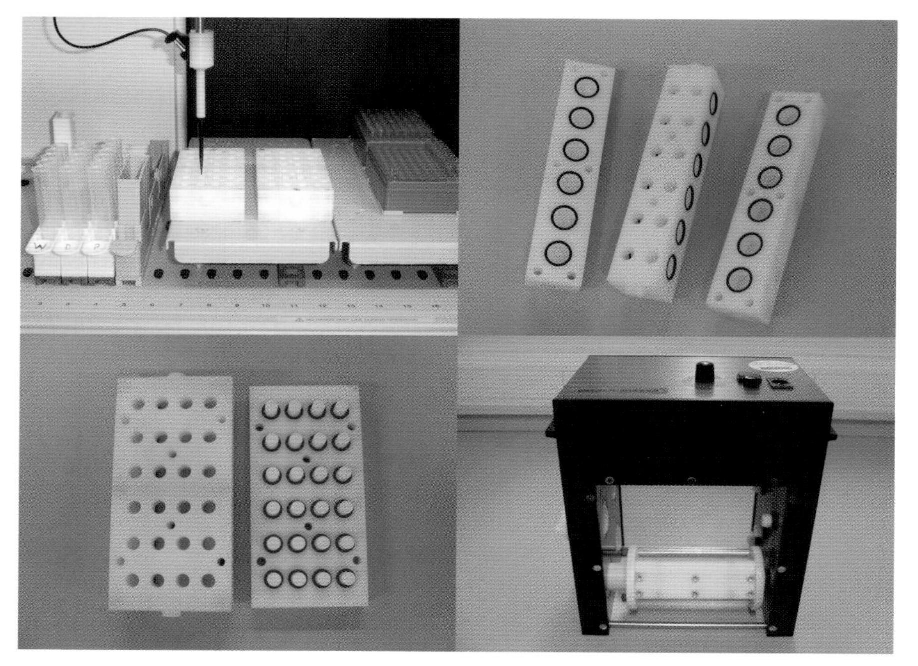

Fig. 1. *Automated PPB assay experimental setup*

not arise to any significant extent as shown in *Fig. 2,b*. The data from single-compound experiments and experiments using mixtures show excellent agreement over the entire dynamic range of PPB measurement. Furthermore, statistical analysis (a two-tailed paired t-test) indicates that data obtained when compounds are dialyzed and analyzed as mixtures compared to each compound being dialyzed and analyzed separately are not significantly different at the 95% confidence level. The validation of the use of mixtures in PPB studies has also been recently reported by other researchers [31].

3. Biological Aspects

3.1. *The Constituents of Human Plasma*

Human plasma is known to contain over 60 different proteins, the major component being albumin which comprises *ca.* 60% of the total plasma protein [20]. The next most-abundant and well-characterized protein is α_1-acid glycoprotein. Human serum albumin has a molecular

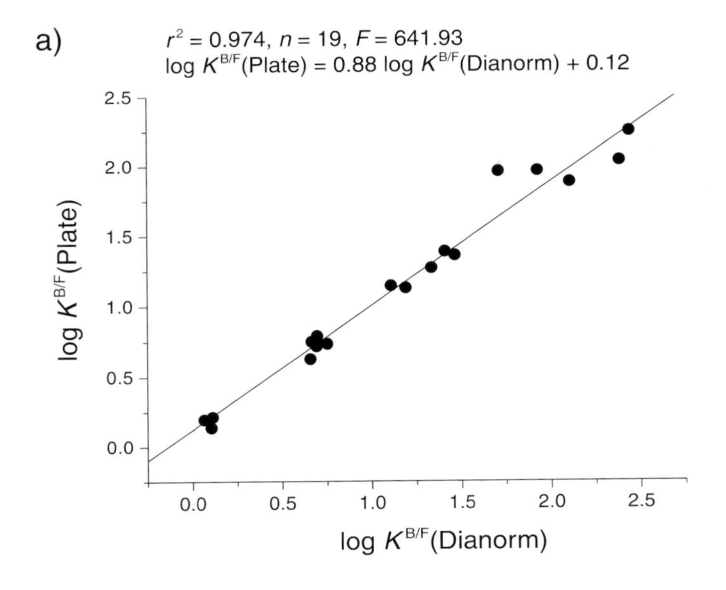

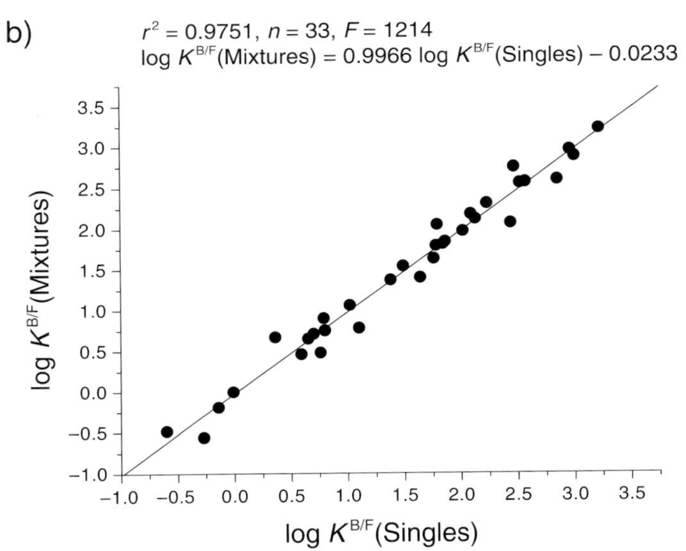

Fig. 2. a) *Log* $K^{B/F}$ *for the custom-built dialysis plate* vs. *data obtained using the commercially available* Dianorm^{TM} *apparatus.* b) *Log* $K^{B/F}$ *from mixtures* vs. *singles.*

weight of 66458 Da and contains 585 amino acid residues. At least 18 different mutations of human serum albumin have been identified and are primarily due to a single amino acid mutation, accounting for distinct protein–ligand binding [18]. The concentration of albumin in a normal

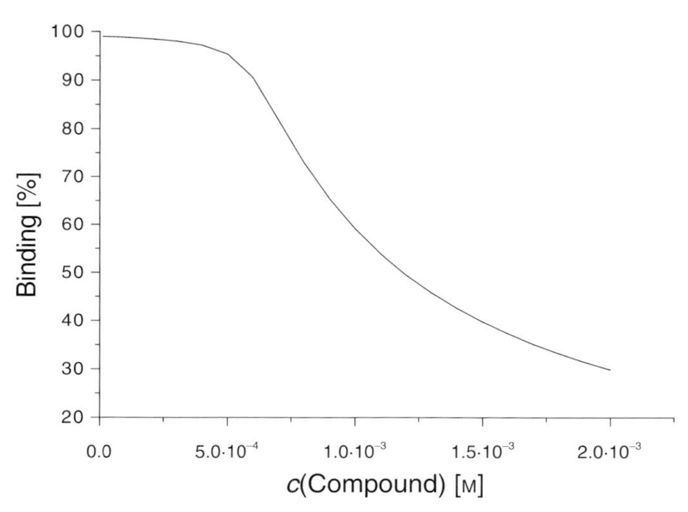

Fig. 3. *Theoretical binding curve of a highly bound compound to albumin* (600 μM)

healthy adult male is typically 43 g l^{-1}, with a range of 35–53 g l^{-1}. Females have a *ca.* 9% lower concentration, 38 g l^{-1} [1][20], and this is argued to account for the gender difference in binding of chlorodiazepoxide and warfarin [1].

In diseased patients, the albumin concentration can be significantly different. For patients with nephrotic syndrome, burns or cirrhosis, the albumin concentration can be less than 10 g l^{-1}, *i.e.*, 20–30% of the normal concentration [32]. In contrast, the normal plasma concentration (human) of α_1-acid glycoprotein is 0.4–1.0 g l^{-1} (10–30 μM) ($M_r = 44000$ Da), and in patients with inflammatory diseases, it can be elevated by up to four- to fivefold [32]. Taking into account the significant variation in protein content and concentration in human plasma (and other species), it is useful to determine PPB using large plasma pools containing a statistically reasonable number of donors [18].

Plasma contains many other globulins (the name of a family of proteins precipitated from plasma or serum by addition of $(NH_4)_2SO_4$). These can be separated into many subgroups, the main ones being α-, β-, and γ-globulins, which differ with respect to the associate lipid or carbohydrate. Immunoglobulins (antibodies) are in the *a* and *b* fractions, lipoproteins are in the *c* and *d* fractions. Other substances in the globulin fractions include macroglobulin, plasminogen, prothrombin, euglobulin, antihemomorphic globulin, fibrinogen, and cryoglobulin. The lipoproteins in plasma, of which α_1-acid glycoprotein is one, can be further classified into very high density (VHDL), high density (HDL), low density (LDL), and very low density lipoproteins (VLDL). The higher the density the lower the lipid content. Lipoproteins are macromolecular complexes displaying characteristic sizes, densities, and compositions. All lipoproteins contain protein components, called apoproteins, and polar lipids (phospholipids) in a surface film surrounding a neutral core (free and esterified cholesterol, triglycerides). The plasma lipoproteins vary in composition with respect to the lipid component, because their principal physiological function is to transport lipids in a water soluble form, but also vary with respect to the polypeptide chain composition. Lipoprotein plasma concentration may vary five- to tenfold. γ-Globulins generally only marginally account for the plasma binding of drugs [33]. Often, it is only when a drug is present at very high concentrations that binding to components other than albumin or α_1-acid glycoprotein occurs [3]. Considerable intersubject variability in the PPB of some compounds (four-to-fivefold variation in fu_p is not uncommon) and in the concentration of proteins exists even within healthy human volunteers [1]. Genetically determined variations in amino acid sequences of human serum albumin can also contribute to variability in binding and cause higher variability in patients with highly bound drugs. The binding affinities of warfarin, salicylate, and diazepam to five known variants of human serum albumin have been studied [34]. The association constants for all three drugs to albumin decreased by a factor of four- to tenfold for some of the mutations relative to each other.

3.2. *Species Differences in Plasma Protein Binding*

The extent of binding of drugs to plasma proteins may differ significantly between animal species and between different strains of a given species [35][36]. This is in contrast to tissue binding which has been shown to be constant across different species [3][35][37]. When comparing PPB data from different laboratories, values may vary for several reasons including buffer composition, drug concentration, pH, the experimental method used to separate free from bound drug (equilibrium dialysis *vs.* ultrafiltration), and temperature. The observed species and strain differences in plasma protein binding may reflect differences in plasma albumin concentration and affinity and/or the number of binding sites on albumin, or other proteins, *e.g.*, α_1-acid glycoprotein, to which a drug may bind [35]. It is not uncommon for fu_p in human compared to different animals to vary by a factor of up to five [1][35]. Occasionally, even more-dramatic variability has been observed. Valproic acid, for example, shows PPB that ranges from $fu_p = 0.05$ in human to $fu_p = 0.82$ in mouse [38], a difference in free fraction of greater than 16-fold. The %PPB of zamifenacin is 99.98%

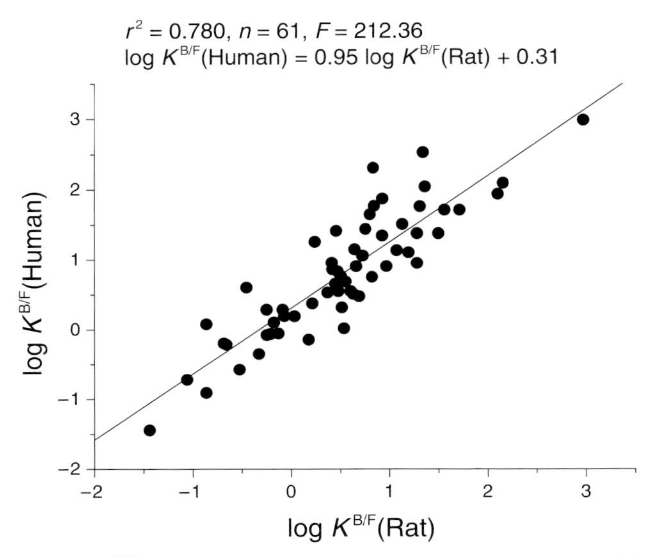

Fig. 4. *Log* $K^{B/F}$ *in human* vs. *rat plasma for a set of marketed oral drugs*

in humans whilst only 99.80% in rat, *i.e.*, a tenfold variability in fu_p [39][40]. *Fig. 4* shows a plot of log $K^{B/F}$ in human *vs.* rat for a range of marketed oral drugs, covering all charge types (acids, bases, and neutral compounds). It can be seen that, although the data from the two species show a general correlation, some of the drugs do have markedly different plasma binding in the two species.

4. The Importance of Plasma Protein Binding in Drug Metabolism and Pharmacokinetic Profiling

4.1. *Effect of PPB on Drug Clearance*

The extent of plasma protein binding is extremely important in its influence on many pharmacokinetic parameters. For drugs with low hepatic clearance, the *in vivo* clearance can be approximated by *Eqn. 7* [41]:

$$CL \cong fu_p \times CL_{int} \tag{7}$$

where CL_{int} is the intrinsic clearance of the drug. Under these conditions, clearance is directly proportional to fu_p. An example of this behavior is given by the clearance of warfarin in male *Sprague–Dawley* rats which is

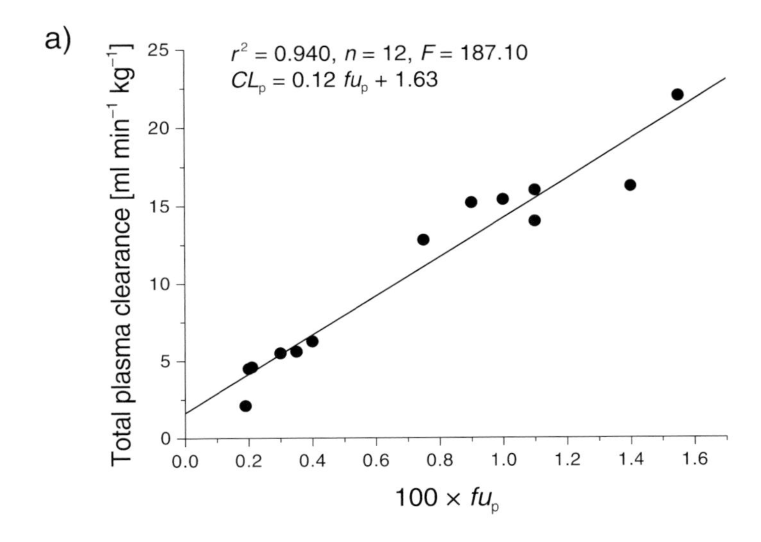

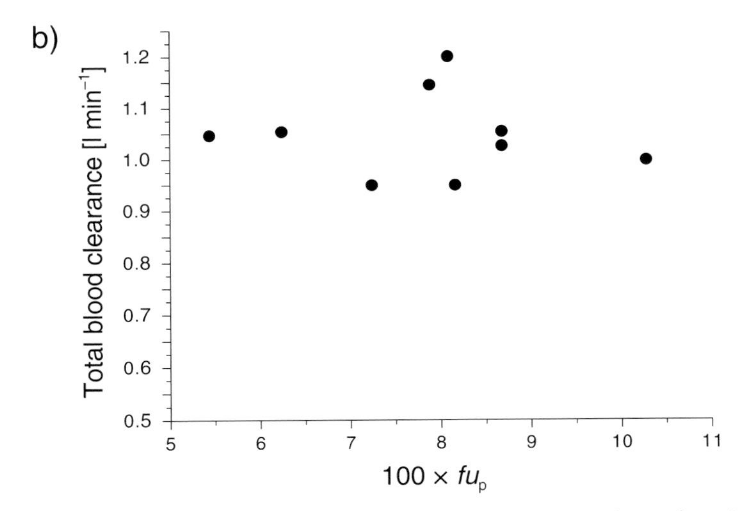

Fig. 5. a) *Total plasma clearance* vs. *the free fraction of warfarin in plasma from individual rats.* b) *Propanolol total clearance* vs. *plasma free fraction in human volunteers.*

proportional to the free fraction of the drug within each individual rat [42][43], as shown in *Fig. 5,a*.

For drugs with high hepatic clearance, the clearance is largely controlled by the liver blood flow, Q_h, and is independent of the extent of plasma binding [41] as shown by *Eqn. 8*:

$$CL \cong Q_h \tag{8}$$

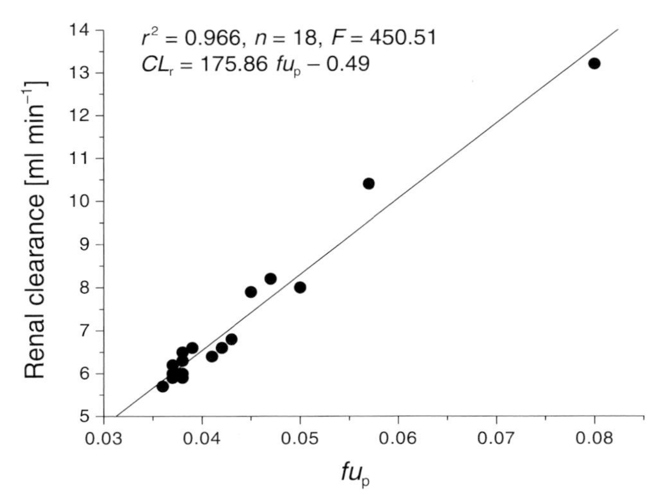

Fig. 6. *Renal clearance* vs. *plasma free fraction of ceftriaxone in human volunteers*

This type of behavior is illustrated by propranolol, where the clearance in humans does not depend on the human free fraction of drug in plasma [44], as shown in *Fig. 5,b.*

The degree of plasma protein binding can not only influence metabolic clearance, but also renal clearance [45]. *Fig. 6* demonstrates how the human renal clearance of ceftriaxone varies with the plasma free fraction, with higher plasma free fraction in individual patients leading to higher renal clearance of the drug.

4.2. *Affect of PPB on Drug Volume of Distribution*

The steady-state volume of distribution, V_{SS}, of a drug is approximately related to its free fraction in plasma by *Eqn. 9* [1][36][46–48]:

$$V_{SS} = V_p + \left[\left(\frac{fu_p}{fu_t}\right) \times V_t\right] \tag{9}$$

where V_p is the volume of the plasma compartment (0.045 l kg^{-1} in humans), fu_t is the free fraction of drug in tissues, and V_t is the volume of the extravascular tissue compartments, *i.e.*, usually viewed as the physical volume outside of the plasma into which a drug distributes. It is evident that V_{SS} is dependent on both plasma and tissue binding. From *Eqn. 9*, the value of V_{SS} for a drug, which is bound more extensively in tissues than in plasma ($fu_t < fu_p$) may be in excess of the actual physical volume into which the drug distributes. For a drug that is able to penetrate cell walls, the distribution space may correspond to total tissue water, whereas, for a drug excluded from cells, it will be equal to the extracellular volume. For very highly plasma-bound compounds with low tissue affinity, V_{SS} is limited by the volume of distribution of plasma proteins. *Ca.* 59% of the total body albumin and 40% of α_1-acid glycoprotein are located in the interstitial fluid of the extravascular space [35][49], and hence plasma proteins have a volume of distribution in excess of V_p (V_{SS} albumin *ca.* 0.1 l kg^{-1} [4]). When a drug is not bound to either plasma or tissue proteins ($fu_p = fu_t = 1$), V_{SS} will be approximately equal to the volume of water, *i.e.*, 40 l (0.57 l kg^{-1}). For drugs with values of $V_{SS} > 0.3$ l kg^{-1}, V_p is small compared with V_{SS}, and, consequently, V_{SS} becomes directly proportional fu_p (*Eqn. 9*).

While tissue binding has been shown to remain approximately constant on changing from one species to another [3][35][37], plasma protein binding of compounds can vary dramatically between species leading to considerable interspecies differences in V_{SS} (*Fig. 7,a*). The antimuscarinic agent zamifenacin is a lipophilic base which has a very high degree of PPB. The %PPB of the drug is 99.8% in rats and 99.98% in humans, and a consequence of this is that V_{SS} is tenfold higher in rats than in humans [39][40].

If we consider V_{SS} of a drug in both human and rat, then using *Eqn. 9* along with the assumption that $V_{SS} > 0.3$ l kg^{-1}, we can write *Eqns. 10* and *11*:

$$V_{SS}(\text{human}) \cong \left[\left(\frac{fu_p(\text{human})}{fu_t(\text{human})}\right) \times V_t(\text{human})\right] \tag{10}$$

$$V_{SS}(\text{rat}) \cong \left[\left(\frac{fu_p(\text{rat})}{fu_t(\text{rat})}\right) \times V_t(\text{rat})\right] \tag{11}$$

If we now assume equivalent tissue binding between the species (fu_t(rat) $= fu_t$(human) $= fu_t$), the combination of *Eqns. 10* and *11* allows elimination of the fu_t term leading to *Eqn. 12*:

$$V_{SS}(\text{human}) \cong \left(\frac{fu_p(\text{human})}{fu_p(\text{rat})}\right) \times V_{SS}(\text{rat}) \tag{12}$$

Eqn. 12, which amounts to the commonly used assumption that unbound volume of distribution is independent of species, gives a simple method for the prediction of human volume of distribution using readily available preclinical data. *Fig. 7,b* shows a plot of the observed logarithm of the human volume of distribution for a set of marketed oral drugs *vs.* the predicted value using *Eqn. 12*. The correlation in *Fig. 7,b* is much tighter

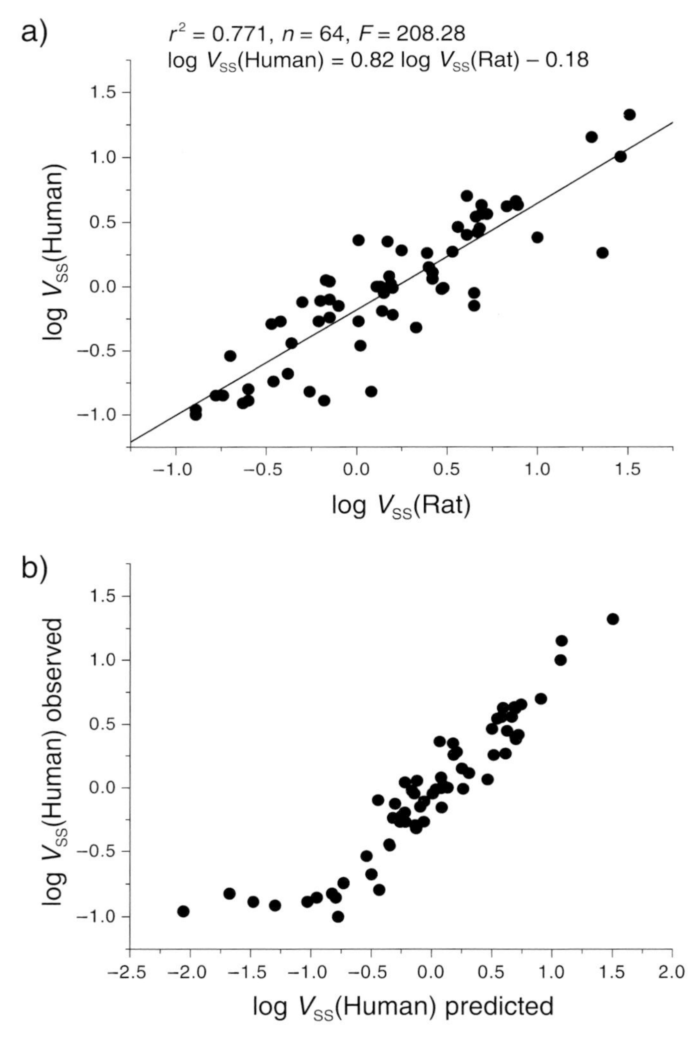

Fig. 7. a) *Log* V_{SS}*(human)* vs. *log* V_{SS}*(rat) for a set of marketed oral drugs.* b) *Observed* vs. *predicted log* V_{SS}*(human)* (*Eqn. 11*) *for marketed oral drugs.*

than that in *Fig. 7,a* showing that the interspecies differences in V_{SS} are largely due to interspecies differences in PPB and that this effect is well modeled by *Eqn. 12*. An exception to this are the compounds to the left of *Fig. 7,b* where V_{SS} in human and rat approaches the volume of distribution of albumin, at which point interspecies differences in PPB will not lead to interspecies differences in V_{SS}. These compounds do not obey the assumptions leading to *Eqn. 12*, and their behavior is better understood with use of the more-complex method given by *Rowland* and *Tozer* [4].

5. Drug–Drug and Disease–Drug Interactions

5.1. *Drug–Drug Interactions*

The obvious question arises as to whether highly plasma-bound drugs can displace each other from plasma proteins. The potential does exist, but as most drugs have therapeutic concentrations far below the plasma concentration of albumin and often below that of α_1-acid glycoprotein, proteins are rarely at the point of saturation, and so competition for binding sites (and hence displacement) will not occur [41] [50].

Although the literature is replete with examples of drug–drug interactions suggested to be due to PPB interactions, many of these have now been attributed to other mechanisms [50]. The drug–drug interactions are often due to cytochrome P450 inhibition, or by co-administered drugs acting as competitive substrates, although a few may still be attributable to pure displacement. The *Table* lists some drugs for which other mechanisms are in fact responsible for drug–drug interactions, other than the original suggestion of PPB displacement.

The fact that high drug concentrations are required to induce significant displacement interactions is illustrated by studies on the effect of salicylic acid on the PPB of ibuprofen [25], which show that very high concentrations of salicylic acid (1500 μM) have only a moderate effect on the free concentration of ibuprofen, as exemplified in *Fig. 8*. The possibility of significant albumin displacement interactions *in vivo* is therefore only a reality when one of the interacting drugs has a very low potency (or short half-life) and consequently is dosed in such a way that very high plasma concentrations are achieved. Since plasma concentrations of α_1-acid glycoprotein are *ca.* 40-fold lower than that of albumin, displacement interactions at this protein could be possible at more-typical therapeutic

Table. *Drug–Drug Interactions Originally Suggested to Be Attributable to Plasma Protein Displacement, and the Actual Mechanism Responsible* [51]

Drug	Displacing drug	Mechanism responsible
Methotrexate	Salicylate	Inhibition of renal clearance
Phenytoin	Valproate	Inhibition of metabolism
Tolbutamide	Phenylbutazone	Inhibition of metabolism
	Salicylates	Pharmacodynamic
	Sulfonamides	Inhibition of metabolism
Warfarin	Phenylbutazone	Inhibition of metabolism
	Clofibrate	Pharmacodynamic
	Choral Hydrate	Possible pure displacement
	Sulphamethoxazole	Inhibition of metabolism
	Sulphinpyrazole	Inhibition of metabolism

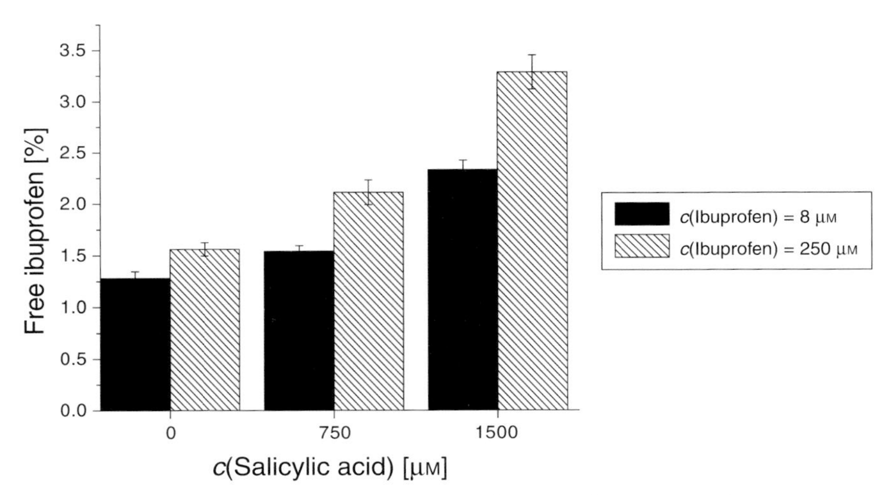

Fig. 8. *Extent of binding of ibuprofen to human plasma in the presence of varying concentrations of salicylic acid.* Error bars are 95% confidence intervals.

drug concentrations. For example, bupivacaine and mepivacaine both bind with a high affinity to α_1-acid glycoprotein, and with lower affinity to albumin [30]. Bupivacaine (9 µM) caused a 110% increase in the free fraction of mepivacaine (3 µM) in a 15 µM solution of α_1-acid glycoprotein. In serum, however, the lower affinity binding to the much higher high capacity albumin dampens out such a displacement such that the increase in free fraction of mepivacaine was lower at 65%.

The importance of changes in plasma protein binding still appears to be a concern of many clinicians, regulators, and industrial drug developers in relation to drug–drug interactions which could lead to increased free concentrations of drugs in the plasma or blood, even though the literature demonstrates this is of little clinical importance. The concern is usually based on the intuitive belief that when a drug is displaced from its plasma binding protein, the increased free drug concentrations will cause an increase in drug effect and potential toxic effects [41]. However, even if a displacement does occur, its effect should not be considered in isolation, since the resulting increased free fraction of displaced drug can influence the clearance and volume of distribution. The pharmacokinetic half-life is defined by *Eqn. 13*:

$$t_{1/2} = \ln 2 \times \frac{V_{SS}}{CL} \tag{13}$$

For high-extraction-ratio drugs with $V_{SS} > 0.3\ \mathrm{l\,kg^{-1}}$, combining *Eqns. 8* and *9* leads to *Eqn. 14*, which shows that $t_{1/2}$ will be proportional to fu_p. Hence, the increased free fraction of displaced drug will in turn lead to an increase in half-life.

$$t_{1/2} \cong \frac{\left[\ln 2 \times \left(\frac{fu_p}{fu_t}\right) \times V_t\right]}{Q_{organ}} \tag{14}$$

For low-extraction-ratio drugs, for which $V_{SS} > 0.3\ \mathrm{l\ kg^{-1}}$, combining *Eqns. 7* and *9* leads to *Eqn. 15* [41], hence $t_{1/2}$ is expected to be independent of fu_p:

$$t_{1/2} \cong \frac{\left[\ln 2 \times \left(\frac{V_t}{fu_t}\right)\right]}{CL_{int}} \tag{15}$$

It is evident therefore that, depending upon the intrinsic clearance of a drug, its volume of distribution, clearance, and half-life may or may not be dependent on changes in PPB.

Exposure is a measure of the drug levels a patient experiences after a dose or series of doses. It is a measure of concentration integrated over time, often referred to as the area under the curve (AUC). It is generally accepted that the pharmacological and toxicological effects of a drug are related to the exposure of the unbound drug concentrations, AUC_U. Unbound AUC from an oral dose is given by *Eqn. 16*, where F is the oral bioavailability:

$$AUC_U = \frac{fu_p \times F \times \mathrm{Dose}}{CL} \tag{16}$$

It has been shown [41] that the unbound oral AUC as defined above can be rewritten as *Eqn. 17*:

$$AUC_U = \frac{F_{abs} \times \mathrm{Dose}}{CL_{int}} \tag{17}$$

where F_{abs} is the fraction of the oral dose absorbed. Clearly, this expression for unbound exposure is independent of extent of plasma binding, hence any changes in plasma binding caused by drug interactions are not expected to lead to any significant changes in pharmacological or toxicological effects.

5.2. *Disease–Drug Interactions*

Many diseases cause significant changes in the concentrations of plasma proteins. An extreme example is that of analbumenic rats which have a plasma albumin concentration of only 0.6 μM. In normal rat plasma, warfarin was shown to be 98.8% bound, whereas in analbumenic rat plasma it is 64% bound [51]. The 1000-fold decrease in albumin concentration leads to only a 30-fold increase in fu_p indicating that the reduction in binding due to decreased albumin concentration is to some extent offset by simultaneous binding to other plasma proteins.

A study of the pharmacokinetics of naproxen in patients with rheumatoid arthritis compared to healthy volunteers [52] highlighted statistically significant differences in CL/F and V_{SS}/F, but importantly there were no significant differences in half-life. The patients with the disease had a lower plasma albumin concentration compared to healthy volunteers, resulting in the former having an average %PPB of 99.70% and the latter an average of 99.92%. Naproxen has low hepatic clearance, hence a statistically significant increase in clearance in diseased patients was found as expected from *Eqn. 7*. The increase in fu_p in diseased patients also led to a statistically significant increase in V_{SS}, as expected from *Eqn. 8*. However, no change in $t_{1/2}$ was found as predicted by *Eqn. 15*.

A later study [53] again examined naproxen pharmacokinetics in patients with rheumatoid arthritis, but now determining the pharmacokinetic parameters during active disease and then, in the same patients, during a period of disease improvement. During active disease, the patients had lower plasma concentrations of human serum albumin compared to when they had undergone improvement, and hence a decreased extent of PPB. Again, this manifested itself in significantly increased values of CL/F and V_{SS}/F during active disease but no change in $t_{1/2}$.

In summary, it is important to understand that any changes in PPB arising from drug or disease interaction cannot be regarded as being in isolation from the pharmacokinetic parameters of AUC, AUC_U, clearance, volume of distribution, and half-life. Depending on the properties of the drug, fu_p can have a range of effects on these parameters, and little effect is expected on the total exposure to free drug [41].

6. Quantitative Structure–Property Relationships in Plasma Protein Binding

It is of fundamental importance to understand, and hence be able to control, the factors that affect the degree of plasma binding of compounds. Armed with such an understanding, it should be possible for chemists to design and synthesize structures that exhibit the desired extent of plasma binding within a particular chemical series. A quantitative structure–human serum albumin (HSA) binding relationship has been reported for a wide range of drug-like compounds [7]. The extent of HSA binding was determined using a fast-gradient HPLC method. It was also found that for the range of drugs used in the study, HSA binding correlated very strongly with measured binding to whole plasma. The degree of HSA binding correlated positively with lipophilicity, as indexed by calculated log $D_{7.4}$. However, the degree of binding was higher for acidic compounds than for non-acids (basic, neutral, and zwitterionic compounds) for a given log $D_{7.4}$. The observation of enhanced binding of acids has also been reported elsewhere [8][9]. A number of other studies on selected drugs have reported linear correlations between log $K^{B/F}$ (human, rat, rabbit, and mouse) and log P or log $D_{7.4}$ [2–11].

We have generated in-house human plasma protein binding data for a large and diverse set of compounds covering all charge types. *Fig. 9,a* shows the relationship between log $K^{B/F}$ and lipophilicity as indexed by experimentally determined log $D_{7.4}$ for non-acids and by log P for acids (defined as compounds with an acidic pK_a < 7.4). It can be seen that the higher the lipophilicity of a compound, the higher the extent of binding. The root-mean-square error (RMSE) in describing the PPB data using lipophilicity alone is 0.53 log units. On the %-binding scale, this equates to a mean error of 3.4-fold in the prediction of fu_p. The good correlation with log P for acids and with log $D_{7.4}$ for compounds of other charge types suggests that both the neutral and ionized forms of acids bind to plasma proteins (mostly to albumin), while the neutral form of compounds of other charge types is the predominantly bound species.

To further understand the physicochemical properties that control the extent of PPB, a PLS (partial least squares) model was generated. During model construction, the human PPB data set was divided into a training set (75% of the entire data set) with which the model was fitted, and a test set (25% of the entire data set), with which the robustness and predictivity of the model was assessed. *Fig. 9,b* shows a plot of observed log $K^{B/F}$ *vs.* predicted log $K^{B/F}$ using the PLS model. The PPB data are modeled well with the test set (RMSE = 0.42 log units, *i.e.*, a 2.6-fold mean error in the prediction of fu_p). The PLS model offers considerable improvement over

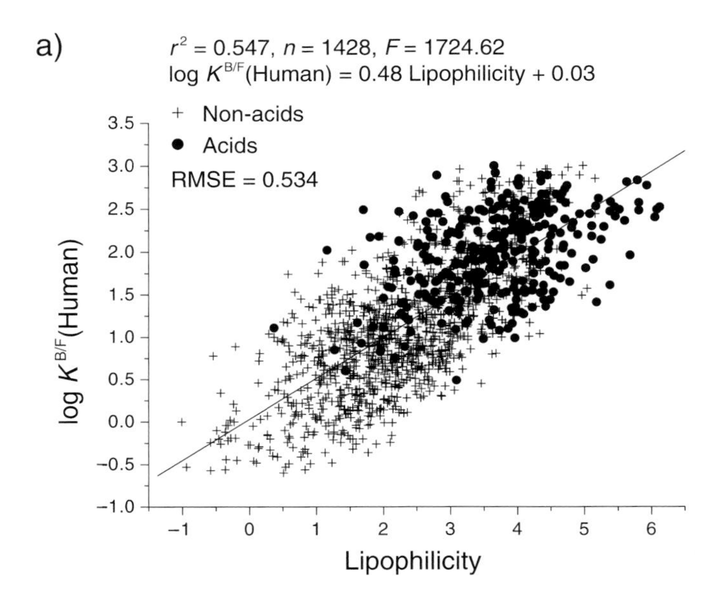

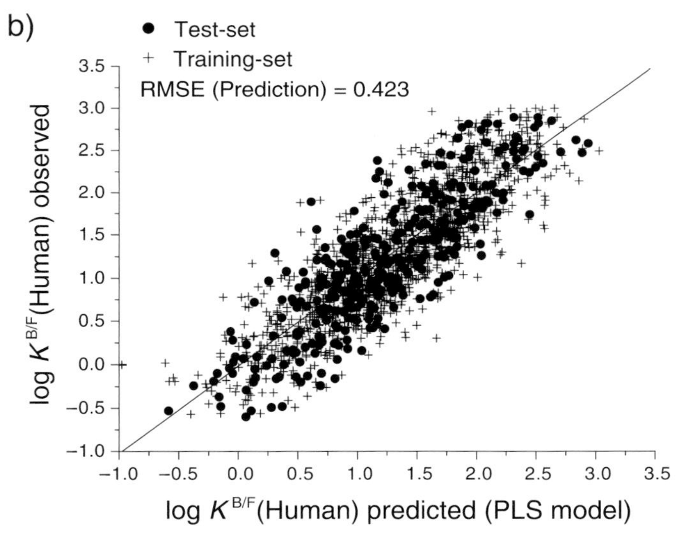

Fig. 9. a) *Log* $K^{B/F}$*(human) vs. lipophilicity for a set of* AstraZeneca *compounds.* b) *Observed* vs. *predicted log* $K^{B/F}$*(human) from PLS model using the same set of compounds.*

lipophilicity alone as indicated by the lower RMSE in prediction (0.42 *vs.* 0.53 log units). This clearly highlights the dependence of plasma binding on a variety of structural and physical properties in addition to lipophi-

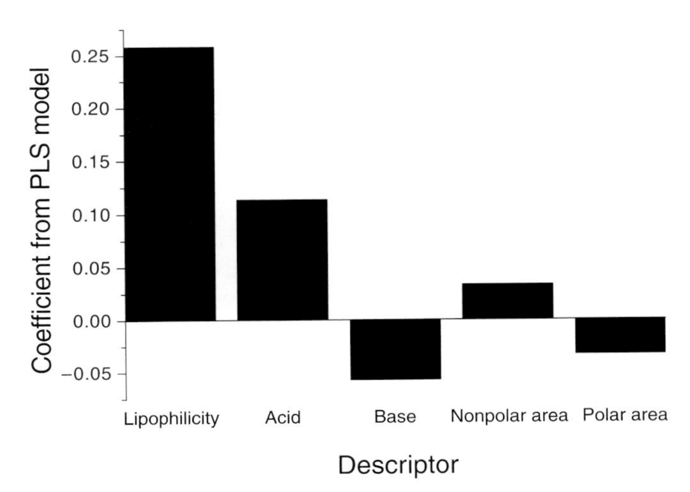

Fig. 10. *PLS Coefficients from the human PPB QSPR model*

licity. Other than lipophilicity, important descriptors in the PLS model are the pK_a value of ionizing groups within the molecule, various size-related descriptors and descriptors relating to the H-bonding properties of the molecules. This is illustrated by *Fig. 10*, which shows the PLS coefficients of some of the most significant descriptors included in the model. Moving from left to right in *Fig. 10*, the descriptors become less important in describing the PPB data, while the positive or negative nature of each coefficient indicates the directional influence of each descriptor on PPB. Lipophilicity, for example, has a positive coefficient, and hence increasing lipophilicity will increase the degree of PPB, while polar surface area has a negative coefficient, and increasing this property will decrease the extent of PPB.

7. Conclusions

The ability to measure the extent of plasma binding of candidate drugs, along with an understanding of the influence that plasma binding has on many aspects of a drug's behavior, is of fundamental importance to the drug discovery process. Further, an understanding of the structural and physical properties of compounds that control the extent of plasma binding allows manipulation of the plasma binding properties of candidate drugs, and this is an integral part of the modern optimization process. Protein binding for drugs can vary considerably from one species to another, and even between healthy human volunteers. The extent of plasma binding has an important effect on *in vivo* efficacy and toxicity and also influences pharmacokinetic parameters such as metabolic and renal clearance and volume of distribution. PPB has less of an effect on drug–drug interactions than has been suggested by some of the older literature in this area. Instead, co-administration of drugs more often causes interaction through cytochrome P450 inhibition. The extent of binding is largely controlled by the lipophilicity of a drug. Higher log $D_{7.4}$ for non-acids and higher log P for acids generally leads to an increased extent of binding. Among other properties, which can be manipulated by chemists to optimize PPB, are extent of ionization, the size of the molecule, and the H-bonding properties of functional groups within a molecule.

REFERENCES

[1] G. R. Wilkinson, *Drug Metab. Rev.* **1983**, *14*, 427.
[2] M. Laznicek, A. Laznickova, *J. Pharm. Biomed. Anal.* **1995**, *13*, 823.
[3] J. P. Tillement, G. Houin, R. Zini, S. Urien, E. Albengres, J. Barre, M. Lecomte, P. D'Athis, B. Sebille, in 'Advances in Drug Research, Vol. 13', Ed. B. Testa, Academic Press, London, 1984, p. 59–93.
[4] M. Rowland, T. N. Tozer, 'Clinical Pharmacokinetics – Concepts and Applications', 3rd edn., Lippincott Williams & Wilkins, London, 1995.
[5] B. B. Brodie, H. Kurtz, L. J. Schanker, *J. Pharmacol. Exp. Ther.* **1960**, *130*, 20.
[6] F. Herve, S. Urien, E. Albengres, J. Duche, J. P. Tillement, *Clin. Pharmacokinet.* **1994**, *26*, 44.
[7] K. Valko, S. Nunhuck, C. Bevan, M. H. Abraham, D. Reynolds, *J. Pharm. Sci.* **2003**, *92*, 2236.
[8] H. van de Waterbeemd, D. A. Smith, B. C. Jones, *J. Comput.-Aided Mol. Des.* **2001**, *15*, 273.
[9] A. M. Davis, R. Riley, D. R. Flower, in 'Drug Design: Cutting Edge Approaches', Ed. D. R. Flower, Royal Society of Chemistry, Cambridge, UK, 2002, p. 106–123.
[10] S. Toon, M. Rowland, *J. Pharmacol. Exp. Ther.* **1983**, *225*, 752.
[11] M. Laznicek, J. Kvetina, J. Mazak, V. Krch, *J. Pharm. Pharmacol.* **1987**, *39*, 79.
[12] M. Lobell, V. Sivarajah, *Mol. Divers.* **2003**, *7*, 69.
[13] J. Koch-Weser, E.M. Sellers, *Med. Intelligence* **1976**, *294*, 311.
[14] I. M. Klotz, *Acc. Chem. Res.* **1974**, *7*, 162.
[15] I. M. Klotz, D. L. Hunston, *J. Biol. Chem.* **1975**, *250*, 3001.
[16] I. M. Klotz, D. L. Hunston, *Proc. Natl. Acad. Sci. U.S.A.* **1977**, *74*, 4959.
[17] H. G. Weder, J. Schildknecnt, P. Kesselring, *Am. Lab.* **1971**, 15.
[18] I. Kariv, H. Cao, K. R. Oldenburg, *J. Pharm. Sci.* **2001**, *90*, 580.
[19] I. M. Klotz, F. M. Walker, R. B. Pivan, *J. Am. Chem. Soc.* **1946**, *68*, 1486.
[20] J. H. Lin, D. M. Cocchetto, D. E. Duggan, *Clin. Pharmacokinet.* **1987**, *12*, 402.
[21] P. B. Rheberg, *Acta Phys. Scand.* **1943**, *5*, 305.
[22] Z. Liu, F. Li, Y. Huang, *Biomed. Chromatogr.* **1999**, *13*, 262.
[23] S. P. Colowick, F. C. Womack, *J. Biol. Chem.* **1968**, *244*, 774.
[24] S. Hall, M. Rowland, *J. Pharmacol. Exp. Ther.* **1983**, *227*, 174.
[25] L. Aarons, D. M. Grennan, M. Siddiqui, *Eur. J. Clin. Pharmacol.* **1983**, *25*, 815.
[26] J. Yuan, D. C. Yang, J. Birkmeier, J. Stolzenbac, *J. Pharmacokinet. Biopharm.* **1995**, *23*, 41.
[27] D. S. Hage, J. Austin, *J. Chromatogr., B* **2000**, *739*, 39.

[28] C. F. Chignell, 'Actualités de Chimie Thérapeutique, Vol. 8', Société de Chimie Thérapeutique, Paris, 1981, p. 291–302.
[29] C. Bertucci, P. Salvadori, E. Domenici, in 'The Impact of Stereochemistry on Drug Development and Use', Eds. H. Y. Aboul-Enein, I. W. Wainer, Wiley, New York, 1997, p. 521–543.
[30] C. T. Hartrick, W. E. Dirkes, D. E. Coyle, P. Prithvi Raj, D. D. Denson, *Clin. Pharmacol. Ther.* **1984**, *36*, 546.
[31] E. N. Fung, Y. Chen, Y. Y. Lau, *J. Chromatogr., B* **2003**, *795*, 187.
[32] T. F. Blaschke, *Clin. Pharmacokinet.* **1977**, *2*, 32.
[33] K. M. Piafsky, *Clin. Pharmacokinet.* **1980**, *5*, 246.
[34] U. Kragh-Hansen, S. Brennan, M. Galliano, O. Sugita, *Mol. Pharmacol.* **1990**, *37*, 232.
[35] B. Fichtl, A. v. Nieciecki, K. Walter, in 'Advances in Drug Research, Vol. 20', Ed. B. Testa, Academic Press, London, 1991, p. 117–166.
[36] F. Lombardo, R. S. Obach, M. Y. Shalaeva, F. Gao, *J. Med. Chem.* **2002**, *45*, 2867.
[37] S. Zhou, J. W. Paxton, P. Kestell, M. D. Tingle, *J. Pharm. Pharmacol.* **2001**, *53*, 463.
[38] W. Loscher, *J. Pharmacol. Exp. Ther.* **1978**, *204*, 255.
[39] K. C. Beaumont, P. V. Macrae, C. M. Miller, D. A. Smith, P. Wright, *Br. J. Clin. Pharmacol.* **1994**, *38*, 179.
[40] D. A. Smith, B. C. Jones, D. K. Walker, *Med. Res. Rev.* **1996**, *16*, 243.
[41] L. Z. Benet, B. Hoener, *Clin. Pharmacol. Ther.* **2002**, *71*, 115.
[42] G. Levy, A. Yacobi, *J. Pharm. Sci.* **1974**, *63*, 805.
[43] A. Yacobi, G. Levy, *J. Pharm. Sci.* **1975**, *64*, 1995.
[44] G. H. Evans, A. S. Nies, D. G. Shand, *J. Pharmacol. Exp. Ther.* **1973**, *186*, 114.
[45] K. Stoeckel, P. J. McNamara, R. Brandt, H. Plozza-Nottebrock, W. H. Ziegler, *Clin. Pharmacol. Ther.* **1981**, *29*, 650.
[46] S. Oie, T. N. Tozer, *J. Pharm. Sci.* **1979**, *68*, 1203.
[47] F. Lombardo, R. S. Obach, M. Y. Shalaeva, F. Gao, *J. Med. Chem.* **2004**, *47*, 1242.
[48] R. S. Obach, J. G. Baxter, T. E. Liston, B. M. Silber, B. C. Jones, F. Macintyre, D. J. Rance, P. Wastall, *J. Pharmacol. Exp. Ther.* **1997**, *283*, 46.
[49] H. Kurz, B. Fichtl, *Drug Metab. Rev.* **1983**, *14*, 467.
[50] P. E. Rolan, *Br. J. Clin. Pharmacol.* **1994**, *37*, 125.
[51] J. Hirate, C. Zhu, I. Horikoshi, S. Nagase, *Int. J. Pharm.* **1990**, *65*, 149.
[52] F. A. van den Ouweland, M. J. A. M. Franssen, L. B. A. van de Putte, Y. Tan, C. A. M. van Ginneken, F. W. J. Gribnau, *Br. J. Clin. Pharmacol.* **1987**, *23*, 189.
[53] F. A. van den Ouweland, F. W. J. Gribnau, C. A. M. van Ginneken, Y. Tan, L. B. A. van de Putte, *Clin. Pharmacol. Ther.* **1988**, *43*, 79.

In vivo Pharmacokinetic Profiling of Drugs

by **Heidi Wunderli-Allenspach**

Institute of Pharmaceutical Sciences, Dept. of Chemistry and Applied Biosciences,
ETH Hönggerberg, CH-8093 Zurich
(e-mail: heidi.wunderli@pharma.ethz.ch)

Abbreviations
AUC: Area under the curve; BW: body weight; *c*: plasma concentration of a drug; *c*(*t*) plasma concentration of a drug at time *t*; $c_{ss,av}$: average plasma concentration of a drug at steady state; *CL*: total clearance of drug from plasma; CL_H: hepatic clearance of drug from plasma; CL_{int}: intrinsic clearance (capacity); CL_{NR}: nonrenal clearance of drug from pasma; CL_R: renal clearance of drug from plasma; *D*: dose; *E*: extraction ratio; E_H: hepatic extraction ratio; E_p: extraction ratio related to plasma; $E_{p,H}$: hepatic extraction ratio related to plasma; *F*: bioavailability of drug; f_a: fraction of drug absorbed; f_e: fraction of drug absorbed that is excreted unchanged in urine; f_{fp}: fraction systemically available after first liver passage; f_u: fraction of drug unbound in plasma; f_{uT}: fraction of drug unbound in tissue; GF: glomerular filtration; *GFR*: glomerular filtration rate; *H*: hematocrit; Q_H: hepatic blood flow; $Q_{p,H}$: hepatic plasma flow; R_{ac}: accumulation ratio; τ: dosing interval; $t_{1/2}$: half-life (first-order kinetics); $t_{1/2(\lambda_Z)}$: terminal half-life; *V*: volume of distribution (apparent volume of distribution); V_p: plasma volume; V_T: physiological volume of water outside plasma into which a drug distributes (*i.e.*, extracellular to total body water minus plasma volume).

1. Introduction

The pharmacokinetic behavior of a drug results from the superposition of its absorption, distribution, biotransformation (metabolism), and excretion in the body. These processes are usually summarized under the acronym of ADME. The pharmacokinetic behavior is mainly determined by the physicochemical characteristics of the compound such as molecular weight, pK_a value, and lipophilicity. Although population studies reveal much inter- and intraindividual pharmacokinetic variability, specific pharmacokinetic core parameters can be defined for each drug in healthy young individuals (*Table 1*). All of them can be determined by either *in vitro* experiments (f_u) or by clinical studies comprising plasma and, in the case of f_e, plasma and urine data.

Table 1. *The Pharmacokinetic Core Parameters*

Core parameter	Symbol	Dimension	Method for determination
Volume of distribution	V	[l] or [l kg^{-1}]	clinical studies: plasma concentration–time curves
Terminal half-life (in most cases corresponding to the biological half-life)	$t_{1/2(\lambda_Z)}$	[h]	clinical studies: plasma concentration–time curves, compartment models
Fraction of absorbed drug excreted unchanged in urine	f_e	none	clinical studies: plasma concentration–time curves and cumulative-urine curves
Fraction unbound in plasma	f_u	none	*in vitro* experiments, *e.g.*, equilibrium dialysis or ultracentrifugation
Bioavailability (formulation specific)	F	none	clinical studies: plasma concentration–time curves after intra- and extravascular application

2. The ADME *in vivo* Profiling Flow Chart

The pharmacokinetic core parameters can be used to establish a flow chart for the *in vivo* ADME profiling of compounds (*Fig. 1*). The relevant calculations are summarized in *Table 2*. Based on this analysis, predictions can be made on the pharmacokinetic behavior of a drug and its potential interactions at the level of each ADME process. Simplest assumptions are made in a first approach which can be revised at a later stage if necessary. The step-by-step procedure is described below.

2.1. *Distribution*

The distribution of a compound is primarily dependent on its physicochemical parameters and its affinity to plasma and tissue proteins, to receptors, as well as to other components such as nucleic acids and glycosylated proteins and lipids. It has been shown that predictions can be made for the apparent volume of distribution of a new compound by means of empirical equations comprising the apparent partition coefficient under physiological conditions (pH 7.4) and the fraction unbound in plasma [1–3]. Exceptions to the rule are found with compounds that exhibit affinities to a specific binding site, *e.g.*, digoxin that has a high affinity for Na^+/K^+-ATPase, which leads to a relatively high volume of

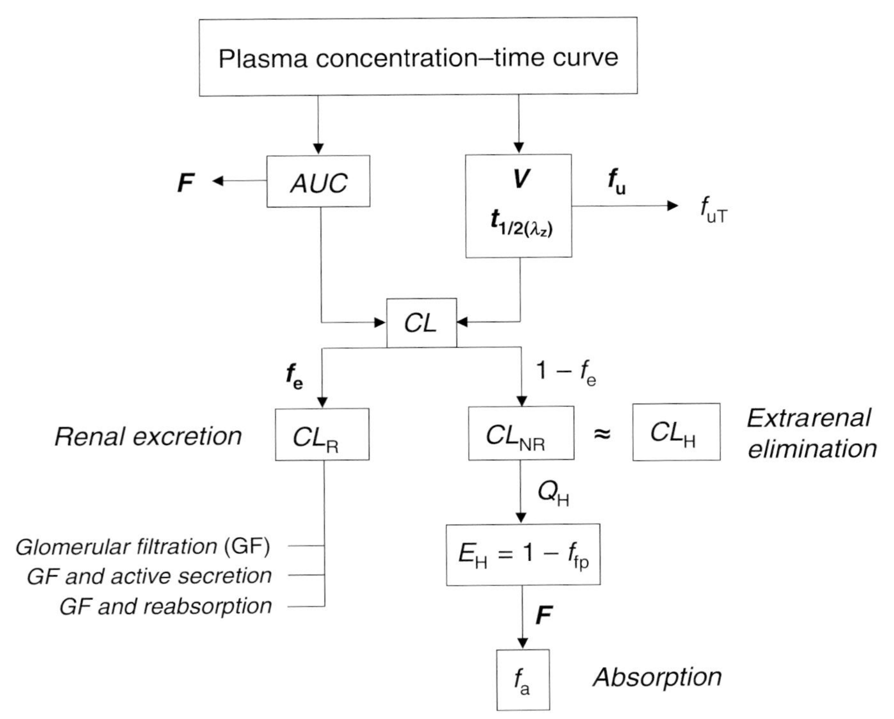

Fig. 1. *The flow chart of* in vivo *pharmacokinetic profiling of drugs. AUC*, area under the plasma concentration–time curve from $t=0$ to $t \to \infty$; *V*, volume of distribution; f_u, unbound fraction of drug in plasma; f_{uT}, unbound fraction of drug in tissue; *F*, bioavailability (bioavailable fraction of dose); $t_{1/2(\lambda_z)}$, terminal half-life (in most cases identical to the biological half-life, *i.e.*, the half-life relevant for therapy); *CL*, total body clearance; f_e, fraction of drug absorbed that is excreted unchanged into the urine; CL_R, renal clearance; CL_{NR}, nonrenal clearance; CL_H, hepatic clearance (for a first estimate $\approx CL_{NR}$); Q_H, hepatic blood flow; *E*, extraction ratio; f_{fp}, fraction of dose absorbed available systemically; f_a, fraction of dose absorbed. Core parameters in boldface.

distribution (*ca.* 6 l kg^{-1}) as compared to the more-lipophilic digitoxin (*ca.* 0.6 l kg^{-1}).

To estimate the fraction of unbound drug in tissue (f_{uT}), and thus get an idea on the overall distribution of the compound in the body, a two-compartment model, *i.e.*, plasma and tissue, can be used [4]. This leads to the following relations (*Eqns. 1* and *2*):

$$V = V_p + V_T \left(\frac{f_u}{f_{uT}}\right) \tag{1}$$

$$f_{uT} = \left(\frac{f_u}{V - V_p}\right) V_T \tag{2}$$

Table 2. *Calculations for the* in vivo *Profiling of Drugs*

Parameter	Calculations/Estimates
f_{uT}	$V = V_p + V_T\left(\frac{f_u}{f_{uT}}\right) \rightarrow f_{uT} = \left(\frac{f_u}{V - V_p}\right) V_T$
F	$F = \frac{AUC_{extravascular}\, D_{i.v.}}{AUC_{i.v.}\, D_{extravascular}}$ (CL constant)
CL	$CL = k_{10} V = \left(\frac{\ln 2}{t_{1/2}}\right) V$
CL_R	$CL_R = \frac{Ae(\infty)}{AUC} = \frac{f_e\, F\, D}{AUC} = f_e\, CL$
CL_{NR}	$CL_{NR} = (1 - f_e)\, CL$
	Assumption: only renal and hepatic elimination[a])
	$CL_{NR} \approx CL_H = Q_{p,H}\;\; E_{p,H} = Q_{p,H} \frac{f_u CL_{int}}{Q_{p,H} + f_u CL_{int}}$
	$E_{p,H} = \frac{CL_H}{Q_{p,H}}$ [b])
f_{fp}	$f_{fp} = 1 - E_{p,H}$
f_a	$F = f_a\, f_{fp} = f_a\,(1 - E_{p,H})$
	$f_a = \frac{F}{f_{fp}} = \frac{F}{1 - E_{p,H}}$

[a]) This assumption must be revised, for example when $E > 1$. [b]) The core parameters with the exception of f_u stem from plasma data. Accordingly, plasma flow through liver ($Q_{p,H} = Q_H (1 - H)$) is used in the calculations instead of blood flow (Q_H), and thus the balance relates to extraction from the plasma (E_p). The hematocrit H equals 0.44.

where V_p represents the plasma volume, which corresponds to *ca.* 4% of body weight (BW), *i.e.*, *ca.* 3 l for a 70-kg person. V_T is estimated as the physiological volume of the drug that corresponds to a maximum of the total body water minus plasma volume (*ca.* 56% BW, *i.e.*, *ca.* 39 l for a 70-kg person). When a drug is not distributed throughout the body fluids, *i.e.*, when $V < 42$ l, V_T equals $(V - V_p)$, and as a consequence f_u equals f_{uT}. Using the mass balance, it can be calculated in percentage of dose to what extent a drug is found in the plasma and tissue, respectively (*Table 3*).

To estimate the risk of displacement from the same binding partner upon comedication with another compound, four points need to be taken into consideration. Significant displacement can be expected if: *a*) plasma protein binding is high ($f_u < 0.1$), *b*) the apparent volume of distribution is relatively small (< 0.2 l kg^{-1}), *c*) the therapeutic range of the drug at risk of displacement is narrow, and *d*) the molar therapeutic concentration of the displacing drug is in the same range as the molar concentration of the

Table 3. *Distribution of Selected Drugs in Plasma and Tissue*

Parameter	Thiopental	Streptomycin	Warfarin	Diazepam
V [l]	161	18.2	7.7	140
V_p [l]	3	3	3	3
V_T [l]	39	15.2	4.3	39
f_u	0.15	0.66	0.005	0.03
f_{uT}	0.0375	0.66	0.005	0.009
plasma free[a])	0.45 c (0.3%)	1.98 c (10.9%)	0.015 c (0.2%)	0.09 c (0.06%)
plasma bound[a])	2.55 c (1.6%)	1.02 c (5.6%)	2.985 c (38.8%)	2.91 c (2.08%)
tissue free[a])	5.85 c (3.6%)	10.03 c (55.1%)	0.024 c (0.3%)	1.23 c (0.88%)
tissue bound[a])	152.26 c (94.6%)	5.17 c (28.4%)	4.677 c (60.7%)	135.77 c (96.98%)

[a]) Calculated with mass balance: $A_{tot} = A_p + A_T$ and $c\ V = c\ V_p + c_T\ V_T$; amount in plasma free: $f_u\ V_p\ c$; amount in plasma bound: $(1 - f_u)\ V_p\ c$; amount in tissue free: $f_{uT}((f_u/f_{uT})\ V_T)\ c$; amount in tissue bound: $(1 - f_{uT})((f_u/f_{uT})\ V_T)\ c$.

binding partner. Thus, the affinity constant to human serum albumin (HSA) of all acidic drugs is so high that, at therapeutic levels, HSA represents their main binding protein in plasma. Physiological concentrations of HSA are in the range of 500–700 μM. A classical and clinically relevant displacement takes place upon comedication of warfarin (displaced drug, M_r 308.3; association constant $K = 2.1 \times 10^5\ \text{M}^{-1}$; therapeutic concentration range of 3–32 μM) and salicylic acid (displacing drug, M_r 138.1; association constant $K = 3.8 \times 10^5\ \text{M}^{-1}$; therapeutic concentration range of 150–2200 μM). Interestingly enough, clinically relevant interactions at the level of protein binding are less frequent than anticipated. This is mainly due to the fact that the unbound fraction of drug is not only relevant for the effect, but also directly influences elimination in the kidneys and liver (see calculations in *Table 2*). An increase in f_u thus leads to an increased clearance and therefore to an internal regulation of the drug level within a short period of time.

The pH-partition hypothesis is a useful tool to estimate the potential drug concentrations in different body fluids (*Fig. 2*), being based solely on the physicochemical characteristics of the compound. The concentration ratios are calculated with the *Henderson–Hasselbalch* equation for two aqueous compartments with differing pH values such as plasma (pH 7.4) and milk (pH 6.6), or plasma and stomach (pH 1–3.5).

The following equations are used for weak acids (*Eqn. 3*) and weak bases (*Eqn. 4*), respectively:

$$\frac{c_1}{c_2} = \frac{1 + 10^{\mathrm{pH}_1 - \mathrm{p}K_a}}{1 + 10^{\mathrm{pH}_2 - \mathrm{p}K_a}} \qquad (3)$$

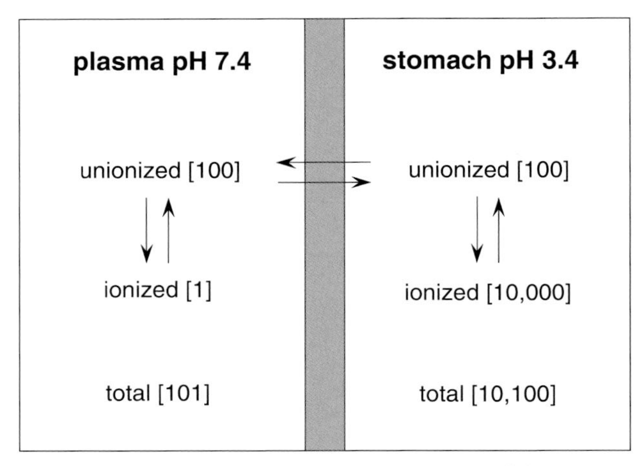

Fig. 2. *The pH-partition hypothesis: schematic representation of the partitioning of a weak base* (pK_a 5.4) *between two aqueous phases with pH 7.4* (plasma) *and pH 3.4* (stomach) *at equilibrium*. A lipophilic barrier separates the two phases (dark grey) that is mainly permeable for the nonionized, more-lipophilic species. Of the molecules, 99% are ionized at pH 3.4 and only 1% at pH 7.4. At equilibrium, this results in a 100-fold enrichment of compound in the acidic environment.

$$\frac{c_1}{c_2} = \frac{1 + 10^{pK_a - pH_1}}{1 + 10^{pK_a - pH_2}} \tag{4}$$

Based on these calculations, it can be predicted that the milk/plasma ratio is higher for weak bases and lower for weak acids. If estimates are made with the pH-partition hypothesis, differences can occur as compared to *in vivo* data. This can be due to the fact that equilibrium is not necessarily reached or that active transport of drug into milk occurs. Another particularly important aspect is that differences in protein binding between the two compartments under consideration will shift the equilibrium and thus influence concentration ratios (*Table 4*).

2.2. *Renal Excretion*

Kidneys are both regulation and elimination organs. The renal clearance of a drug can comprise different overlapping processes, in particular glomerular filtration (GF), active secretion, passive or active reabsorption, and biotransformation. Depending on the drug, not all of these processes are of equal importance. Thus, GF occurs continuously. About one tenth of the renal blood flow is deviated through the filters of the glomeruli (*ca.* 120 ml min^{-1} in healthy adults). As with all other elimi-

Table 4. *pH-Dependent Urinary Excretion of Drugs* (selection)

Renal clearance	Drug (pK_a)[a]
increased in *acidic* urine	amphetamine (9.7 B) chloroquine (8.4 B) codeine (8.3 B) imipramine (9.5 B) morphine (8.1 B) nicotine (9.7 B) procaine (9.0 B)
increased in *alkaline* urine	amino acids (A) barbiturates (A) nalidixic acid (6.0 A) nitrofurantoin (7.2 A) probenecid (3.4 A) salicylic acid (3.0 A) sulfonamides (A)

[a]) A = acidic pK_a; B = basic pK_a.

nation processes, only unbound drug is cleared. Provided that the affinity of a drug for a transport protein is high enough, the compound can also be cleared by active secretion. Whether or not a compound is excreted in urine depends on its physicochemical characteristics. Passive reabsorption occurs as a function of the apparent partition coefficient, the pK_a value of the drug, and the urinary pH (pH range of 4.5–8.0). Estimates for the partitioning at equilibrium can be made with the *Henderson–Hasselbalch* equations (*Eqns. 3* and *4*). Renal reabsorption of weak electrolytes by passive diffusion can be influenced by changing the urinary pH. A significant change can be expected when the acidic drug has a pK_a between 3.0 and 7.5 and the basic drug a pK_a between 7.0 and 11.0. Some examples of drugs with pH-dependent renal clearance are listed in *Table 4*.

The total clearance, CL, can be calculated from the pharmacokinetic core parameters V and $t_{1/2(\lambda_z)}$, and together with f_e one obtains the renal clearance CL_R. For a drug that is only excreted through GF, CL_R corresponds to the product of the GF rate (GFR) and f_u. When CL_R is larger than the product of GFR and f_u, this indicates that net active secretion occurs. In contrast, when CL_R is smaller than the product of GFR and f_u, net reabsorption can be postulated. Renal insufficiency has a direct impact on the GFR. Dosage adjustments should be considered for compounds with $f_e > 0.3$ when the GFR falls below *ca.* 25% of the normal rate. For compounds with $f_e < 0.3$, adjustments may be adequate for drugs that are transformed to active metabolites with high f_e.

2.3. *Nonrenal Elimination*

The nonrenal clearance (CL_{NR}) is obtained as the difference between CL and CL_R. In a first approximation, CL_{NR} is assumed to equal CL_H. This is obviously a simplification which has to be verified at a later step of the analysis (see below).

Making use of the physiological or hemodynamic perfusion model [5], the hepatic extraction ratio $E_{p,H}$ can be calculated from CL_H with the help of the hepatic plasma flow Q_H (*Eqn. 5*):

$$E_{p,H} = \frac{CL_H}{Q_{p,H}} \tag{5}$$

Accordingly, the fraction of the absorbed dose remaining systemically available after the first liver passage, f_{fp}, equals $(1 - E_{p,H})$.

Hepatic clearances below *ca.* 200 ml min^{-1} reflect an extraction ratio of $E < 0.3$ indicative of capacity-limited biotransformation. In this case, the enzyme capacity (also called intrinsic clearance, CL_{int}) of the liver is rate-limiting for elimination. This means that changes in the enzyme capacity through enzyme induction or enzyme inhibition directly affect hepatic clearance. As only the unbound fraction of drug undergoes biotransformation, f_u also has a direct influence on the hepatic clearance (*Eqn. 6*):

$$\text{For} \quad E_H \rightarrow 0: \quad CL_H \approx f_u\, CL_{int} \tag{6}$$

With high extraction compounds ($E > 0.7$), the enzyme capacity is high and not rate limiting. These are the so-called first-pass drugs. Here, the blood flow represents the limiting factor, whereas changes in enzyme capacity (*i.e.*, enzyme induction or inhibition) do not influence the hepatic clearance (*Eqn. 7*):

$$\text{For} \quad E_H \rightarrow 1: \quad CL_H \approx Q_H \tag{7}$$

First-pass drugs show high inter- and intraindividual variability in plasma concentration–time curves.

When calculations with *Eqn. 5* result in $E > 1$, the assumption that the nonrenal clearance is identical to the hepatic clearance has to be revised. In this case, presystemic clearance (*e.g.*, in the intestinal wall and/or the portal vein) has to be assumed in addition to the clearance occurring upon hepatic first passage.

2.4. *Reduced Bioavailability: Limited Absorption or First-Pass Effect?*

In contrast to the other core parameters, bioavailability F is not constant for a given compound but depends on its specific formulation. F results from two contributions, the fraction of the absorbed dose (f_a) and the systemically available fraction of the absorbed dose (f_{fp}) after a possible first-pass extraction (*Eqn. 8*):

$$F = f_a f_{fp} \tag{8}$$

With F known for a specific formulation and E calculated as sketched above, rough estimates can be made whether a reduced bioavailability originates from limited absorption or is the result of a first-pass effect. From the examples shown in *Table 5*, both propranolol and morphine are high-extraction drugs with possible presystemic elimination, whereas penicillin G and sulfasalazine are low-extraction drugs with limited absorption.

Table 5. *Reduced* F – *Limited Absorption or First-Pass Effect?*

Parameter	Propranolol	Morphine	Penicillin G	Sulfasalazine
F	0.35[a])	0.4[a])	0.3	0.3
$E_{p,H}$[b])	0.9–1.5[c])	0.25–0.65	0.14	0.24
f_a[b])	~1	~1	~0.3	~0.4

[a]) Variable. [b]) Parameters calculated with data from [4] and [6]. [c]) Possible presystemic elimination.

3. Considerations about Dosage Regimens

The pharmacokinetic core parameters are also useful for a rough estimate of the plasma concentration–time curve produced by multiple dosing [4]. For a dosage regimen with constant dose and constant dosing interval, a steady state is reached (>95% of the maximum plasma concentration) after *ca.* 5 $t_{1/2}$. The average plasma concentration ($c_{ss,av}$) at steady state depends on the dose (D) applied and the dosing interval (τ) in relation to the half-life of the respective drug (*Eqn. 9*):

$$\frac{F D}{\tau} = CL \; c_{ss,av} = V \; \frac{\ln 2}{t_{1/2}} \; c_{ss,av} \tag{9}$$

Given the therapeutic concentration range to be reached for a given drug, the appropriate D and τ values can be chosen using its V and $t_{1/2}$ values.

An accumulation ratio, R_{ac}, has been defined that corresponds to the ratio of the maximum plasma concentration at steady state over the maximum plasma concentration after a single dose. R_{ac} depends only on $t_{1/2}$ and τ (*Eqn. 10*):

$$R_{ac} = \left(1 - e^{-\frac{\ln 2}{t_{1/2}}\tau}\right)^{-1} \tag{10}$$

Changes in half-life, *e.g.*, due to renal insufficiency or to enzyme induction or inhibition, directly influence both the time when the steady state is reached and the R_{ac} and thus the plasma concentrations at steady state.

4. Conclusions

In vivo pharmacokinetic profiling of drugs based on their core parameters is a useful instrument in designing drug therapies. Although it is clear that refinements are needed for more than a rough classification, the estimates presented here provide important clues about tissue distribution, renal excretion, and biotransformation in the liver. They also allow predictions on possible drug interactions upon comedication, as well as on the plasma concentration–time curve to be expected upon multiple dosing.

REFERENCES

[1] S. Oie, T. N. Tozer, *J. Pharm. Sci.* **1979**, *68*, 1203.
[2] W. A. Ritschel, C. V. Hammer, *Int. J. Clin. Pharmacol. Ther. Toxicol.* **1980**, *18*, 298.
[3] F. Lombardo, R. S. Obach, M. Y. Shalaeva, F. Gao, *J. Med. Chem.* **2004**, *47*, 1242.
[4] M. Rowland, T. N. Tozer, 'Clinical Pharmacokinetics. Concepts and Applications', 3rd edn., Lippincott Williams & Wilkins, Philadelphia, PA, 1995.
[5] G. R. Wilkinson, D. G. Shand, *Clin. Pharmacol. Ther.* **1975**, *18*, 377.
[6] W. A. Ritschel, G. L. Kearns, 'Handbook of Basic Pharmacokinetics Including Clinical Applications', 5th edn., American Pharmaceutical Association, Washington DC, 1999.

Part III. Physicochemical Strategies

Automated Parallel Synthesis in Support of Early Drug Discovery: Balancing Accessibility of Chemistry with the Design of Drug-Like Libraries
Carmen M. Baldino

New Insights into the Lipophilicity of Ionized Species
*Giulia Caron** and *Giuseppe Ermondi*

Physicochemical and Biological Profiling in Drug Research. Elog$D_{7.4}$ 20,000 Compounds Later: Refinements, Observations, and Applications
*Franco Lombardo**, *Marina Y. Shalaeva*, *Brian D. Bissett*, and *Natalya Chistokhodova*

Lipid Bilayers in ADME: Permeation Barriers and Distribution Compartments
Stefanie D. Krämer

High-Throughput Solubility, Permeability, and the MAD PAMPA Model
Alex Avdeef

Correlations between PAMPA Permeability and log *P*
Karl Box, *John Comer**, and *Farah Huque*

Predicting the Intestinal Solubility of Poorly Soluble Drugs
Alexander Glomme, *J. März*, and *Jennifer B. Dressman**

Accelerated Stability Profiling in Drug Discovery
*Edward H. Kerns** and *Li Di*

Physicochemical Characterization of the Solid State in Drug Development
Danielle Giron

Automated Parallel Synthesis in Support of Early Drug Discovery: Balancing Accessibility of Chemistry with the Design of Drug-Like Libraries

by **Carmen M. Baldino**

ArQule Inc., 19 Presidential Way, Woburn, MA 01801-5140, USA
(e-mail: carmen.baldino@bioduro.com)

Abbreviations
AMAP™: Automated molecular-assembly plant; DOS: diversity-oriented synthesis; ELSD: evaporative light-scattering detector; HPLC: high-performance liquid chromatography; HT: high-throughput; IMDAF: intramolecular *Diels–Alder* reaction of furans; LC/MS: inline liquid chromatography and mass spectrometer used for analytical characterization of molecules; *MapMaker*™: an integrated suite of computational chemistry and informatics tools used for library design; MS: mass spectrometer; SAR: structure–activity relationships.

---

1. Introduction

Automated parallel synthesis provides chemistry muscle to early-stage drug discovery programs enabling the efficient and comprehensive exploration of desirable chemistry space [1]. A discovery strategy that involves the high-throughput screening of a compound collection with a parallel synthesis heritage will fuel the rapid expansion of SAR from validated chemical procedures. This approach, however, relies heavily on the accessibility of synthetic methods to prepare the molecules of interest. Combinatorial chemistry techniques have been plagued with the challenge of developing methods for the synthesis of the right compounds and not just those that are chemically feasible. This formidable obstacle can be overcome only by integrating multiple disciplines and technologies. Herein, we provide details of our approach with applications in support of drug discovery programs.

2. Chemical Technology Platform

Our integrated high-throughput parallel synthesis platform enables the rapid development of solution-phase-chemistry protocols for the efficient preparation of large libraries (1000–5000 members) used for lead generation, as well as smaller (50–500 members) library sets used for rapid follow-up on primary screening hits or for lead optimization. The technology platform can be viewed as a lever that transforms traditional medicinal chemistry into an effective high-throughput science capable of addressing the productivity gap in drug discovery. Herein, the breadth and flexibility of the automated molecular-assembly plant (*AMAP*™) system is described.

ArQule's philosophy in developing the *AMAP*™ has focused on a modular unit operation approach similar to that practiced in typical chemical engineering disciplines. We have established unit operations that mimic those available to the medicinal chemist, but are performed in parallel on the system. The modular workstations are designed to perform a specific task or set of closely related tasks. This approach allows the chemist to use the various unit operations in an appropriate fashion based on the library reaction scheme to synthesize the desired compounds.

The foundation of this strategy is the reaction block (*Fig. 1*), which is an anodized aluminum block that conforms to a 24- or 96-well microtiter plate footprint. Inserting glass vials into the wells of the reaction block creates 24- or 96-jacketed reaction vessels. Each modular workstation is then designed to incorporate this reaction block into the automated process. This standardized format allows for multiple reaction steps to be performed on a scale from 50 to 500 μmol, which, when followed by HPLC

Fig. 1. ArQule *reaction blocks: the standard work unit footprint used on the* AMAP™ *system.* The aluminum 24- or 96-well reaction blocks used with glass vial inserts allow for library chemistry to move seamlessly through multiple automated synthetic, workup, and purification steps in a parallel fashion.

purification, typically yield between 20 and 100 μmol. This format not only meets the current synthetic needs of pharmaceutical research and development, but also allows the employment of practical, commercially available automation, enabling the development of new tools and applications to expand that capabilities going forward.

3. Chemist-Friendly Library Design Tools

A fundamental problem in the design of the optimal molecules is that it requires the combination of the skill set of a medicinal chemist and an available computational chemistry resource both of which are in short supply in the industry. We have addressed this issue by integrating several software tools into a single web-based platform (*MapMaker*™) accessible to medicinal chemists for library design [2] [3] (*Fig. 2*). This tool allows for the automated creation of exportable files with a convenient template enabling chemists to interactively visualize the impact of modifications on the original library design. The true value of this system is the efficiency

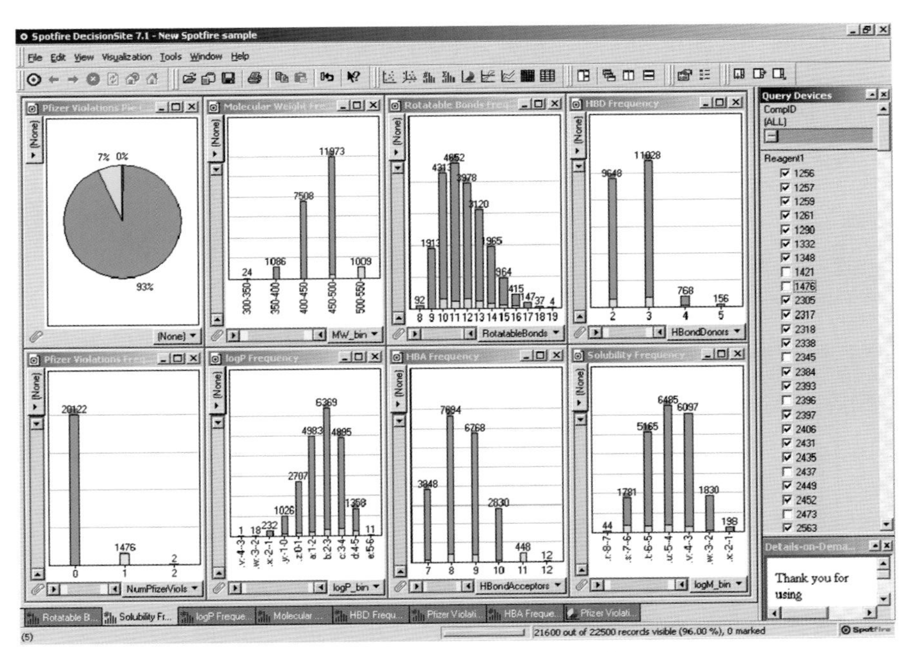

Fig. 2. *Optimized virtual library profile using the* MapMaker™ *platform.* An example of an optimized virtual library profile from the *MapMaker*™ platform, which is an integrated computational chemistry and informatics platform that uses a genetic algorithm to optimize for specific physicochemical properties and pharmacophore fit while maximizing diversity.

gains obtained by enabling the chemists to be responsible for their own designs ensuring that the final purified compounds embody the design intent of the original idea.

4. Standardized Chemistry Development Process

Technology alone does not allow for the efficient development of automated synthetic methods. We have found that the translation of bench-level know-how to the automated synthesis of libraries requires a standardized chemistry development approach that accounts for both chemical feasibility and automated processing details. We have addressed this problem by using specific groups of scientists focusing in on critical areas pertaining to the development of synthetic or analytical methods followed by the translation of those methods to an automated process. The key process points include *Early Development & Optimization* (optimization of synthetic route and conditions), *Analytical Test Plate* (optimization of purification and characterization conditions), *Reagent Qualification* (validation of reagent performance), *Pilot Synthesis* (*Pilot Test Plate* and *Pilot 1st Run*, validation of performance under automated processing conditions) and *High-Throughput Synthesis* (library synthesis). This strategy ensures that the right scientists are focused on the development process at each stage, which is summarized in *Fig. 3*.

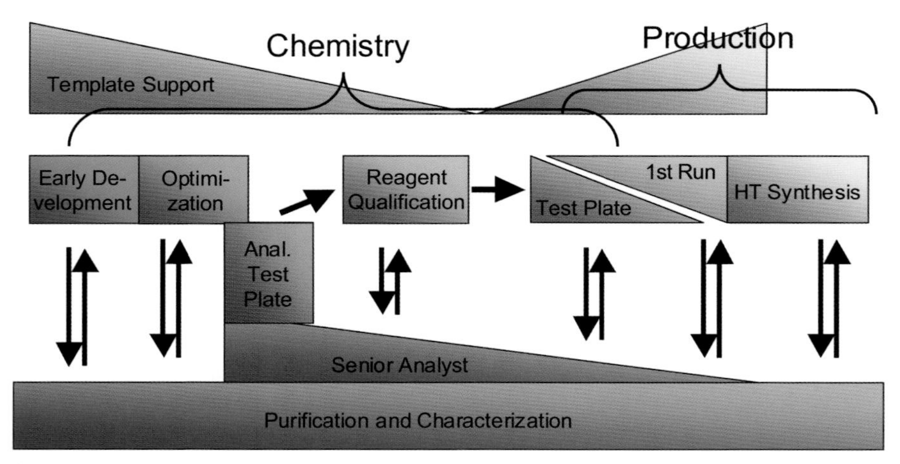

Fig. 3. ArQule*'s standardized chemistry development strategy.* This is a graphical representation illustrating our automated approach to chemistry development for parallel synthesis, which relies on focusing specific groups of synthetic chemists on key elements of the process while providing the appropriate level of support from analytical chemistry.

5. High-Throughput Analytical Chemistry

A successful HT chemistry strategy requires a comparable emphasis on high-throughput analytical chemistry. Quality has always been the cornerstone criteria for medicinal chemists, which is unlikely to change. Therefore, the capabilities and capacities in high-throughput analytical chemistry must keep pace with those in synthesis. With this in mind, *ArQule* has developed and successfully operated a high-throughput HPLC purification and LC/MS characterization process supporting both lead generation and lead optimization programs [4]. The purification approach relies on MS-triggered fraction collection with a one-to-one correlation of reaction and collection vessels (*Fig. 4*). This strategy is coupled with a 5-min reversed-phase gradient, which has provided over 375,000 purified samples during 2003.

After purification, every sample is analyzed for purity using a 2.5-min reversed-phase HPLC/UV/ELSD/MS method [5–8] (*Fig. 5*). This allows for the characterization of diverse collections of compounds by using three detectors in parallel. The speed and separation efficiency of the method has allowed for the characterization of over one million samples during 2003.

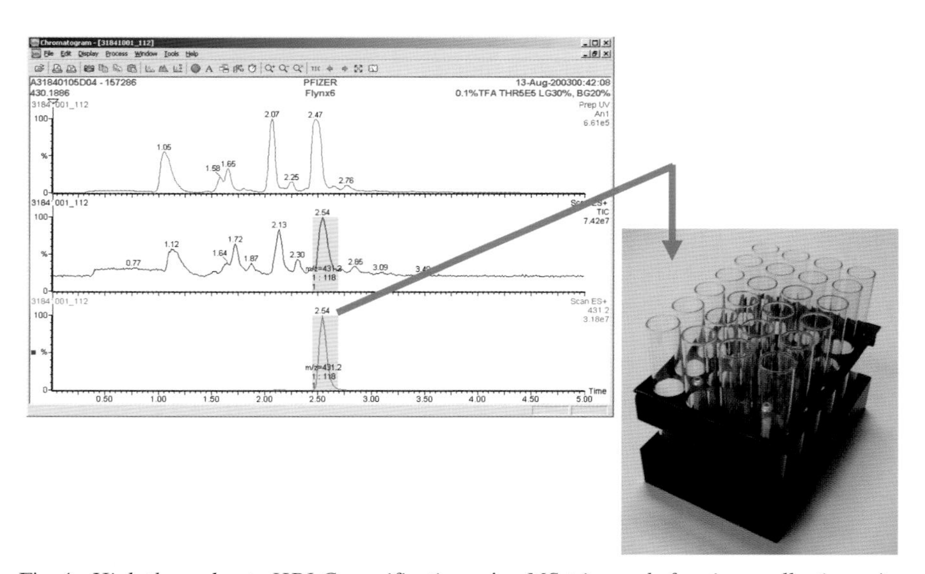

Fig. 4. *High-throughput HPLC purification* via *MS-triggered fraction collection.* An example of a high-throughput purification process using reversed-phase HPLC separation followed by the MS-triggered collection of a single fraction per reaction.

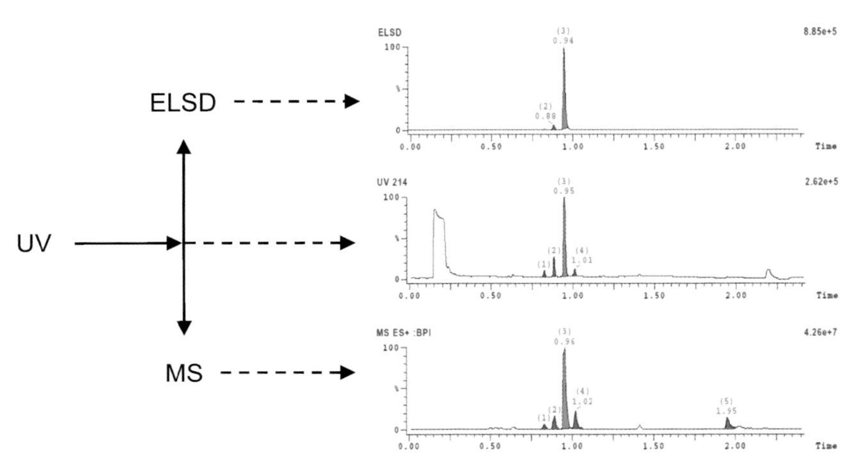

Fig. 5. *High-throughput HPLC/UV/ELSD/MS characterization approach.* An example of a purified sample having been characterized by three distinct detectors including ultraviolet (UV) spectroscopy, evaporative light-scattering detector (ELSD), and mass spectrometer (MS).

6. Applications

ArQule's integrated chemistry technology platform including an automated parallel synthesis system, library design tools, high-throughput analytical chemistry, and a standardized development process have been detailed. The question that still needs to be answered is 'What can a chemist really do with this system?' The following examples of diversity-oriented synthesis (DOS) illustrate the complexity of chemotypes that are possible through the use of this approach [9].

Scheme 1 describes the *Pictet–Spengler* cyclization of a tryptamine and methyl 2-furly-α-ketoester in an automated step providing the intermediate disubstituted 1-(2-furyl)-1-carboxymethyl-β-tetrahydrocarbolines **1**. *N*-Acylation of **1** followed by intramolecular *Diels–Alder* cycloaddition affords the functionalized noryohimban analogs **2** and **3**. In addition, the noryohimban analogs **2** and **3** are rich in functional group density, which allows for the further elaboration of those scaffolds providing access to a number of novel chemotyps as described in *Scheme 2*.

Scheme 1. *Functionalized Noryohimbans by an Intramolecular* Diels–Alder *Reaction of Furans* (IMDAF). An example of a synthetic strategy providing access to a natural-product-like scaffold and a library of related noryohimban analogs.

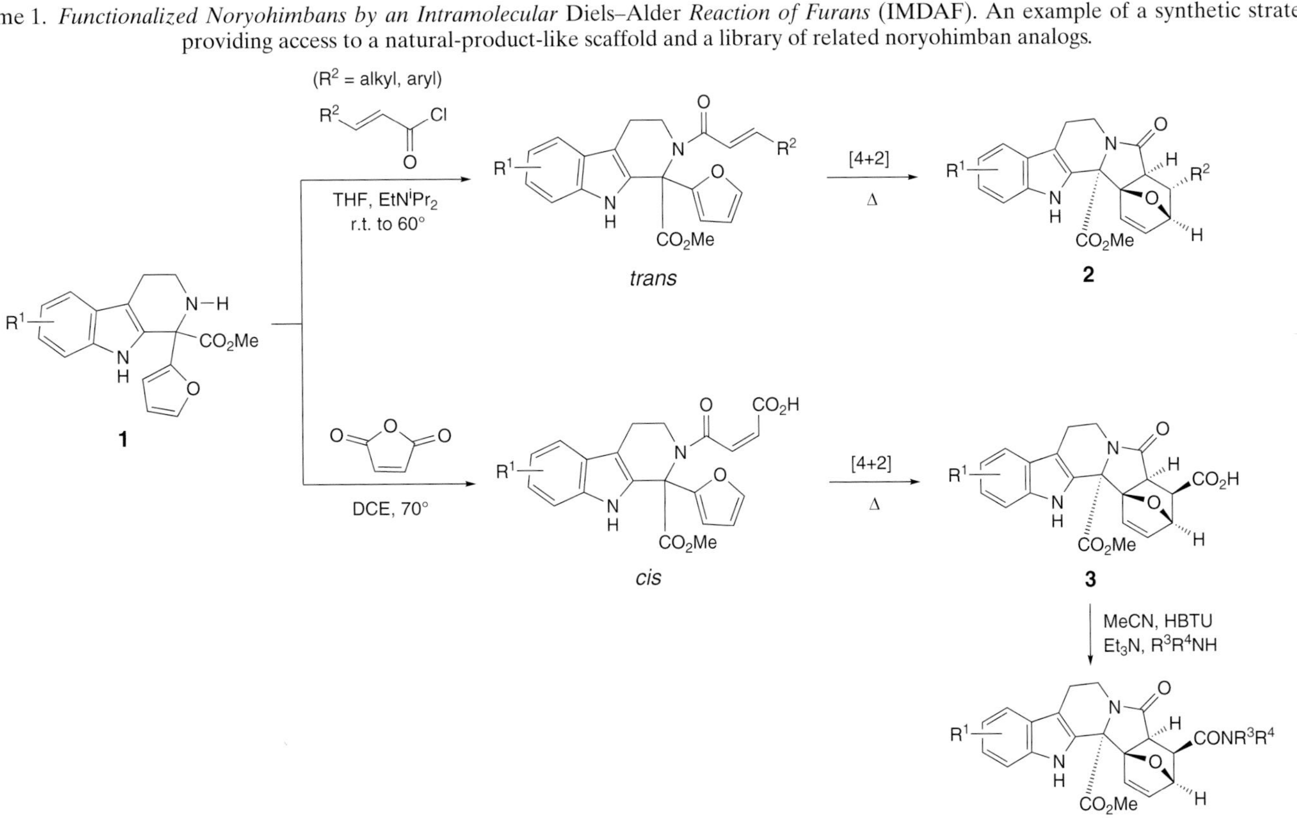

Scheme 2. *Exploiting the Functionality of the Cycloadducts.* An example of sampling the available diversity in a highly functionalized natural-product-like scaffold through the synthesis of libraries of further elaborated chemotypes.

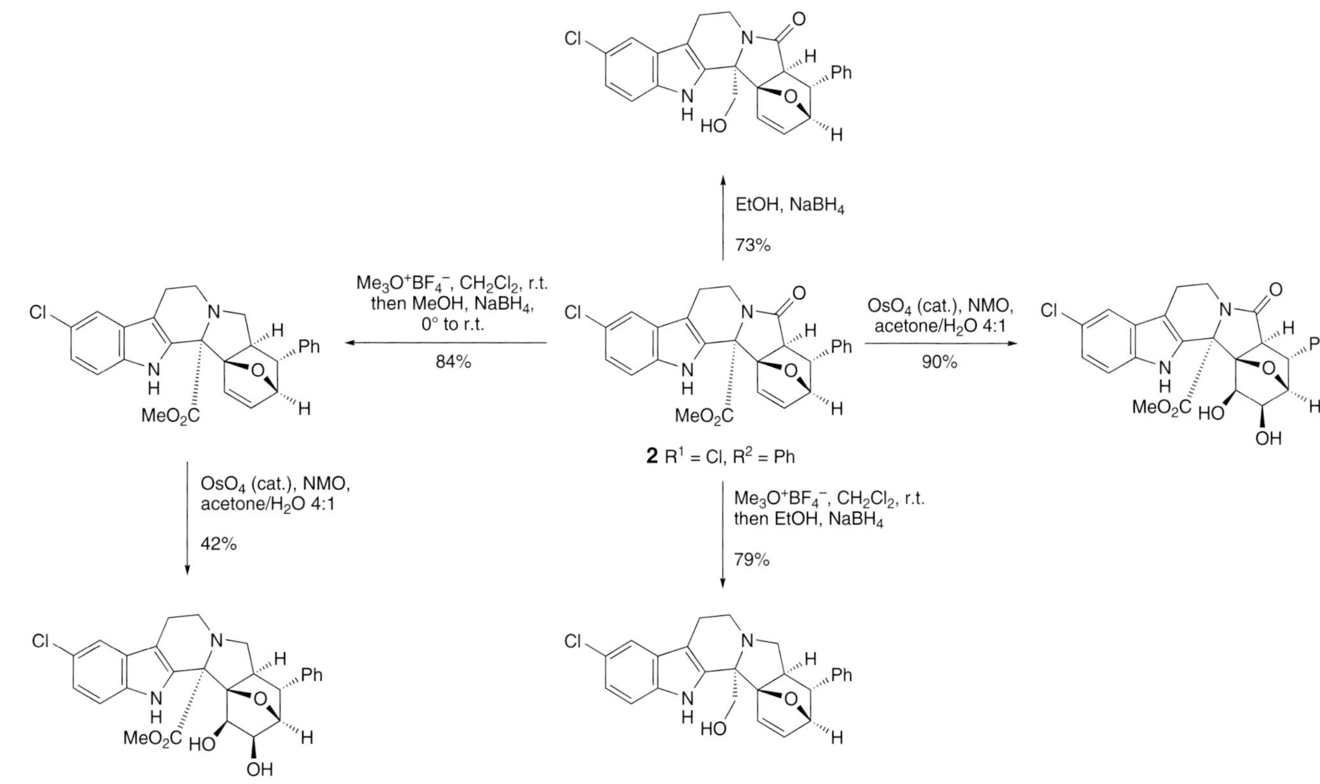

7. Conclusions

In conclusion, the effectiveness of the *ArQule* automated parallel synthesis strategy in support of drug discovery has been described and illustrated through several examples. It is our belief that the true benefits of the advances made over the last five years by *ArQule* and many other scientists focused on a variety of high-speed chemistry approaches have yet to be fully realized. The final assessment must then wait to determine if these efficiency gains will translate into healthier pipelines of high-quality clinical candidates for the pharmaceutical industry.

REFERENCES

[1] C. M. Baldino, *J. Combinat. Chem.* **2000**, *2*, 89.
[2] S. Gallion, L. Hardy, A. Sheldon, *Curr. Drug Disc.* **2002**, January, 25.
[3] A. Yasri, D. Hartsough, *J. Chem. Inf. Comput. Sci.* **2001**, *41*, 1218.
[4] J. N. Kyranos, H. Cai, B. Zhang , W. K. Goetzinger, *Curr. Opin. Drug Disc. Dev.* **2001**, *4*, 719.
[5] J. N. Kyranos, H. Cai, D. Wei, W. K. Goetzinger, *Curr. Opin. Biotechnol.* **2001**, *12*, 105.
[6] J. N. Kyranos, J. C. Hogan Jr., *Modern Drug Disc.* **1999**, 73.
[7] J. N. Kyranos, J. C. Hogan Jr., *Anal. Chem.* **1998**, *70*, 389A.
[8] W. K. Goetzinger, J. N. Kyranos, *Am. Lab.* **1998**, *30*, 27.
[9] D. Fokas, J. E. Patterson, G. Slobodkin, C. M. Baldino, *Tetrahedron Lett.* **2003**, *44*, 5173.

New Insights into the Lipophilicity of Ionized Species

by **Giulia Caron*** and **Giuseppe Ermondi**

Dipartimento di Scienza e Tecnologia del Farmaco, via Giuria 9, I-10125 Torino
(e-mail: giulia.caron@unito.it)

Abbreviations
ASA: Water-accessible surface area of all atoms; ASA+: water-accessible surface area of all atoms with positive partial charge; ASA−: water-accessible surface area of all atoms with negative partial charge; ASA_H: contribution of all hydrophobic atoms to the ASA; %B(DBA): percent albumin binding; CD: cyclodextrin; ClogP: log P calculated by the MedChem method; CSD: *Cambridge Structural Database*; DCE: 1,2-dichloroethane; IAM: immobilized artificial membranes; LFER: linear solvation–free energy relationship; log K'_{HSA}: logarithm of the equilibrium constant of binding to human serum albumin; log $D^{7.4}$: logarithm of the distribution coefficient at pH 7.4; log P: logarithm of the partition coefficient (in general); log P^I: logarithm of the partition coefficient of an ionized species I; NPOE: nitrophenyl octyl ether; PEOE: partial equalization of orbital electronegativities; PCA: principal component analysis; QMD: quenched molecular dynamics; QSAR: quantitative structure–activity relationship.

---

1. Introduction

Chemicals can be categorized according to their electrical features [1] into neutral molecules (total electrical charge = 0), unionized molecules (which contain no ionized group), charged molecules (total electrical charge different from zero), and ionized molecules (containing one or more ionized groups).

How different in their lipophilicity are unionized and ionized compounds? Is it possible to generalize the effect of charges on lipophilicity? Is it sufficient to modify lipophilicity rules valid for unionized compounds to investigate the lipophilicity of ionized solutes? This chapter tries to answer these questions by critically discussing *a*) the changes that a positive or negative electrical charge induces in the unionized structure of compounds in a lipophilicity-related perspective, *b*) the role of electrostatic interactions in a few well-known biphasic systems, and by giving some preliminary indications about a series of molecular descriptors well suited to rationalize the lipophilicity of ions.

2. The Solute

2.1. *The Extent of Ionization Governs Lipophilicity in Biphasic Systems*

Weak acids and bases ionize in solutions to varying extents depending on their pK_a values and on pH (the human physiological pH is 7.4 but in some compartments can have values as low as 1 or as high as 8 [2]).

It is not sufficient for a compound to have an ionizable center to be considered as ionized at a given pH, because it is the extent of ionization that determines its behavior in biphasic systems. We propose that compounds must be ionized more than 90% for the charge to influence strongly its partitioning behavior in a biphasic environment. The screen at 90% is due to the observation that, when 90% of the ionized species is present, the error made in neglecting $\log P^I$ when estimating $\log D^{pH}$ becomes important (*Fig. 1*) [3].

For zwitterionic ampholytes (ampholytes for which the relation $pK_a^{acidic} < pK_a^{basic}$ is true [4]), the presence of ions is also governed by the ratio of concentrations of the two neutral microspecies (zwitterion/neutral) expressed by the equilibrium constant of tautomerism, K_Z.

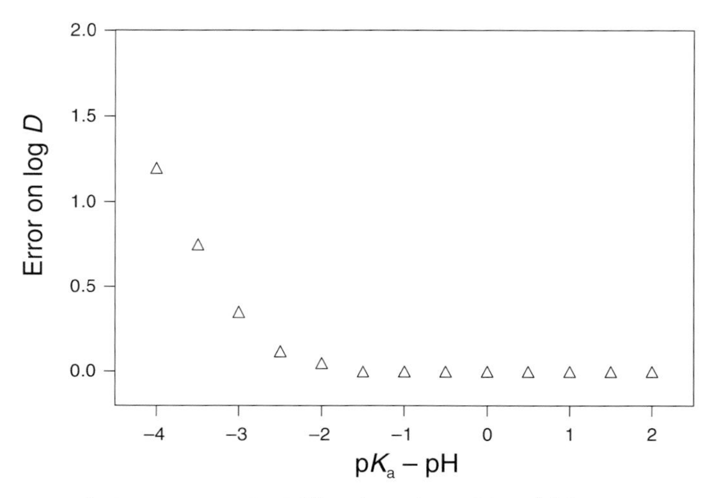

Fig. 1. *Error made in estimating log* D^{pH} *with neglect of log* P^I [3]: *the case of a generic monoacid for which* $pK_a = 3$, *log* $P^N = 4$ *and log* $P^I = 1.5$

2.2. *Does a Representative Data Set for Ions Exist?*

A well-distributed data set of 80 structurally diverse neutral model compounds with which to generate comparable LFER equations was defined [5]. Could a similar data set also be designed for charged molecules? This appears doubtful mainly because ions should be 'biologically interesting', with the consequence that their chemical variability cannot be optimized.

According to *Bergström et al.* [6], it is a good strategy to work with small data sets (15–20 compounds) of reliable biological data, although the structural variability of the chemicals must be checked chemometrically. An easy and clever method based on principal component analysis (PCA) has been presented [6]. However, charge-related parameters (ASA+ and ASA−, see below for definition and applications) should also be included as descriptors besides traditional descriptors such as ClogP, water-accessible surface area, number of rotatable bonds, number of aromatic bonds, number of H-, C-, N-, and O-atoms, number of H-bond donors and acceptors.

In this chapter, we use two data sets. In both, the chemical variability of the dominant (>90%) species at physiological pH was confirmed by a PCA analysis performed with the SIMCA software [7]. *Fig. 2* gives the scores of the first two principal components (t1 and t2), which describe 66% (*Data Set 1*, *Table 1*, and *Fig. 2,a*) and 63% (*Data Set 2*, *Table 2*, *Fig. 2,b*) of the diversity in the space of descriptors. The data sets were found to cover all four quadrants of the PCA plot, indicating that they are heterogeneous and, thus, significant.

3. Lipophilicity Systems

Lipophilicity systems are usually classified by the characteristics of the more-lipophilic phase. When it is an organic solvent (*e.g.*, octanol), we are in the presence of isotropic systems; when it is a suspension (*e.g.*, liposomes), we are in the presence of anisotropic systems, and when it is a stationary phase in liquid chromatography, we are in the presence of anisotropic chromatographic systems.

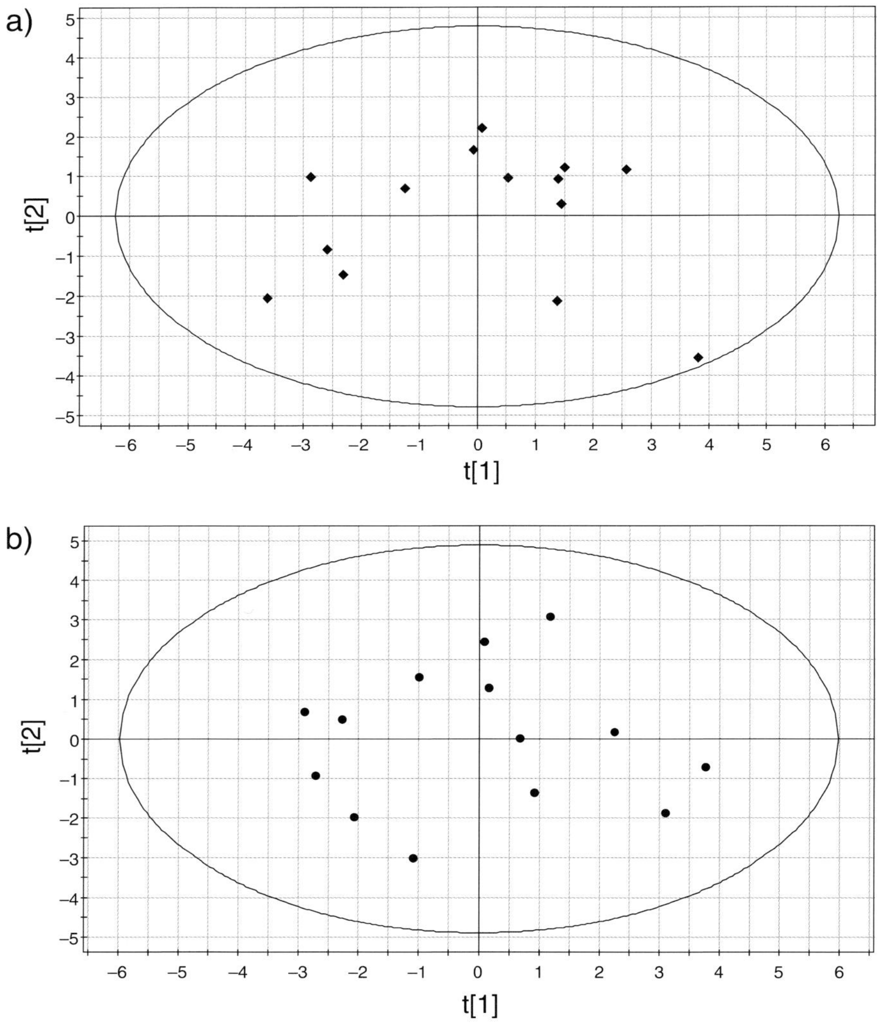

Fig. 2. *Principal component analysis* (PCA) *to investigate the heterogeneity of the data sets discussed. a*) *Data Set 1* (compounds in *Table 1*). *b*) *Data Set 2* (compounds in *Table 2*).

3.1. *Intermolecular Interactions Governing Lipophilicity*

Based on LFER results, lipophilicity is commonly factorized into nonpolar terms positively related to lipophilicity, and polar terms negatively related to lipophilicity (*Eqn. 1*):

$$\log P = vV - \Lambda \qquad (1)$$

Table 1. Data Set 1: *Albumin Binding Experimental Data and Theoretical Descriptors Calculated by the Molecular Package MOE* [8]

Drug	pK_a	Species[a])	%B(DAB)[b])	log K'_{HSA}[c])	ne_ASA+[d])	ion_ASA+[e])	*diff+*[f])	ne_ASA−[g])	ion_ASA−[h])	*diff−*[i])
Acebutolol	9.52	c	0	−0.21	434.00	444.86	−10.86	175.13	157.45	17.68
Amlodipine	9.03	c	65.4	–	451.98	475.64	−26.66	206.67	185.02	21.65
Chloroquine	8.10, 9.94	c	8.7	–	368.45	383.48	−15.03	216.53	202.44	14.10
Furosemide	3.65, 10.24	a	87.2	−0.13	246.59	229.69	16.90	256.59	274.85	−18.26
Hydrochlorthiazide	8.78, 9.96	n	50.8	−0.42	174.13	174.13	0.00	248.89	248.89	0.00
Indomethacin	4.5	a	91.1	0.47	317.73	309.65	8.08	259.92	260.70	−0.78
Methotrexate	3.76, 4.83, 5.60	n	49.9	−0.77	457.63	457.63	0.00	250.95	250.95	0.00
Naproxen	4.18	a	>95	0.25	278.52	262.22	16.30	175.51	190.28	−14.77
Nicardipine	7.17	n	93.4	–	463.85	463.85	0.00	240.94	240.94	0.00
Prazosin	6.5	n	46.4	0.06	525.39	525.39	0.00	128.33	128.33	0.00
Quinidine	4.46, 8.52	c	27.8	0.44	407.27	422.43	−15.16	143.96	124.81	19.16
Ranitidine	8.48	c	3.9	−0.1	435.03	483.47	−48.44	158.40	113.64	44.76
Tenoxicam	4.95	a	89.6	–	234.58	211.79	22.80	242.90	262.95	−20.05
Tetracycline	3.3, 7.7, 9.5	n	35.5	−0.08	370.59	370.59	0.00	227.37	227.37	0.00

[a]) Dominant (>90%) species at physiological pH: n = neutral or ampholyte, a = anion, c = cation. [b]) Ultracentrifugation data taken from [9]. [c]) Chromatographic data taken from [10]. [d]) The water-accessible surface area of all atoms with positive partial charge calculated on the neutral species of the molecule. [e]) The water-accessible surface area of all atoms with positive partial charge calculated on the electrical state of the molecule dominant at physiological pH. [f]) [d]) minus [e]). [g]) The water-accessible surface area of all atoms with negative partial charge calculated on the neutral species of the molecule. [h]) The water-accessible surface area of all atoms with negative partial charge calculated on the electrical state of the molecule dominant at physiological pH. [i]) [g]) minus [h]).

Table 2. Data Set 2: *Lipophilicity Descriptors and Theoretical Descriptors Calculated by the Molecular Package MOE* [8]

Drug	pK_a	Species[a])	$\log D_{lip}^{7.0}$ [b])	$\log k_{IAMw}^{7.0}$ [c])	ASA_H[d])	ne_ASA+[e])	ion_ASA+[f])	*diff*+[g])
Aspirin	3.50 [2]	a	1.60 [2]	−0.95 [11]	235.18	193.19	183.82	9.37
Clonidine	8.11 [12]	n	1.29 [12]	1.36 [12]	392.58	224.34	224.34	0.00
Diazepam	3.45 [12]	n	3.58 [12]	2.34 [12]	443.67	271.51	271.51	0.00
Diclofenac	3.99 [2]	a	2.66 [2]	2.43 [11]	418.50	199.99	194.20	5.79
Ibuprofen	4.45 [2]	a	1.94 [2]	1.12 [11]	381.57	297.29	272.86	24.43
Imipramine	9.34 [12]	c	2.83 [12]	3.30 [12]	529.27	351.25	402.52	−51.27
Nicotine	3.23, 8.00 [12]	n	2.30 [12]	0.78 [12]	353.20	281.53	281.53	0.00
Pheno-barbital	7.20 [12]	n	2.15 [12]	0.81 [12]	296.74	231.66	231.66	0.00
Phenytoin	7.94 [12]	n	3.05 [12]	1.86 [12]	363.52	233.14	233.14	0.00
Procaine	9.03 [12]	c	1.62 [12]	1.02 [12]	414.05	322.04	335.79	−13.74
Propranolol	9.53 [12]	c	2.69 [12]	2.44 [12]	492.22	354.32	364.05	−9.73
Rilmenidine	9.22 [12]	c	2.11 [12]	1.03 [12]	367.48	304.18	315.81	−11.63
Tetracaine	8.5 [13]	c	2.26 [2]	1.75 [13]	508.62	399.43	438.77	−39.34
Warfarin	5.0 [14]	a	1.4 [15]	2.7 [15]	432.19	301.10	271.19	29.91

[a]) Dominant (>90%) species at physiological pH: n = neutral or ampholyte, a = anion, c = cation. [b]) Logarithm of the distribution coefficient at pH = 7.0 in the phosphatidylcholine liposomes/water system. Data were obtained either by dialysis or by potentiometry according to the cited reference. [c]) Capacity factor determined on IAM.PC.DD2 HPLC column at pH = 7.0. [d]) The contribution of the water-accessible surface area of all hydrophobic atoms calculated on the electrical state of the molecule dominant at physiological pH. [e]) The water-accessible surface area of all atoms with positive partial charge calculated on the neutral species of the molecule. [f]) The water-accessible surface area of all atoms with positive partial charge calculated on the electrical state of the molecule dominant at physiological pH. [g]) [e]) minus [f]).

where v is a constant, V the molar volume, and where Λ accounts for the polarity of the molecule.

When expressed by partition coefficients measured in traditional isotropic systems, the polar interactions encoded in lipophilicity are H-bonds plus *Keesom* (orientation) and *Debye* (induction) forces, whereas the nonpolar interactions are the *London* (dispersion) forces and hydrophobic interactions [16]. Interestingly, *Eqn. 1* fails to encode some important recognition forces, most-notably ionic bonds, which are of particular importance when modeling the interaction of ionized compounds with anisotropic media. It has, thus, been proposed that *Eqn. 1* be modified to *Eqn. 2* [1], in which I accounts for ionic interactions:

$$\log P = vV - \Lambda + I \quad (2)$$

Eqn. 2 can be applied to all partitioning systems. The value of I is different from zero as long as two conditions are verified: *a*) the solute is ionized, and *b*) the more-lipophilic phase bears well-localized charges (see below).

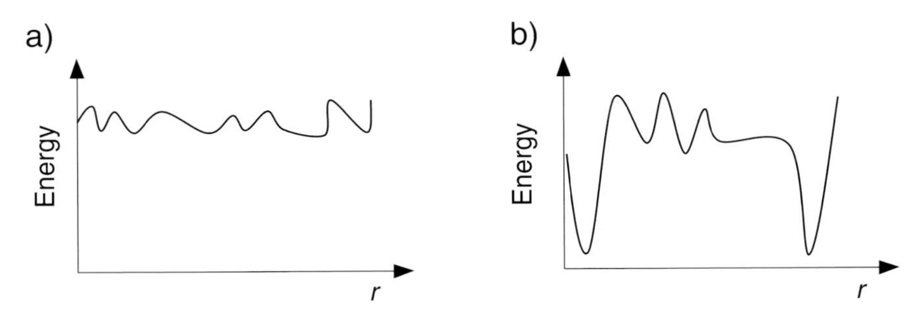

Fig. 3. *Schematic representations of the free energy landscape. a*) Partitioning mechanisms. *b*) Binding mechanisms.

3.2. *Binding and Partitioning*

Energetic criteria allow to distinguish between partitioning and binding mechanisms: in the case of partitioning, the free energy landscape of the complex shows broad, overlapping, poorly defined, and shallow minima (*Fig. 3,a*). Conversely, the free energy landscape of the complex in binding systems is such that it contains one or a very few narrow, relatively deep, and clearly defined energy minima (*Fig. 3,b*). In contrast, binding and partitioning cannot, thus, be distinguished by the nature of the intermolecular interactions involved [17]: the same intermolecular forces that govern partitioning–lipophilicity systems also govern binding–affinity systems. Given this clarification, the use of *ad hoc* binding systems to investigate ionic interactions also operating in partitioning systems appears reasonable.

4. Molecular Descriptors to Investigate the Intermolecular Forces Governing Lipophilicity

4.1. *Solvatochromic Descriptors*

Linear solvation–free energy relationships (LFERs), the most-powerful tool available to unravel the intermolecular forces governing lipophilicity [18] [19], are not (yet) able to deal with ions. The factorization of $\log P$ using LFERs is based on the well-known solvatochromic parameters (α, β, π^*, and V_w), which were highly innovative at the period of their discovery, but are limited to neutral compounds, as also confirmed by their experimental determination in which the apolar CCl_4 solvent is used despite its inability to solubilize most compounds in their ionized form.

Nevertheless, LFERs are often used in the literature to rationalize *in vitro* results (*e.g.*, permeability) obtained for data sets containing compounds almost completely ionized at the experimental pH. This is clearly illogical, as illustrated by ibuprofen. It is clear that, at pH 7.4, ibuprofen (pK_a *ca.* 4.5) has completely lost the H-bonding donor properties of its neutral COO group, and, thus, its α value cannot be assumed to be 0.59 (calculated by Absolv [20]) in any LFER study applied to experiments performed at physiological pH.

Thus, it has become crucial to modify the solvatochromic parameters to take electrical charges into account. To a first approximation, the α and β properties of ions can be calculated by adding or subtracting the contribution of acidic H-atoms and doublets from corresponding data of the unionized species. However, this appears to be an oversimplification, because electronic rearrangement (for example charge delocalization) must be considered. In other words, correcting solvatochromic descriptors for the contribution of ionized species is not trivial. It is, therefore, advisable to look for other parameters better suited to handle these problems. In particular, the required descriptors should be calculated for each electrical state of the molecule and be able to take three-dimensional effects into account.

4.2. *ASA Descriptors*

Descriptors that combine molecular surface area and partial atomic charge are of great interest in this connection [21]. These descriptors depend not only on the partial charges of the molecules but also on their conformation. According to the positive or negative sign of the charge, we distinguish between ASA+ and ASA− descriptors. The former indicates the water-accessible surface area of all atoms with partial positive charge, whereas ASA− is the water-accessible surface area of all atoms with partial negative charge. Because of their definition, standard molecular-modeling tools can easily help visualize the ASA descriptors.

For a given ionizable compound with a basic (or acidic) pK_a, the difference between ASA+ (or ASA−) of the neutral species and ASA+ (or ASA−) of the ionized species is called *diff+* (or *diff−*) and yields information on the local and global changes that the positive (or negative) electrical charge induces on the neutral structure. These changes are a function of the topographical (how molecular charges are exposed on the molecular surface) and chemical features of the charge itself, and, because of their definition, they are also sensitive to conformational effects. The higher the absolute value of *diff+* (or *diff–*) is, the stronger are the changes induced on the molecular properties of the neutral species by protonation (or deprotonation).

Let us examine the influence of protonation on two well-known calcium antagonists, lercanidipine (*Fig. 4*) and amlodipine (*Fig. 5*), whose pharmacological behavior is discussed below. *Figs. 4,a* and *4,b* show the minimum energy conformer of neutral and cationic lercanidipine, respectively. For both structures, ASA is grey, and the zones in which ASA is positive are blue. No huge difference in blue content is observable when comparing *Figs. 4,a* and *4,b*. The basic N-atom is in fact sterically hindered and, thus, poorly expressed on the ASA. The reverse is true for amlodipine for which the protonated amino group is fully expressed on the ASA and, thus, considerably enhances the blue content in *Fig. 5,b* (cationic species) compared to *Fig. 5,a* (neutral species). The *diff+* values express numerically the different influence of protonation on the two drugs: – 16.94 for lercanidipine and – 29.63 for amlodipine.

The ASA descriptors are very flexible, since formal charges can be calculated at any level of theory, the radius of the probe sphere in ASA calculations can be varied to the operator's convenience, and any conformational analysis tool can be used to generate conformers. In this chapter, we have used PEOE (partial equalization of orbital electronegativities) charges, 1.4 Å as the radius probe, and quenched molecular dynamics (QMD). Work is in progress to refine ASA descriptors by testing different combinations of computational tools.

5. Lipophilicity of Ionized Solutes in Isotropic Systems

During the *logP2000* Symposium [22], many scientists pointed out that the logarithm of the partition coefficient *P* of an ionized form in a given solvent system $(\log P^{I}_{\mathrm{solv}})$ is a good lipophilicity descriptor for ions (see below).

5.1. *How to Obtain log* P^{I}

5.1.1. *Experimental Methods*

Because of the accelerating development of new experimental techniques, few methods are available to measure the lipophilicity of ionized species in isotropic systems. Potentiometry [23] is considered the standard

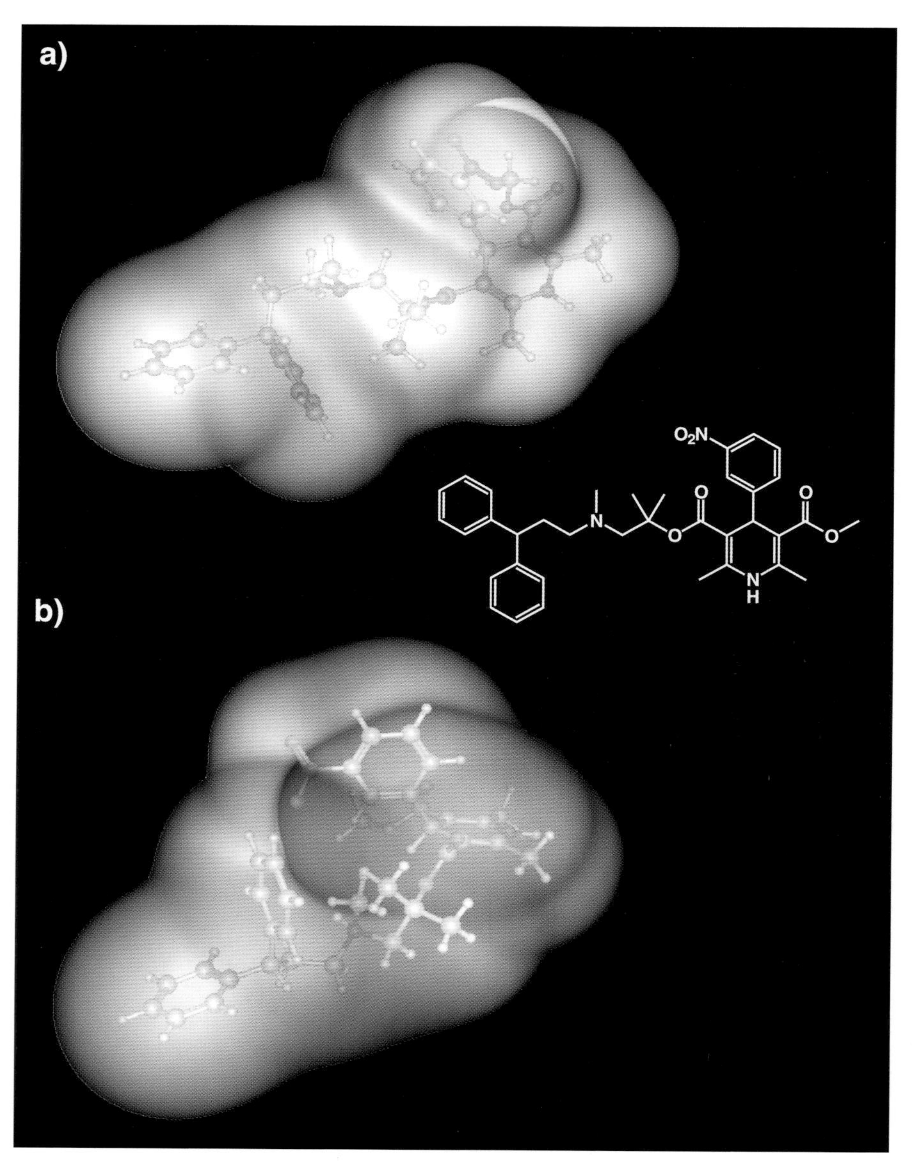

Fig. 4. *Water-accessible surface area* (ASA) *for lercanidipine.* ASA is colored by partial charges (calculated by the PEOE method): blue indicates positively charged regions, grey areas are uncharged regions. Negatively charged zones are not shown. *a*) The most-stable conformer (obtained by QMD) for neutral lercanidipine. *b*) The most-stable conformer (obtained by QMD) for cationic lercanidipine; positively charged regions are poorly exposed.

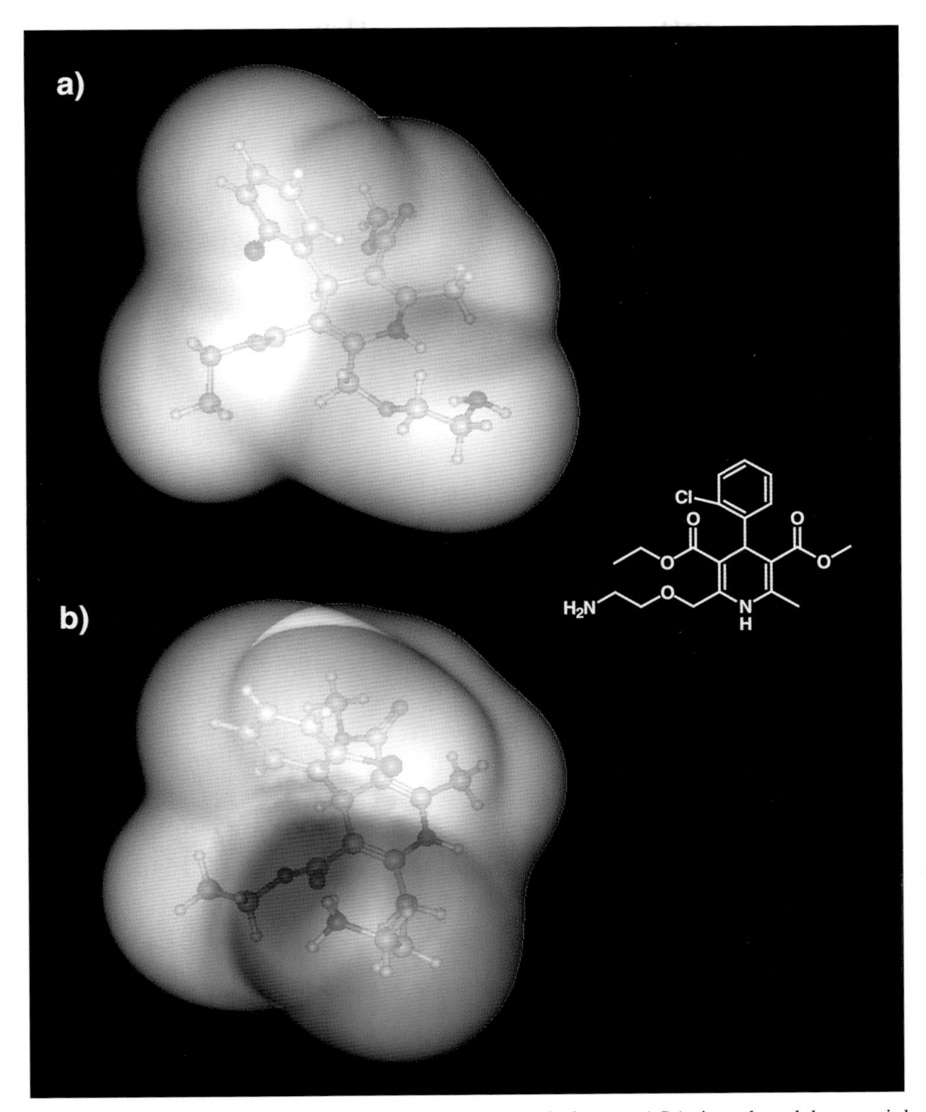

Fig. 5. *Water-accessible surface area* (ASA) *for amlodipine.* ASA is colored by partial charges (calculated by the PEOE method): blue indicates positively charged regions, grey areas are uncharged regions. Negatively charged zones are not shown. *a*) The most-stable conformer (obtained by QMD) for neutral amlodipine. *b*) The most-stable conformer (obtained by QMD) for cationic amlodipine; positively charged regions are well exposed.

tool, because a variety of solvents can be used and a broad range of log P values measured. Recently, *Girault et al.* [24] showed that cyclic voltammetry (CV) is the only method, which can yield the intrinsic log P^I (called

$\log P^{0,i}$), since $\log P^{I}$ values obtained by the shake-flask or pH-metric method are strongly influenced by experimental conditions and particularly by phase volumes and the nature of counter-ions. The CV approach is also valid for zwitterions [25]. In addition, a new electrochemical method based on a liquid layer immobilized between two aqueous compartments has also been reported [26]. This procedure is based on 96-well microfilter plates and allows high-throughput applications. Even if the electrochemical approach is able to guarantee a very high number of reliable data, it must be stressed that these methods cannot be routinely applied in medicinal chemistry because of their serious limitations, particularly the very few usable organic solvents (1,2-dichloroethane (DCE) and nitrophenyl octyl ether (NPOE)) and the resulting inaccessibility of most biphasic systems such as the standard octanol/water.

5.1.2. *Computational Tools*

No theoretical method based on parameterization is available to calculate $\log P^{I}$. Rough predictions of $\log P^{I}$ can be obtained by subtracting a given value from $\log P^{N}$ [3]. This is possible because in isotropic systems, charges are carried by salt buffers and solutes [1] and, thus, have no defined location (*Fig. 6,a*). As a result, no 'specific' electrostatic interaction occurs, and ions partition less than the corresponding neutral species, the decrement depending on the partitioning system, the compound's characteristics, the nature of counter-ions, and the ionic strength. In octanol/water, the average value is *ca.* 3.5 [27]. For the DCE/

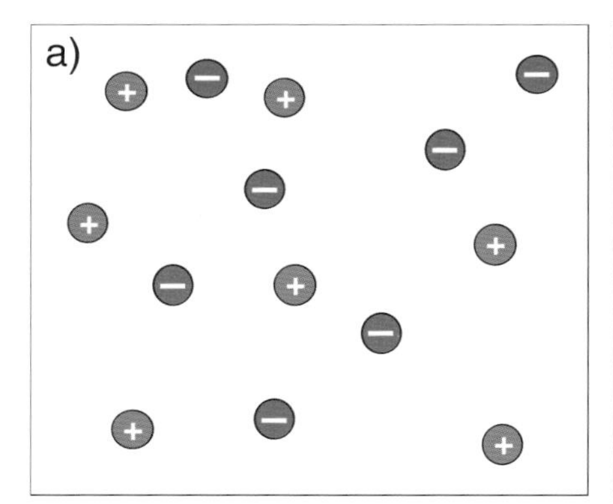

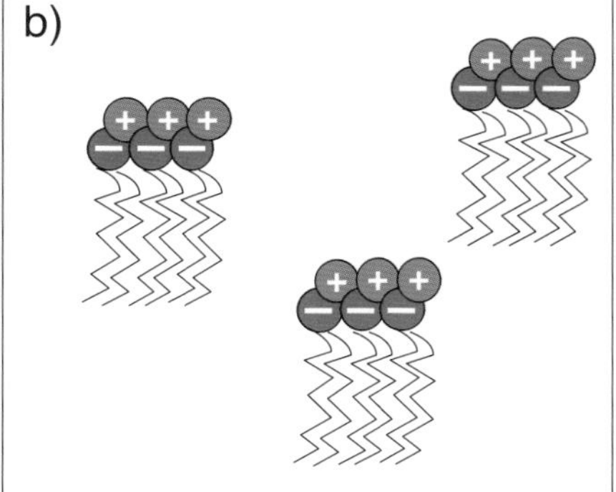

Fig. 6. *The location of charges governs the existence or absence of ionic interactions* (see text for details). *a*) Undefined location: charges are carried by salt buffers. *b*) Well-defined location: charges are located on the lipidic phase (*e.g.*, the polar head of purified phospholipids).

water system, the difference in log P between neutral and ionized species ranges from 3 to 6 [28].

5.2. *Log* P[I] *Information Content*

It has been argued that ion partitioning in isotropic systems has little relevance with biological systems [2]. Even if this assumption is true, the analysis of the lipophilicity of ionized solutes in isotropic systems with different polar properties (*e.g.*, octanol/water and alkane/water) is a validated tool [29][30] to obtain structural information about intramolecular interactions [17], because the tendency of solutes to form internal H-bonds in octanol and in water is usually comparable, while nonpolar solvents (*e.g.*, alkane) strongly favor internal H-bonds.

β-Blockers are a good series of drugs to show the influence of a positive charge on a molecular structure. The multiple internal H-bonding pattern of cationic acebutolol shown in *Fig. 7* has been demonstrated by combining experimental lipophilicity with molecular-modeling tools [29]. This particular feature does not affect most compounds belonging to the series.

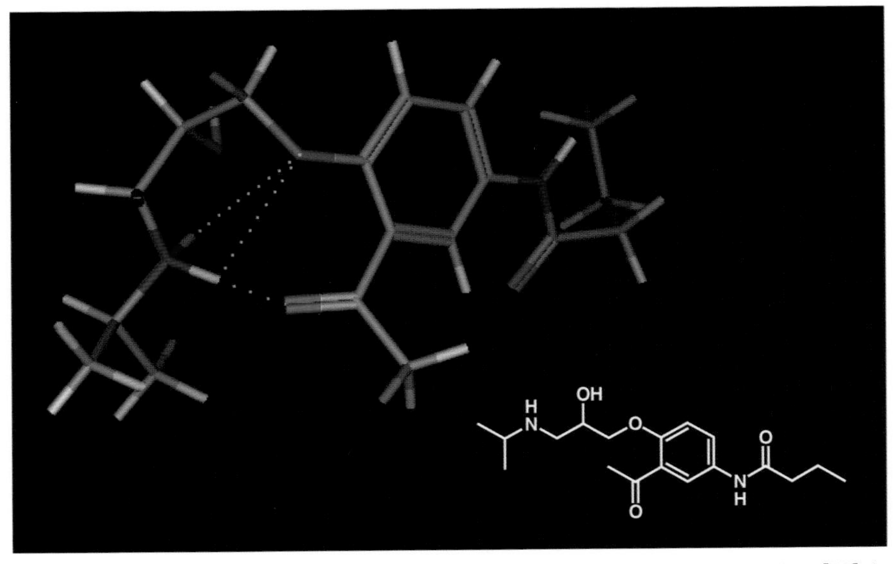

Fig. 7. *Multiple internal H-bonding pattern of cationic acebutolol as obtained in* [29] *by combining experimental lipophilicity with QMD simulations*

6. Lipophilicity of Ionized Solutes in Anisotropic Systems

6.1. *How to Obtain log* P^I*?*

A number of different experimental approaches are able to yield log P^I values, but their review is beyond the scope of this chapter and has already been addressed in [1]. Conversely, no computational tool exists, and, thus, log P^I cannot yet be predicted *in silico*.

6.2. *Log* P^I *Information Content*

Anisotropic systems resemble biological systems more than do isotropic ones. As a general principle, *Eqn. 2* holds also for anisotropic systems where the presence of electrostatic interactions (the I term in *Eqn. 2*) is due to the well-defined location (*Fig. 6,b*) of ionic charges [1][12][31–33], which are, thus, able to form ionic bonds with ionized solutes. Electrostatic forces, however, vary in features and strength with the system investigated. As a consequence of their variability, ionic interactions may represent the predominant interaction force between the solute and the solvent system, whereas, in other cases, hydrophobicity and/or polar forces govern the interaction [1].

A number of examples will be given below to illustrate the role of electrostatic interactions and the relevance of ASA descriptors in various anisotropic biphasic systems.

7. The Contribution of Ionic Interactions in Nonchromatographic Systems

7.1. *Partitioning Systems*

Membranes and artificial membranes such as liposomes are often used as the lipidic phase in anisotropic systems. An electrostatic interaction with phospholipids has been demonstrated for cations [34]. Lercanidipine (*Fig. 4*) and amlodipine (*Fig. 5*) are long-acting calcium channel blockers of the 1,4-dihydropyridine type, a feature that results from their high and comparable affinity for membranes (the log K_p(mem) of lercanidipine is 5.5, that of amlodipine is 4.3 [35]. However, the lipophilicity of the two drugs in octanol/water is very different, log $D^{7.4}_{oct}$ being *ca.* 6 for lercanidipine [36] and 1.6 for amlodipine [37].

The extent of ionization and the balance between ionic and hydrophobic interactions may be the crucial features that differentiate the membrane-binding mode of amlodipine and lercanidipine. Lercanidipine is ionized to *ca.* 50% at pH 7.0 (pK_a *ca.* 7.0 [36]), whereas amlodipine is completely cationic (pK_a *ca.* 9 [37]). A strong electrostatic interaction with phospholipids has been demonstrated for amlodipine [38], whereas hydrophobic interactions have been hypothesized to dominate for lercanidipine [35]. ASA Descriptors can confirm this hypothesis. Indeed, the basic N-atom of cationic lercanidipine is hidden by hydrophobic moieties and is, therefore, not sufficiently exposed to generate electrostatic interactions with the negative charge of the phospholipids (in *Fig. 4,b*, the blue region due to the protonated amine is small). The opposite is true for amlodipine, where the primary amine is like a freely movable flag (in *Fig. 5,b*, the blue region due to the protonated amine is large).

7.2. *Binding Systems*

Cyclodextrin/water is a good example of a biphasic system (a macromolecular dispersion) governed by binding mechanisms but recalling partitioning systems. The less-polar phase, in fact, provides hydrophobic regions well-separated from hydrophilic zones, which can also bear electrical charges. The solvent-accessible surface area (ASA) of the crystal structure of a β-cyclodextrin taken from the *Cambridge Structural Database* (CSD) was calculated and is reported in *Fig. 8* with polar regions in yellow and hydrophobic regions in blue. The snapshot shows that the blue and yellow regions are well separated.

The role of charge in substrate–cyclodextrin complexation was investigated by comparing the binding of neutral and charged substrates to a neutral cyclodextrin (CD) and a negatively charged one [39]. For the negatively charged CD, the complexation constants decreased upon substrate ionization, but the extent of the decrease depended on the positive or negative nature of the charge. In particular, cationic papaverine and prazosin more or less maintained their complexation strength, while anionic naproxen and warfarin showed decreased complexation. These experimental findings may be ascribed to additional ionic attractive (or repulsive) forces between the charged sulfonates and the positive (or negative) guest molecules. In particular, the bases papaverine and prazosin appear to be similarly attracted, while the acidic naproxen is much more strongly repelled than warfarin. The results for the two acids are shown in *Fig. 9*, indicating that the accessibility of the charge on the ASA is the critical feature governing the interaction. In fact, the negative

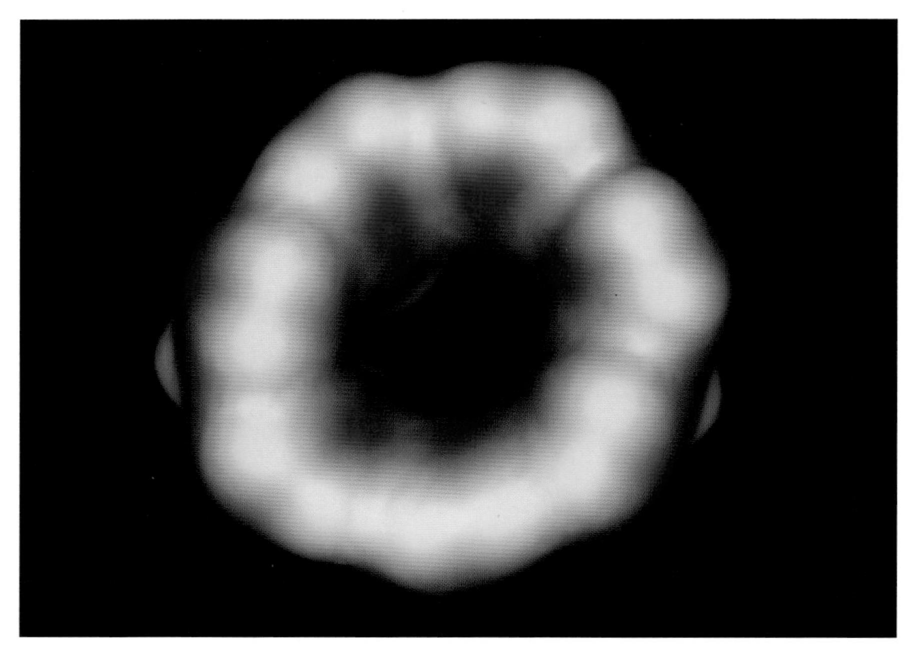

Fig. 8. *Water-accessible surface area* (ASA) *calculated for the X-ray structure of a β-cyclodextrin.* Hydrophobic regions are blue and polar regions are yellow.

charge of anionic naproxen (*Fig. 9,a*) is more exposed than the negative charge of warfarin (*Fig. 9,b*).

Albumin/water is an anisotropic system for which the interactions between drug and receptor occur in regions that are more clearly defined and more easily saturated than the interaction regions in liposomes. It has been demonstrated [9] that the linear correlation between the percentage of drug bound (%B(DAB)) and $\log D^{7.4}_{\text{oct}}$ is significant for neutral compounds and cations, but not for anions, where albumin binding is probably governed by electrostatic interactions (*Table 1*). In particular, all acidic drugs investigated (furosemide, indomethacin, naproxen, and tenoxicam) have high albumin-binding values despite their large differences in $\log D^{7.4}$. The features of the negative charge on the anions (*Table 1*) and the reference compound warfarin (*Table 2*) were investigated in terms of their ASA− and *diff*− values. No particular trend was found, making it reasonable to assume that the mere presence of a negative charge, rather than its features, allows acids to bind to albumin.

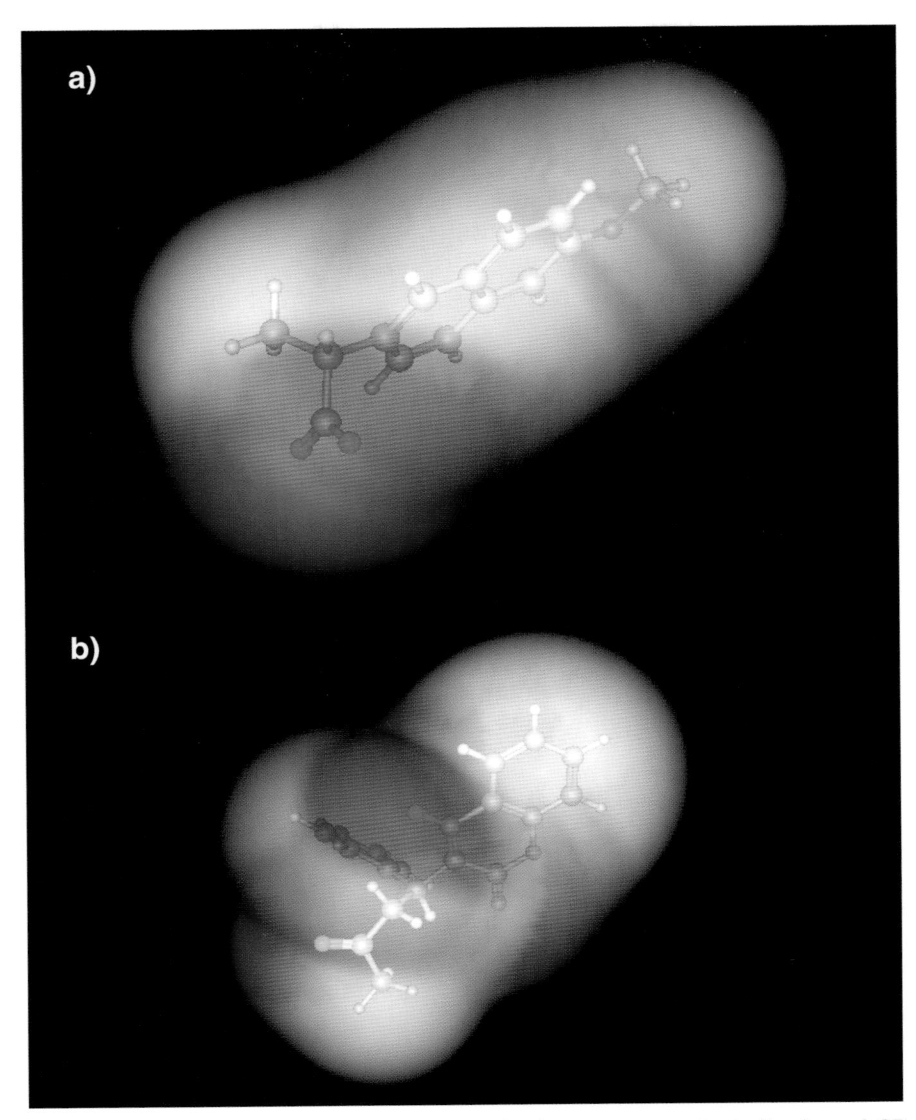

Fig. 9. *The different binding strengths of charged substrates to an anionically charged CD depend on how molecular charges are exposed on the molecular surface.* The major exposure of the negative charge of anionic naproxen (*a*; average conformation) explains its lower complexation ability compared with the anionic warfarin (*b*; average conformation). ASA is colored by partial charges: red indicates negatively charged regions, grey areas are uncharged regions. Positively charged zones are not shown.

8. Lipophilicity of Ionized Solutes in Anisotropic Chromatographic Systems

Because of the ready availability of data and many known linear relationships between log P and retention parameters, chromatographic methods are often used to replace traditional lipophilicity measurements [40]. This is reasonable when neutral species are considered, but it is a source of severe inaccuracy when examining the lipophilicity of ions, because the balance between electrostatic and hydrophobic interactions is different in chromatographic and nonchromatographic systems, as discussed below for two chromatographic methods.

Many studies have used immobilized artificial membranes (IAM) [13][41–44]. *Table 2* brings together the IAM and potentiometric data of *Data Set 2* obtained at pH 7.0. The two sets of values are poorly correlated ($r^2=0.27$, *Fig. 10,a*), but IAM data are correlated to the water-accessible surface area of all hydrophobic atoms (ASA_H) ($r^2=0.70$, graph not shown). This is not the case for log $D_{\text{lip}}^{7.0}$ ($r^2=0.10$, graph not shown).

High-performance affinity chromatography equipped with an immobilized albumin column has recently been proposed to determine albumin binding [10]. The correlation between chromatographic retention factors (log K_{HSA}) and albumin binding determined by ultracentrifugation (%B(DAB)) is again poor (*Fig. 10,b*, $r^2=0.10$; data and relative references in *Table 1*).

9. Conclusions

The lipophilicity of neutral species has been extensively studied. Conversely, the lipophilicity of ionized species is a more-recent field of interest, which, besides using the same equipment as neutral species, also requires additional tools to check modifications induced in a molecular structure by the introduction of a charge (*i.e.*, the ionization process *per se*), and to handle the electrostatic interactions, which control many drug/biphasic systems.

In this connection, preliminary but encouraging results have been obtained using ASA descriptors (molecular parameters that combine information related to both surface area and partial atomic charges). These results appear to indicate that, in most (but not all) biphasic systems, ions elicit electrostatic interactions with the nonaqueous phase as long as their charge is sufficiently exposed on their accessible surface. Work is in

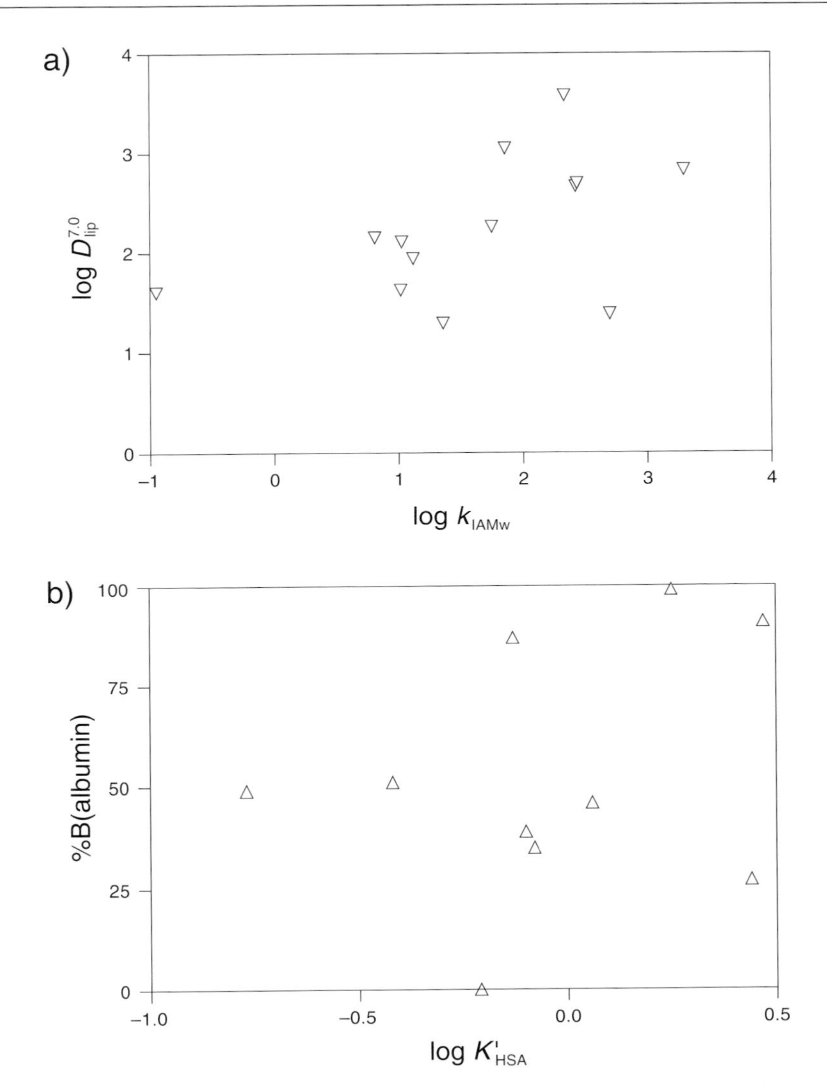

Fig. 10. *Experimental data obtained by nonchromatographic and chromatographic methods are not interchangeable when ionized solutes are included in the data sets. a*) Interaction with liposomes measured by potentiometry (log $D^{7.0}_{lip}$, direct method) and log $k^{7.0}_{IAMw}$ for *Data Set 2*. *b*) Albumin binding measured by ultracentrifugation (%B(DAB), direct method) and log K'_{HSA} for *Data Set 1*.

progress to confirm these preliminary results and to extend the study to a larger number of biphasic systems.

G. C. and *G. E.* are indebted to *Bernard Testa* for useful suggestions.

REFERENCES

[1] G. Plempler van Balen, C. a Marca Martinet, G. Caron, G. Bouchard, M. Reist, P. A. Carrupt, R. Fruttero, A. Gasco, B. Testa, *Med. Res. Rev.* **2003**, *24*, 299.
[2] A. Avdeef, 'Absorption and Drug Development. Solubility, Permeability, and Charge State', Wiley, New York, 2003.
[3] G. Caron, G. Ermondi, *Mini-Rev. Med. Chem.* **2003**, *3*, 821.
[4] A. Pagliara, P. A. Carrupt, G. Caron, P. Gaillard, B. Testa, *Chem. Rev.* **1997**, *97*, 3385.
[5] A. Pagliara, E. Khamis, A. Trinh, P. A. Carrupt, R. S. Tsai, B. Testa, *J. Liq. Chromatogr.* **1995**, *18*, 1721.
[6] C. A. S. Bergström, M. Strafford, L. Lazorova, A. Avdeef, K. Luthman, P. Artursson, *J. Med. Chem.* **2003**, *46*, 558.
[7] SIMCA-P, Vers. 10.0.2.0, *Umetrics*, Umeå, 2003.
[8] MOE, Vers. 2002.03, *Chemical Computing Group Inc.*, Montreal, Canada, 2002.
[9] G. Ermondi, M. Lorenti, G. Caron, *J. Med. Chem.* **2004**, *47*, 3949.
[10] G. Colmenarejo, A. Alvarez-Pedraglio, J.-L. Lavandera, *J. Med. Chem.* **2001**, *44*, 4370.
[11] F. Barbato, M. I. La Rotonda, F. Quaglia, *J. Pharm. Sci.* **1997**, *86*, 225.
[12] A. Taillardat-Bertschinger, C. a Marca Martinet, P. A. Carrupt, M. Reist, G. Caron, R. Fruttero, B. Testa, *Pharm. Res.* **2001**, *19*, 729.
[13] M. Amato, F. Barbato, P. Morrica, F. Quaglia, M. I. La Rotonda, *Helv. Chim. Acta* **2000**, *83*, 2836.
[14] C. Ottiger, H. Wunderli-Allenspach, *Pharm. Res.* **1999**, *16*, 643.
[15] A. Taillardat-Bertschinger, A. Galland, P. A. Carrupt, B. Testa, *J. Chromatogr., A* **2002**, *953*, 39.
[16] B. Testa, G. Caron, P. Crivori, S. Rey, M. Reist, P. A. Carrupt, *Chimia* **2000**, *54*, 672.
[17] B. Testa, P. A. Carrupt, P. Gaillard, R. S. Tsai, in 'Lipophilicity in Drug Action and Toxicology', Eds. V. Pliska, B. Testa, H. van de Waterbeemd, VCH Publishers, Weinheim, 1996, p. 49–71.
[18] M. J. Kamlet, R. W. Taft, *J. Am. Chem. Soc.* **1976**, *98*, 377.
[19] R. W. Taft, M. J. Kamlet, *J. Am. Chem. Soc.* **1976**, *98*, 2886.
[20] Absolv, Vers. 1.4.05.63, *Sirius Analytical Instruments Ltd.*, 2000.
[21] D. T. Stanton, P. C. Jurs, *Anal. Chem.* **1990**, *62*, 2323.
[22] 'Pharmacokinetic Optimization in Drug Research. Biological, Physicochemical, and Computational Strategies', Eds. B. Testa, H. van de Waterbeemd, G. Folkers, R. H. Guy, Verlag Helvetica Chimica Acta, Zürich, 2001.
[23] A. Avdeef, in 'Lipophilicity in Drug Action and Toxicology', Eds. V. Pliska, B. Testa, H. van de Waterbeemd, VCH Publishers, Weinheim, 1996, p. 109–139.
[24] F. Reymond, V. Gobry, G. Bouchard, H. H. Girault, in 'Pharmacokinetic Optimization in Drug Research. Biological, Physicochemical, and Computational Strategies', Eds. B. Testa, H. van de Waterbeemd, G. Folkers, R. H. Guy, Verlag Helvetica Chimica Acta, Zürich, 2001, p. 327–349.
[25] G. Bouchard, A. Pagliara, P. A. Carrupt, B. Testa, V. Gobry, H. H. Girault, *Pharm. Res.* **2002**, *19*, 1150.
[26] S. M. Ulmeanu, H. Jensen, G. Bouchard, P. A. Carrupt, H. H. Girault, *Pharm. Res.* **2003**, *20*, 1317.
[27] H. Kubinyi, 'QSAR: Hansch Analysis and Related Approaches', VCH Publishers, Weinheim, 1993.
[28] G. Bouchard, P. A. Carrupt, B. Testa, V. Gobry, H. H. Girault, *Chem.–Eur. J.* **2002**, *8*, 3478.
[29] G. Caron, G. Steyaert, A. Pagliara, F. Reymond, P. Crivori, P. Gaillard, P. A. Carrupt, A. Avdeef, J. E. Comer, K. J. Box, H. H. Girault, B. Testa, *Helv. Chim. Acta* **1999**, *82*, 1211.
[30] V. Chopineaux-Courtois, F. Reymond, G. Bouchard, P. A. Carrupt, B. Testa, H. H. Girault, *J. Am. Chem. Soc.* **1999**, *121*, 1743.
[31] S. D. Krämer, C. Jakits-Deiser, H. Wunderli-Allenspach, *Pharm. Res.* **1997**, *14*, 827.
[32] G. V. Betageri, J. A. Rogers, *Int. J. Pharm.* **1988**, *46*, 95.

[33] G. M. Pauletti, H. Wunderli-Allenspach, *Eur. J. Pharm. Sci.* **1994**, *1*, 273.
[34] S. D. Krämer, A. Braun, C. Jakits-Deiser, H. Wunderli-Allenspach, *Pharm. Res.* **1998**, *15*, 739.
[35] L. Herbette, M. Vecchiarelli, A. Leonardi, *J. Cardiovasc. Pharmacol.* **1997**, *29*, S19.
[36] A. Leonardi, E. Poggesi, C. Taddei, L. Guarnieri, P. Angelico, M. R. Accomazzo, S. Nicosia, R. Testa, *J. Cardiovasc. Pharmacol.* **1997**, *29*, S10.
[37] U. Franke, A. Munk, M. Wiese, *J. Pharm. Sci.* **1999**, *88*, 89.
[38] L. Herbette, D. G. Rhodes, R. Preston Mason, *Drug Des. Delivery* **1991**, *7*, 75.
[39] V. Zia, R. A. Rajewski, V. J. Stella, *Pharm. Res.* **2001**, *18*, 667.
[40] A. Nasal, D. Siluk, R. Kaliszan, *Cur. Med. Chem.* **2003**, *10*, 381.
[41] F. Barbato, M. I. La Rotonda, F. Quaglia, *Eur. J. Med. Chem.* **1996**, *31*, 311.
[42] L. H. Alifrangis, I. T. Christensen, A. Berglund, M. Sandberg, L. Hovgaard, S. Frokjaer, *J. Med. Chem.* **2000**, *43*, 103.
[43] R. Kaliszan, A. Nasal, A. Bucinski, *Eur. J. Med. Chem.* **1994**, *29*, 163.
[44] A. Taillardat-Bertschinger, P. A. Carrupt, F. Barbato, B. Testa, *J. Med. Chem.* **2003**, *46*, 655.

Physicochemical and Biological Profiling in Drug Research. ElogD$_{7.4}$ 20,000 Compounds Later: Refinements, Observations, and Applications

by **Franco Lombardo***, **Marina Y. Shalaeva**, **Brian D. Bissett**, and **Natalya Chistokhodova**

Molecular Properties Group, *Pfizer Global Research and Development*, Groton Laboratories, Groton, CT 06340, U.S.A.
(e-mail: franco.lombardo@pfizer.com)

Abbreviations
ADME: Absorption, distribution, metabolism, and excretion; DDE: dynamic data exchange; HBA: H-bond acceptor; HBD: H-bond donor; HPLC: high-performance liquid chromatography; HT: high throughput; t_R: retention time; VBA: visual basic for applications; V_D: volume of distribution.

1. Introduction

Several methods exist for the determination of the partition (or distribution) coefficient for drugs or other compounds, whether in a low-, medium-, or fairly high-throughput fashion. Early in our efforts, we chose to look at RP-HPLC methods, since we thought they offered the best chance of being amenable to automation and, at the same time, no additional instrumentation would have been required, over and above HPLC instruments already present in most laboratories.

RP-HPLC methods however, like any other we are aware of, are not devoid of complications, which range from the need for a judicious choice of mobile phase, to column lifetime issues, to data acquisition and analysis protocol. Another potentially significant problem is the frequency and impact of the observed nonlinearity [1], which manifests itself as a curvature in a plot of capacity factor (log k') *vs.* the amount of organic solvent present (typically MeOH) in the mobile phase. The linear extrapolation of the capacity factor to a pure aqueous mobile phase thus becomes an issue.

Our long-term plan was to start the development of an internal database, with potentially tens of thousands of compounds, and to apply the data generated to the development of predictive ADME models, whether experimental or computational. Thus, the potential for a good degree of accuracy was another factor we considered.

Our initial efforts led to the development of the $ElogD_{7.4}$ method, published in 2001 [2], which is used for high-throughput lipophilicity measurements and, more specifically, for the determination of the octanol/water distribution coefficient at pH 7.4 for neutral and basic drugs. We discuss our experience, issues, and solutions encountered in the determination of the distribution coefficient for over 20,000 compounds, and the application of the data generated by means of this method to the prediction of volume of distribution in human [3][4] as well as its potential use to extract H-bonding information.

2. Significance of log *D* and log *P*

It may seem redundant, in the context of a chapter written for the *logP2004 Lipophilicity Symposium*, to discuss the significance of log *P* and log *D* determinations, especially since the present symposium was preceded by two very successful international meetings on the same topic, in 1995 [5] and 2000 [6]. An enormous body of information is, of course, available on this subject, and the introductory section of essentially every publication on this topic will discuss the significance of log *P* or log *D*. However, for the sake of completeness, we will cite some important aspects of ADME that rely on lipophilicity, and we will discuss the application of our own data in a later section, while stating that our discussion will be limited to the more 'classical' octanol/water (buffer) system, although arguably lipophilicity and octanol/water partition (or distribution) coefficients are not necessarily synonyms in the context of ADME phenomena.

A review of the available methods for the experimental determination of lipophilicity is beyond the scope of this chapter, and we refer the reader to the work of *Kerns et al.* [7], especially for what concerns the high-throughput profiling of drug compounds, or to the more-recent review article by *Caron* and *Ermondi* [8], as well as the ample discussion in the proceedings of the previous symposia [5][6]. We note, however, that many authors in the field seem to divide these methods in 'high(er) accuracy–low(er) speed', such as the potentiometric and the classical shake-flask methods, and 'low(er) accuracy–high(er) speed', such as HPLC or similar chromatographic methods. We submit that this 'partition' of the partition coefficient determination methods is not necessarily as definite as it may be thought to be, and that environment- and operator-specific aspects, such as training and understanding of the experimental caveats as well as the actual experimental conditions including compound purity, will dictate the final outcome to a very large extent.

The determination of the apparently 'simple' partition (neutral species) or distribution (charged and neutral species) coefficients is not trivial at all, and the literature is replete with examples of very high data variability for a given compound. These examples range from the use of radioactively labeled material without a radiochemical (and general) purity check, to the determination of very high ($\geq 10^5$) or very low ($\leq 10^{-2}$) partition or distribution coefficients by shake-flask techniques without due attention paid to the solvent ratios and mutual presaturation of the phases used. The choice of the method will, of course, depend heavily upon the stage of use, but some commonly encountered generalizations may be questionable.

A condensed yet informative table, detailing the involvement and the 'importance' of lipophilicity, was reported by *Smith et al.* [9], who used 'check-marks' to show the relative importance of lipophilicity in absorption, clearance, volume of distribution, and other aspects. There is, of course, a higher level of complexity in the interpretation of lipophilicity data *vs. in vivo*, or even *in vitro* membrane model data. Therefore, the partition (or distribution) of a solute between two homogeneous and immiscible phases cannot be expected to model exactly the very complex and often poorly understood physiological phenomena underlying ADME. That level of information is nevertheless a very useful starting point, showing how ubiquitous lipophilicity is in the behavior of drugs *in vivo*.

3. Computational *vs.* Experimental Methods

It is well known that the computational approach, from structure only, is a highly desirable tool, and there is a plethora of computational packages available, commercially or free of charge through the internet, for the calculation of log *P*. However, we note that the same is not true when the calculation of log *D* (the logarithm of the distribution coefficient) is desired. It is, of course, possible to calculate the pK_a and log *P* by means of several different packages, but the user should be mindful of the caveats and errors involved in each calculation.

One of the most-recent review articles, written by *Caron* and *Ermondi* [8], discusses the problem of the variability in calculated values and its causes, and we will briefly touch upon them in this section. A general problem is that fragmental constants methods, which represent the vast majority of approaches, cannot deal well with long-distance conformational and H-bonding effects, while they can reproduce reasonably well inductive, steric, and resonance effects encoded in the generally large data sets used. Thus, and unbeknownst to the user, they may fail without warning with any structural class, and the user should be mindful of that, since, more often than not, he does not know the range of data and structures used in the training of the various models.

So, what possible 'defenses' exist? If the user is not attempting to score a virtual library, and (some of) the compounds are therefore available, it would be advisable to run determinations on a few analogs, even though the choice of 'sample' compounds, based on some similarity algorithm or chemical intuition, is often not a trivial matter. A rugged and trusted experimental method, familiar to the operator and coupled with the understanding of potential caveats (solubility may be a problem, for example, in potentiometric titrations) should lead to the generation of good data for a subset of a class of analogs or a library. The comparison of the experimental results with computational estimates may offer some help in accepting or rejecting the latter, although there will be no guarantees regarding the accuracy of the computed data when flexibility, or other significant structural variations such as the presence of tautomeric equilibria, will be introduced in otherwise 'close' analogs.

What if the compounds are not available? In those cases, the only recourse is the use of multiple programs and, perhaps, the average values with attention to an agreed upon standard deviation. However, and even for relatively similar compounds and keeping the caveats on 'similarity' expressed above in mind, it is not rarely seen that variations of 2 log units among packages occur in dealing with fairly complex structures as encountered in drug research. This outcome will leave the user wondering about the usefulness of such calculations. Unfortunately, it has been our experience that such cases are encountered fairly often, and it is difficult to rationalize these behaviors, even when the user is quite knowledgeable about the caveats of such calculations.

Perhaps, the only way to attempt to decrease these errors is the development of in-house log P/log D computational prediction programs. This approach would also offer training capabilities and the flexibility of a model not dependent on a vendor for training set expansion and further improvements. Needless to say, the data used in the training of internal model should be at the very least self-consistent and possess a fairly good level of accuracy. If that is the case, it is likely that many of the functional groups and scaffolds used in further synthetic efforts will be well represented in the training set used in the modeling efforts, and that will likely translate in an improved performance.

4. ElogD$_{7.4}$: A General Presentation

4.1. *ElogD$_{7.4}$: General Features and Performance*

We have developed a fairly rugged and automated method for the determination of log $D_{7.4}$, which we have termed ElogD, but which is not suitable for acidic compounds that would be significantly or completely ionized at pH 7.4, yielding an anion. Although this is an obvious limitation, the judicious choice of structural filters, and the fact that the vast majority of compounds in drug research are either neutral or basic ones, contribute to reduce the impact of this limitation on the usefulness of the method and on the automated compound submission protocol needed for its efficient application in an industrial setting. We have not developed quantitative rules based on the fraction ionized (which varies with the amount of methanol), and it is also possible that 'soft' anions, where the charge is fairly delocalized, might be suitable for this system. However, our criteria for the exclusion of an acidic compound is based on generally known pK_a values in water which translates into the exclusion of say, carboxylic acids, but not of hydroxamic acids or phenols, even though structural variations may significantly alter the expected pK_a.

We begin with a brief description of the general features of our method and its capabilities, together with a realistic assessment of the latter based on the large numbers of runs we have performed since its development. Throughout this chapter, we will use log D in place of log $D_{7.4}$.

The method relies on a series of isocratic runs, in the range of 15–70% MeOH, grouped in three ranges: a low range comprising 25, 20, and 15% MeOH, a medium range comprising 50, 45, and 40% MeOH, and a high range comprising 70, 65, and 60% MeOH. The flow rate varies from 0.5 to 2 ml min^{-1}, going from the low to the high range and using 1 ml min^{-1} for the medium range. The column, in all cases, is a *Supelcosil LC-ABZ* with a polar embedded functionality.

The first equation shown (*Eqn. 1*) was based on 90 compounds and correlates the capacity factor extrapolated to 0% MeOH or log k'_w from 3 isocratic determinations in each range with log D data obtained either by shake-flask measurements or potentiometric titrations. We customarily placed compounds in one of the ranges on the basis of computed log $D_{7.4}$ values and on the basis of our experience and statistical work performed on several hundreds of compounds. This work allowed us to develop 'threshold' values to help ensure that compounds have been run in a suitable range, *i.e.*, low lipophilicity compounds are run in the low range which comprises 25, 20, and 15% MeOH.

The correlation shown below yielded good statistics and a slope close to unity, which in itself is a good indication of the similarity in the balance of forces underlying log D and as it was demonstrated by the analysis of the solvation parameters in our previously reported ElogP method [10].

$$\log D_{\mathrm{oct}} = 1.12\ (\pm 0.023)\ \log k'_{\mathrm{w}} + 0.212\ (\pm 0.043) \qquad (1)$$

$$n = 90,\ r^2 = 0.964,\ s = 0.309,\ F = 2337,\ q^2 = 0.962$$

Since the ElogP and ElogD methods differed only by the presence of decylamine in the mobile phase, necessary to run basic compounds, and the $\log k'_{\mathrm{w}}$ values obtained for neutral compounds under both protocols were essentially identical, we concluded that the same balance of forces underlying log P values was underlying our ElogP and ElogD methods as well.

We have since made some modifications to the routine preparation of mobile phase as well as to the data analysis protocols, and we have expanded the data set used for the correlation to 163 nonproprietary compounds, which yielded the correlation shown in *Eqn. 2*, while *Fig. 1* shows a plot of the data.

$$\log D_{\mathrm{oct}} = 1.08\ (\pm 0.020)\ \log k'_{\mathrm{w}} + 0.200\ (\pm 0.040) \qquad (2)$$

$$n = 163,\ r^2 = 0.949,\ s = 0.369,\ F = 3000,\ q^2 = 0.948$$

It can be seen that the correlation is still excellent, with a slightly lower slope (closer to 1) and intercept, albeit the standard deviation is slightly higher, using a nearly doubled number of compounds. This is, in itself, a good indication of the ruggedness of the method. The variability introduced by the new compounds, a heterogeneous set of structures with data from various sources, even though composed of well-known (and likely 'well-behaved') drugs, seems fairly low. And, in our hands, the results are usually very close to potentiometric and/or shake-flask data generated for proprietary compounds, the vast majority of them generated by different laboratories. This is not to say, of course, that we do not observe, in some cases, significant disagreements among methods and laboratories, which are *always* reason for further understanding of the strengths and weaknesses of our method, and are also used to help maintain our database at the highest possible level of accuracy.

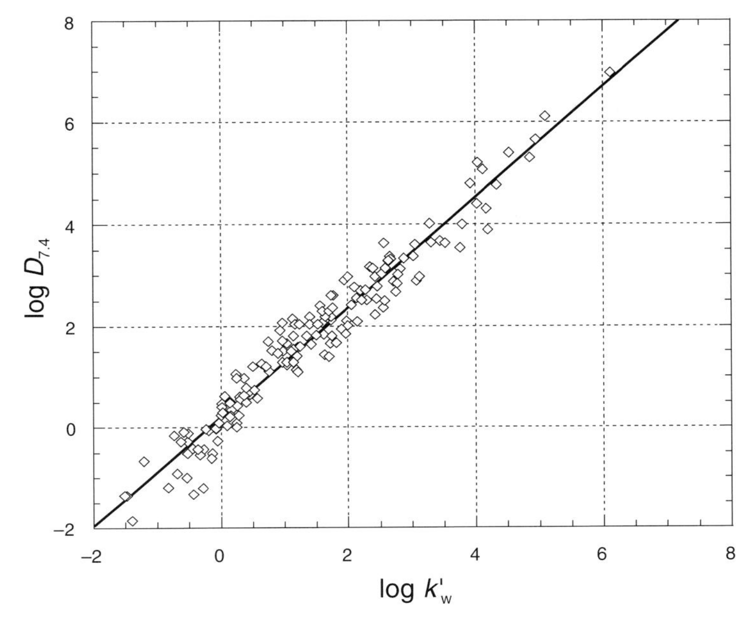

Fig. 1. *Plot of log* $D_{7.4}$ vs. *log* k'_w *data generated according to the ElogD protocol for 163 neutral and basic solutes.* The log $D_{7.4}$ data were either generated in-house or taken from the literature. The line of fit corresponds to *Eqn. 2*.

This performance does, of course, come at a cost in terms of speed. In general, using several HPLC instruments and with one person performing the runs and the data analysis, we can reach a maximum of four 96-well plates per week, or *ca.* 360 compounds, considering the wells needed for control standards. These values, to some researchers, may not seem high-throughput, but a careful scrutiny of the results and a fairly detailed data analysis are an integral part of these efforts to ensure accuracy and self-consistency against curvature phenomena, seemingly discrepant results for relatively close analogs and the possibility of confounding impurities (multiple peaks of similar intensities). We consider these efforts fundamental to ensure the quality of the data set and to achieve as little variability as possible, stemming from column to column, instrument to instrument, and compound quality. The ambitious goal of developing a reliable in-house computational model is another driving force of these efforts, and that is equivalent to investigating the nature of the forces involved in these partition phenomena and the ability of computed descriptors to capture them. Thus, the accuracy and reliability of the experimental data generated and stored cannot be overemphasized, and it is our belief that accuracy should be balanced with speed.

4.2. *ElogD$_{7.4}$: Data Acquisition and Analysis Software*

When designing any automated data analysis system, a balance must be struck between these three factors: *1*) ease of use, *2*) ease of maintenance, and *3*) complexity. Ease of maintenance and ease of use become dominant factors when the software is utilized across sites at different locations within the organization. In such an instance, the software must be intuitive enough that the users are able to learn how to use it from reading a manual. Similarly, the software must be upgradeable by the people using it across different sites without the need for intervention by the person who has written or is maintaining the software. The first two factors are not as critical when the software is utilized at a single site and the developer is on site for support purposes. The final factor complexity is usually inversely proportional to the ease of use factor.

The system in place is relatively easy to maintain and use, and is currently utilized across three *Pfizer* sites, Groton CT, La Jolla CA, and Ann Arbor MI. The system is a combination of *Agilent* HPLC Chemstation macros on the front end and *Microsoft* Excel Visual Basic for Applications (VBA) macros on the back end (*Fig. 2*). The Chemstation macros function to extract necessary parameters from the HPLC Chemstation system tables (such as area, retention time, compound number, *etc.*). These parameters are then checked for validity (peaks resultant from DMSO or other known factors removed) and then the information is placed into a known comma delimited file where it can be imported into nearly any application, but in this instance it is imported into *Microsoft* Excel.

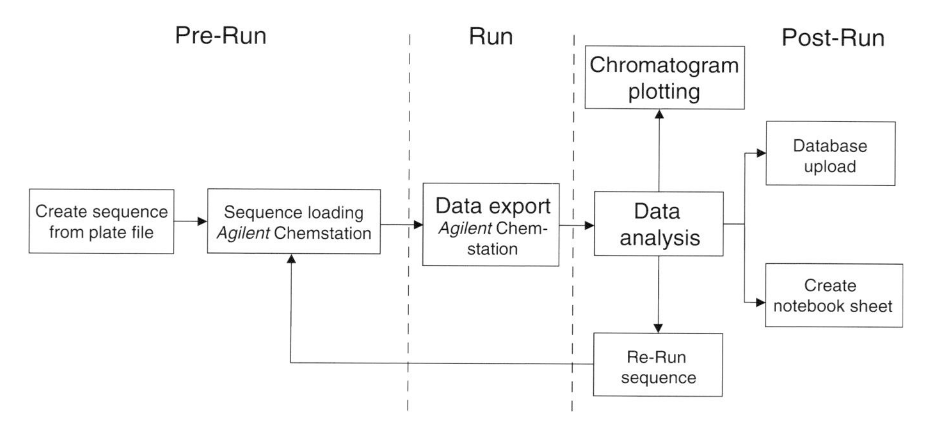

Fig. 2. *Flow chart of core automation macros*

Once the data is imported into *Microsoft* Excel, VBA routines are run on the data. The VBA routines in *Microsoft* Excel then analyze the data and calculate the ElogD of the compounds (if possible), and prepare a custom report for inspection by the scientist. Compounds for which a satisfactory ElogD cannot be calculated are flagged, and automatically slated for analysis by a different method. Similarly, standard comments are added to the report for upload to the corporate database for the following conditions: 'exceeds reasonable range of measurements', 'multiple peaks observed, largest peak taken', 'no peaks observed', and 'very low intensity'. For compounds yielding questionable results, it is possible to set up a print queue to print out all the relevant data (peaks, areas, retention times, and chromatograms) from the *Agilent* Chemstation software directly from *Microsoft* Excel using DDE. This provides the scientist with the flexibility to examine the results of the analysis in great detail and make an educated determination as to what could have gone wrong, what peak should have been chosen, *etc.* A scientist examines the automated report, makes the necessary corrections, and a flat file for upload to the corporate database can be generated directly in *Microsoft* Excel.

The automated data analysis system currently in place is a good compromise between complexity and ease of use. The software currently returns a correct ElogD value for *ca.* 70% of the compounds processed in the ElogD screen without requiring intervention from the scientist.

4.3. *ElogD*$_{7.4}$*: Problems and Solutions*

The quality of the method and its performance has been constantly monitored, and we describe here some of the modifications implemented since the publication of our original ElogD work [2].

The original protocol, set up to minimize the time spent in preparing the mobile phase while assuring a constant supply of it, called for a detailed procedure executed by a contract laboratory in larger batches for a monthly supply of *ca.* 20 l of the aqueous component. This is not a trivial aspect in industrial HT laboratories, since a great deal of emphasis is placed on the efficiency of screens and on the minimization of manual steps performed by scientists. However, we later discovered that some impurity was leading to the darkening and severe clogging of the HPLC-purge-valve polytetrafluoroethylene (PTFE) frit causing unacceptably frequent shutdowns. We reasoned that the problem could be due to the degradation of the air-sensitive decylamine, which may not be tolerating the fairly large span of time (*ca.* two months) intervening between the preparation and use of mobile phase. The solution involved a modification of the protocol, which now calls for the preparation of the mobile phase in smaller batches in our laboratory from concentrated buffer batches made by a contract laboratory, followed by addition of decylamine and adjustment of pH. This modification, however simple, has some impact on the time, but the net outcome is favorable.

Another question, perhaps of more-general interest, is whether there is much sense in dealing with compounds that have log *P*/log *D* values in excess of, say, 5 since these compounds are likely to have undesirable ADME properties, ranging from high metabolic rates to poor solubility. The accurate determination of ElogD values above this threshold are possible with our method, but we adopted the practice of excluding these compounds, through a single run at 75% MeOH, on the basis of their retention time (t_R) *vs.* t_R of the highly lipophilic drug amiodarone, run in parallel with every plate. When the t_R threshold is exceeded the compound is assigned a value >5 and reported as such, without further screening. The compound can, of course, be run in the high range at a later stage, against a specific request, which may for example stem from the need of high range values for computational purposes.

Finally, we wish to comment, albeit in fairly general terms, on the question of 'curvature', *i.e.*, nonlinearity in the extrapolation to 0% MeOH. In general, we have observed that the three-point extrapolation yields a variable range of r^2 values, which is considered acceptable above 0.98 for the high range (70, 65, and 60% MeOH), and above 0.95 for the medium range (50, 45, and 40% MeOH). The low range (25, 20, and 15% MeOH) yields a highly variable and often very low r^2, especially for negative ElogD values such as the one observed for atenolol (ElogD $= -1.5$). In the latter case, the r^2 value is not considered a reliable indication of the quality of the run, and it is not used as a 'guide'.

For some compounds, such as metoclopramide, spiperone, and haloperidol, a curvature (nonlinear behavior across a wide range of MeOH amounts) has been reported [1]. However, using our protocol, we have not observed such phenomenon for these compounds. Conversely, digoxin has shown a high curvature. We do not have firm understanding of this phenomenon, which may be difficult to unravel for complex drug-like molecules, but we note that a relatively high number of H-bond donor (HBD) and/or H-bond acceptor (HBA) atoms (HBD $=6$ and HBA $=14$ for digoxin) seems to result in a high(er) propensity for curvature. This may form the basis for the extraction of *intramolecular* HBD or HBA data from log k'_w determined at different ranges and/or from the comparison between log k'_w and log $k'_{\%}$, the latter parameter being the logarithm of the capacity factor at a given percentage of MeOH (see *Sect. 2.6*). How well this approach may work and what degree of resolution it may yield, especially when dealing with not too dissimilar compounds in terms of HBD/HBA capabilities, remains to be established.

At the same time, it is important to note here that most of the observations regarding fairly large curvature across ranges ($\geq$1 log unit difference between values obtained at different ranges, say high and medium) made during 'routine' (HT) runs were not confirmed or were much smaller when the same proprietary compounds were run separately in small sets, allowing for a lower column 'burden' compared with the long duration of the HT runs. A good deal of the fluctuation observed may be due to column degradation (or other fluctuations) possibly resulting from the compression of a large number of consecutive runs ($>$800) needed to maintain a fairly high speed of analysis without reconditioning of the column, and not, as it may be thought, due to drastic changes in the balance of forces between ranges, at least for most compounds. The large number of runs is likely to have a detrimental impact on the column performance, for example due to impurities and/or traces of DMSO in the samples and the pH value used, which is close to the maximum recommended pH for the type of column. In the 'off-line' measurements, using newer columns and a much smaller sets, the impact of several of these factors may be largely decreased, and one recourse is to monitor the performance of the column with standards prior and during HT runs.

4.4. *ElogD$_{7.4}$: the Extraction of H-Bonding Information*

It has been previously reported [11][12] that log P_{oct} is completely insensitive to the solute H-bond acidity, while the log k'_{30-70} values seem influenced by it, even though no generalization can be made as to which system to use, and MeOH may not be the cosolvent of choice, for the purpose described below. *Valko et al.* [12] suggested using H-bond acidity as well as several other terms to improve the correlation between RP-HPLC derived chromatographic hydrophobicity indices (determined in MeCN/buffer mixtures) and log *P*. We have shown [10] that the ElogP method relies on the same balance of forces as classical 'shake-flask' method and, since the acquisition of log k'_{15-75}data (range of MeOH content: 15–70%) is part of the ElogP (or ElogD) method, it may be useful and potentially rewarding to explore the possibility of extracting H-bonding data from these comparisons as a part of the same determination.

The approach we used here was to determine whether there was merit in finding an expression for ElogD in terms of a subset of the parameters of the *Abraham*'s solvation equation [13]:

$$SP = c + eE + sS + aA + bB + vV \qquad (3)$$

where a and b are the H-bond acidity and basicity coefficients, respectively, and where SP would be represented by the difference (Δ) between $\log k'_w$ and $\log k'_n$, and $n = \%$ of MeOH.

However, our preliminary findings show that the use of MeOH as a cosolvent for $\log k'_{15-25}$ measurements does not allow for H-bond acidity determination, especially since the absolute value of the coefficient of the parameter A is low and relatively invariant among conditions. The larger differences, between $\log k'_w$ and $\log k'_n$, were observed in terms of the coefficients b and v, that is, in terms of HBA character and volume of the solute. We only had the parameters, from literature, for a small number of drugs and the test may not be very significant, but the use of very different mobile phases may be what is needed in these efforts, as well as larger data sets.

Mobile phases comprising MeCN may be much more suitable for the purpose at hand, as in *Valko*'s work [12], and they will be considered.

4.5. *ElogD$_{7.4}$: Application to the Prediction of Volume of Distribution*

The volume of distribution (V_D) is a proportionality constant, which is often devoid of a physical meaning since the measured volume of distribution in many cases far exceeds the physical volume in which a drug may dissolve, and which is generally accepted to be (as total body water) in the range of 0.6–0.8 l kg^{-1}. Nevertheless, its knowledge is of paramount importance to predict the half-life of a drug and, therefore, its dosing regimen. The latter has obviously important implications in the establishment of a suitable therapeutic regimen, which includes the aspect of patient compliance.

It may seem intuitive that a large distribution into the body may be related, at least in part and assuming passive diffusion, to lipophilicity. However, we also note that limited work in this area has shown that the presence of a higher amount of fat tissues (as in the obese) did not seem to translate into an increase in V_D in human [14], and the amount of partition in the fat tissue may not be an overwhelming determinant of V_D, especially when different classes of compounds are considered, even though it is an important one [15]. Furthermore, most of the correlations existing in the literature focused on fairly small data sets and/or on the use of analogs. It seemed useful, in attempting to predict V_D, to derive a larger correlation, armed with the possibility of generating fairly accurate ElogD values in a reasonable short time, for neutral and basic compounds.

We have recently reported in detail the derivation and application of our method to the prediction of V_D in human [3][4], which is based on the use of measured parameters, but without the use of any kind of *in vivo* data. This is, of course, desirable to spare animal resources but also operator time and the amount of drug needed, which is often a precious commodity at early stages of research.

The model relies on three parameters, ElogD, the fraction of drug ionized at pH 7.4 ($f_{i7.4}$, from a measured pK_a), and the fraction unbound in plasma (f_u, from plasma protein binding determinations) with the advantage that all these parameters can be measured *in vitro* and at a fairly high throughput [7][16][17]. The resulting equation, shown below (*Eqn. 4*) [4], yields a very good set of statistics, especially considering the variability embedded in such a heterogeneous set of literature V_D and f_u data.

$$\log f_{ut} = 0.008\ (\pm\ 0.075) - 0.229\ (\pm\ 0.041)\ \text{ElogD} - 0.931\ (\pm\ 0.078)\ f_{i(7.4)} + 0.888\ (\pm\ 0.096)\ \log f_u \quad (4)$$

$n = 120$; $r^2 = 0.866$; rmse $= 0.366$; $F_{3,116} = 250.9$; $q^2 = 0.854$; p-value < 0.0001; mean-fold error for the prediction of training set $V_{D_{ss}} = 2.08$

This method involves the calculation of f_{ut}, the fraction unbound in tissues, from which the calculation of V_D is possible by means of the *Øie–Tozer* equation [18]. It is not suitable, however, for the calculation of V_D values for acidic compounds, since *i*) ElogD is not suitable for the determination of distribution coefficients for acidic compounds, *ii*) the determination of V_D and f_u for acidic compounds is quite difficult given the large amount of plasma protein binding and, *iii*) almost invariably very small V_D values are observed for acidic compounds. This class of compounds represents quite a challenge in terms of discriminating compounds having small values often well below 1 l kg^{-1}. Work is in progress in our group to address these aspects computationally as well as experimentally, and lipophilicity remains an important parameter.

5. Conclusions

We have attempted to cover some of the general and some of the particular aspects of the determination and use of lipophilicity which, in the mind and work of a great number of scientists in the field, is inextricably linked to the octanol/water (buffer) partition (or distribution) coefficient, proposed some 40 years ago by *Hansch* [19].

We have, of course, also referred to more-specific aspects and to our own method of log D determination as well as to its possible applications, namely the estimation of H-bond capability of a solute and the application of ElogD to the calculation of V_D.

Many questions, however, remain about the use of log P whether from octanol/water or other binary mixtures. One, in particular, comes to mind: after having practiced this approach for 40 years, where are we now, especially computationally?

The question is not a rhetorical one, since there are more and more log P calculation methods continuously appearing in the literature, but which most users (and the authors of this chapter) would be skeptical about, given the similarity in the data set used and the fact that drug-like molecules have experienced an increase in complexity and lipophilicity with time. 'Ready-made' data sets are very attractive, and an argument may be that an improvement in the statistics with an identical data set may be worth of notice. We have observed, fairly frequently, deviations of 2 log units from an 'accepted' value, and we have noted above that in-house efforts may be the appropriate avenue of progress, when a method is expected to deal with a variety of functional groups and interactions present in fairly complex and lipophilic drug-like compounds, such as the ones encountered in industrial drug research.

What about the data from various low-, medium-, to high-throughput methods that populate the literature? And what about the ones that are used to populate the industry in-house data sets? The answer is not easy, but it should be borne in mind that no method is truly and fully automated without extensive validation, constant monitoring, and refinement. This should include frequent and as accurate as possible controls on self-consistency and method performance, using data generated by multiple methods and multiple laboratories, whenever possible. If that is not the case, much of an improvement in the understanding of the lipophilic behavior of drugs cannot be expected, nor can a significant improvement of computational packages be achieved, if that is the goal. And, as experimentalists, we would not have fulfilled our responsibility of populating the literature and in-house data sets with reliable and therefore useful data. Last but not least, educational efforts aimed at the medicinal chemistry community will greatly benefit from reliable and reproducible data.

REFERENCES

[1] N. El Tayar, H. van de Waterbeemd, B. Testa, *J. Chromatogr.* **1985**, *320*, 293; N. El Tayar, H. van de Waterbeemd, B. Testa, *J. Chromatogr.* **1985**, *320*, 305.
[2] F. Lombardo, M. Y. Shalaeva, K. A. Tupper, F. Gao, *J. Med. Chem.* **2001**, *44*, 2490.
[3] F. Lombardo, R. S. Obach, M. Y. Shalaeva, F. Gao, *J. Med. Chem.* **2002**, *45*, 2867.
[4] F. Lombardo, R. S. Obach, M. Y. Shalaeva, F. Gao, *J. Med. Chem.* **2004**, *47*, 1242.
[5] M. S. Tute, in 'Lipophilicity in Drug Action and Toxicology', Eds. V. Pliska, B. Testa, H. van de Waterbeemd, Wiley-VCH, Weinheim, 1996, p. 7–26.
[6] J. Comer, K. Tam, in 'Pharmacokinetic Optimization in Drug Research. Biological, Physicochemical, and Computational Strategies', Eds. B. Testa, H. van de Waterbeemd, G. Folkers, R. Guy, VHCA, Zürich, 2001, p. 275–304.
[7] E. H. Kerns, *J. Pharm. Sci.* **2001**, *90*, 1838.
[8] G. Caron, G. Ermondi, *Mini-Rev. Med. Chem.* **2003**, *3*, 821.
[9] D. A. Smith, H. van de Waterbeemd, D. K. Walker, 'Pharmacokinetics and Metabolism in Drug Design', Wiley-VCH, Weinheim, 2001, p. 135.
[10] F. Lombardo, M. Y. Shalaeva, K. A. Tupper, F. Gao, *J. Med. Chem.* **2000**, *43*, 2922.
[11] A. J. Leo, in 'Biological Correlations – The Hansch Approach. ACS Advances in Chemistry Series 114', Ed. R. F. Gould, American Chemical Society, Washington, D.C., 1972, p. 51–60.
[12] K. Valko, C. M. Du, C. Bevan, D. P. Reynolds, M. H. Abraham, *Curr. Med. Chem.* **2001**, *8*, 1137.
[13] Y. H. Zhao, J. Le, M. H. Abraham, A. Hersey, P. J. Eddershaw, C. N. Luscombe, D. Boutina, G. Beck, B. Sherborne, I. Cooper, J. A. Platts, *J. Pharm. Sci.* **2001**, *90*, 749.
[14] G. Cheymol, J.-M. Poirier, P.-A. Carrupt, B. Testa, J. Weissenburger, J.-C. Levron, E. Snoeck, *Br. J. Clin. Pharmacol.* **1997**, *43*, 563.
[15] S. Bjorkman, *J. Pharm. Pharmacol.* **2002**, *54*, 1237.
[16] I. Kariv, C. Cao, K. R. Oldenburg, *J. Pharm. Sci.* **2001**, *90*, 580.
[17] M. J. Banker, T. H. Clark, J. A. Williams, *J. Pharm. Sci.* **2003**, *92*, 967.
[18] S. Øie, T. N. Tozer, *J. Pharm. Sci.* **1979**, *68*, 1203.
[19] T. Fujita, J. Iwasa, C. Hansch, *J. Am. Chem. Soc.* **1964**, *86*, 5175.

Lipid Bilayers in ADME: Permeation Barriers and Distribution Compartments

by **Stefanie D. Krämer**

Institute of Pharmaceutical Sciences, Department of Chemistry and Applied Biosciences, ETH Zürich, CH-8093 Zürich
(e-mail: skraemer@pharma.ethz.ch)

Abbreviations
BLM: Black lipid membranes; CAD: cationic amphiphilic drug; CPP: cell-penetrating peptide; EPR: electron paramagnetic resonance; log *D*: logarithmic value of the distribution or apparent partition coefficient; log *P*: logarithmic value of the absolute partition coefficient; NMR: nuclear magnetic resonance; P_{app}: apparent permeation coefficient; Pgp: P-glycoprotein; PhC: egg phosphatidylcholine; PhI: phosphatidylinositol from a natural source.

---

1. Introduction

Cell membranes consist of a vast variety of lipids and proteins [1]. The several hundred lipid species [2] form bilayers of a few nanometer thickness with the lipid acyl chains pointing towards the center plane and the polar headgroups facing the aqueous phases on either side of the membrane. According to our current understanding, cell membranes consist of dynamic domains with distinct lipid and protein compositions and arrangements. In addition, the two faces of the cell membranes display different lipid and glycosylation patterns. An illustrative sketch of such a membrane was drawn by *Kinnunen* [3].

In terms of ADME (absorption, distribution, metabolism, excretion), the lipid bilayers act both as barriers between aqueous compartments and as distribution compartments for lipophilic drugs, determining their pharmacokinetic behavior. Focusing on the cellular level and considering the heterogeneous distribution of lipids and proteins in the cellular membrane pool, the distribution of a lipophilic drug can only be heterogeneous, and its translocation between the opposite lipid leaflets of a membrane region must strongly depend on the characteristics of the particular domain. In other words, regarding membrane-associated lipophilic drug molecules as 'xeno-lipids', it becomes obvious that they will show an inhomogeneous distribution and flip-flop behavior among the different membrane domains.

The octanol/buffer partition system, which is frequently used to predict the body distribution and barrier passage of a drug, can hardly provide sufficient information to describe the partitioning of drug compounds at the cellular level and their behavior on the route along and across the membrane domains to reach their respective targets.

Much information has accumulated during the last years on how different lipids influence the membrane affinity of drugs [4][5]. Less is known on lipid-bilayer permeation of drug compounds. More data are needed on permeation kinetics across lipid bilayers of different composition mimicking membrane domains such as the apical plasma membrane of barrier cells or the vicinity of target proteins. Techniques which have been used so far to determine permeation kinetics of drugs or drug like compounds across lipid bilayers will be summarized in this article and a novel assay to investigate lipid-bilayer permeation of aromatic carboxylic acids will be discussed. Systems which are not strictly using lipid bilayers, such as PAMPA, are discussed by *Avdeef* in the following Chapter of these *Proceedings* [6].

2. Relationship between Membrane Affinity and Permeation

A simple model of membrane partitioning and permeation of an amphiphilic molecule is shown in *Fig. 1*. The presumably most favored positions during the permeation process are depicted, *i.e.*, the hydrated states in the two aqueous compartments and the positions in the two lipid leaflets with the hydrophilic part of the solute in the region of the lipid headgroups and the hydrophobic moiety in the acyl chain environment. Based on this model, permeation can be described by two equations (for a symmetrical membrane and aqueous compartments with identical conditions): one for the partitioning between the aqueous compartments and the lipid leaflets with the partition coefficient as equilibrium constant, the other one for the flip-flop between the two leaflets. The rate constants of both equilibria are decisive for the permeation rate and each of them can be rate-limiting depending on the physicochemical characteristics of the solute and the lipid bilayer.

Considering the two equilibria, it becomes clear why high membrane affinity does not necessarily mean fast permeation. Protonated, charged propranolol, as an example, has a much higher affinity to membranes containing negatively charged lipids than neutral propranolol [7][8].

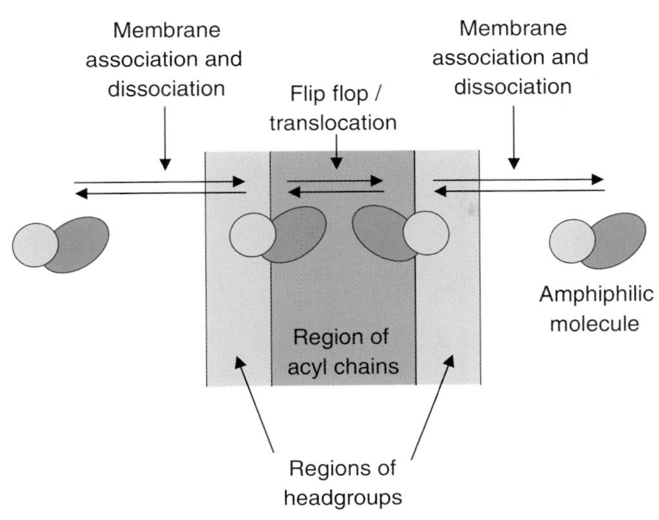

Fig. 1. *Model of lipid bilayer partitioning and permeation of an amphiphilic molecule.* The polar headgroup and the hydrophobic acyl chain regions of a lipid bilayer are indicated in light and dark grey, respectively. Four favored positions of an amphiphilic molecule are shown. In the membrane, its hydrophilic moiety (light grey) prefers the lipid headgroup region and the hydrophobic part (dark grey) assembles with the hydrophobic acyl chains. The molecule partitions between the aqueous phase (no color) and the two lipid leaflets and, in addition, translocates between the two lipid leaflets. The corresponding equilibria are indicated. The equilibrium constant of membrane association and dissociation is the partition coefficient.

However, there is strong evidence that only neutral solutes show significant permeation across lipid bilayers (see below). It could even be argued that a very high affinity of ionized compound lowers the total pool of neutral compound and therefore has a negative effect on permeation.

3. Drug Partitioning between Lipid Bilayers and Water

Partitioning between lipid bilayers and water has been extensively discussed in the *Proceedings of the 2nd Symposium on Lipophilicity, LogP2000* [4]. In our laboratory, we use unilamellar liposomes in an equilibrium dialysis system to investigate the affinity of drugs to membranes of different lipid compositions. The addition of negatively charged phosphatidylinositol (PhI) or oleic acid to phosphatidylcholine (PhC) membranes led to a striking increase in the membrane affinity of positively charged propranolol. The log P value of the protonated base was 1.5 log units higher using pure PhI liposomes than PhC liposomes, and 12% (mol/mol) oleic acid in PhC membranes enhanced the value by 0.8

log units. Distribution profiles followed a *Henderson–Hasselbalch* function considering the ionization states of the solute and the lipids. Log D values were highest in the pH range where the lipids were negatively charged and the drug was positively charged. In the PhC/PhI liposome system, the distribution coefficient D at pH 6 and 8 was exponentially dependent on the molar fraction of PhI in the bilayer. In PhC/oleic acid liposomes, the free fatty acid changed its protonation state around physiological pH causing a strong pH dependence of the membrane affinity of propranolol in this pH range [7][8]. Based on these findings a heterogenous and pH-dependent distribution of drugs and other solutes can be expected in the cell.

Here, a partitioning phenomenon is described in more detail that could elucidate how α-tocopherol (vitamin E) reverses phospholipidosis, a side-effect of so-called cationic amphiphilic drugs (CADs). Phospholipidosis denotes the accumulation of lipids and drug molecules in lysosomes resulting in microscopically visible lamellar bodies in the cells [9]. The current model is that CADs follow the pH gradient from the cytosol (pH 7.2) to the lysosomes (pH 4.5) where they associate with the phospholipids and directly or indirectly inhibit their metabolism by phospholipases. It was observed *in vivo* and in cell culture that treatment with α-tocopherol reversed phospholipidosis, and it was hypothesized that the vitamin alters the binding affinity between the phospholipids and the CADs [10].

Using PhC liposomes and PhC/PhI/cholesterol liposomes mimicking the lysosomal membranes in partition experiments, we found that α-tocopherol increased the log P of the neutral CAD desipramine resulting in an increase in membrane affinity (log D) at pH 7.4 rather than the expected decrease. However, α-tocopherol had no influence on the log P of the protonated CAD and on the log D at pH 4.5, which corresponds to the pH of the lysosomes (*Fig. 2*). From these findings, we concluded that treatment with α-tocopherol *in vivo* could lead to a redistribution of desipramine from the lysosomal lipid/drug accumulations at low pH to nonlysosomal membranes and lipoproteins around pH 7.4 [5].

The chromanol OH group of α-tocopherol was indispensable for the effect as α-tocopherol acetate, and cholesterol had no influence on desipramine partitioning. 2,2,5,7,8-Pentamethyl-6-chromanol, which lacks the phytyl side-chain of α-tocopherol but has otherwise an identical structure, showed the same effect as α-tocopherol.

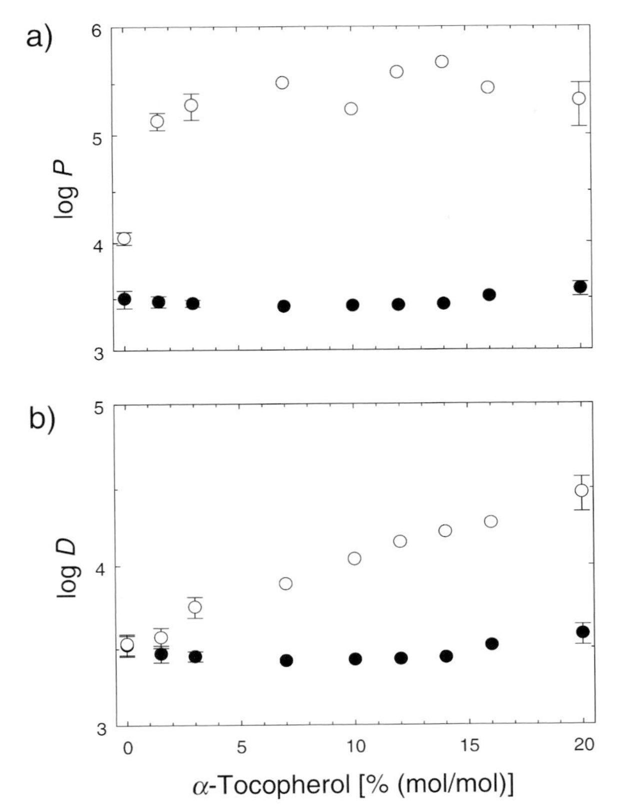

Fig. 2. *Concentration-dependent effect of α-tocopherol on the partitioning of desipramine in PhC liposomes.* Distribution coefficients D between liposomal membranes and buffer were determined by equilibrium dialysis at 37° and 0.21M ionic strength. Log P values were fitted from the pH-dependent distribution. *a*) ○, log P of neutral desipramine; ●, log P of protonated desipramine. *b*) ○, log D at pH 7.4; ●, log D at pH 4.5. Reprinted from [5] with permission from *Elsevier*.

4. Drug Permeation across Lipid Bilayers

4.1. *The Permeation Coefficient*

The most frequently used measure for permeation is the apparent permeation coefficient, P_{app}, which reflects the clearance (CL) per surface area (A). It is calculated from the initial permeation rate, $(dc/dt)_{t=0}$, according to *Eqn. 1* (*e.g.*, [11]),

$$P_{app}\left[\frac{cm}{s}\right] = \frac{CL}{A} = \left(\frac{dc}{dt}\right)_{t=0} \cdot \frac{1}{\Delta c_{t=0}} \cdot \frac{V}{A}\left[\frac{cm^3}{s \cdot cm^2}\right] \tag{1}$$

where V denotes the volume of the aqueous compartment with the concentration c and $\Delta c_{t=0}$ the concentration gradient between the concentrations at the start and at equilibrium. The term $(dc/dt)_{t=0} \cdot (1/\Delta c_{t=0})$ equals the first derivative of the exponential concentration/time curve in the respective aqueous compartment.

4.2. *Factors Influencing Lipid Bilayer Permeation*

Studies on the pH dependence of lipid bilayer permeation revealed an effect of the ionization states of the solute and membrane components, respectively (*e.g.*, [12][13]). The neutral species permeates much faster than the charged species, resulting in a strong pH dependence of the permeation rates. Experimentally determined permeation coefficients are usually corrected for the molar fraction of neutral solute in the aqueous compartment.

Permeation coefficients are very sensitive to temperature changes. In the studies of *Frézard* and *Garnier-Suillerot* [13] and *Finkelstein* [14], permeation coefficients increased by factors of 5 to 10 when the temperature was raised by 10°. The temperature was always above the transition temperature of the lipids.

The lipid composition of the membrane strikingly influences permeation rates. Permeation coefficients of butane-1,4-diol, acetamide, or formamide varied by a factor of close to 100 between PhC membranes and membranes consisting of sphingomyelin and cholesterol. Permeation across PhC/cholesterol membranes was intermediate [14]. In general, the addition of *ca.* 20–40% or more cholesterol to the lipid bilayers leads to a decrease in permeation coefficients (*e.g.*, [13] and [15]). The water permeability of membranes mimicking the exofacial leaflet of apical plasma membranes of epithelial cells, *i.e.*, containing cholesterol, PhC, sphingomyelin and glycosphingolipids, was 10 times lower than the permeability of membranes consisting of phosphatidylethanolamine, phosphatidylserine, PhI, and cholesterol, mimicking the cytoplasmic leaflet [16].

A relationship between permeation and lipophilicity has been postulated at least since the work of *Overton* on the osmotic characteristics of the cell [17]. Early work with planar lipid bilayers confirmed the positive correlation between permeation coefficients and partition coefficients in the octanol/water or other partition systems [18–21].

4.3. *Permeation Studies with Planar Lipid Bilayers*

Unlike the shake-flask method to measure octanol/water partition coefficients or equilibrium dialysis to study membrane affinities, there is no reference technique to determine lipid bilayer permeation which would be applicable for all kinds of solutes. The two most frequently used systems are planar lipid bilayers (black lipid membranes, BLM) and liposomes. The *Table* gives a selection of published permeation coefficients across lipid bilayers determined by different techniques.

Planar lipid bilayers form when a lipid solution is spread over a hole of, *e.g.*, 200 μm diameter in a *Teflon* partition of *e.g.*, 25 μm thickness, or when a lipid monolayer formed at a water/air interface is mounted on either side over the hole [16]. The bilayer thus separates two aqueous compartments, which are accessible for probes and sampling. Conditions can be changed in both compartments during the experiment. The technique even allows the formation of asymmetric bilayers (*e.g.*, [16]). The drawbacks are the relative instability of the membranes and the distances between membrane and probe containing the unstirred water layer representing a relevant diffusion layer of 100–200 μm width for permeating solutes [12][22].

In 1973 and 1984, *Gutknecht* and *Tosteson* [12], and *Walter* and *Gutknecht* [24] studied the flux of radiolabeled carboxylic acids across planar lipid bilayers. From their apparent permeation coefficients at different pH values, they concluded that the unstirred water layer contributed considerably to the permeation barrier. Based on a model, where the charged solute diffuses through the unstirred water layer while only the uncharged solute permeates the lipid bilayer, and using the apparent permeation coefficients at various pH values, they fitted the permeation coefficients of the neutral compounds across the bilayer alone. This led to relatively high permeation coefficients as compared to other studies (see the *Table*).

The model, where only neutral solutes cross membranes and ionize at the membrane/water interface depending on pH and pK_a, was used by several groups to determine the permeability of weak acids and bases. Permeation rates were calculated from the pH gradient, which developed across the membrane upon permeation [20][23–25].

4.4. *Permeation Studies with Liposomes*

Liposomes are vesicles consisting of one or several concentric lipid bilayers. Like planar lipid bilayers, liposomal membranes separate the

Table. *A Selection of Published Permeation Coefficients across Lipid Bilayers*

Permeant[a])	System[b])	Technique[c])	Permeation coefficient [cm s^{-1}]	Ref.
Water	Egg PhC, BLM, 25°		$2 \cdot 10^{-3}$	[20]
ATP (pH 8)	DMPhC/DPPhC 50:50, Lip, 32°	Separation	$2 \cdot 10^{-10}$	[26]
Acetic acid	Egg PhC, BLM, 25°	Recording pH changes	$2.4 \cdot 10^{-4}$	[20]
	Egg PhC, Lip, 7°	NMR	$5 \cdot 10^{-4}$	[27]
	Egg PhC, BLM	Radioactive tracer	$6.6 \cdot 10^{-3}$	[28]
Hexanoic acid	Egg PhC, BLM	Radioactive tracer	1.1	[28]
Benzoic acid	Egg PhC, BLM	Radioactive tracer	0.55	[28]
Salicylic acid	Egg PhC, BLM	Radioactive tracer	0.77	[12][28]
	Egg PhC, Lip, 20°	Tb^{3+} assay	$1.3 \cdot 10^{-4}$	[15]
Salicylate (deprotonated)	Egg PhC, BLM	Radioactive tracer	$< 10^{-7}$	[12]
4-(Hydroxymethyl)benzoic acid	Egg PhC, Lip, 25°	Separation	$1.6 \cdot 10^{-3}$	[21]
Substituted 4-methylbenzoic acids	Egg PhC, Lip, 25°	Separation	$4.1 \cdot 10^{-5}$ to $1.1 \cdot 10^{0}$	[21]
Substituted 4-methylhippuric acids	Egg PhC, Lip, 25°	Separation	$9.9 \cdot 10^{-9}$ to $4.9 \cdot 10^{-4}$	[21]
Pirarubicin	Egg PhC/egg PhA/Chol 75:5:20, Lip, 20°, pH 6.0	Quenching by DNA	$2.6 \cdot 10^{-3}$	[13]
Daunorubicin	Egg PhC/egg PhA/Chol 75:5:20, Lip, 20°, pH 6.0	Quenching by DNA	$6 \cdot 10^{-4}$	[13]
	Egg PhC/egg PhA/Chol 60:20:20, Lip, 20°, pH 7.4	Quenching by DNA	$4 \cdot 10^{-5}$	[13]
Doxorubicin	Egg PhC/egg PhA/Chol 60:20:20, Lip, 20°, pH 7.4	Quenching by DNA	$2 \cdot 10^{-5}$	[13]

[a]) Uncharged species except when stated otherwise. [b]) BLM: black lipid membranes (planar lipid bilayers); Lip: liposomes; DMPhC: dimyristoyl-PhC; DPPhC: dipalmitoyl-PhC; PhA: phosphatidic acid; Chol: cholesterol. [c]) See text for details.

aqueous phase in different compartments. For permeation studies, it is convenient to distinguish the extraliposomal (outer) and the intra-liposomal (inner) compartments. The latter is the sum of all aqueous compartments entrapped by the vesicles. Liposomes are more stable than planar lipid bilayers. Unilamellar liposomes are in the size range between *ca.* 25 nm and several micrometer in the case of giant liposomes. Stable preparations with low curvature stress have typically diameters of one or a few hundred nanometers [4]. Disadvantages compared to planar lipid bilayers are that the compartments are not accessible for sampling without a further separation step. The intraliposomal compartment cannot be reached by physical probes and its conditions cannot easily be changed during the experiments. In addition, the formation of asymmetric bilayers is not straightforward. Unlike with planar lipid bilayers, the unstirred water layer can be neglected. Liposome diameters and the distances between liposomes are generally much smaller than the unstirred water layers in experiments with planar lipid bilayers.

4.4.1. *Separation Techniques*

Most techniques used in permeation studies with liposomes require a concentration gradient between the inner and outer aqueous phases. To establish a concentration gradient from the inner to the outer compartment, liposomes are prepared in the presence of the permeant or incubated with it after the preparation. After equilibrium is established, the liposomes are separated from the extraliposomal permeant using a size-exclusion column. The efflux of permeant along the established gradient already begins during the exchange of the outer aqueous phase.

If permeation kinetics are determined from the concentrations at different time points in one or both compartments, the compartments have either to be separated at each time point or the inner and outer permeant molecules have to show distinguishable characteristics, which can be used for quantification. If separation is required, the concentration gradient usually points from inside to outside, and the concentration is determined in the outer compartment after size-exclusion chromatography, ultrafiltration, or ultracentrifugation [21] [26] [29]. These methods are only appropriate if permeation is much slower than the separation steps. Liposomes are usually cooled during the separation procedures to slow down membrane passage.

To overcome the technical limits for fast permeating solutes, *Eytan et al.* [30] used multilamellar liposomes, to estimate permeation half-lives. The permeant was incubated with multilamellar liposomes, which were then separated from the outer aqueous phase at different time intervals to determine the concentration of liposome-associated permeant. After a fast step of binding, which was assigned to the association to the outermost lipid leaflet, a slower increase in association was observed and interpreted as the binding kinetics to the inner layers (which is directly proportional to permeation). To separate the liposomes from the outer buffer phase, the samples were layered under a sucrose density gradient with ether as top layer. Upon centrifugation at high *g* forces the liposomes migrated to the ether phase, where they dissolved, allowing the liposome-associated drug to be quantified. The outer aqueous phase containing non-membrane-associated solute remained in the lower part of the tube. The method was used to compare the permeation of P-glycoprotein (P-gp) substrates and modulators. P-gp modulators showed faster permeation than substrates of P-gp transport [30][31].

4.4.2. *Permeation Experiments Using Fluorescence Techniques*

Fluorescent probes with self-quenching characteristics have been used to study the permeability of liposomes under different conditions such as lipid composition, pH, type of ions and ionic strength, or temperature. Leakage of carboxyfluorescein or calcein, which are quenched at high concentrations out of liposomes, is detectable from the increase in fluorescence upon dilution into the outer compartment [32–35].

Rhodamine 123 fluorescence is quenched when the compound associates with a membrane. *Eytan et al.* [30] described a two-step quenching kinetics when rhodamine 123 was added to unilamellar or multilamellar liposomes. The first step, which was too fast to be measured, was again assigned to the partitioning into the outer lipid leaflet. The second step, which was slow enough to be quantified, reflected the translocation from the outer to the inner lipid leaflet(s) and was used as a measure for the permeation rate.

Fluorescence quenching in one of the aqueous phases can also be achieved by the interaction with a nonpermeating reaction partner. *Regev* and *Eytan* [36] loaded liposomes with doxorubicin and added DNA to the sample to bind and quench doxorubicin which desorbed from the membrane. After the first fast step, the slower exponential step was again used as a measure of permeation rate. *Frézard* and *Garnier-Suillerot* [13] encapsulated DNA in the liposomes and added anthracycline derivatives to the outer phase. A fluorescence quenching assay was also used to determine the permeation kinetics of tastants [37].

Fluorescence enhancement upon interaction of the permeant with an entrapped compound was used to characterize membrane passage of tetracycline. Tetracycline repressor protein, which enhanced tetracycline fluorescence, was entrapped in liposomes which were then incubated with the antibiotic [38].

Alternatively, the permeant could alter the fluorescence of a reaction partner in either of the aqueous compartments. In this case, permeation and membrane association of the fluorescent agent should be negligible compared to the permeation of the compound studied. The number of candidates is small. Examples are carboxyfluorescein, which is pH sensitive, terbium^{3+} (Tb^{3+}), which increases its luminescence upon ligation or complexation with excited aromatic carboxylic acids (see *Sect. 4.5*), and 1-aminonaphthalene-3,6,8-trisulfonic acid (ANTS), which is quenched by 1,1′-[1,4-phenylenebis(methylene)]bispyridinium dibromide (DPX), both with low membrane permeability [39]. Carboxyfluorescein was used to determine the permeation of free fatty acids across vesicle membranes. The assay was based on the same principle as ΔpH measurements with planar lipid bilayers: only the neutral acid permeates the lipid bilayer carrying a proton from the outer compartment and releasing it into the liposomal lumen [40].

4.4.3. *Permeation Measurements Based on Changes in UV/VIS Absorbance*

Permeation can be determined from changes in UV/VIS absorbance following a chemical reaction between the permeant and a nonpermeating agent in one of the aqueous compartments. One example is the entrapment of $[FeSCN]^{2+}$ in liposomes, which absorbs at 450 nm. After addition of F^- to the liposomes and permeation of either of the reaction partners, F^- displaces SCN^- from the complex leading to a decrease in absorbance [41].

Allicin permeation was determined using GSH-loaded vesicles. Influxing allicin reacted with the SH group and the product was determined by HPLC [42].

4.4.4. *Permeation Studies Using Enzymatic Reactions in the Acceptor Compartment*

Permeation of entrapped glucose was determined using an enzymatic assay resulting in a colored product [43]. Using an enzymatic reaction to measure permeation kinetics is only possible if it is not the rate-limiting step.

4.4.5. *NMR and EPR in Permeation Studies*

Taking advantage of the low membrane permeation of paramagnetic ions used as shift reagents in NMR, *Alger* and *Prestegard* [27] and *Xiang* and *Anderson* [44–47] applied NMR ^{1}H-line broadening employing Pr^{3+} to determine permeation coefficients of various carboxylic acids across lipid bilayers. Electron paramagnetic resonance (EPR) was applied to determine transmembrane translocation of spin-labeled steroids using 6-phenylascorbic acid as extraliposomal reducing agent [48].

4.4.6. *Changes in Light-Scattering Characteristics upon Permeation*

The geometry of a liposome is determined by the relation between the entrapped volume and the inner membrane surface. At highest volume/surface ratio, the vesicle forms a sphere which collapses at decreasing volume/surface ratios. Collective changes in the inner volumes of a liposome sample can be detected by changes in the light-scattering characteristics of the liposome preparation. This was used by *Neitchev et al.* [49] to determine water permeation across liposomal membranes following an osmotic gradient. A similar method was used to study permeation of solutes assuming that permeation builds up an osmolarity gradient [50]. A related approach was chosen by *Hill et al.* [51] using carboxyfluorescein quenching to detect changes in the entrapped volumes of the liposomes upon permeation of a solute.

4.4.7. *Tb^{3+} Assay to Study Permeation of Aromatic Carboxylic Acids*

We recently established a method to study permeation kinetics of aromatic carboxylic acids across liposomal membranes [52]. Using this technique, we currently investigate the influence of permeant characteristics as well as the influence of the membrane composition on permeation. The first results are described as a free communication in the accompanying CD-ROM [15].

The assay is based on the luminescence of the rare earth ion Tb^{3+}, which is enhanced upon chelation or ligation by an excited aromatic system. Tb^{3+} has been used before in combination with dipicolinic acid to study liposome fusion [53]. For our studies, liposomes were prepared by extrusion in the presence of Tb^{3+} and separated from extraliposomal Tb^{3+} using a size-exclusion column. The liposomes were immediately incubated with the aromatic carboxylic acid to be tested. The Tb^{3+} luminescence increased in a bi-exponential manner as shown in *Fig. 3,a* and *b* for salicylic acid. No leakage of the rare earth ion was detectable under the experimental conditions. Luminescence of the complex itself was linearly dependent on the solute concentration in the estimated concentration range in the liposomal lumen. Stopped-flow experiments revealed that the complex between an aromatic carboxylic acid and Tb^{3+} was formed within a few seconds. This is fast enough to be distinguishable from the permeation kinetics of the studied carboxylic acids, which displayed permeation half-lives in the range of minutes at pH 7.0 and 20°.

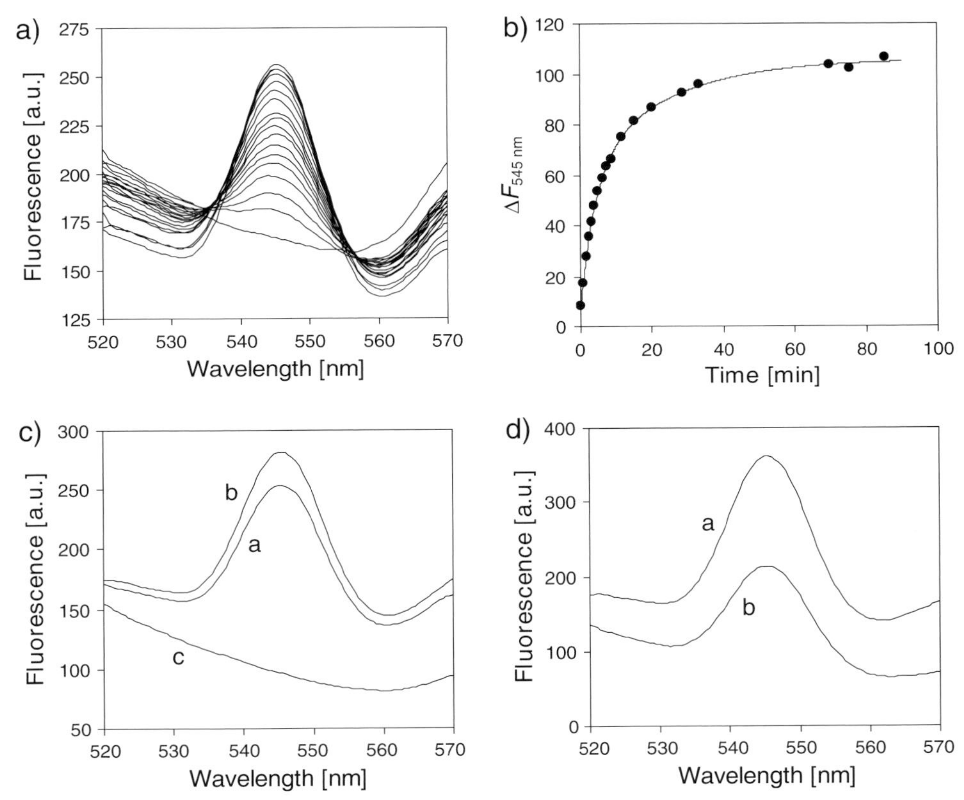

Fig. 3. *Entry of salicylic acid into Tb^{3+}-containing PhC liposomes.* Tb^{3+}-containing PhC liposomes were incubated with salicylic acid, and fluorescence scans were monitored at λ_{ex} 318 nm. *a*) Fluorescence scans before addition of salicylic acid, (no peak at 545 nm) and at different time points between 0 and 85 min after salicylic acid addition. *b*) Peak heights ΔF_{545nm} as calculated from the spectra in *a*. (—) bi-exponential fit of the data. *c*) After incubation of liposomes with salicylic acid for 85 min (*Line a*), EDTA was added to displace salicylic acid from extraliposomal Tb^{3+} (*Line b*). After EDTA addition, liposomes were lysed with *Triton X-100* (*Line c*). *d*) After incubation of liposomes with salicylic acid for 50 min (*Line a*), liposomes were lysed with *Triton X-100* in the absence of EDTA (*Line b*). a.u., Arbitrary units. Modified from [52] with permission from *Elsevier*.

To investigate whether the luminescent complex was inside or outside the liposomes, an excess of EDTA was added to the liposomes after the incubation with salicylic acid. EDTA has a higher affinity to Tb^{3+} than salicylic acid and would displace salicylic acid from extraliposomal complexes. EDTA did not extinguish the luminescence, except when the liposomes were lysed with *Triton X-100* (*Fig. 3,c*). Without EDTA, luminescence persisted also after lysis (*Fig. 3,d*).

As luminescence measurements are very sensitive, permeation studies can be performed at low permeant concentrations, *e.g.*, 5 µM salicylic acid with *ca.* 5 mM lipids. The low permeant concentration precludes effects on the membrane characteristics [4]. We currently investigate whether the assay can be adapted to a broad pH range.

4.4.8. *The Tb^{3+} Assay to Study Lipid Bilayer Permeation of Cell-Penetrating Peptides*

Therapeutic macromolecules such as peptides, proteins, oligonucleotides, or polysaccharides show poor membrane permeation if they do not bind to a receptor to enter the cell by endocytosis. A new concept of barrier permeation was introduced with the discovery of cell-penetrating peptides (CPPs) [54]. It was shown in several *in vivo* and *in vitro* studies that CPPs are able to carry macromolecules into cells. Based on studies with cell cultures, it was proposed that at least some CPPs are able to permeate lipid bilayers without the need of any energy-consuming mechanism. Using the Tb^{3+} liposomal assay, we could show that the CPP TAT(44–57), which was linked *via* a cystein to the aromatic carboxylic acid *N*-(4-carboxy-3-hydroxyphenyl)maleimide, did not permeate liposomal membranes of different lipid compositions [52]. In an assay using MDCK cells, a cell line of kidney origin with an epithelial phenotype [55], we found that fluorescein-labeled TAT(44–57) was unable to enter living cells. *Fig. 4* shows MDCK cells incubated with ethidium homodimer-1, which stains the nucleus of cells with impaired plasma membranes, and with the fluorescein-labeled TAT(44–57) peptide. Living cells showed accumulation of the peptide at the basal side but no peptide uptake. Only after cells started to deteriorate as seen from the ethidium staining, was TAT-fluorescein detectable in the cells. From this, we concluded that TAT(44–57) permeation across the lipid bilayer does not occur unassisted and that its cell-penetrating capability is not independent of cell type.

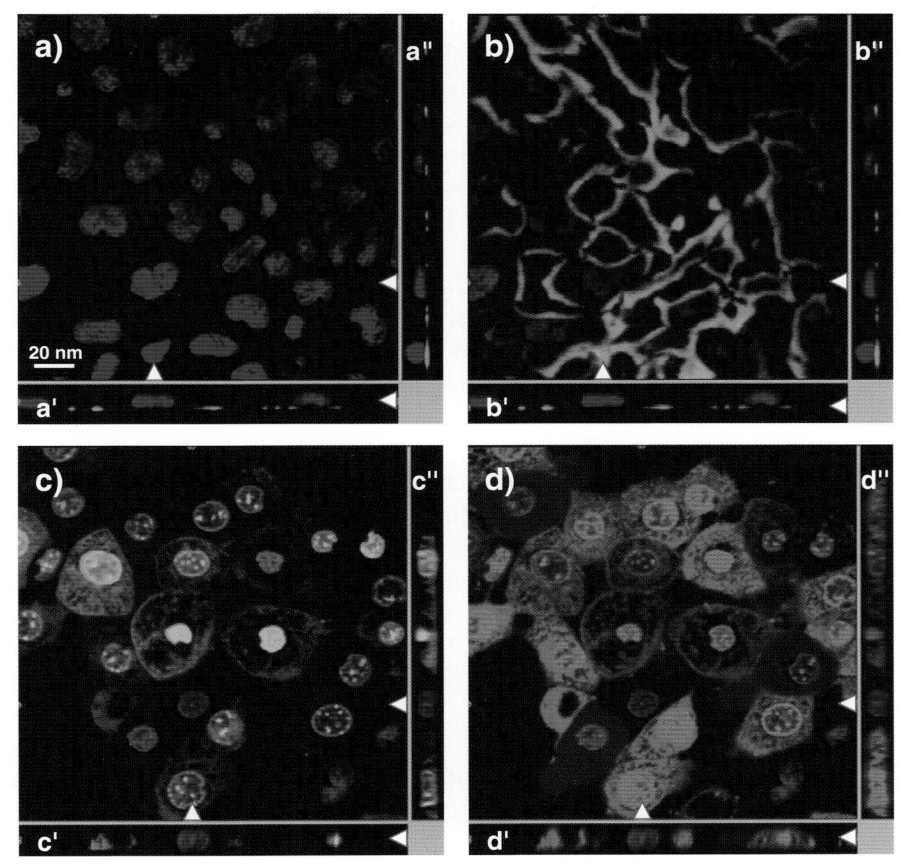

Fig. 4. *Lack of entry of the CPP TAT(44–57) into living MDCK cells.* Cells grown for 1–2 days were incubated for 20 min with 1 μM Cys-TAT(44–57)-maleimido-fluorescein, *Hoechst 33342* (nuclei), and ethidium homodimer-1 at 37°. Cells were mounted without washing or fixation. *a*)–*c*) *Hoechst 33342* (blue) and peptide fluorescence (green); *d*) *Hoechst 33342* (blue) and ethidium homodimer-1 fluorescence (red). *a*) *x,y* Optical section through the center of the nuclei immediately after mounting and *b*) through a lower part of the same cells close to the cover slip. a′, b′ and a″, b″: *x,z* and *y,z* projections. Arrowheads indicate the positions of the projections. *c*) *x,y* Optical section through the center of the nuclei of cells with peptide uptake 15 min after mounting. *d*) Ethidium homodimer-1 staining of the respective area. Turquoise and magenta staining of the nuclei indicates co-localization of *Hoechst 33342* with either peptide (*c*) or ethidium homodimer-1 (*d*). Reprinted from [52] with permission from *Elsevier*.

4.5. *Relevance of Lipid Bilayer Permeation for Passage across* in vivo *Barriers*

The concept that lipid bilayer permeation along a concentration gradient represents the major route for lipophilic drugs to cross *in vivo* barriers has recently been challenged, based on the ever increasing number of transporter proteins that keep being published. *Al-Awqati* postulated that most drugs are substrates of a transporter and that only a minority of drugs simply permeates lipid bilayers [56].

However, the fact that most drugs show linear pharmacokinetics implies that nonspecific mechanisms are more relevant for the ADME processes than specific mechanisms. If transporters would be relevant for the pharmacokinetics of most drugs, nonlinear pharmacokinetics should occur much more frequently due to the saturability of transporter systems. In addition, the pharmacokinetic behavior of most drugs correlates with their physicochemical characteristics. This would not be the case if their ADME processes would be controlled by different transport systems. Of course, many outliers exist, and, for some of them, relevant transport systems have been identified.

Lipid bilayer permeation is even critical for drugs transported by P-gp, one of the most-intriguing efflux system in drug therapy. There is evidence that fast-permeating drugs cannot be expelled to a sufficient extent from cells by the multidrug resistance transporter P-gp, but can act as P-gp modulators by inhibiting the efflux of other drugs [30].

5. Conclusions

With our growing understanding of cell membrane domains and microenvironments of target membrane proteins and transporters, the partition and permeation behavior of drugs and drug candidates attracts new interest. Phenomena such as reversibility of phospholipidosis, bilayer permeation of CPPs or P-gp modulation have been elucidated using partition and permeation experiments with model membranes. Efforts are needed to develop simple reference systems to measure permeation kinetics across lipid bilayers.

Assays on membrane affinity and permeation will progressively fill the gaps between test-tube experiments in early drug discovery and the pharmacological effects and pharmacokinetic behavior *in vivo*.

I thank *Heidi Wunderli-Allenspach* for very interesting discussions and her helpful input to this manuscript. I am grateful to *Bernard Testa*, *Maja Günthert*, and *Anita Thomae* for carefully reading it.

REFERENCES

[1] O. G. Mouritsen, *Chapt. 3*, p. 49–70.
[2] W. Dowhan, *Annu. Rev. Biochem.* **1997**, *66*, 199.
[3] M. Edidin, *Nat. Rev. Mol. Cell Biol.* **2003**, *4*, 414.
[4] S. D. Krämer, in 'Pharmacokinetic Optimization in Drug Research: Biological, Physicochemical, and Computational Strategies', Eds. B. Testa, H. van de Waterbeemd, G. Folkers, R. Guy, VHCA, Zürich, 2001, p. 401–428.
[5] M. Marenchino, A. L. Alpstäg-Wöhrle, B. Christen, H. Wunderli-Allenspach, S. D. Krämer, *Eur. J. Pharm. Sci.* **2004**, *21*, 313.
[6] A. Avdeef, *Chapt. 14*, p. 221–241.
[7] S. D. Krämer, C. Jakits-Deiser, H. Wunderli-Allenspach, *Pharm. Res.* **1997**, *14*, 827.
[8] S. D. Krämer, A. Braun, C. Jakits-Deiser, H. Wunderli-Allenspach, *Pharm. Res.* **1998**, *15*, 739.
[9] R. Lüllmann-Rauch, in 'Lysosomes in Applied Biology and Therapeutics', Eds. J. T. Dingle, P. J. Jacques, I. H. Shaw, North Holland Publishers, New York, 1979, p. 49–130.
[10] U. E. Honegger, I. Scuntaro, U. N. Wiesmann, *Biochem. Pharmacol.* **1995**, *49*, 1741.
[11] S. Paula, A. G. Volkov, D. W. Deamer, *Biophys. J.* **1998**, *74*, 319.
[12] J. Gutknecht, D. C. Tosteson, *Science* **1973**, *182*, 1258.
[13] F. Frezard, A. Garnier-Suillerot, *Biochim. Biophys. Acta* **1998**, *1389*, 13.
[14] A. Finkelstein, *J. Gen. Physiol.* **1976**, *68*, 127.
[15] A. V. Thomae, H. Wunderli-Allenspach, S. D. Krämer, free communication in the accompanying CD-ROM.
[16] A. V. Krylov, P. Pohl, M. L. Zeidel, W. G. Hill, *J. Gen. Physiol.* **2001**, *118*, 333.
[17] E. Overton, *Vierteljahrsschrift Naturforsch. Ges. Zürich* **1899**, *44*, 88.
[18] A. Finkelstein, *J. Gen. Physiol.* **1976**, *68*, 127.
[19] M. Poznansky, S. Tong, P. White, J. Milgram, A. Solomon, *J. Gen. Physiol.* **1976**, *67*, 45.
[20] J. Wolosin, H. Ginsburg, W. Lieb, W. Stein, *J. Gen. Physiol.* **1978**, *71*, 93.
[21] P. T. Mayer, T. X. Xiang, B. D. Anderson, *AAPS PharmSci* **2000**, *2*, E14.
[22] P. Pohl, S. M. Saparov, Y. N. Antonenko, *Biophys. J.* **1998**, *75*, 1403.
[23] P. Pohl, T. I. Rokitskaya, E. E. Pohl, S. M. Saparov, *Biochim. Biophys. Acta* **1997**, *1323*, 163.
[24] V. Y. Evtodienko, D. I. Bondarenko, Y. N. Antonenko, *Biochim. Biophys. Acta* **1999**, *1420*, 95.
[25] V. Y. Erukova, O. O. Krylova, Y. N. Antonenko, N. S. Melik-Nubarov, *Biochim. Biophys. Acta* **2000**, *1468*, 73.
[26] P. A. Monnard, D. W. Deamer, *Orig. Life Evol. Biosphere* **2001**, *31*, 147.
[27] J. R. Alger, J. H. Prestegard, *Biophys. J.* **1979**, *28*, 1.
[28] A. Walter, J. Gutknecht, *J. Membr. Biol.* **1984**, *77*, 255.
[29] T. Xiang, Y. Xu, B. D. Anderson, *J. Membr. Biol.* **1998**, *165*, 77.
[30] G. D. Eytan, R. Regev, G. Oren, Y. G. Assaraf, *J. Biol. Chem.* **1996**, *271*, 12897.
[31] G. D. Eytan, R. Regev, G. Oren, C. D. Hurwitz, Y. G. Assaraf, *Eur. J. Biochem.* **1997**, *248*, 104.
[32] J. N. Weinstein, S. Yoshikami, P. Henkart, R. Blumenthal, W. A. Hagins, *Science* **1977**, *195*, 489.
[33] H. Komatsu, P. L. Chong, *Biochemistry* **1998**, *37*, 107.
[34] J. Davidsen, O. G. Mouritsen, K. Jorgensen, *Biochim. Biophys. Acta* **2002**, *1564*, 256.
[35] D. Marathe, K. P. Mishra, *Radiat. Res.* **2002**, *157*, 685.
[36] R. Regev, G. D. Eytan, *Biochem. Pharmacol.* **1997**, *54*, 1151.
[37] I. Peri, H. Mamrud-Brains, S. Rodin, V. Krizhanovsky, Y. Shai, S. Nir, M. Naim, *Am. J. Physiol., Cell Physiol.* **2000**, *278*, C17.
[38] A. Sigler, P. Schubert, W. Hillen, M. Niederweis, *Eur. J. Biochem.* **2000**, *267*, 527.
[39] M. A. Requero, F. M. Goni, A. Alonso, *Biochemistry* **1995**, *34*, 10400.
[40] R. Busche, J. Dittmann, H. D. Meyer zu Duttingdorf, U. Glockenthor, W. von Engelhardt, H. P. Sallmann, *Biochim. Biophys. Acta* **2002**, *1565*, 55.
[41] S. Kaiser, H. Hoffmann, *J. Colloid Interface Sci.* **1996**, *184*, 1.

[42] T. Miron, A. Rabinkov, D. Mirelman, M. Wilchek, L. Weiner, *Biochim. Biophys. Acta* **2000**, *1463*, 20.
[43] S. C. Kinsky, J. A. Haxby, D. A. Zopf, C. R. Alving, C. B. Kinsky, *Biochemistry* **1969**, *8*, 4149.
[44] T. X. Xiang, B. D. Anderson, *J. Pharm. Sci.* **1995**, *84*, 1308.
[45] T. X. Xiang, B. D. Anderson, *Biophys. J.* **1997**, *72*, 223.
[46] T. X. Xiang, B. D. Anderson, *Biochim. Biophys. Acta* **1998**, *1370*, 64.
[47] T. X. Xiang, B. D. Anderson, *Biophys. J.* **1998**, *75*, 2658.
[48] P. Muller, A. Herrmann, *Biophys. J.* **2002**, *82*, 1418.
[49] V. Neitchev, E. Kostova, M. Goldenberg, L. Doumanova, *Int. J. Biochem. Cell Biol.* **1997**, *29*, 689.
[50] C. Dordas, P. H. Brown, *J. Membr. Biol.* **2000**, *175*, 95.
[51] W. G. Hill, R. L. Rivers, M. L. Zeidel, *J. Gen. Physiol.* **1999**, *114*, 405.
[52] S. D. Krämer, H. Wunderli-Allenspach, *Biochim. Biophys. Acta* **2003**, *1609*, 161.
[53] J. Wilschut, N. Duzgunes, R. Fraley, D. Papahadjopoulos, *Biochemistry* **1980**, *19*, 6011.
[54] M. Lindgren, M. Hallbrink, A. Prochiantz, U. Langel, *Trends Pharm. Sci.* **2000**, *21*, 99.
[55] J. A. McRoberts, M. Taub, M. H. Saier, in 'Functionally Differentiated Cell Lines', Ed. G. Sato, Liss, New York, 1981, p. 117–139.
[56] Q. Al-Awqati, *Nat. Cell Biol.* **1999**, *1*, E201.

High-Throughput Solubility, Permeability, and the MAD PAMPA Model[1])

by **Alex Avdeef**

*p*ION Inc., 5 Constitution Way, Woburn, MA 01801, USA
(tel: +1 781 935 8939, ext. 22; fax: +1 781 935 8938; e-mail: aavdeef@pion-inc.com)

Abbreviations
A: Area of the PAMPA filter [cm^2]; AP: absorption potential; BA: bioavailability; BBB: blood–brain barrier; BCS: biopharmaceutics classification system; BE: bioequivalence; c_A, c_D: acceptor and donor aqueous solute concentration, resp. [mol cm^{-3}]; $c_m(x)$: solute concentration inside of a membrane, at position x [mol cm^{-3}]; c_o: aqueous concentration of the uncharged species [mol cm^{-3}]; D_{aq}: aqueous diffusivity [cm^2 s^{-1}]; D_m: diffusivity of a solute inside a membrane [cm^2 s^{-1}]; D_{mw}: pH-dependent membrane/water distribution coefficient; GIT: gastrointestinal tract; h: membrane thickness [cm]; HIA: human intestinal absorption; J: flux across a membrane [mol cm^{-2} s^{-1}]; k_a: absorption rate constant [min^{-1}]; MAD: maximum absorbable dose [mg]; PAMPA: parallel artificial membrane permeability assay; P_{app}: apparent permeability coefficient, based on rat absorption rate constants [cm s^{-1}]; P_e: effective artificial membrane permeability, pH-dependent [cm s^{-1}]; pK_a: ionization constant (negative log form); P_o: intrinsic artificial membrane permeability, pH-independent [cm s^{-1}] ; P_{ow}: octanol/water partition coefficient for an uncharged species; S: solubility [M, μg ml^{-1}, or mg ml^{-1}]; S_o: intrinsic solubility, that is, the solubility of the uncharged species; UWL: unstirred water layer; V_L: volume of luminal fluid (250 ml); α: empirical constant (usual values 0.7–1.0); ε_a: apparent microfilter porosity (defined by *Eqn. 13*); ε: nominal microfilter porosity, as specified by the manufacturer; ν: stirring speed [rpm].

1. Introduction

1.1. *Absorption*

It is well accepted that physicochemical properties of oral drug candidates can be used to predict their biological activity. This chapter discusses how human intestinal absorption (HIA) of ionizable compounds can depend simultaneously on three key properties: solubility, permeability, and pK_a. This relationship is exemplified by the absorption potential (AP) [1], the biopharmaceutics classification system (BCS) [2–4], and the

[1]) Contribution No. 14 in the 'PAMPA – a Drug Absorption Model' series from *p*ION.

maximum absorbable dose (MAD) calculations [5][6]. *Fick*'s laws of diffusion [7] and the pH-partition hypothesis [8] underlie these relationships.

For ionizable test compounds, the pH at the site of absorption affects the rate of absorption. The pH-partition hypothesis suggests that permeability will be highest at the pH where the molecule is least charged. *This is also the pH where the molecule is least soluble.* Hence, at the site of absorption, the amount of the uncharged species and the tendency of the neutral species to cross the phospholipid membrane barrier are both important predictors of absorption. The *intrinsic* permeability coefficient, P_o, characterizes the membrane transport of the *uncharged* species, whose concentration in saturated solution (excess solid present) is determined by the *intrinsic* solubility constant, S_o, which puts a limit on the concentration of the *uncharged* form of the test compound, c_o. The pK_a of a compound, together with the pH at the site of absorption, can be used to calculate the amount of uncharged species present in solution.

In this chapter, the relationship between solubility, permeability, and pK_a will be reviewed, with PAMPA (parallel artificial membrane permeability assay) [9] chosen as a prototypical measure of permeability. It has been demonstrated that PAMPA can predict biological measures of permeability: Caco-2 permeability [10], rat intestinal permeability [10], and human jejunal permeability [11]. Primary application focus here will be on how passive absorption can be quickly and conveniently approximated by the MAD model – not with the rigor and precision of complex *in silico* calculations, *e.g.*, GastroPlus™ [12], but with precision suitable to make earliest classifications of discovery molecules. Usually, MAD calculations use rat intestinal absorption rate constants [6], which are often determined during candidate-optimization stages of discovery. So, the traditional MAD analysis is limited by a very low-throughput *in vivo* assay. We will propose a high-throughput *in vitro* variant of MAD. Namely, PAMPA will be used as a substitute for rat data, to allow MAD calculations to be made much earlier in the discovery process, and at a greatly reduced assay cost.

1.2. *Solubility and Permeability in* Fick*'s First Law*

Fick's first law applied to homogeneous membranes at steady state [7][11] may be stated as (*Eqn. 1*):

$$J = D_m \frac{dc_m}{dx} = D_m[c_m(0) - c_m(h)]/h \qquad (1)$$

where J is the flux [mol cm^{-2} s^{-1}], $c_m(0)$ and $c_m(h)$ are the concentrations [mol cm^{-3}] of solute *within* the membrane at the two water–membrane boundaries (at positions $x=0$ and $x=h$, where h is the thickness of the membrane [cm], and D_m is the diffusivity of the solute within the membrane [cm^2 s^{-1}]. At steady state, the concentration gradient, dc_m/dx within the membrane is linear, hence the difference may be used in the right side of *Eqn. 1*.

The limitation of *Eqn. 1* is that measurement of concentrations of solute within different parts of a membrane is not feasible. However, since one can estimate (or possibly measure) the *apparent* (pH-dependent) partition coefficients (in modern terms, the distribution coefficients) between bulk water and the membrane, D_{mw}, one can convert *Eqn. 1* into a more-accessible form (*Eqn. 2*):

$$J = D_m\, D_{mw}\, (c_D - c_A)/h \tag{2}$$

where the bulk water concentrations in the donor and acceptor compartments (c_D and c_A, resp.) can be readily measured. It is common practice to combine D_m, D_{mw}, and h into one composite quantity, called 'effective permeability', P_e (*Eqn. 3*):

$$P_e = D_m\, D_{mw}/h \tag{3}$$

The link of *Eqn. 2* to solubility comes in the concentration terms. Consider 'sink' conditions, where c_A is essentially zero. *Eqn. 2* reduces to the following simple flux equation:

$$J = P_e\, c_D \tag{4}$$

c_D may be ideally equal to the dose of the drug, unless the dose exceeds the solubility limit, in which case it is equal to the solubility. If only the uncharged molecular species permeates membrane barriers appreciably, then *Eqn. 4* may be restated as *Eqn. 5*:

$$J = P_o\, c_o \leq P_o\, S_o \tag{5}$$

where P_o and c_o are the intrinsic permeability and concentration of the uncharged species, respectively. The intrinsic permeability does not depend on pH (thin lines in *Fig. 1*), but its cofactor in the flux equation, c_o, does (dashed lines in *Fig. 1*). The concentration of the uncharged species is always equal to or less than the intrinsic solubility of the species, S_o (*Eqn. 5*).

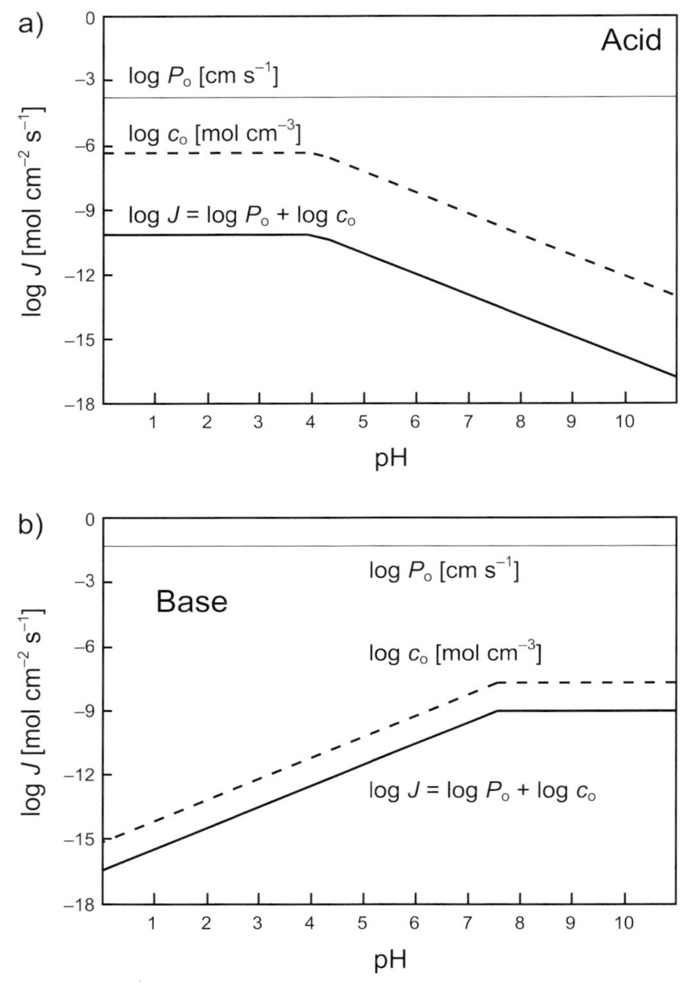

Fig. 1. *Logarithm of flux* vs. *pH plots for* a) *an acid* (ketoprofen, 75 mg dose) *and* b) *a base* (verapamil, 180 mg dose), *calculated at dose concentrations* [13]

In solutions that are saturated at some pH, the plot of log c_o *vs.* pH for an ionizable molecule is simply a combination of straight segments, joined at a point of discontinuity indicating the boundary between the saturated state and the state of complete dissolution. The pH of these junction points is dependent on the dose level used in the calculation, and the maximum value of c_o is equal to S_o in a saturated solution (*Eqn. 5*).

Fig. 1,a shows that, for an acid, log c_o is a dashed horizontal line (log c_o = log S_o) in the saturated solution (at low pH), and decreases with a slope of -1 in the pH domain, where the solute is completely dissolved (pH $>$ 4). *Fig. 1,b* shows that, for a base, the plot of log c_o *vs.* pH is a dashed horizontal line at high pH ($>$7.5) in a saturated solution and is a line with a slope of $+1$ for pH values less than the pH of the onset of precipitation. The log flux curves are indicated by the thick lines in *Fig. 1*.

1.3. *pH at the most Probable Site of Absorption*

The properties of the gastrointestinal tract (GIT) have been reviewed extensively (*e.g.*, [11]). *Ca.* 99% of the surface area available for absorption is mainly concentrated in the jejunum and ileum, and the transit time of luminal content in these intestinal segments is usually not longer than 270 min. At the beginning of the journey in the small intestine, the proximal jejunal pH can be as low as 5 and the distal ileal pH can be as high as 8. In a fasted state, the average pH of the small intestine is *ca.* 6.5; *ca.* 1 to 2 h after food intake, the small intestine pH may drop to 5.

1.4. *Absorption Potential*

Dressman et al. [1] proposed the AP function as a simple basis for predicting the HIA fraction. Performing a dimensional analysis of the factors likely to be associated with absorption, neglecting luminal degradation and first-pass metabolism, the authors proposed the following function (*Eqn. 6*):

$$\mathrm{AP} = \log\left[\left(D_{\mathrm{mw}}^{6.5} S_{\mathrm{o}}^{37} V_{\mathrm{L}}\right)/X_{\mathrm{o}}\right] \tag{6}$$

The membrane/water distribution coefficient at pH 6.5 ($D_{\mathrm{mw}}^{6.5}$) was approximated by measurements in the octanol/water system, since neither the PAMPA nor the Caco-2 permeability model was known at that time. The intrinsic solubility at 37° is represented by S_{o}^{37} [mg ml^{-1}]; the luminal intestinal volume V_{L} was estimated as 250 ml; X_{o} is the dose in mg. The AP function is dimensionless. It can be seen that the numerator in *Eqn. 6* is proportional to the flux function in *Eqn. 5*. *Eqn. 6* implies a relationship between solubility, permeability, and pK_{a}. Seven compounds were used by the original authors to illustrate the AP concept. Examples of the use of the AP function may be found in the recent literature [14].

1.5. *The Biopharmaceutics Classification System* (BCS)

The transport model considered in this chapter, based on permeability and solubility, is also incorporated in the BCS proposed by the *FDA* as a bioavailability-bioequivalence (BA/BE) regulatory guideline [3]. The BCS allows estimation of the likely contributions of three major factors: dissolution, permeability, and solubility in the pH range from 1 to 7.5, which affect oral drug absorption from immediate-release solid oral products. *Fig. 2* shows the four BCS classes, based on high and low designations of solubility and permeability. If a molecule is classed as highly soluble, highly permeable (*Class 1*), and does not have a narrow therapeutic index, it may qualify for a waiver of the very expensive BA/BE clinical testing. Such a classification can save a pharmaceutical company considerable costs.

The solubility scale is defined in terms of the volume (ml) of water required to dissolve the highest dose strength at the lowest solubility in the pH range from 1 to 7.5, with 250 ml being the dividing line between high and low. So, high solubility refers to complete dissolution of the highest dose in 250 ml in the pH range from 1 to 7.5.

Permeability is the major rate-controlling step when absorption kinetics from the GIT is controlled by drug biopharmaceutical factors and

Biopharmaceutics Classification System

	HIGH SOLUBILITY	LOW SOLUBILITY
HIGH PERMEABILITY	CLASS 1 (amphiphilic) diltiazem, phenylalanine, labetolol, antipyrine, captopril, glucose, enalapril, L-dopa, metoprolol, propranolol 1	CLASS 2 (hydrophobic) flurbiprofen, ketoprofen, naproxen, desipramine, diclofenac, itraconazole, piroxicam, verapamil, carbamazepine, phenytoin 2
LOW PERMEABILITY	CLASS 3 (hydrophilic) famotidine, atenolol, cimetidine, acyclovir, ranitidine, nadolol, hydrochlorothiazide 3	CLASS 4 terfenedine, furosemide, cyclosporin 4

Fig. 2. *Biopharmaceutics classification system*. If the highest dose of a drug is fully dissolved in 250 ml over the pH interval 1–7.5, the solubility is classed high. If the permeability is classed 'high', a drug has a human jejunal permeability equal to or greater than 10^{-4} cm s^{-1}. With *Class 1* drugs, the rate of dissolution limits the *in vivo* absorption rate. With *Class 2* drugs, solubility limits absorption flux. With *Class 3* drugs, permeability is rate determining. With *Class 4* drugs, no *in vitro–in vivo* correlation is expected.

not by formulation factors. Permeability in the BCS is based on human jejunal permeability values, with 'high' being $P_{app} > 10^{-4}$ cm s^{-1} and 'low' being below that value. Values of well-known drugs have been determined *in vivo* at pH 6.5 [15]. The high permeability class boundary is intended to identify drugs that exhibit nearly complete absorption (>90% of an administered oral dose) from the small intestine.

1.6. *Maximum Absorbable Dose*

Medicinal chemists, charged with making new compounds in a discovery project, can be imagined to often ask the question: 'How soluble do I need to make my drug candidates, so that they are well-enough absorbed?' *Johnson* and *Swindell* [5] and *Curatolo* [6] described a relatively simple approach that can serve as a conceptual tool for understanding the quantitative aspects of drug absorption (*Eqn. 7*):

$$\mathrm{MAD} = S^{6.5}\, V_{\mathrm{L}}\, k_{\mathrm{a}}\, t \tag{7}$$

The units of MAD are mg, referring to the expected drug quantity absorbed during the transit time t (270 min). The solubility at pH 6.5 is indicated by $S^{6.5}$ [mg ml^{-1}]. The estimated luminal volume is V_{L} (250 ml). The transintestinal absorption rate constant is indicated by k_{a} [min^{-1}].

Metaphorically, MAD is easy to visualize. Imagine a slightly 'leaky' tube (representing the small intestine), whose internal volume is 250 ml. A slurry is made in a beaker, by placing enough drug into 250 ml of pH 6.5 buffer, so that a saturated solution forms. The dissolution process is allowed to reach equilibrium. The number of mg that dissolve is equal to $S^{6.5}\, V_{\mathrm{L}}$. The drug suspension is then poured into the leaky (permeable) tube, and the timer is started. The absorption rate constant, k_{a}, describes how quickly the dissolved quantity of drug 'leaks' out of the tube each minute. If the process is allowed to proceed for 270 min, then the total amount of 'leakage' is the estimate of the amount of drug that should be intestinally absorbed. If that estimated quantity exceeds the likely dose for the drug candidate, human intestinal absorption should be 100%. This is a very simple model, and yet it is thought to be quite useful to medicinal chemists for estimating a target value for the solubility of their candidate compounds [16].

The transintestinal absorption rate constant in *Eqn. 7* becomes available after rat intestinal perfusion experiments are done. This usually takes place during lead optimization in the discovery project. We will propose to substitute k_{a} values with PAMPA-based permeability, to make the MAD calculation possible at a much earlier time in discovery, long before the expensive *in vivo* studies. Absorption rate constants and membrane permeability are related by the expression (*Eqn. 8*):

$$k_a = P_{app} \cdot A/V_L \tag{8}$$

where P_{app} is the *in vivo* apparent permeability coefficient and A the absorption surface area. In a typical rat perfusion experiment, the length of the subjected intestine is estimated to be 100 cm, and the radius is 0.178 cm [10]. Hence, the P_{app} is estimated by multiplying the absorption rate constant by one-half the radius. To perform early MAD calculations, a link needs to be made between P_{app} based on rat data and P_e, the effective permeability based on PAMPA measurement (see *Sect. 4*).

1.7. Hansch–Fick *Solubility–Permeability Relationship*

Fig. 3,a shows the empirical relationship (*Eqn. 9*) between the intrinsic solubility (log S_o, with S_o in mol cm^{-3}) and the octanol/water partition coefficient (log P_{ow}). Such *Hansch*-type analysis suggests at least a trend between solubility and lipophilicity:

$$\log S_o = -0.63 \log P_{ow} - 2.0 \tag{9}$$

$$n = 101;\ r^2 = 0.56;\ s = 1.1$$

Fig. 3,b shows the empirical relationship (*Eqn. 10*) between the intrinsic permeability, log P_o, and the octanol/water partition coefficient:

$$\log P_o = +1.18 \log P_{ow} - 5.7 \tag{10}$$

$$n = 164;\ r^2 = 0.68;\ s = 1.5$$

The two equations may be combined to eliminate the lipophilicity term:

$$\log P_o = -1.87 \log S_o - 15.0 \tag{11}$$

It must be emphasized that the above relationship holds only for passive transport. From *Eqn. 11*, it can be concluded that very insoluble compounds must be very permeable. So, how is it possible to have a BCS Class 4? This will be further discussed in *Sect. 5*.

Taking S_o [mol cm^{-3}], *Eqns. 5* and *11* may be combined to produce the empirical relationship for a saturated solution (*Eqn. 12*):

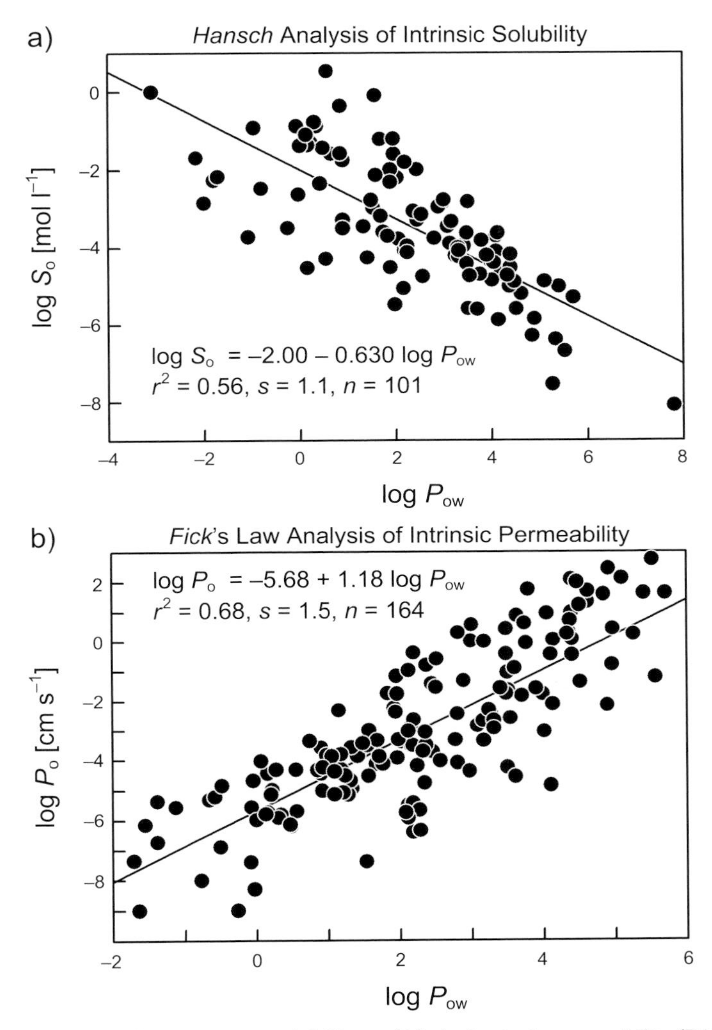

Fig. 3. *Logarithm of* a) *the intrinsic solubility and* b) *the intrinsic permeability* (PAMPA) vs. *logarithm of the octanol/water partition coefficient for a series of common drug molecules, largely taken from* [11]

$$J = 10^{-15}/S_o^{0.87} \tag{12}$$

Consider a very soluble molecule with the intrinsic solubility of 1000 μM. According to the above equation, the flux is expected to be *ca.* 1.6×10^{-10} mol cm^{-2} s^{-1}. It is interesting to note that, were the compound very insoluble, say 1 μM, the calculated flux according to *Eqn. 12* would be 420 times *greater* than the value calculated for the very soluble compound.

2. Recent Advances in the PAMPA Technology

The acronym PAMPA, or parallel artificial membrane permeability assay, was coined by *Kansy* and co-workers in their widely read 1998 paper [9]. The published *Hoffmann-La Roche* PAMPA method involves creating a filter-immobilized artificial membrane by infusing a lipophilic microfilter with 10% *w/v* egg lecithin dissolved in dodecane. The filter membrane is used to separate an aqueous solution containing a test compound from an aqueous buffer initially free of the molecule (*Fig. 4*). PAMPA enables the kinetics of transport by diffusion to be studied in this permeation cell. Microtitre plate technology allows 96 permeation cells to be simultaneously formed, increasing the speed while lowering the cost.

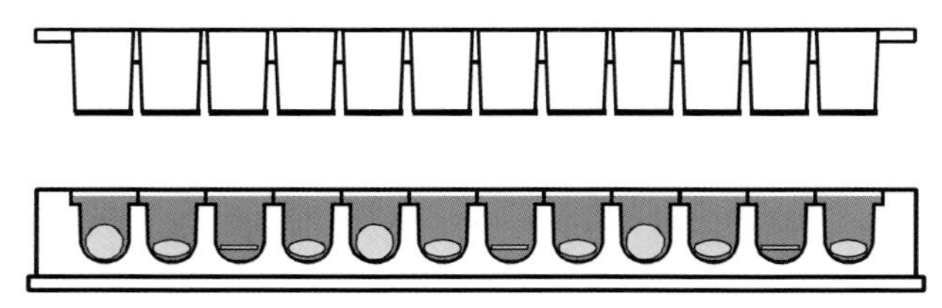

Fig. 4. *PAMPA Sandwich: perspective view and cross sectional view.* Individual well stirrers (lower view) are magnetically driven.

In the two-year period after the original publication, several companies developed their own versions of the assay. During this period, a commercial instrument was launched by *p*ION Inc. At the *logP2000 Conference* organized by *Bernard Testa* in Lausanne in March 2000, *Kansy et al.* [17], *Avdeef* [18], and *Faller* and *Wohnsland* [19] discussed the emerging PAMPA technology. Since then, several other PAMPA papers have appeared, *e.g.*, *Avdeef et al.* [20], *Wohnsland* and *Faller* [21], *Sugano et al.* [22–25], *Zhu et al.* [26], *Veber et al.* [27], and *Di et al.* [28]. The first international symposium on the topic was held in San Francisco in July 2002 (www.pampa2002.com), where nearly all researchers known to be involved with the new technique presented papers and posters. Several reviews with PAMPA coverage have been published, citing the literature between 1998 and 2003 [13][29–35], and a book by *Avdeef* [11] devotes a large portion to the topic of PAMPA.

2.1. *Membrane Retention and pH Gradients*

Wohnsland and *Faller* [19][21] and *Sugano et al.* [22–25] defined effective permeability equations used in PAMPA assuming zero loss of test compound to the lipid phase and to the plastic surfaces of the wells (mass balance assumed to be confined to the aqueous phase). The equations presented by *Avdeef* [11][34] directly incorporate the additional effects of *a*) membrane retention (complete mass balance) and *b*) pH gradients.

2.2. *Filter Porosity*

Faller's group was the first to factor filter porosity into their permeability equation by multiplying the filter area by the nominal porosity of 0.2. This made their permeability scale five times greater than that of others who neglect the porosity correction when using filters of the same porosity (in effect, assuming porosity to be unit value).

Nielsen and *Avdeef* [36] introduced the concept of apparent filter porosity. If more lipid is deposited on the filter than can be accommodated by the volume of the pores, the apparent porosity (ε_a) is different from the nominal porosity (ε). For example, *Faller*'s group [19][21] deposited 0.75 μl of hexadecane on top of 10 μm thick polycarbonate filters, which had the nominal porosity $\varepsilon = 0.2$. The lipid volume substantially exceeded the pore volume. The resulting thickness of the excess lipid layer is 29 μm, giving a total apparent membrane thickness of 39 μm. The excess lipid significantly alters the contribution of the pores to the overall resistance in the membrane barrier. Analysis of the geometry suggests that, instead of using the nominal porosity $\varepsilon = 0.2$, it would have been appropriate to apply the apparent porosity $\varepsilon_a = 0.50$, a 150% increase over the nominal value. In such cases, the apparent porosity is calculated from *Eqn. 13* [36]:

$$\varepsilon_a = \frac{V/A + h(1 - \varepsilon)}{[V/A + h(1/\varepsilon - \varepsilon)]} \tag{13}$$

where V [cm^3] is volume of lipid deposited, A [cm^2] is the filter area, h [cm] is the filter thickness, and ε is the nominal filter porosity. Most PAMPA practitioners use the metrics and IPVH filter suggested by *Kansy*, *i.e.*, $V = 4$ μl, $A = 0.3$ cm^2, and $h = 125$ μm. The nominal filter porosity of Millipore's IPVH filters is $\varepsilon = 0.7$ according to the manufacturer. The resulting thickness of the excess lipid layer is 46 μm, giving a total apparent membrane thickness of 171 μm. Applying these metrics to *Eqn. 13* yields $\varepsilon_a = 0.76$, a 9% increase from the nominal value.

2.3. *PAMPA Lipids*

In PAMPA experiments, test compounds need to transfer by passive diffusion through the membrane environment created on the filter plate. Since pure phospholipids are solids that will not disperse into filters, a nonpolar solvent is normally used to dissolve the phospholipid prior to filter coating. This allows experimenters flexibility in choice of lipid, provided the lipid can be dissolved in an inert solvent. *Faller*'s group [19][21] has demonstrated that some solvents alone (*e.g.*, hexadecane) can provide adequate results for simple permeability testing. Further attempts to replicate *in vivo* conditions using highly biomimetic lipid compositions have also been made. *Sugano* and co-workers [22–25] demonstrated the use of complex lipid combinations similar to those found *in vivo* [11]. The high cost of these lipids may prohibit their widespread adoption, but the *Sugano* lipid formulation demonstrates the open system nature of PAMPA. In searching for the ideal PAMPA model to predict human jejunal permeability, *Avdeef* [11] reported the evaluation of *ca.* 50 different lipid compositions. A lecithin-based lipid combination, referred to as the *Double-Sink*™ formulation, has been described by *Avdeef* and co-workers [10][11][34][36][37]. Excellent correlations between PAMPA permeability based on this membrane and several absorption parameters have been demonstrated. Furthermore, this new lipid is very cost effective.

2.4. *Problems of Low Solubility*

The properties of buffer solutions used in the donor wells are very important to the experiment. A key problem found when implementing PAMPA is the low solubility of many research compounds, some soluble only in the low micromolar range. This fact is often not fully appreciated by experimenters since PAMPA papers tend to emphasize results based on catalog drug compounds with good aqueous solubility. These drugs can be assayed at concentrations up to 500 μM [22–25] without difficulty, but when low solubility compounds are encountered, the analytics may become problematic. Two approaches have been suggested to overcome this problem, namely the use of excipients or cosolvents.

Kansy, who used solutions of glycocholic acid in pH 6.5 buffer to solubilize compounds, first described the use of excipients in PAMPA experiments [17]. Other solubilizing agents have been tested, including cosolvents, to overcome the problems of low sample solubility. *Ruell et al.* [38] described a cosolvent procedure, based on the use of 20% *v*/*v* MeCN in a universal pH buffer. For the first time, it was possible to measure the permeability of compounds like cyclosporin A, paclitaxel, and amiodarone (6 *nanograms* ml^{-1} intrinsic solubility [11]). Their measured intrinsic permeability coefficients are 6.2×10^{-4}, 3.2×10^{-3}, and 13 cm s^{-1}, respectively. A procedure was devised, where values determined in 20% MeCN are extrapolated to cosolvent-free conditions.

2.5. *Lipophilicity Gradient* (Sink for Bases in the Acceptor Compartment)

The composition of the acceptor well solution plays an equally important role in the outcome of permeability experiments. In many reported literature studies, donor and acceptor solutions are of the same composition. This is contrary to the *in vivo* GIT conditions where compounds, after passing through the cells of the intestinal wall, are immediately removed from the receiver site by blood flow assisted by their binding to serum proteins. This sink state maintains the largest possible concentration gradient across the membrane and thus hastens the transfer across the intestinal barrier. In PAMPA experiments, adding carrier proteins and other agents that bind compounds in aqueous solution can be used. The detection method most often applied in PAMPA is direct UV spectroscopy, so proteins and other agents having swamping UV absorbance need to be avoided. *Avdeef* recently described a nonselective binding agent added to the acceptor well buffer to create a sink condition simulating the presence of serum proteins and blood flow [11][34]. Acceptor solution agents that strongly bind the test compounds can greatly reduce the time needed for permeability experiments.

2.6. *Stirring*

Because of the efficient mixing near the surface of the GIT, the *in vivo* unstirred water layer (UWL) is estimated to be 30–100 μm thick [39]. The UWL in the blood–brain barrier (BBB) is < 1 μm, given that the diameter of the capillaries is *ca.* 6 μm and the tight fit of the distorted circulating erythrocytes gives efficient mixing [40]. However, in unstirred *in vitro* permeation cells, the UWL can be 1500–4000 μm thick, depending on permeation cell geometry and dimensions [11][21][36][37]. If the assays ignore the UWL effect with lipophilic test compounds, the resulting permeability values will not correctly indicate the *in vivo* conditions of permeability, and will merely reveal properties of water rather than membrane permeation.

When the solution in a permeation cell is stirred, the thickness of the unstirred water layer decreases [10][11][20][21][41][42], and the UWL plays a lesser role in the overall *in vitro* transport process. To mimic the *in vivo* UWL, stirring must reduce the nascent thickness from 1500–4000 μm to less than 100 μm. The hydrodynamic equation [41] relating thickness of the UWL to the stirring speed is (*Eqn. 14*):

$$h = (D_{aq}/K)\ \nu^{-\alpha} \tag{14}$$

where D_{aq} is the aqueous diffusivity of the test compound, ν is the stirring speed [rpm], and K and α are empirical hydrodynamic constants. *Adson et al.* [41] reported values of $K = 4.1 \times 10^{-6}$ cm s^{-1} and $\alpha = 0.8$ based on data for testosterone in a stirred Caco-2 assay. *Avdeef et al.* [42] reported the values $K = 23 \times 10^{-6}$ cm s^{-1} and $\alpha = 0.71$ for PAMPA, based on the average behavior of 13 different lipophilic molecules, stirred from 49 to 622 rpm.

A most welcome aspect of stirring the solutions in PAMPA is that lipophilic molecules can be characterized with 15 min permeation assay times, a notable decrease from the 15 h originally used by *Kansy et al.* [9].

2.7. *Future Direction*

Although PAMPA has been used primarily in discovery projects, *Liu et al.* [43] have described a procedure for screening solubilizing excipients, signalling the entry of PAMPA into preformulation applications.

3. High-Throughput Solubility Measurement by the Self-Calibrating UV Method

The high-throughput direct-UV microtitre plate method used to characterize solubility–pH profiles of test compounds, using the μSOL Discovery/Evolution instruments (*p*ION), does not require serial dilution calibration curves [18]. Each sample serves as its own calibrant by the unique method.

Two identical 96-/384-well microtitre plates are filled with a constant capacity universal buffer (mixture of zwitterionic buffers, free of phosphate and boric acid), where each well may have its own pH adjusted automatically in the interval pH 3 to 10 (or 0.1M HCl may be used to produce gastric pH). To each well of the two plates, an aliquot of 10–30 mM DMSO stock solution of sample is added, typically to produce 70–200 μg ml^{-1} aqueous solutions, with 0.5–1% DMSO in the wells. One of the two plates serves as a 'reference' plate (see later *Fig. 5*) and the other serves as a 'sample' plate. Other concentration ranges may be used, and DMSO may be entirely eliminated in some applications.

A substantial amount of a nonvolatile water-miscible cosolvent is added quickly to each well in the reference plate, to inhibit/suppress any precipitation. The UV spectra are taken of the 'reference' plate as quickly as possible. On the other hand, the 'sample' plate is allowed to sit undisturbed, during which time no cosolvent is added to it. *Ca.* 15 h later, the excess solid in the sample plate is either filtered or removed by centrifugation. At this step, the same amount of cosolvent as was added to the 'reference' plate is added to the solid-free solutions in the 'sample' plate, and the UV spectra are taken.

Considerable attention to detail is pursued in the processing of the spectral data, with area-under-the-curve used for quantitation purposes, usually in the wavelength interval 240–500 nm. Blank spectra are taken, to subtract the absorption contributions of DMSO, buffer, and the plastic UV plate. Spectral shape analysis is performed to ensure the absence of decomposition and other artifacts, such as those caused by compounds that act as 'promiscuous inhibitors' [44]. The software running the assay has the capability to correct the solubility data for the effects of DMSO binding, excipients, and aggregation (*e.g.*, 'promiscuous inhibitor' effect), provided that reliable pK_a values of the compounds are known or can be estimated [18].

Although the assay seems to be slow (15 h usually), it is actually quite high in throughput, since microtitre-plate format is used, and it is possible to process *ca.* 1500–3000 wells per day. A variant of the assay, called 'kinetic solubility', is available, where the 15 h incubation time is reduced to less than 10 min. The results are usually different from the normal 'equilibrium' assay, and best match those obtained by the popular turbidity-based assays [16].

Fig. 5 shows the measured absorption spectra of phenazopyridine ('reference' and 'sample'). As precipitation takes place to varying degrees at different pH values, the spectra of the sample solutions change in optical densities, according to *Beer*'s law. This can be clearly seen in *Fig. 5,a* for the sample spectra, where the sample spectra have the lowest OD values at pH 9.0 and systematically show higher OD values as pH is lowered, a pattern consistent with that of a weak base. The changing OD values indicate that solubility changes with pH. *Fig. 6* shows the measured solubility–pH profile for phenazopyridine.

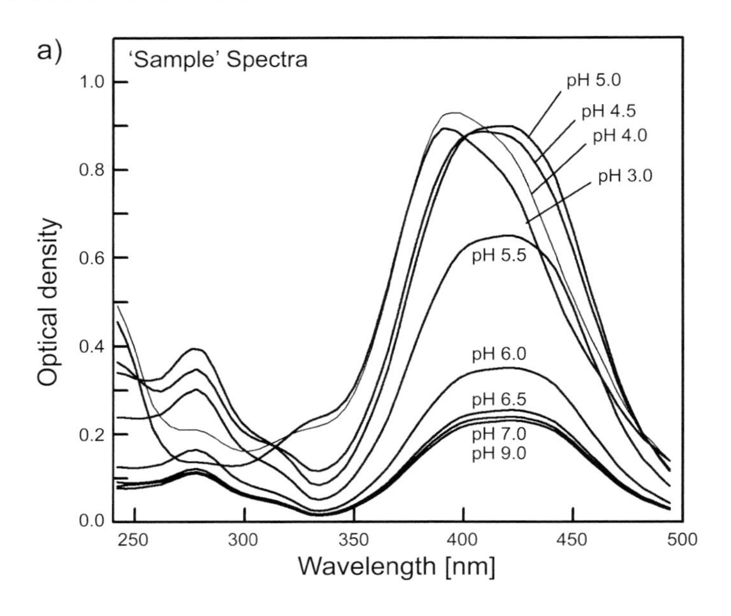

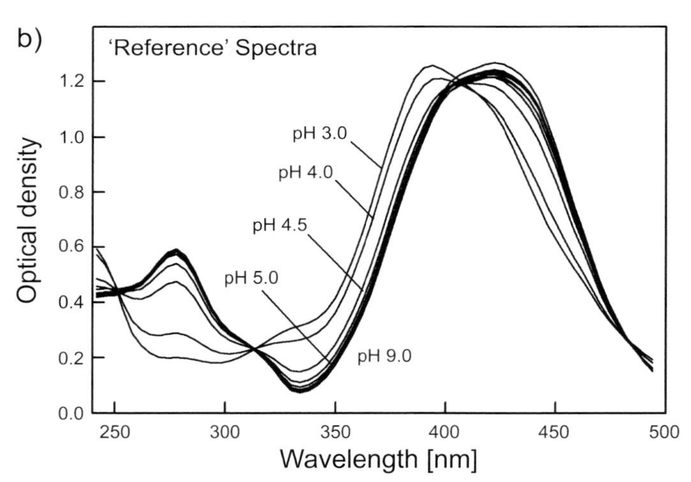

Fig. 5. *UV Spectra of phenazopyridine as a function of pH. a*) 'Sample' wells after filtration of precipitate, *b*) 'reference' wells with solutions prepared in such a way that precipitation was suppressed.

4. MAD PAMPA for Early Discovery

Bermejo et al. [10] described remarkable correlations between PAMPA, rat *in situ* intestinal perfusion, and Caco-2 permeability data based on 17 fluoroquinolones, including three congeneric series with

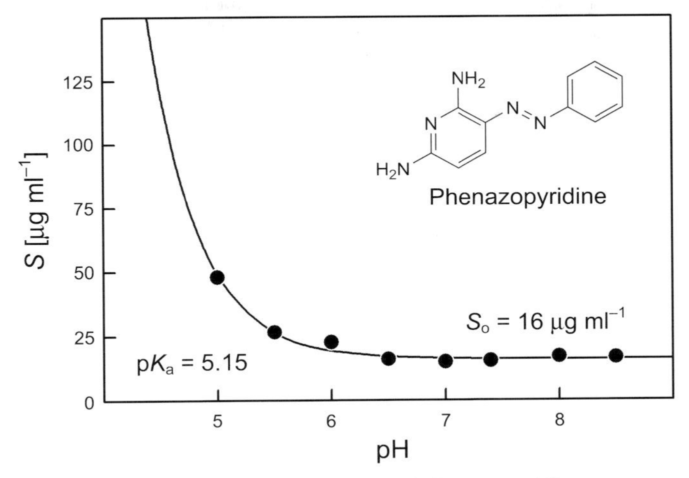

Fig. 6. *Solubility–pH Profile of phenazopyridine.*

systematically varied alkyl chain length at the distal N-atom ($N^{4\prime}$-position) of the piperazine residue. *Fig. 7* shows the correlation between the rat data and PAMPA [10]. The following simple correlation equation was proposed by the investigators:

$$\log P_{\mathrm{app}}^{\mathrm{Rat}} = -2.33 + 0.438 \log P_{\mathrm{o}}^{\mathrm{PAMPA}} \tag{15}$$

$$n=17;\ r^2=0.87;\ s=0.14;\ F=103.$$

From *Eqn. 8*, $P_{\mathrm{app}}=0.00148\,k_{\mathrm{a}}$, given the dimensions of the rat intestine. By combining *Eqns. 7, 8*, and *15*, MAD may be approximated on the basis of *Eqn. 16*:

$$\begin{aligned}\mathrm{MAD\ [mg]} &= 4.56 \cdot 10^4 \cdot S^{6.5}\ [\mu\mathrm{g\ ml^{-1}}] \cdot 10^{(-2.33+0.438\ \log\ P_{\mathrm{o}})} \\ &= S^{6.5}\ [\mu\mathrm{g\ ml^{-1}}] \cdot 10^{(+2.33+0.438\ \log\ P_{\mathrm{o}})}\end{aligned} \tag{16}$$

The *Table* attempts to answer the medicinal chemist's question: 'How soluble does a drug candidate need to be?' The *Table* lists the MAD analysis for seven fluoroquinolones, with data taken from [10]. The last three columns are the calculated target solubility values at pH 6.5, in units of µg ml^{-1}, corresponding to three levels of expected potency, corresponding to 0.1, 1, and 10 mg kg^{-1} dose levels. If the measured solubility exceeds the target values in the *Table*, then intestinal aborption is expected to be complete (100%). *Lipinski* and colleagues at *Pfizer* often refer to

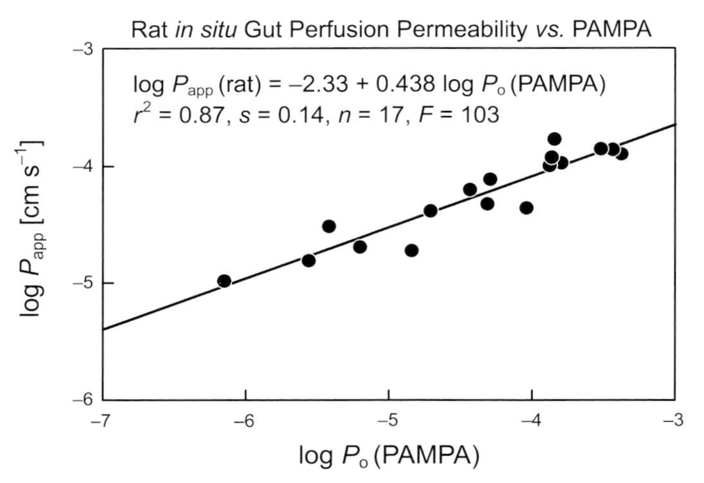

Fig. 7. *Rat* in situ *perfusion permeability compared to PAMPA intrinsic permeability for 17 substituted fluoroquinolones* (adapted from the work of *Bermejo et al.* [10])

three classes of permeability, implicitly based on rat absorption data (in 10^{-6} cm s^{-1}) [6][16], namely:

- low permeability: $P_{app} < 10$ or PAMPA < 1
- average permeability: $P_{app} = 10$ to 40 or PAMPA $= 1$ to 20
- high permeability: $P_{app} > 40$ or PAMPA > 20.

If a compound is expected to be formulated with an average dose of 70 mg and is showing average permeability based on PAMPA measurements, *Eqn. 16* suggests that the target solubility should be greater than

Table. *MAD PAMPA Calculation for Fluoroquinolones*

Compound	P_{app}[a]) obs.	PAMPA P_o[b])	P_{app}[c]) calc.	MAD/S[d])	Target solubility[e]) at pH 6.5 [μg ml^{-1}]		
					dose = 7 mg	dose = 70 mg	dose = 700 mg
Ciprofloxacin	16	3	17	0.77	9	90	904
$N^{4'}$-Methylciprofloxacin	63	37	53	2.44	3	29	287
$N^{4'}$-Propylciprofloxacin	119	137	95	4.33	2	16	162
$N^{4'}$-Butylciprofloxacin	141	302	134	6.12	1	11	114
Ofloxacin	20	6	25	1.12	6	62	624
Sarafloxacin	19	14		1.59	4	44	439
Sparfloxacin	44	92		3.63	2	19	193

[a]) Measured rat *in situ* intestinal perfusion permeability [10^{-6} cm s^{-1}] [10]. [b]) Measured double-sink PAMPA permeability [10^{-6} cm s^{-1}] [10]. [c]) Calculated rat permeability, using *Eqn. 15* [10^{-6} cm s^{-1}]. [d]) The permeability contribution to MAD (see *Eqn. 16*). [e]) The target solubility is the lowest value of solubility which would yield a MAD value equal to the projected dose.

roughly 54 μg ml^{-1} (average of the first four compounds in the penultimate column of the *Table*) to observe 100% intestinal absorption. If both the potency and permeability are high, then solubility can be as low as *ca.* 2 μg ml^{-1} (average of the last three compounds in the *Table*) to observe 100% intestinal absorption. If very high doses need to be formulated, and if permeability is poor, than solubility needs to be greater than 1000 μg ml^{-1} to observe complete absorption.

5. Outlook

The *Hansch–Fick* analysis in *Sect. 1.7* suggested that BCS *Class 4* could not be valid if passive permeability were the only mechanism of transport. The compounds categorized as *Class 4* in *Fig. 2* (furosemide, cyclosporin, terfenadine) are low in permeability because they are substrates for P-glycoprotein efflux transport. When the compounds are extremely low in solubility, it is a common practice to use low concentrations (*ca.* 1 μM) in Caco-2 assays. (Clinically relevant concentrations may be as high as 100 μM.) The low assay concentrations may be below the saturable level of active transporters, and thus passive processes may not be dominant in transport. The BCS approach is still evolving, and interprations of the underlying solubility–permeability–pK_a properties have been furthered by methods such as PAMPA.

The PAMPA method has demonstrated its versatility in many instances since 1998. It is a remarkable 'open-system' approach, where practitioners may formulate their own lipid barriers for any number of different applications, not all focused on permeability screening. The method can be low cost, very fast, and a particularly helpful add-on to cellular permeability assays such as Caco-2. It readily provides information about passive transport permeability, not complicated by other mechanisms such as active transport and metabolism. Low solubility of test compounds is not an obstacle to permeability measurement, as demonstrated with the cosolvent method. Continuing improvements in PAMPA will make it the method of choice for primary screening of permeability. *Eqn. 15* has been shown to hold for fluoroquinolones, which are ampholytes, slightly negatively charged at pH 7.4. Other equations for acids and bases are currently being developed in our laboratory, and progress in generalizing MAD PAMPA is soon expected. These are still the early days of PAMPA, in some ways resembling those of HPLC in the 1980s.

We thank Drs. *Manfred Kansy*, *Holger Fischer*, *Bernard Faller*, *Kiyohiko Sugano*, *Ed Kerns*, *Li Di* and Profs. *Marival Bermejo* and *Norman Ho* for valuable discussions during the formative stages of the PAMPA model.

REFERENCES

[1] J. B. Dressman, G. L. Amidon, D. Fleisher, *J. Pharm. Sci.* **1985**, *74*, 588.
[2] G. L. Amidon, H. Lennernäs, V. P. Shah, J. R. Crison, *Pharm. Res.* **1995**, *12*, 413.
[3] 'Guidance for Industry, Waiver of *in vivo* Bioavailability and Bioequivalence Studies for Immediate Release Solid Oral Dosage Forms Based on a Biopharmaceutics Classification System', FDA, Washington D.C., 2000.
[4] E. Rinaki, G. Valsami, P. Macheras, *Pharm. Res.* **2003**, *20*, 1917.
[5] K. Johnson, A. Swindell, *Pharm. Res.* **1996**, *13*, 1795.
[6] W. Curatolo, *Pharm. Sci. Tech. Today* **1998**, *1*, 387.
[7] T. F. Weiss, 'Cellular Biophysics. Vol. I: Transport', The MIT Press, Cambridge, MA, 1996.
[8] L. S. Schanker, D. J. Tocco, B. B. Brodie, C. A. M. Hogben, *J. Am. Chem. Soc.* **1958**, *123*, 81.
[9] M. Kansy, F. Senner, K. Gubernator, *J. Med. Chem.* **1998**, *41*, 1007.
[10] M. Bermejo, A. Avdeef, A. Ruiz, R. Nalda, J. A. Ruell, O. Tsinman, I. González, C. Fernández, G. Sánchez, T. M. Garrigues, V. Merino, *Eur. J. Pharm. Sci.* **2004**, *21*, 429.
[11] A. Avdeef, 'Absorption and Drug Development', Wiley-Interscience, New York, 2003, p. 128–246.
[12] *Simulations Plus Inc.*, www.simulations-plus.com.
[13] A. Avdeef, *Curr. Topics Med. Chem.* **2001**, *1*, 277.
[14] K. Balon, B. W. Mueller, B. U. Riebesehl, *Pharm. Res.* **1999**, *16*, 882.
[15] S. Winiwarter, N. M. Bonham, F. Ax, A. Hallberg, H. Lennernäs, A. Karlen, *J. Med. Chem.* **1998**, *41*, 4939.
[16] C. A. Lipinski, in 'Annual Meeting of the Society of Biomolecular Screening, Sept. 24, 2002', Society of Biomolecular Screening, The Hague, The Netherlands, 2002.
[17] M. Kansy, H. Fischer, K. Kratzat, F. Senner, B. Wagner, I. Parrilla, in 'Pharmacokinetic Optimization in Drug Research. Biological, Physicochemical, and Computational Strategies', Eds. B. Testa, H. van de Waterbeemd, G. Folkers, R. Guy, Verlag Helvetica Chimica Acta, Zürich, 2001, p. 447–464.
[18] A. Avdeef, in 'Pharmacokinetic Optimization in Drug Research. Biological, Physicochemical, and Computational Strategies', Eds. B. Testa, H. van de Waterbeemd, G. Folkers, R. Guy, Verlag Helvetica Chimica Acta, Zürich, 2001, p. 305–326.
[19] B. Faller, F. Wohnsland, in 'Pharmacokinetic Optimization in Drug Research. Biological, Physicochemical, and Computational Strategies', Eds. B. Testa, H. van de Waterbeemd, G. Folkers, R. Guy, Verlag Helvetica Chimica Acta, Zürich, 2001, p. 257–274.
[20] A. Avdeef, M. Strafford, E. Block, M. P. Balogh, W. Chambliss, I. Khan, *Eur. J. Pharm. Sci.* **2001**, *14*, 271.
[21] F. Wohnsland, B. Faller, *J. Med. Chem.* **2001**, *44*, 923.
[22] K. Sugano, H. Hamada, M. Machida, H. Ushio, K. Saitoh, K. Terada, *Int. J. Pharm.* **2001**, *228*, 181.
[23] K. Sugano, H. Hamada, M. Machida, H. Ushio, *J. Biomol. Screen.* **2001**, *6*, 189.
[24] K. Sugano, N. Takata, M. Machida, K. Saitoh, K. Terada, *Int. J. Pharm.* **2002**, *241*, 241.
[25] K. Sugano, Y. Nabuchi, M. Machida, Y. Aso, *Int. J. Pharm.* **2003**, *257*, 245.
[26] C. Zhu, L. Jiang, T. M. Chen, K. K. Hwang, *Eur. J. Med. Chem.* **2002**, *37*, 399.
[27] D. F. Veber, S. R. Johnson, H. Y. Cheng, B. R. Smith, K. W. Ward, K. D. Kopple, *J. Med. Chem.* **2002**, *45*, 2615.
[28] L. Di, E. H. Kerns, K. Fan, O. J. McConnell, G. T. Carter, *Eur. J. Med. Chem.* **2003**, *38*, 223.

[29] E. H. Kerns, *J. Pharm. Sci.* **2001**, *90*, 1838.
[30] E. H. Kerns, L. Di, *Curr. Topics Med. Chem.* **2002**, *2*, 87.
[31] L. Di, E. H. Kerns, *Curr. Opin. Chem. Biol.* **2003**, *7*, 402.
[32] E. H. Kerns, L. Di, *Drug Disc. Today* **2003**, *8*, 316.
[33] J. A. Ruell, *Mod. Drug Disc.* **2003**, *6*, 28.
[34] A. Avdeef, in 'Drug Bioavailability. Estimation of Solubility, Permeability, Absorption and Bioavailability', Eds. H. van de Waterbeemd, H. Lennernäs, P. Artursson, Wiley-VCH, Weinheim, 2003, p. 46–70.
[35] K. A. Youdim, A. Avdeef, N. J. Abbott, *Drug Disc. Today* **2003**, *8*, 997.
[36] P. E. Nielsen, A. Avdeef, *Eur. J. Pharm. Sci.* **2004**, *22*, 33.
[37] J. A. Ruell, K. L. Tsinman, A. Avdeef, *Eur. J. Pharm. Sci.* **2003**, *20*, 393.
[38] J. A. Ruell, O. Tsinman, A. Avdeef, *Chem. Pharm. Bull.* **2004**, *52*, 561.
[39] H. Lennernäs, *J. Pharm. Sci.* **1998**, *87*, 403.
[40] W. M. Partridge, 'Peptide Drug Delivery to the Brain', Raven, New York, 1991, p. 52–88.
[41] A. Adson, P. S. Burton, T. J. Raub, C. L. Barsuhn, K. L. Audus, N. F. H. Ho, *J. Pharm. Sci.* **1995**, *84*, 1197.
[42] A. Avdeef, P. Nielsen, O. Tsinman, *Eur. J. Pharm. Sci.* **2004**, *22*, 365.
[43] H. Liu, C. Sabus, G. T. Carter, C. Du, A. Avdeef, M. Tischler, *Pharm. Res.* **2003**, *20*, 1820.
[44] S. L. McGovern, E. Caselli, N. Grigorieff, B. K. Shoichet, *J. Med. Chem.* **2002**, *45*, 1712.

Correlations between PAMPA Permeability and log *P*

by **Karl Box**[a]), **John Comer***[a]), and **Farah Huque**[b])

[a]) *Sirius Analytical Instruments Ltd.*, Riverside, Forest Row Business Park, Forest Row, East Sussex, RH18 5DW, UK (e-mail: John.Comer@sirius-analytical.com)
[b]) Department of Chemistry, Cardiff University, P.O. Box 912, Cardiff CF10 3TB, UK

Abbreviations
DOPC: Dioleylphosphatidylcholine; GI: gastrointestinal; PAMPA: parallel artificial membrane permeation assay; P_e: effective permeability; P-gp: P-glycoprotein; P_o: intrinsic permeability; P_u: permeability through the unstirred water layer; pK_a^{flux}: apparent pK_a in unstirred PAMPA experiment.

---

1. Introduction

It is useful in drug research and development to know whether and to what degree drugs are absorbed and transported in the human body. One way to find out is to measure drug absorption *in vivo* in humans, for it certainly tells people what they want to know – that the drug was absorbed. These measurements are made during the late-stage development of drugs that are likely to come to market.

When investigating the properties of newly discovered drugs, it is also useful to assess whether they may be absorbed. However, *in vivo* experiments during the discovery process are out of the question for a number of reasons – not enough drug is available, the cost of experiments is very high, and the toxicity of the drug has not yet been assessed.

A plausible alternative is to assess the drug's ability to permeate membranes, as those drugs that readily permeate are more likely to be absorbed and transported around the body. There are several ways to make this assessment. One method is to measure the permeability through a membrane *in vitro*, as exemplified by Caco-2 and PAMPA measurements [1]. Another approach is to measure lipophilicity (as expressed for example by the log *P* parameter) in a solvent/water partition experiment, as log *P* values are known to correlate with permeability for many drugs

[2]. Finally, methods exist for interpreting chemical structures *in silico* to predict absorption, permeability, lipophilicity, and related properties. In this chapter, we will examine the relationship between measured PAMPA permeability and measured lipophilicity.

2. Permeability

2.1. *Membranes Made from Cells*

A widely used method for measuring permeability is to measure the passage of drugs across membranes made from cultured cells such as Caco-2. Drugs can permeate through Caco-2 membranes by one or more of the following mechanisms: passive diffusion, active transport, or paracellular transport. They may also be rejected by the membranes by P-gp mediated efflux processes. There is a good deal of literature on Caco-2 and other methods involving membranes made from cultured cells [3]. These measurements have great utility, but success calls for a combination of skills, including cell culture, robotics, and analytical chemistry. Inter-laboratory comparison of results calls for standardization of methodology and practice. It is helped by the near-universal adoption of just two or three cell lines, though there has been debate about how well these cells mimic real human membranes [4].

2.2. *Membranes Made from Lipids and Alkanes*

Another method for measuring permeability is to measure the passage of drugs through artificial membranes made from lipids or organic solvents such as alkanes (C_{10} or higher) [5][6]. It is likely that drugs can permeate through such membranes only by the mechanism of passive diffusion. Permeability experiments are done in closed vessels in which measured volumes of aqueous solution are separated by membranes – the so-called sandwich. In the automated PAMPA, sandwiches are made in specially designed 96-well plates. At the start of the experiment, the drug should be in solution on one side of the membrane only (the donor side). If the drug is permeable, it will appear after some time on the acceptor side. After a fixed incubation time, the sandwich is separated and sample concentrations are measured in the donor and acceptor cells. Permeability is calculated from these measured concentrations, taking into account the volumes of the solutions and the incubation time. PAMPA experiments provide two classes of results: a yes/no answer to the question 'did the molecule cross the membrane?' and a number that expresses the rate of permeability in units such as cm s^{-1}. Until 2003, all published PAMPA studies were done in cells that contained aqueous solutions with the same composition on either side of the membrane – the so-called iso-pH method. However, the development of PAMPA methodology is at an earlier stage than Caco-2, and the use of many different membranes and solution compositions have been reported [5–8].

3. A Short Note on log *P*

Because log *P* measurements are made using pure solvents under equilibrium conditions, inter-laboratory comparison of results is generally excellent, and they approach closely to the ideal of physicochemical constants. While the use of log *P* in octanol/water systems is popular for getting a general idea of pharmacokinetics from a single number, log *P* values obtained in alkane/water systems have also been used to predict the behavior of molecules in biological systems [9]. Some workers have proposed that the rate-limiting step in passive diffusion through a lipid bilayer is the diffusion through the lipid core of the membrane, rather than ionic and H-bonding interactions at the membrane's surface. Following this idea, a PAMPA has been described which uses membranes made from hexadecane, leading to a good correlation between PAMPA permeability measured in this manner and alkane/water partition coefficients, as well as with measured human GI absorption [6]. More recently, a method of measuring log $P_{\text{alkane/water}}$ has been described which uses the same apparatus as for PAMPA measurements [10]. It does not seem surprising that PAMPA results obtained using hexadecane membranes correlate with log $P_{\text{alkane/water}}$ values. However, it is interesting to speculate to what extent PAMPA results obtained with membranes made from 2% DOPC dissolved in dodecane would correlate with log *P*.

4. Experimental Outline

4.1. *Measurement of Permeability*

The measurement of permeability is illustrated by our recent work using the iso-pH method described by *Ruell et al.* [11]. In this work, PAMPA permeability of 60 small organic molecules, mostly drugs, was studied using 16-hour incubation times without stirring at 12 pH values between 3 and 10. Details of the experimental method have been reported elsewhere [12]. The membranes were made by pasting a 2% solution of DOPC in dodecane on to filter supports. It has been suggested that membranes made by this method form bilayers of DOPC on the filter support [13]. Presumably such bilayers would be surrounded by a good deal of dodecane.

The selected molecules were ionizable, with the pK_a value that defined the neutral species lying within the pH range of the assay – this made it possible to calculate P_o, the intrinsic permeability of the neutral form. The pK_a values used to convert the measured P_e values to P_o values were obtained from the literature, or by measurements with the *Sirius GLpKa* instrument. The PAMPA measurements were made on a *p*ION *Evolution* system with pH mapping, run on a *Tecan Genesis* robot and *Molecular Devices Spectramax* UV plate reader, and with preliminary data analysis performed by *p*ION's PAMPA Evolution software. Samples in 96-well plates were placed manually on the instrument and in the UV plate reader, but all other operations were automated. Seven samples and one standard were analyzed on each plate at 12 different pH values – four plates (28 samples) could be analyzed during a working day. Many more samples could be analyzed at a single pH.

Fig. 1 illustrates a PAMPA result for tetracaine, a basic drug with pK_a values of 2.34 and 8.45. Note the similar shapes of the UV spectra measured in the reference, donor, and acceptor cells. This similarity confirms that the same substance was present in each cell. Note how the absorption measured in the acceptor cell (and therefore the concentration) is lower than that measured in the donor – this indicates that the permeation process was still under way at the time the PAMPA sandwich was separated. This is a sign of a well-designed experiment, as the rate of permeation cannot be derived if the measurement went to full equilibrium and the concentration in the donor and acceptor are equal. Note also that the sum of the maximum absorbance in the donor and acceptor cells is similar to the maximum absorbance in the reference – this is a sign that very little compound was retained in the membrane. The profile of P_e *vs.* pH is shown at the bottom right. It shows that the rate of permeability is highest at high pH where the molecule is unionized. This observation accords with pH–partition theory, which suggests that drugs must be unionized when they cross lipid or alkane membranes by passive diffusion.

While *Fig. 1* shows P_e measured at 12 different pH values, the sample could have been shown to be permeable if just one experiment had been done only at pH 10, because it is known to be a base, and the pK_a values of 2.34 and 8.45 indicate that it would be unionized at pH 10. However, it is difficult to choose an appropriate pH for an experimental sample before the measurement is made. As a compromise, to save time and to get higher

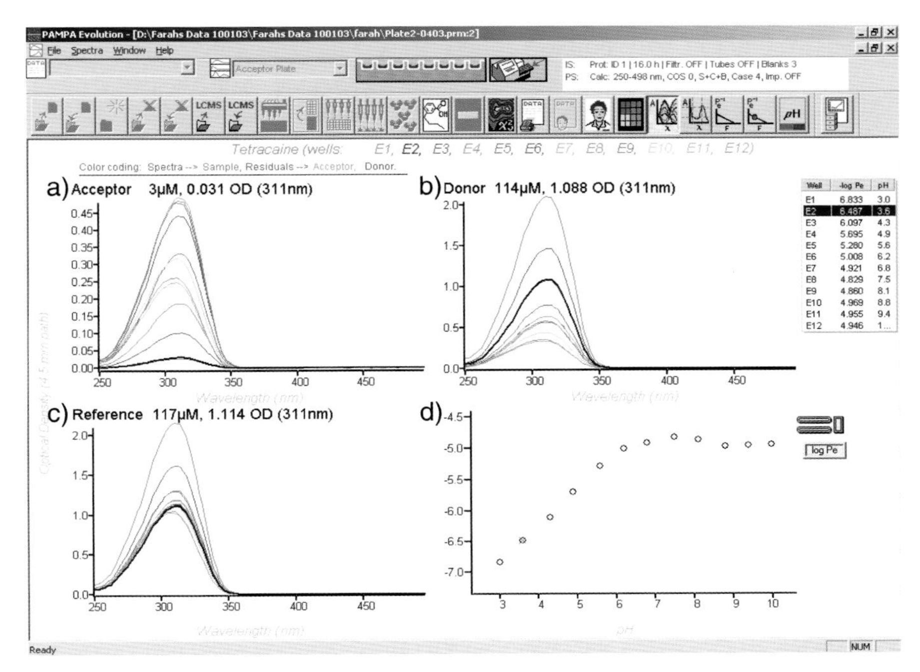

Fig. 1. *Screen display from pION Evolution software showing permeability of tetracaine through a membrane made from 2% DOPC in dodecane. Graphs a–c* show the UV absorption of tetracaine at 12 pH values in acceptor, donor, and reference cells. *Graph d* shows P_e at 12 pH values.

throughput, some laboratories running PAMPA assays measure at three pH values – a low pH at which acidic samples will be unionized, a high pH at which basic samples will be unionized, and a mid-range pH at which ampholytes will probably be unionized. Nonionizable substances should have the same permeability across the whole pH range.

To pass from the bulk solution in the donor cell to the bulk in the acceptor, the dissolved sample not only has to cross the membrane itself, but also the unstirred water layer on either side of the membrane. Passage through the unstirred water layer proves to be a slow step, but it can be speeded up by stirring. Until now, PAMPA in most laboratories has probably been done without stirring, because it has proved difficult to get reliable, reproducible stirring of the PAMPA sandwich during incubation. However, recent introduction of 96-well magnetic stirring looks promising [14].

While useful for comparing a result with those from other samples and standards within a single laboratory, a P_e value measured at a single pH without stirring is not a physicochemical constant, as the value obtained is

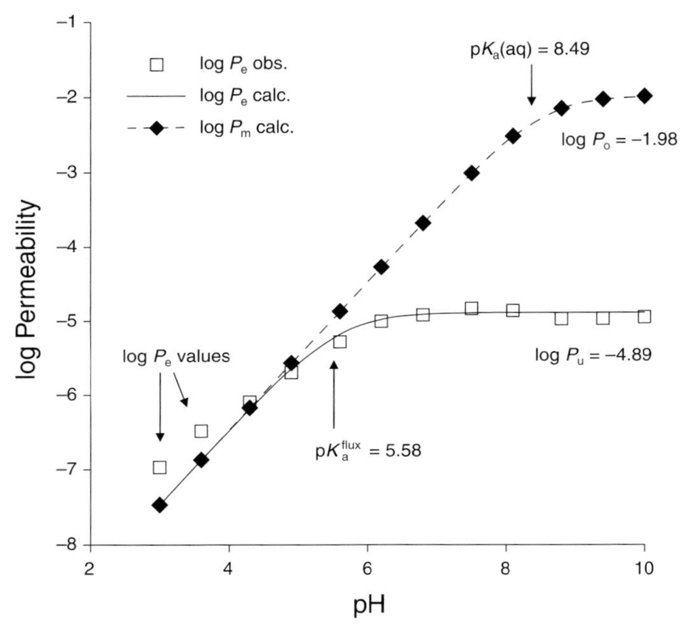

Fig. 2. *Graph showing the correction of* P_e *to* P_o *for tetracaine.* P_e is measured at 12 pH values. The line of best fit to these P_e values is used to estimate log P_u, the permeability through the unstirred water layer. P_u is lower than P_o for lipophilic molecules. The aqueous pK_a value is used in the calculation of P_o.

too condition-dependent. A more universally useful value would be the P_o value obtained if passage through the unstirred water layer were not the rate-limiting step. While P_e may approach P_o in cells with effective stirring, P_o may also be calculated for ionizable compounds in unstirred experiments by taking advantage of pK_a differences. *Fig. 2* shows the P_e of tetracaine measured at 12 pH values. The apparent pK_a in the unstirred experiment (pK_a^{flux}) is shown in the graph. It is several units lower than the true pK_a of the sample in aqueous solution. If the aqueous pK_a is known, P_o can be calculated from the profile of P_e *vs.* pH, using published equations [12].

As reported in our earlier work [12], P_o values were calculated for 43 of the compounds in the study, and are shown in the *Table*. Of the other compounds studied, concentrations were equal in the donor and acceptor compartments for four samples (2-iodophenol, 3-chlorophenol, phenol, thymol), which made it impossible to calculate P_e values. There are two possible explanations for these equal concentrations. Equilibration may have occurred if the samples were very permeable. On the other hand, these four samples were all low-molecular-weight phenols, and they may have attacked and disrupted the membranes. Four compounds (atenolol, famotidine, moxonidine. triamterene) produced strong UV spectra in the donor compartments but no spectra in the acceptor compartments. The significance of this result will be discussed later.

Of the remaining compounds, seven (bepredil, miconazole, nitrofurantoin, terfenadine, prazosin, prochloraz, and tryptamine) produced noisy data that suggested they may not have been fully dissolved in the donor compartment at certain pH values – perhaps they could have been successfully measured if the aqueous buffers had contained water-miscible solvents to enhance their solubility. Two compounds (atropine and ephedrine) absorbed UV too weakly to produce useful signals – perhaps they could have been measured if a more-sensitive detection method had been used.

4.2. *Measurement of log* P

Measured log $P_{octanol/water}$ values are shown in the *Table*. Six were taken from published sources, while the remainder were measured at *Sirius* by the pH-metric method, as described below.

It has been shown that log P values measured in cyclohexane are more or less identical to values measured in dodecane [15]. Moreover, *Abraham* equations derived from measured log P in cyclohexane/water and hexadecane/water are very similar [16]. Therefore, log P values measured in different alkane/water systems are expected to be similar. Of the log $P_{alkane/water}$ values reported in the *Table*, one log $P_{cyclohexane/water}$ and two log $P_{heptane/water}$ values were found in the literature. Eight log $P_{dodecane/water}$ values measured potentiometrically were kindly provided by *Glynn et al.* [15], while 32 dodecane/water and one cyclohexane/water values were measured in-house. Log $P_{dodecane/water}$ values for the remaining compounds were too low to measure.

Measurements were done by the pH-metric method using the *Sirius GLpKa* instrument. This method has already been validated for the measurement of log $P_{octanol/water}$ [17–19], but it can also be used with other partition solvents [20]. Aqueous pK_a values were obtained from the literature or were measured during this study, either pH-metrically or using the D-PAS UV accessory for *GLpKa*.

On a practical note, dodecane is probably the best alkane to use in pH-metric experiments. Cyclohexane and heptane are so volatile that significant amounts of solvent evaporate during the titrations, hindering calculation of results. The higher alkanes are viscous, making them difficult to handle. Dodecane is neither volatile nor viscous – it floats on water,

Table. *Measured and Calculated Properties of the Compounds in This Study*. Results for pK_a^{flux} and $\log P_o$ were measured using the *p*ION *Evolution* system under iso-pH conditions, with membranes made from 2% DOPC solution in dodecane. Sources for measured pK_a and $\log P$ results and for the calculated parameters E, S, A, B, and V are indicated.

No.	Name	M_r	pK_a [a][b]	pK_a^{flux}	$\log P_{oct}$ [b]	$\log P_{alk}$ [b][c]	$\log P_o$	E	S	A	B	V	Source
	Acids												
1	Aspirin	180.16	3.49 [21]	3.8	0.90 [21]	−1.52 [22][d]	−5.56	0.78	0.8	0.49	1	1.2879	[e]
2	3-Chlorophenol	128.56	8.80		2.50 [23]	−0.17		0.96	0.97	0.65	0.27	0.9	[f]
3	Chlorthalidone	338.77	9.04	9.72	0.76		−5.97	2.65	3.17	1.25	1.96	2.1747	[f]
4	2,4-Dichlorophenoxyacetic acid	221.04	2.64 [21]	2.93	2.78 [21]	−0.86	−4.38	1.21	1.18	0.77	0.6	1.3761	[e]
5	Diclofenac	296.15	3.99 [24]	6.38	4.51 [24]	1.72 [25]	−2.58	1.84	1.58	0.9	0.83	2.025	[f]
6	3,5-Dimethoxyphenol	154.16	9.09	9.79	1.73	−1.06	−4.12	0.88	1.11	0.53	0.71	1.1743	[f]
7	Flumequine	261.25	6.27 [21]	6.4	1.72 [21]	−1.37	−5.18	1.55	1.92	0.59	1.32	1.791	[f]
8	Flurbiprofen	244.26	4.03 [26]	6.11	4.07	1.49	−2.55	1.3	1.38	0.59	0.59	1.839	[f]
9	4-Hydroxybenzoic acid	138.12	4.32	4.96	1.58 [23]	−0.75	−5.73	0.93	0.9	0.81	0.56	0.9904	[e]
10	Ibuprofen	206.28	4.35 [21]	6.88	3.97 [21]	2.08 [25]	−2.46	0.73	0.97	0.6	0.7	1.7771	[e]
11	Indomethacin	357.79	4.02	5.67	4.20	0.98	−3.38	2.24	2.85	0.4	1.08	2.5299	[e]
12	2-Iodophenol	220.01	8.44 [27]		2.65 [23]	0.80		1.34	0.96	0.37	0.28	1.03	[f]
13	2-Naphthoic acid	172.18	3.92 [28]	5.47	3.30	0.72	−3.4	1.51	1.32	0.59	0.54	1.3007	[f]
14	Nifuroxime	156.1	9.52	9.7	1.34	−1.1	−5.08	1.07	1.09	0.34	0.7	0.9669	[f]
15	4-Nitrophenol	139.11	6.97	7.09	1.91 [23]	−2.22 [22][g]	−5.25	1.07	1.72	0.82	0.26	0.9493	[e]
16	Phenol	94.11	9.74		1.46 [23]	−0.86		0.81	0.87	0.53	0.31	0.78	[f]
17	Phenylacetic acid	136.15	4.06	4.4	1.41 [23]	−1.6	−4.99	0.73	0.97	0.6	0.61	1.0726	[e]
18	4-Propoxybenzoic acid	180.2	4.54 [29]	5.83	3.09 [30]	0.31	−3.55	0.88	1.19	0.59	0.46	1.4131	[f]
19	Theophylline	180.16	8.55 [24]	9.16	0.00 [24]	−1.70 [22][d]	−6.77	1.5	1.6	0.54	1.34	1.2223	[e]
20	Thymol	150.22	10.50 [31]		3.30 [32]	1.40		0.83	0.76	0.53	0.35	1.34	[f]
21	Warfarin	308.33	4.94	6.83	3.39	0.05	−2.9	2.19	2.96	0.34	1.42	2.3077	[f]
	Bases												
22	Aniline	93.13	4.59 [33]	3.34	0.90 [23]	−0.23	−3.36	0.96	0.96	0.26	0.41	0.8162	[e]
23	Atenolol	266.34	9.54 [24]		0.22 [24]			1.4	2.13	0.75	1.91	2.18	[f]
24	Chlorpromazine	318.87	9.24 [21]	3.36	5.40 [21]	4.37 [25]	0.94	2.44	1.37	0.06	1.08	2.4056	[20]
25	*p*-F-Deprenyl	205.27	7.42 [24]	3.31	3.06 [24]	2.64	−0.73	0.83	1.05	0.08	0.84	1.7342	[f]
26	Desipramine	266.39	10.08	6.03	4.21	3.38[g]	−0.83	1.99	1.38	0.07	1.23	2.2606	[20]
27	Diltiazem	414.52	8.02 [34]	5.78	2.88	1.37	−2.67	2.34	2.65	0	2.11	3.1365	[f]
28	Diphenhydramine	255.36	9.08	5.07	3.44	2.67	−0.95	1.39	1.43	0	1.04	2.1872	[f]
29	Eserine	275.35	8.17 [35]	7.64	1.73	−0.46	−4.62	1.58	1.87	0.27	1.73	2.1411	[f]
30	Famotidine	337.45	6.76		−0.81 [36]			2.56	2.56	1.36	2.41	2.26	[f]
31	Fluoxetine	309.33	10.09	5.07	4.61	3.10 [25]	0.07	1.16	1.38	0.1	0.85	2.2403	[f]
32	Haloperidol	375.87	8.43	5.73	3.67	1.53	−2.15	1.92	2.3	0.38	1.62	2.798	[f]

Table (cont.)

No.	Name	M_r	pK_a a)b)	pK_a^{flux}	$\log P_{oct}$ b)	$\log P_{alk}$ b)c)	$\log P_o$	E	S	A	B	V	Source
33	Lidocaine	234.34	7.95 [24]	5.61	2.44 [24]	1.00 [25]	−2.51	1.23	1.43	0.01	1.39	2.0589	[20]
34	Metoprolol	267.37	9.56 [24]	8.3	1.95 [24]	−0.38	−4.07	1	1.69	0.31	1.63	2.2604	e)
35	Moxonidine	241.68	7.54 [37]		0.70 [37]			1.68	1.68	0.76	2.09	1.67	f)
36	Papaverine	339.39	6.39 [24]	4.61	2.95 [24]	0.75	−3.06	2	2.21	0	1.38	2.5914	f)
37	Penbutolol	291.43	9.92 [24]	4.56	4.62 [24]	3.06	0.3	1.26	1.29	0.3	1.38	2.5158	f)
38	Phenazopyridine	213.24	5.07	2.98	3.31	0.42	−2.8	2.46	2.28	0.49	0.96	1.6393	f)
39	4-Phenylbutylamine	149.24	10.51	7.39	2.35	1.00	−2.13	0.81	0.87	0.1	0.72	1.3798	e)
40	Procaine	236.31	9.04 [24]	7.96	2.14 [24]	−0.67 [25]	−3.99	1.14	1.36	0.25	1.41	1.9767	e)
41	Propranolol	259.35	9.53 [24]	6.54	3.48 [24]	1.75 [25]	−2.13	1.71	1.8	0.31	1.26	2.148	e)
42	Quinine	324.42	8.55 [21]	7.68	3.50 [21]	−0.39	−4.14	2.47	1.23	0.37	1.97	2.5512	e)
43	Tetracaine	264.37	8.49 [24]	5.58	3.51 [24]	1.54 [25]	−1.98	1	1.34	0.2	1.43	2.2585	f)
44	Tramadol	263.38	9.49	6.16	2.65	1.82	−1.49	1.13	1.32	0.38	1.45	2.234	f)
45	Triamterene	253.27	6.28		1.18			3.11	3.12	0.74	1.32	1.83	f)
46	Verapamil	454.61	8.72	5.42	3.96	2.27	−1.54	1.66	2.73	0	1.96	3.7861	f)
	Ampholytes												
47	Carazolol	298.38	9.52 [24]h)	8.46	3.73 [24]	0.05	−3.98	2.08	2.02	0.68	1.68	2.378	f)
48	Carbendazim	191.19	4.48i)	3.95	1.74		−4.84	1.64	1.82	0.77	0.74	1.3613	f)
49	Fenoterol	303.35	8.16j)	7.91	0.95		−6.24	2.09	2.23	1.93	1.78	2.3633	f)
50	Furosemide	330.74	3.60k)	3.55	2.56 [36]	−0.6	−4.96	2.09	2.75	1.3	1.26	2.1032	f)
51	Sulfacetamide	214.24	5.22l)	5.87	−0.18		−6.68	1.5	2.74	0.96	1.24	1.4944	f)

a) Only the pK_a that defines the neutral species has been listed. b) All values measured at *Sirius*, unless otherwise stated. c) All values measured in dodecane/water, unless otherwise stated. d) Measured in heptane/water. e) *Abraham* database, licensed to *Sirius Analytical Instruments Ltd.*, Sussex, UK. f) Calculated using the ABSOLV software (*Sirius Analytical Instruments Ltd.*, Sussex, UK). g) Measured in cyclohexane/water. h) Basic pK_a used (acidic pK_a ~13.9). i) Basic pK_a used (acidic pK_a ~10.6). j) Basic pK_a used (acidic pK_as at 9.2, 10.2, 10.9). k) Acidic pK_a used (basic pK_a ~9.8). l) Acidic pK_a used (basic pK_a ~1).

mixes well on stirring, separates quickly without forming micro-emulsions, and is easy to work with.

The method of sample preparation is important. Many samples are poorly soluble in both water and dodecane – unless they are prepared correctly for analysis, they may precipitate out at the interface between the two phases, or on to the walls of the titration vessel. To get clean, reproducible data, the sample should be weighed into the titration vessel, after which dodecane and water are added, and the pH is adjusted to where the sample is ionized. The sample will dissolve in the aqueous layer, and will remain in solution throughout the experiment.

It is also worth noting that no tendency for partitioning of ionized species into alkane was observed in any of the experiments. Ion-pair partitioning into octanol is commonly observed for lipophilic compounds in partitioning experiments with aqueous phase of high ionic strength (*e.g.*, 0.15M KCl). However, ionized species of drugs cannot form stable ion pairs with hydrophilic counter-ions in the extremely hydrophobic alkanes. The practical consequence of this observation is that only one titration is required to measure log $P_{\text{alkane/water}}$ values of monoprotic acids and bases, whereas two or three are required in octanol/water to provide sufficient data to calculate log P for both neutral and ionized species.

5. Correlations between Physicochemical Properties

5.1. *Correlation between PAMPA and log* P

P_0 Values through membranes made from 2% DOPC in dodecane are plotted against log $P_{\text{octanol/water}}$ in *Fig. 3*. There is a broad correlation, but it is not very good. In contrast, a strong correlation with a good linear fit emerges when plotting permeabilities against log $P_{\text{alkane/water}}$ values (*Fig. 4*). The quality and reliability of this fit draws attention to some interesting issues.

The PAMPA membranes were made from a purchased reagent described as a 2% solution of DOPC in dodecane. A good deal of PAMPA work has been done using membranes made from 1–2% solutions of lipids in dodecane [8][38]. The PAMPA results reported here suggest that the membranes containing 2% DOPC were behaving as pure dodecane. Could it be that there was insufficient DOPC to form a complete membrane on the filter, or that the DOPC formed a thin membrane but the rate of permeation was governed by the passage through a much thicker dodecane layer? Recently, workers have reported PAMPA results obtained with membranes made from 20% lipid solutions in 80%

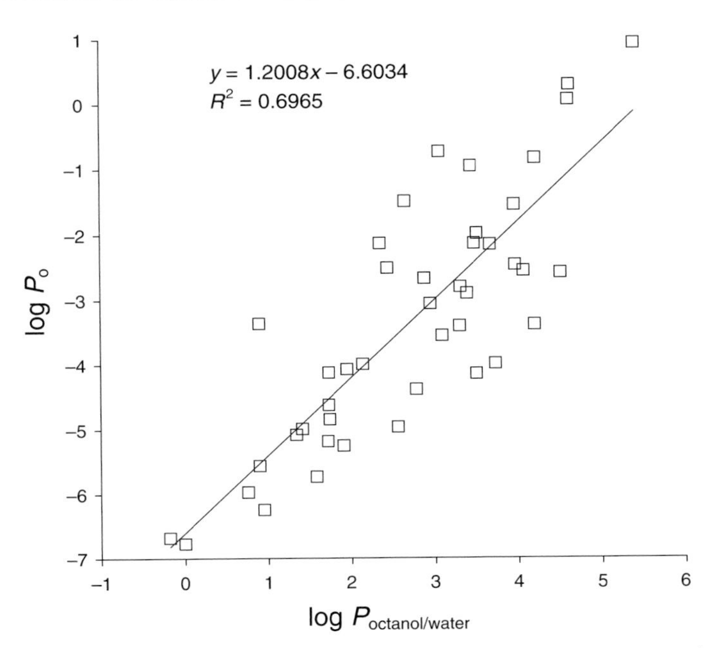

Fig. 3. *Correlation of log* P_o *measured by the iso-pH method using membranes made from 2% DOPC in dodecane* vs. *log* $P_{octanol/water}$

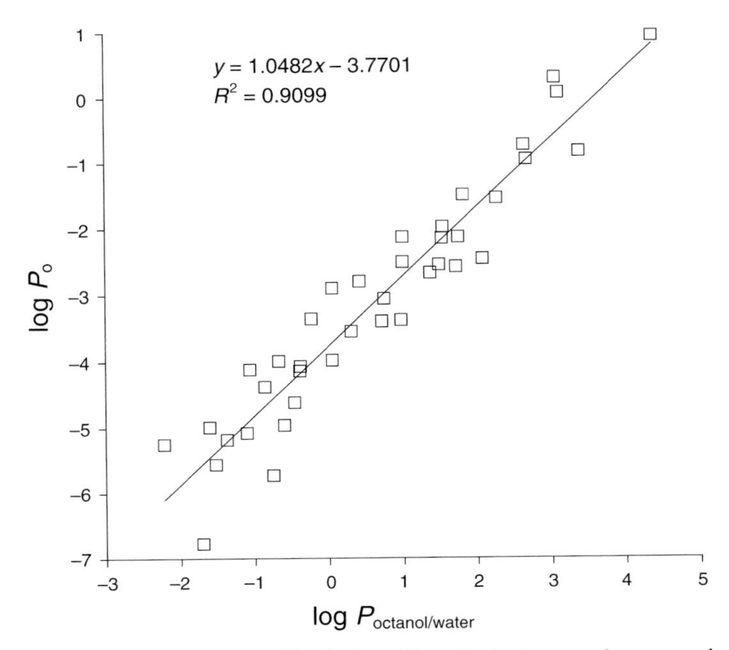

Fig. 4. *Correlation of log* P_o *measured by the iso-pH method using membranes made from 2% DOPC in dodecane* vs. *log* $P_{alkane/water}$

dodecane [39]. It would be interesting to repeat the permeability measurements using these new membrane formulations, and see whether the correlation with log $P_{alkane/water}$ is still apparent.

5.2. *Molecules That Did not Fit the Model*

Log $P_{alkane/water}$ values are not adequate to explain PAMPA permeability of all compounds in membranes made from 2% DOPC in dodecane. Though they had good UV spectra, four compounds (atenolol, famotidine, moxonidine, triamterene) could not be detected in the acceptor wells after 16-hour incubation, suggesting that their PAMPA permeability was very low. Data for famotidine, which shows this behavior, are shown in *Fig. 5*. The log $P_{alkane/water}$ values for these four compounds were also very low. Nevertheless, all these compounds are marketed drugs, and published pharmacokinetic data suggests that they are well absorbed, with reported 30–90% human GI absorption [40]. Their molecular weights are relatively low – perhaps they are absorbed paracellularly or actively transported. Perhaps they are absorbed well enough in the GI tract even though their PAMPA result suggests otherwise, as the absorbing area of the intestine is *ca.* 4 million times greater than the membrane in a PAMPA well, and the gut is well stirred.

Perhaps P_o values could be measured if different membranes or buffers were used in PAMPA, which were better able to interact with molecules with high capacity to form H-bonds. Note that the log $P_{octanol/water}$ values for these four nonpermeable molecules were well within the range of many other compounds that are known to be absorbed. This is not surprising, given the capacity for octanol to form H-bonds. These molecules have a high capacity for H-bond formation, as indicated by the sum of their descriptors A and B listed in the *Table*. Other molecules with high A and B (*e.g.*, chlorthalidone, fenoterol, furosemide) also had low P_o and low log $P_{alkane/water}$ values.

6. Conclusions

Researchers would like to be able to predict drug absorption from physicochemical data. Because it looks at the appearance of compound that has successfully traversed a membrane, PAMPA is a physicochemical surrogate for a biological system, and, although it cannot model paracellular transport, active transport, or efflux processes, it works much faster and for a lower running cost than permeability measurement using

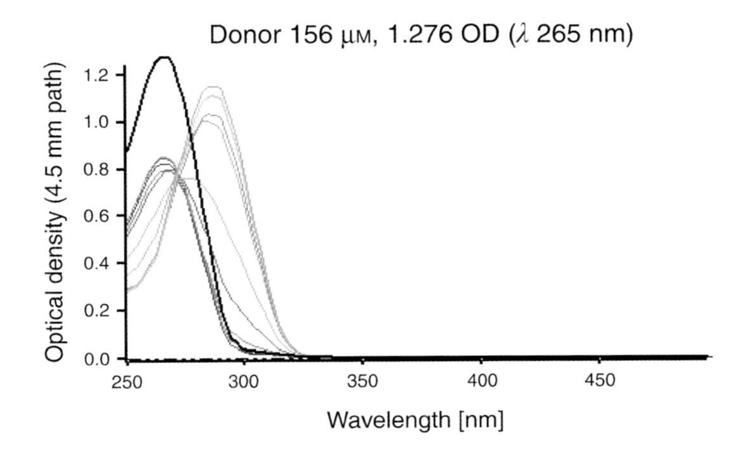

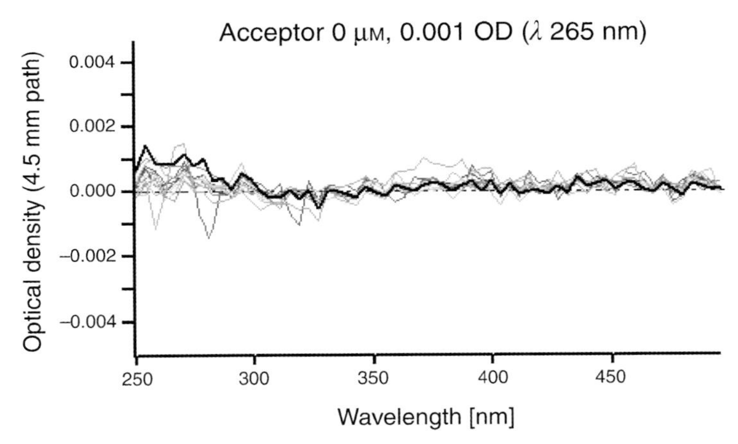

Fig. 5. *Spectra produced in PAMPA experiment for famotidine, a compound that did not appear in the acceptor compartment after a 16-h incubation*

Caco-2 or other cultured cell lines. We have clearly shown that PAMPA P_o values correlate with log $P_{alkane/water}$ for a wide range of small organic molecules and drugs. The effective experimental PAMPA permeability, P_e, depends on pH, which controls the fraction of the more-permeable unionized form of the molecule, and the thickness of the unstirred water layer. If it is true that PAMPA predicts absorption, then absorption could also be predicted by log $P_{alkane/water}$ and the pK_a.

While log $P_{alkane/water}$ tells you that the molecule *should* go through the membrane, PAMPA tells you that the molecule *does* go through the membrane – this is an advantage of the PAMPA technique. On the other hand, log P measured in solvent/water systems is a clearly defined physical property whose value will be similar in a wide range of measurement techniques such as pH-metric or shake flask, while PAMPA methodologies and membrane systems are still under development. From a practical point of view, workers need to choose what parameters to measure. The choice probably depends on the budget available and the throughput requirement, as well as the needs for inter-laboratory comparison.

We thank *Rebeca Ruiz Guevara* for her measurements of log *P* values, and *Jamie Platts* (Department of Chemistry, Cardiff University) and *Jon Mole* (*Sirius Analytical Instruments Ltd.*) for helpful discussion.

REFERENCES

[1] P. Artusson, K. Palm, K. Luthman, *Adv. Drug Delivery Rev.* **2001**, *46*, 27.
[2] G. Camenisch, G. Folkers, H. van de Waterbeemd, *Pharm. Acta Helv.* **1996**, *71*, 309.
[3] G. L. Amidon, P. I. Lee, E. M. Topp, 'Transport Processes in Pharmaceutical Systems', Dekker, New York and Basel, 2000.
[4] H. van de Waterbeemd, D. A. Smith, K. Beaumont, D. K. Walker, *J. Med. Chem.* **2001**, *44*, 1313.
[5] M. Kansy, F. Senner, K. Gubernator, *J. Med. Chem.* **1998**, *41*, 1007.
[6] F. Wohnsland, B. Faller, *J. Med. Chem.* **2001**, *44*, 923.
[7] K. Sugano, H. Hamada, M. Machida, H. Ushio, K. Saitoh, K. Terada, *Int. J. Pharm.* **2001**, *228*, 181.
[8] L. Zhu, T. M. Jiang, K. K. Chen, H. Wang, *Eur. J. Med. Chem.* **2002**, *37*, 399.
[9] B. Testa, P. Crivori, M. Reist, P. A. Carrupt, *Perspect. Drug Disc. Des.* **2000**, *19*, 179.
[10] X. Briand, B. Faller, Poster presented at the *LogP2004 Symposium*, Zürich, 29.2.–4.3.2004, Abstract A-18.
[11] J. A. Ruell, K. L. Tsinman, A. Avdeef, *Eur. J. Pharm. Sci.* **2003**, *20*, 393.
[12] F. T. T. Huque, K. Box, J. A. Platts, J. Comer, *Eur. J. Pharm. Sci.* **2004**, *23*, 223.
[13] M. Thompson, R. B. Lennox, R. A. McLelland, *Anal. Chem.* **1982**, *54*, 76.
[14] Gut Box™, *p*ION Inc., Woburn, MA.
[15] D. P. Edwards, E. Glynn, W. Arbuckle, Y. Lamont, K. Box, Poster presented at the *LogP2004 Symposium*, Zürich, 29.2.–4.3.2004, Abstract A-22.
[16] M. H. Abraham, H. S. Chadha, G. S. Whiting, R. C. Mitchell, *J. Pharm. Sci.* **1994**, *83*, 1085.
[17] B. Slater, A. McCormack, A. Avdeef, J. Comer, *J. Pharm. Sci.* **1994**, *83*, 1280.
[18] J. Comer, K. Chamberlain, A. Evans, *SAR QSAR Environ. Res.* **1995**, *3*, 307.
[19] K. Takács-Novák, A. Avdeef, *J. Pharm. Biomed. Anal.* **1996**, *14*, 1405.
[20] A. M. Zissimos, M. H. Abraham, M. C. Barker, K. J. Box, K. Y. Tam, *J. Chem. Soc., Perkin Trans. 2* **2002**, 1.
[21] *Sirius Tech. Appl. Notes* **1994**, *1*.
[22] A. Leo, C. Hansch, D. Elkins, *Chem. Rev.* **1971**, *71*, 525.
[23] T. Fujita, J. Iwasa, C. Hansch, *J. Am. Chem. Soc.* **1964**, *86*, 5175.
[24] *Sirius Tech. Appl. Notes* **1995**, *2*.
[25] E. Glynn, *Organon*, UK, personal communication.
[26] A. Avdeef, C. M. Berger, C. Brownell, *Pharm. Res.* **2000**, *17*, 85.
[27] C. van Hooidonk, L. Ginjaar, *Recl. Trav. Chim. Pays-Bas* **1967**, *86*, 449.
[28] J. F. J. Dippy, S. R. C. Huges, J. W. Laxton, *J. Chem. Soc.* **1954**, 1470.
[29] B. Jones, J. C. Speakman, *J. Chem. Soc.* **1944**, 19.
[30] Y. Z. Da, K. Ito, H. Fujiwara, *J. Med. Chem.* **1992**, *35*, 3382.

[31] 'Tables of Rate and Equilibrium Constants of Heterolytic Organic Reactions', Ed. V. A. Palm, Moscow, 1975.
[32] C. Hansch, S. Anderson, *J. Org. Chem.* **1967**, *32*, 2583.
[33] J. L. Jensen, A. T. Thibeault, *J. Org. Chem.* **1977**, *42*, 2168.
[34] A. Avdeef, C. M. Berger, *Eur. J. Pharm. Sci.* **2001**, *14*, 281.
[35] K. Y. Tam, K. Takács-Novák, *Anal. Chim. Acta* **2001**, *434*, 157.
[36] A. Avdeef, 'Absorption and Drug Development. Solubility, Permeability and Charge State', Wiley, Hoboken, NJ, 2003.
[37] C. White, *Lilly*, UK, personal communication.
[38] A. Avdeef, M. Strafford, E. Block, M. P. Balogh, W. Chambliss, I. Khan, *Eur. J. Pharm. Sci.* **2001**, *14*, 271.
[39] M. Bermejo, A. Avdeef, A. Ruiz, E. Nalda, J. A. Ruell, O. Tsinman, I. Gonzalez, C. Fernandez, G. Sanchez, T. M. Garrigues, V. Merion, *Eur. J. Pharm. Sci.* **2004**, *21*, 429.
[40] Banque de Données Automatisées sur les Médicaments, Paris, www.biam2.org.

Predicting the Intestinal Solubility of Poorly Soluble Drugs

by **Alexander Glomme**[a][b][1], **J. März**[a], and **Jennifer B. Dressman**[*b]

[a] *Merck KGaA*, Dept. Medicinal Chemistry, Frankfurter Str. 250, D-64271 Darmstadt
[b] Institute of Pharmaceutical Technology, Biocenter Johann Wolfgang Goethe-University, Marie Curie Str. 9, D-60439 Frankfurt (Main) (phone: +49 69 7982 9680; fax: +49 69 8982 9724; e-mail: dressman@em.uni-frankfurt.de)

Abbreviations
BCS: Biopharmaceutics Classification System; CMC: critical micelle concentration; Δlog *SR*: difference in log *SR* between ionized and nonionized form of the compound; lec: lecithin; mixed micelles: micelles composed of bile salts and lecithin; m.p.: melting point; NaTC: sodium taurocholate; *SR*: solubility ratio, defined as the quotient of the solubility capacities of the compound per mol of bile salt and per mol water, resp.

1. Introduction

Solubility is one of the most-important parameters for the absorption of a drug after oral application, since only compounds that are in solution at the site of absorption can penetrate the intestine wall and be transported into the systemic circulation. If the solubility of the drug is insufficient, only a part of the dose can achieve systemic availability. Particularly for poorly soluble drugs, it is important to establish which concentrations can be generated under intestinal conditions. For these purposes, it is not sufficient to simply measure the aqueous solubility, since parameters important to solubility such as pH value and presence of surfactants are not addressed by solubility measurements in water.

In particular, the bile salts play an important role as a solubilizer in the human digestive tract. Their ability to build micelles which can solubilize lipophilic compounds leads to an increase in solubility of these substances and often to a faster dissolution rate. For six steroids, *Mithani* [1] described

[1] Current address: *F. Hoffmann-La Roche Ltd.*, Pharm. & Analytical R&D, CH-4070 Basel (e-mail: alexander.glomme@roche.com).

the following correlation between the log *P*, the water solubility, and the solubility in sodium taurocholate solutions (*Eqn. 1*):

$$\log SR = 0.61 \cdot \log P + 2.217 \quad (1)$$

where log *SR* is the solubility ratio defined as the quotient of the solubility capacities of the compound per mol of bile salt and per mol water, respectively, and log *P* is the well-known octanol/water partition coefficient.

According to this relationship, the more lipophilic a compound, the larger its solubility ratio. This means that at high lipophilicity, which is usually associated with poor aqueous solubility, the solubility in a bile salt micelle is substantially higher than the solubility in water. From the physiological point of view, the solubilization of lipophilic food components by bile salt micelles enhances their solubility and hence availability for absorption across the gut wall. When bile salt levels are depleted, *e.g.*, in short bowel syndrome [2], the assimilation of fats from the diet is seriously curtailed, leading to chronic weight loss. In a similar way, bile salt solubilization assists in the assimilation of lipophilic drugs. Especially after food intake, when intestinal levels of bile salts are substantially increased, it can be expected that lipophilic compounds of class II in the Biopharmaceutics Classification System (BCS), *i.e.*, poorly soluble but highly permeable drugs, will be better solubilized and absorbed due to the elevated bile salt concentration. A knowledge of the solubility under simulated intestinal conditions can be instrumental in the decision as to whether a special formulation of the drug is needed to achieve complete dissolution in the gastrointestinal (GI) tract.

In this chapter, recent advances in understanding of the solubility-increasing effects of the bile salts for poorly soluble drugs are presented. The afore-mentioned correlation between log *P* and the solubility ratio has been extended for a wider range of acids, bases, and neutral compounds by measuring solubilities at different sodium taurocholate (NaTC) concentrations and at different pH values. Further, a separate correlation has been established for mixed micelles, *i.e.*, those containing lecithin as well as bile salt, using a bile salt/lecithin (lec) ratio of 4 : 1. This combination is more relevant to *in vivo* conditions since lecithin is always cosecreted with bile salts in the bile [3]. The influence of food intake on the solubility can be roughly estimated by measuring the solubility at bile salt concentrations of 3.75 mM NaTC/lec (preprandial) and at 15 mM NaTC/lec (postprandial). From these results, it was possible to develop an algorithm for the calculation of the solubility in NaTC/lec micelles. This algorithm can be used to estimate solubility in the small intestine on the basis of the physicochemical properties of the compound and average concentrations of bile in the fed and fasted states and further to predict food effects on the bioavailability of poorly soluble compounds.

2. Methodology

2.1. *Compounds Used to Establish the Correlations*

Poorly soluble compounds with a range of physicochemical characteristics were used to confirm the correlation between log *P* and log *SR* for simple bile salt micelles and to investigate the possibility of establishing a correlation in mixed micelles. These included albendazole, beclomethasone dipropionate, betamethasone 17-valerate, danazol, dexamethasone, dipyridamole, dronabinol, felodipine, glyburide, griseofulvin, itraconazole, ketoconazole, ketotifen fumarate, mefenamic acid, miconazole nitrate, phenytoin, testosterone propionate, and *levo*-thyroxine.

2.2. *Media Utilized*

The solubility of the drugs was examined as a function of pH and NaTC concentration over the range typically encountered in the small intestine. Determinations were performed with the miniaturized shake-flask method [4]. The base medium was chosen with an electrolyte concentration which simulates the intestinal conditions in the fasted state. This medium, with the addition of NaTC ($>$98% purity) and egg lecithin, is described in the literature as the FaSSIF medium and has a pH of 6.5 [5]. To determine the log *SR* values, solubilities in variations of this medium with NaTC concentrations of 1, 3.75, 7.5, 15, and 30 mM either alone (to generate simple bile salt micelles) or with a bile salt/lecithin ratio of 4:1 (to generate mixed micelles) were investigated. The two lower bile salt concentrations reflect typical values in the fasted state, while the three higher concentrations cover the observed range in the fed state. To characterize the relative contributions of pH and micellar solubilization to the solubility of poorly soluble acids and bases, solubilities were determined at core pH values of 5–7, and, for selected ionizable compounds, the pH range was extended to 3–8.

Table 1 summarizes the solubility data for all compounds at the various pH/bile salt/lecithin combinations.

Table 1. *pH-Dependent Solubility of the Investigated Compounds at Different Bile Salt Concentrations*

Compound	Final pH	NaTC [mM]	Solubility [μg ml^{-1}] without lecithin	Solubility [μg ml^{-1}] with lecithin (4:1)
Albendazole	7	0	0.95 ± 0.09	0.95 ± 0.09
		7.5	1.79 ± 0.09	4.73 ± 0.34
		15	4.71 ± 0.03	10.4 ± 0.6
		30	11.9 ± 1.1	23.1 ± 0.6
Beclometasone dipropionate	7	0	0.46	0.46
		7.5	0.92 ± 0.12	3.84 ± 0.20
		15	3.26 ± 0.18	9.84 ± 0.34
		30	14.3 ± 0.4	24.6 ± 0.24
Betamethasone	7	0	63 ± 2	65.7 ± 3.0
		0.1	64 ± 1	
		1	63 ± 1	
		3.75	69 ± 3	
		7.5	69 ± 1	88.1 ± 2
		15	86 ± 2	130 ± 1
		30	118 ± 5	185 ± 5
Danazol	7	0	0.21 ± 0.04	0.21 ± 0.04
		1	0.32 ± 0.05	2.18 ± 0.20
		3.75	0.30 ± 0.07	10.3 ± 0.2
		7.50	1.27 ± 0.06	15.4 ± 0.4
		15	7.13 ± 0.43	31.6 ± 0.8
		30	21.3 ± 0.8	50.4 ± 0.9
Dexamethasone	7	0	92 ± 2	92.0 ± 10.5
		3.75	109 ± 2	
		7.5	114 ± 2	203 ± 21
		15	137 ± 3	255 ± 2
		30	193 ± 5	369 ± 42
Dipyridamole	3.5	0	2200 ± 100	
		7.5	2250 ± 380	
		15	3950 ± 220	
		30	7500 ± 420	
	4.2	0	343	
		7.5	701 ± 22	
		15	1000 ± 330	
		30	1610 ± 45	
	5.0	0	54.1 ± 2.2	54.1 ± 2.2
		1	54.2 ± 1.9	65.1 ± 1.1
		3.75	57.6 ± 2.3	90.3 ± 4.4
		7.5	94.9 ± 3.7	136 ± 3
		15	290 ± 19	250 ± 8
		30	1005 ± 12	480 ± 1
	6.0	0	10.5 ± 0.94	10.5 ± 0.94
		1	10.2 ± 0.65	13.8 ± 0.24
		3.75	19.7 ± 0.87	23.3 ± 0.55
		7.5	49.9 ± 0.97	44.6 ± 0.54
		15	160 ± 5	101 ± 12
		30	600 ± 20	286 ± 15
	7.0	0	4.9 ± 0.1	4.9 ± 0.1
		1	5.4 ± 0.4	6.9 ± 0.2
		3.75	11.4 ± 0.8	15.3 ± 0.4

Table 1 (cont.)

Compound	Final pH	NaTC [mM]	Solubility [μg ml^{-1}] without lecithin	Solubility [μg ml^{-1}] with lecithin (4:1)
		7.5	38.6 ± 1.3	32.7 ± 0.5
		15	136 ± 3	94.4 ± 13.1
		30	504 ± 33	222 ± 3
	7.8	0	6.0 ± 0.2	
		7.5	53.6 ± 0.9	
		15	200 ± 4	633 ± 11
		30		
Dronabinol	7	0	0.03	0.03
		7.5	0.2	95.5
		15	39.2 ± 10.6	191
		30	177	257
Felodipine	7	0	0.86 ± 0.03	0.86 ± 0.09
		1	0.67 ± 0.04	10.4 ± 0.6
		3.75	1.30 ± 0.09	57.8 ± 0.7
		7.5	14.0 ± 0.77	96.9 ± 2.3
		15	53.0 ± 2.77	258 ± 4
		30	162 ± 2	481 ± 59
Glyburide	2	0	0.07	
		7.5	0.16 ± 0.07	
		15	0.92 ± 0.11	
		30	2.74 ± 0.09	
	3	0	0.06 ± 0.01	
		7.5	0.13 ± 0.03	
		15	1.25 ± 0.21	
		30	2.70 ± 0.18	
	5	0	0.10 ± 0.06	0.1 ± 0.06
		1	0.10 ± 0.05	0.2 ± 0.01
		3.75	0.18 ± 0.01	1.1 ± 0.2
		7.5	0.30 ± 0.06	0.7 ± 0.1
		15	0.60 ± 0.04	2.8 ± 0.2
		30	1.25 ± 0.04	11.7 ± 2.8
	6	0	0.62 ± 0.15	0.62 ± 0.15
		1	0.68 ± 0.09	0.9 ± 0.04
		3.75	0.75 ± 0.04	2.1 ± 0.1
		7.5	0.86 ± 0.26	1.7 ± 0.1
		15	1.62 ± 0.07	1.8 ± 0.3
		30	3.40 ± 0.09	15.7 ± 0.21
	7	0	5.6 ± 0.7	5.6 ± 0.72
		1	6.5 ± 1.9	7.7 ± 0.22
		3.75	6.3 ± 0.1	10.6 ± 0.4
		7.5	7.2 ± 0.2	11.9 ± 0.3
		15	12.4 ± 0.3	15.0 ± 0.3
		30	22.0 ± 2.0	33.4 ± 1.2
	8	0	51.2 ± 0.3	
		7.5	70.6 ± 6.6	
		15	85.9 ± 0.5	
		30	132 ± 1.2	
	9	0	98.6 ± 0.8	
		7.5	130 ± 1	
		15	158 ± 1	

Table 1 (cont.)

Compound	Final pH	NaTC [mM]	Solubility [μg ml^{-1}] without lecithin	Solubility [μg ml^{-1}] with lecithin (4:1)
		30	234 ± 4	
	11.9	0	532 ± 10	
		7.5	744 ± 28	
		15	969 ± 34	
		30	1710 ± 60	
Griseofulvin	7	0	10.4 ± 0.1	10.4 ± 0.1
		7.5	18.6 ± 0.2	55.9 ± 1.9
		15	31.8 ± 1.6	71.7 ± 11.9
		30	56.8 ± 1.0	128 ± 26
Hydrocortisone	7	0	326 ± 6	
		0.09	312 ± 1	
		0.93	346 ± 8	
		3.75	377 ± 4	
		7.5	425 ± 1	
		15	528 ± 4	
		30	683 ± 6	
Hydrocortisone hemisuccinate	7	0		3170 ± 7
		7.5		3220 ± 7
		15		3290 ± 7
		30		3415 ± 7
Itraconazole	7	0		0.0002
		3.75		0.04 ± 0.01
		7.5		0.06 ± 0.01
		15		0.12 ± 0.02
		30		0.26 ± 0.03
Ketoconazole	5	0	81.1 ± 8.3	81.1 ± 8.3
		1	87.1 ± 3.7	100.2 ± 2.3
		3.75	92.6 ± 3.2	133 ± 5
		7.5	97.6 ± 3.9	246 ± 8
		15	229 ± 3	399 ± 8
		30	444 ± 22	687 ± 101
	6	0	15.9 ± 0.7	15.9 ± 0.7
		1	16.6 ± 0.5	14.7 ± 0.1
		3.75	16.1 ± 1.1	32.5 ± 1.3
		7.5	19.5 ± 1.4	75.2 ± 2.3
		15	65.5 ± 1.8	157 ± 6
		30	162 ± 5	369 ± 10
	7	0	5.98 ± 0.27	5.98 ± 0.27
		1	6.08 ± 0.20	6.64 ± 0.67
		3.75	8.21 ± 0.54	16.8 ± 0.7
		7.5	12.4 ± 0.3	45.3 ± 1.9
		15	38.8 ± 1.3	93.1 ± 2.8
		30	110 ± 2	212 ± 10
	7.8	0	3.64	
		7.5	8.55 ± 0.03	
		15	30.9 ± 0.9	
		30	108	
Mefenamic acid	2	0	0.06 ± 0.01	
		7.5	0.27 ± 0.02	
		15	3.40 ± 0.08	

Table 1 (cont.)

Compound	Final pH	NaTC [mM]	Solubility [μg ml⁻¹] without lecithin	Solubility [μg ml⁻¹] with lecithin (4:1)
		30	15.9 ± 3.9	
	3	0	0.07 ± 0.01	
		7.5	0.23 ± 0.09	
		15	3.78 ± 0.25	
		30	19.0 ± 0.5	
	5	0	0.65 ± 0.03	0.65 ± 0.03
		1	0.69 ± 0.02	2.10 ± 0.10
		3.75	0.60 ± 0.03	8.90 ± 0.35
		7.5	0.52 ± 0.12	11.4 ± 0.2
		15	5.27 ± 0.27	27.2 ± 0.9
		30	22.1 ± 1.0	45.5 ± 19.9
	6	0	7.20 ± 0.15	7.20 ± 0.15
		1	5.87 ± 0.08	9.60 ± 0.10
		3.75	6.44 ± 0.22	18.1 ± 0.2
		7.5	6.58 ± 0.38	24.8 ± 0.3
		15	19.4 ± 0.3	46.6 ± 0.7
		30	58.8 ± 5.5	95.1 ± 4.2
	7	0	67.2 ± 3.3	67.2 ± 3.3
		1	58.5 ± 0.6	80.8 ± 1.1
		3.75	65.1 ± 0.3	94.8 ± 1.8
		7.5	64.0 ± 5.7	141 ± 4
		15	153 ± 2	248 ± 2
		30	343 ± 30	530 ± 11
	8	0	357 ± 3	
		7.5	419 ± 4	
		15	546 ± 5	
		30	838 ± 9	
	9	0	486 ± 3	
		7.5	585 ± 5	
		15	715 ± 8	
		30	1000 ± 9	
Miconazole nitrate	4.2	0	52.1 ± 21.3	
		7.5	40.8 ± 3.3	
		15	241 ± 5	
		30	894 ± 63	
	5	0	12.1 ± 0.2	
		7.5	21.9 ± 0.9	
		15	280 ± 63	
		30	546	
	5.4	0		2.47 ± 0.17
		7.5		253 ± 89
		15		377 ± 34
		30		628 ± 57
	6	0	0.97 ± 0.07	0.97 ± 0.07
		7.5	3.13 ± 0.11	49.3 ± 8.2
		15	70.2 ± 0.9	209 ± 33
		30	174	401 ± 13
	7	0	0.26 ± 0.07	0.26 ± 0.07
		7.5	1.63 ± 0.31	52.9 ± 30.8
		15	50.1 ± 8.2	134 ± 4

Table 1 (cont.)

Compound	Final pH	NaTC [mM]	Solubility [μg ml^{-1}] without lecithin	Solubility [μg ml^{-1}] with lecithin (4 : 1)
		30	122	289 ± 14
	7.8	0	0.18 ± 0.04	0.18 ± 0.04
		7.5	1.70 ± 0.30	107 ± 1
		15	18.1 ± 0.6	123 ± 1
		30	137 ± 2	261 ± 2
Niclosamide	2	0	0.005 ± 0.001	
		7.5	0.33 ± 0.08	
		15	2.97 ± 0.12	
		30	8.21 ± 0.33	
	3	0	0.005 ± 0.001	
		7.5	0.3 ± 0.1	
		15	3.19 ± 0.17	
	4	0	0.011	0.011
		7.5	0.39 ± 0.15	6.87 ± 0.29
		15	3.47 ± 0.09	13.8 ± 0.2
		30	12.7 ± 0.3	33.4 ± 16.3
	5	0	0.03	0.03
		7.5	0.11 ± 0.01	7.10 ± 0.05
		15	1.49 ± 0.22	13.8 ± 0.1
		30	12.2	39.0 ± 0.9
	6	0	0.17 ± 0.07	0.17 ± 0.07
		7.5	0.22 ± 0.03	7.35 ± 0.08
		15	2.55 ± 0.30	14.8 ± 0.2
		30	14.2	40.2 ± 0.7
	7	0	1.01 ± 0.08	1.01 ± 0.08
		7.5	1.48 ± 0.19	9.75 ± 0.29
		15	18.0	21.2 ± 0.5
		30	51.0	45.6 ± 32.1
	8	0	5.1 ± 0.8	5.1 ± 0.8
		7.5	33.0 ± 0.5	42.9 ± 0.3
		15	116 ± 1	66.6 ± 0.9
		30	282 ± 5	188 ± 52
	9	0	5.15	
		7.5	36.9 ± 3.2	
		15	116 ± 2	
		30	345 ± 14	
Phenytoin	7	0	31.0 ± 2.0	31.0 ± 2.0
		1	29.6 ± 0.3	32.5 ± 1.3
		3.75	29.8 ± 0.5	39.6 ± 1.5
		7.5	32.6 ± 0.7	44.5 ± 1.6
		15	48.6 ± 1.3	58.5 ± 1.9
		30	64.3 ± 1.1	88.7 ± 0.7
Testosterone propionate	7	0	1.87 ± 0.07	1.87 ± 0.07
		7.5	2.78 ± 0.42	88.1 ± 1.9
		15	37.6 ± 3.0	105 ± 2
		30	93	240 ± 5
levo-Thyroxine	5	0	0.26 ± 0.09	0.26 ± 0.09
		1	0.40 ± 0.05	1.70 ± 0.22
		3.75	0.50 ± 0.20	4.80 ± 1.01
		7.5	1.22 ± 0.28	10.1 ± 0.7

Table 1 (cont.)

Compound	Final pH	NaTC [mM]	Solubility [µg ml^{-1}] without lecithin	Solubility [µg ml^{-1}] with lecithin (4:1)
		15	5.63 ± 1.77	17.3 ± 9.4
		20	13.3 ± 1.0	
		30	16.9 ± 2.0	115 ± 19
	6	0	0.27 ± 0.10	0.27 ± 0.10
		1	0.39 ± 0.04	1.6 ± 0.2
		3.75	0.53 ± 0.15	5.2 ± 0.2
		7.5	1.38 ± 0.39	11.6 ± 0.8
		15	5.04 ± 1.09	19.7 ± 2.2
		20	12.0 ± 0.5	
		30	16.5 ± 2.4	
	7	0	0.5 ± 0.1	0.5 ± 0.1
		1	0.6 ± 0.1	1.7 ± 0.2
		3.75	0.8 ± 0.1	9.8 ± 7.8
		7.5	1.7 ± 0.5	12.2 ± 1.7
		15	6.4 ± 1.3	21.2 ± 1.9
		20	17.7 ± 1.2	
		30	19.7 ± 1.5	64.7 ± 6.5

2.3. *Calculation of Solubility Ratios*

The solubility ratio of the solubility of the compound in bile salt to the solubility in water was calculated as follows (*Eqn. 2*):

$$\log SR = \log\left(\frac{c_{\mathrm{Sx}}}{c_{\mathrm{So}}}\right) \tag{2}$$

with $c_{\mathrm{Sx}} = c_{\mathrm{compound}}\,[\mathrm{M}]/c_{\mathrm{bile\ salt}}\,[\mathrm{M}]$, $c_{\mathrm{So}} = S_{\mathrm{o}}\,[\mathrm{M}]/c_{\mathrm{water}}\,[\mathrm{M}]$, $S_{\mathrm{o}}\,[\mathrm{M}]$ = solubility of the drug in aqueous buffer, $c_{\mathrm{water}}\,[\mathrm{M}]$ = mol water per liter (55.55M).

The c_{Sx} value results from the slope of the straight line that is generated when the solubility of the drug is plotted against the bile salt concentration, as shown in *Fig. 1* for danazol. This gradient reveals how much compound can be dissolved per mol of bile salt.

Fig. 1 shows the different behavior of the simple micelle and the mixed micelle systems, respectively. Linearity in the simple NaTC micelles is first observed after reaching the critical micelle concentration (CMC) at *ca.* 6 mM NaTC. Therefore, only the solubility values at 7.5, 15, and 30 mM NaTC were used for the regression analysis. By contrast, values at all bile salt concentration were used for the calculations for mixed NaTC/lec micellar media. The CMC of this system is *ca.* 0.6 mM NaTC/lec (4:1).

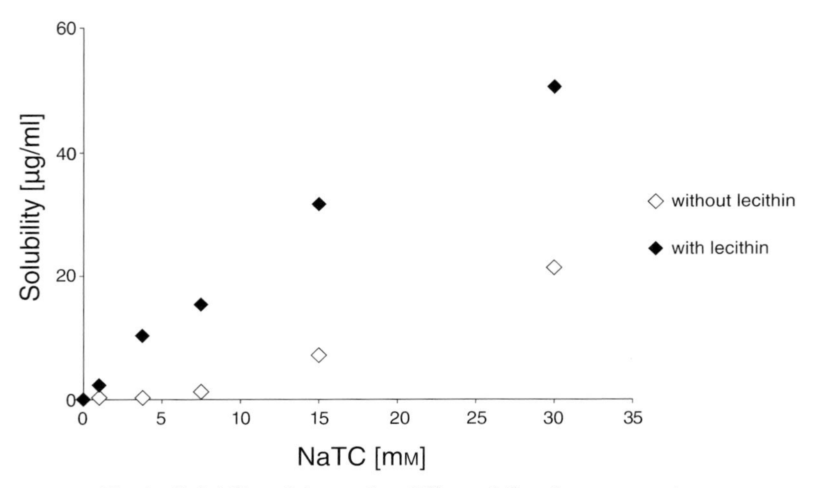

Fig. 1. *Solubility of danazol at different bile salt concentrations*

3. Solubility Ratios for Compounds with pH-Independent Solubility at Typical Intestinal pH

Table 2 summarizes the solubility ratios for compounds which behaved essentially as neutral compounds at typical intestinal pH values. Although some of the compounds are clearly ionizable, their solubilities in simple aqueous buffers and log D values are independent of pH over the range investigated, *i.e.*, pH 5–7.

The log SR values of the mixed micelles are always larger than for the simple NaTC micelles (see *Fig. 2*). Noticeable are the very large error bars of the two compounds itraconazole (log $D = 5.66$ in the pH range studied) and dronabinol (log $D = 6.97$). For these two compounds, the aqueous solubilities could not directly be measured in simple aqueous buffer due to their marked hydrophobicity. In this case, the solubilities had to be measured in presence of cosolvent at several concentrations and the solubility in the corresponding simple aqueous buffer was then estimated by back extrapolation to 0% cosolvent. This resulted in a higher error than the direct determination in aqueous buffer.

For the correlation between the partition coefficients of micelle/water and of octanol/water, the linear *Eqns. 3* and *4* were established. For simple NaTC micelles, the relationship (*Eqn. 3*) was similar to that established by *Mithani* [1] for a series of corticosteroids:

Table 2. *Solubility Ratio* (log *SR*) *and Octanol/Water log* D *of Neutral Compounds*

Compound	Lecithin[a])	log *SR*	log *D*[b])
Albendazole[c])	0	4.43 ± 0.05	3.04 ± 0.10
	w	4.65 ± 0.07	
Beclometasone dipropionate	0	4.87	4.10 ± 0.13
	w	5.00	
Betamethasone[d])	0	3.28 ± 0.02	1.72 ± 0.25
	w	3.57 ± 0.03	
Betamethasone 17-valerate	0	4.40	3.60 ± 0.30
	w	4.97	
Danazol	0	5.38 ± 0.09	4.53 ± 0.32
	w	5.74 ± 0.09	
Dexamethasone[d])	0	3.33 ± 0.02	1.89 ± 0.42
	w	3.73 ± 0.06	
Dronabinol	0	7.17 ± 1.26	6.97 ± 0.44
	w	7.37 ± 1.25	
Felodipine	0	5.64 ± 0.04	4.26 ± 0.42
	w	6.03 ± 0.02	
Griseofulvin	0	3.96 ± 0.01	2.47 ± 0.46
	w	4.30 ± 0.05	
Hydrocortisone[d])	0	3.30 ± 0.02	1.61 ± 0.34
Itraconazole[c])	w	6.36 ± 1.49	5.66 ± 0.46
Phenytoin	0	3.39 ± 0.10	2.47 ± 0.49
	w	3.53 ± 0.03	
Testosterone propionate	0	5.07 ± 0.03	3.72 ± 0.44
	w	5.35 ± 0.06	
Triamcinolone[d])	0	2.97 ± 0.04	1.16 ± 0.13

[a]) 0 = without lecithin (simple bile salt micelles), w = with lecithin (mixed micelles). [b]) For neutral compounds, log D = log P. [c]) Albendazole (pK_a = 3.3) and itraconazole (pK_a = 4.0) are very weak bases but behave in the intestinal pH range like neutral compounds. [d]) log *SR* was determined with measured values from *Bakatselou* [6].

$$\log SR = 0.74 \cdot \log D + 2.01 \qquad n = 14;\; r^2 = 0.95;\; s = 0.26 \tag{3}$$

and for mixed NaTC/lec micelles (ratio 4 : 1):

$$\log SR = 0.74 \cdot \log D + 2.29 \qquad n = 13;\; r^2 = 0.94;\; s = 0.30 \tag{4}$$

The two lines in *Fig. 2* run nearly parallel to each other and the linear equations confirm this determination. The two equations differ only in the constant term, which is *ca.* 0.26 log units larger in the mixed micelles. The only variable parameter of the log *SR* calculation is the gradient of mol

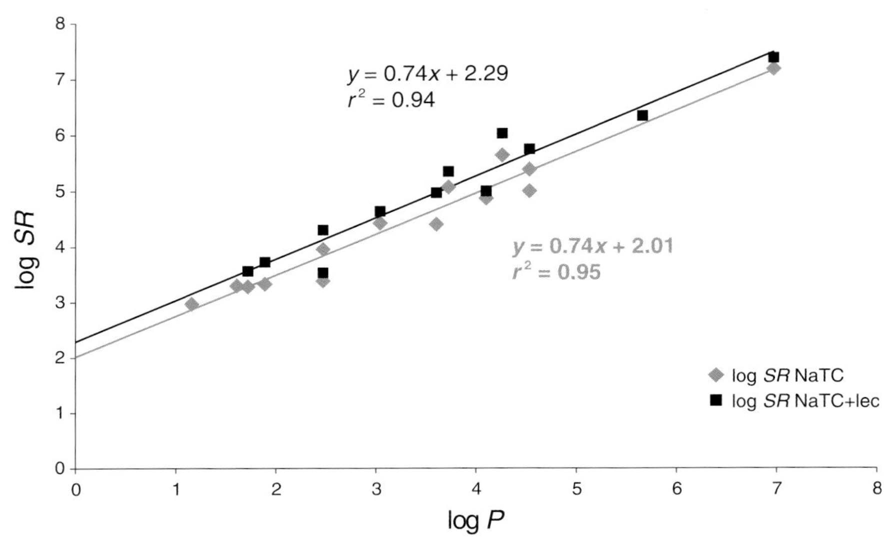

Fig. 2. *Correlation between log* D *and log* SR *values for drugs not ionized in the pH range 5–7.* For neutral compounds, log D = log P.

drug per mol bile salt. From these results, it can be concluded that the mixed micelles have a better solubilization capacity than simple bile salt micelles for neutral drugs. Conversion of the factor 0.26 log units to the linear scale results in a factor of 2. The solubility improvement by the mixed micelles is thus *ca.* twice as high as in the simple micelles. For example, albendazole has a solubility of 1 μg ml^{-1} at 0 mM NaTC and a solubility of 12 μg ml^{-1} at 30 mM NaTC – a twelvefold increase. In the mixed micelles, a 24-fold increase over the same bile salt range is observed (1–23 μg ml^{-1}). Note that this rule of thumb applies only at concentrations above the CMC of the simple NaTC micelles.

An attempt to improve the algorithm by including the melting point was only partially fruitful. The melting point (m.p.) was found to produce a significant improvement in the correlation (r^2 and standard deviation) for the simple NaTC micelles, and the following *Eqn. 5* resulted:

$$\log SR = 0.60 \cdot \log D - 0.004 \cdot \text{m.p.} + 3.24$$
$$n = 14; \; r^2 = 0.97; \; s = 0.2 \quad (5)$$

Both r^2 and the standard deviation improved in relation to the previous algorithm. It would be reasonable that the melting point plays a role in the log SR calculation, because it affects the aqueous solubility of the compound [7]. However, the uptake of the drug into the micelle is affected to a greater degree by lipophilicity. The relative magnitude of the weighting terms (0.6 *vs.* 0.004, resp.) for log D and m.p. confirm this. In contrast to the simple bile salt micelles, the correlation with the melting point did not produce better results for the mixed NaTC/lec micelles. The melting point was not found to be significant. It is difficult to draw a concrete conclusion here, though, since the compound set does not cover a sufficiently wide melting point range (most of the compounds have a melting point $>200°$).

For the mixed micelles, the solubility of a neutral drug at any NaTC/lec concentration can be calculated from the above-mentioned straight line *Eqn. 6*:

$$S_{(\mathrm{NaTC/lec4:1})} = c_{\mathrm{NaTC/lec}} \cdot \frac{S_o}{c_{\mathrm{water}}} \cdot 10^{(0.75 \cdot \log P + 2.27)} + S_o \quad (6)$$

with S_o = solubility in the aqueous buffer, $c_{\mathrm{NaTC/lec}}$ = desired NaTC/lec concentration, $c_{\mathrm{water}} = 55.55\mathrm{M}$.

Only the log P and the melting point are required to calculate the solubility at any bile salt concentration (with NaTC/lec 4 : 1), since solubility in the aqueous buffer can be estimated on the basis of these two parameters. In turn, these two parameters can be calculated by many readily available computer programs, so that a fast estimation of the intestinal solubility can take place without a single laboratory measurement. Since an inaccurately computed value strongly affects the solubility ratio, it is nevertheless recommended to determine the aqueous solubility experimentally.

4. Application to BCS Related Calculations

According to the Biopharmaceutics Classification System (BCS) [8], drugs can be divided into four categories, depending on their permeabilities and solubilities. The class to which the drug is assigned has implications for the approval of oral products containing the drug. For example, fast-dissolving generic products containing a Class I drug can be approved on the basis of dissolution data, without requiring a pharmacokinetic study. The cut-off for solubility is a maximum dose/solubility ratio of 250 ml throughout the pH range 1.2–7.5. In this case, the solubility used is the solubility in simple aqueous buffers. Obviously, if this criterion were to be applied to assess whether a drug can be developed as an oral dosage form, many drugs which today are available on the market as oral dosage forms would never have been developed. For development purposes, it is more reasonable to determine the solubility in the presence of appropriate concentrations of mixed micelles and calculate the maximum dose which could be dissolved under intestinal conditions, assuming volumes of *ca.* 500 ml for the fasted state and 1–1.5 l for the fed state [9].

5. Application to Forecasting Food Effects on Oral Drug Bioavailability

The NaTC/lec concentrations at 3.75 mM and at 15 mM NaTC are of particular importance because they reflect average bile salt concentrations in the intestinal tract under fasted and fed conditions, respectively. A clear increase in solubility at 15 mM bile salt compared to the solubility at 3.75 mM leads to an expectation of a positive effect of ingestion of food on the solubility and consequently on the bioavailability of the compound. In the following table (*Table 3*), measured and calculated solubility ratios for the fasted and the fed state are shown. The solubility increase due to the bile salt concentration was compared with experimental bioavailabilities from the literature using the bioavailability ratio between fasted and fed state for all compounds where such data were available. It is clearly evident that the difference in the bioavailability due to food intake can be estimated by the experimental solubility data as well as by the calculated data.

Table 3. *Comparison of the Bile Salt Induced Increase of Solubility with Food-Dependent Increase of Bioavailability for Selected Compounds*

Compound	Exper. solubility ratio fed/fasted	Calc. solubility ratio fed/fasted	Exper. bioavailability ratio fed/fasted [Ref.]
Danazol	3	4	4 [10]
Felodipine	4	4	2 [11]
Griseofulvin	2	2	2 [12]
Itraconazole	3	4	>2 [13]
Phenytoin	1	2	<2 [14]

In vivo, the food-induced increase in solubility might be even larger because of additional dietary components and their digestive products in the intestinal lumen, *e.g.*, free fatty acids and monoglycerides. In addition, the volume in the small intestine is increased by food and digesting secretions so that more volume is available to dissolve the drug.

6. The Solubility Ratio for Ionizable Compounds

Unlike compounds that are essentially neutral at intestinal pH, the solubility ratio (*SR*) of acids and bases has no linear relationship with their log *P* value. For these compounds, ionization has a major influence on the distribution into micelles. If the log *D* for ionizable compounds is plotted *vs.* the pH value, a sigmoidal curve is obtained similar to the plot of the aqueous solubility as a function of pH. To explore further the similarity in the two curves, the pH range was extended to obtain the limiting values of the two curves.

6.1. *The Bases: Dipyridamole, Ketoconazole, and Miconazole*

Table 4 shows the distribution coefficient (octanol/water) and the solubility ratio (log *SR*) of the examined bases at different pH values. The log *D* value of the bases as well as their log *SR* value show a clear pH dependence: the values at low pH are smaller than at high pH values. At low pH values, the compounds are mainly ionized, resulting in a substantially better aqueous solubility. Because in both definitions of distribution coefficients (log *D* and log *SR*) the aqueous solubility stands in the denominator, a higher concentration of the drug in the aqueous phase will lead to smaller log *D* and log *SR* values.

The expected pH-dependent sigmoidal curve of the log *SR* is confirmed for dipyridamole in *Fig. 3*. At the beginning of both curves, between pH 7 and pH 8, the values run parallel to the *x* axis. Close to the pK_a value ($pK_a = 6.21$) the curves begin to drop as the pH declines. But then a divergence in the curves with further decrease in pH is observed. At pH 4, the log *SR* value begins to run parallel to the *x* axis again, the log *D* value however not until pH 2. Further, it is observed that the difference between the two parallel sections of the curve for log *SR* is smaller than for the log *D* value. Calculation of log *D* as a function of pH indicates that the difference between the distribution coefficients of the ionized form compared to the neutral form is, in general, *ca.* 4, although for bases this difference is often somewhat smaller. However, the $\Delta \log SR$ value (difference between log *SR* of ionized and nonionized form) was in every case investigated much smaller than the $\Delta \log D$ value.

It is noticeable that the solubilities of dipyridamole are higher in the simple NaTC micelles than in the mixed NaTC/lec micelles even at high bile salt concentration. This leads to a lower log *SR* value. On the other hand, miconazole and ketoconazole exhibited behavior similar to that of

Table 4. *Solubility Ratio* (log *SR*) *and Octanol/Water log* D *of Bases at Different pH Values*

Compound	Lecithin[a])	Final pH	log *SR*	log *D*
Dipyridamole	0	3.5	3.77 ± 0.02	1.15 ± 0.48
		4.2	3.82	1.89 ± 0.48
		5.0	4.63 ± 0.06	2.66 ± 0.48
		6.0	5.12 ± 0.08	3.48 ± 0.48
		7.0	5.38 ± 0.06	3.83 ± 0.48
		7.8	5.39 ± 0.04	3.89 ± 0.48
	w	5.0	4.17 ± 0.02	2.66 ± 0.48
		6.0	4.70 ± 0.06	3.48 ± 0.48
		7.0	4.94 ± 0.03	3.83 ± 0.48
Ketoconazole	0	5.0	4.02 ± 0.07	2.50 ± 0.30
		5.5	4.17[b])	2.97 ± 0.30
		6.0	4.35 ± 0.02	3.39 ± 0.30
		7.0	4.61 ± 0.04	3.90 ± 0.30
		7.5	4.65[b])	3.98 ± 0.30
		7.9	4.84 ± 0.07	4.00 ± 0.30
	w	5.0	4.15 ± 0.06	2.50 ± 0.30
		6.0	4.62 ± 0.03	3.39 ± 0.30
		7.0	4.81 ± 0.03	3.90 ± 0.30
Miconazole	0	4.2	4.62 ± 0.24	3.18 ± 0.31
		5.0	5.01 ± 0.09	3.95 ± 0.30
		6.0	5.63 ± 0.05	4.76 ± 0.26
		7.0	6.04 ± 0.14	5.09 ± 0.20
		7.8	6.30 ± 0.15	5.14 ± 0.20
	w	5.6	5.65 ± 0.07	4.40 ± 0.27
		6.0	5.90 ± 0.06	4.76 ± 0.26
		7.0	6.31 ± 0.14	5.09 ± 0.20
		7.8	6.42 ± 0.12	5.14 ± 0.20

[a]) 0 = without lecithin (simple NaTC micelles), w = with lecithin (mixed micelles). [b]) Calculated from experimental data [15].

the neutral compounds, with solubility ratios in the lecithin containing micelles *ca.* 0.2–0.3 log units larger than in simple taurocholic micelles.

6.2. *Acidic Compounds: Glyburide, Mefenamic Acid, and Niclosamide*

For acids, the same general observations can be made as for the bases. In the nonionized state, log *D* and log *SR* are similar. In the proximity of the pK_a value (4.99 for glyburide) the first turning point of the sigmoidal curve can be observed. The difference between the maximum and the minimum log *SR* value amounts to *ca.* 1.5 log units for glyburide (*Fig. 4*). This Δlog *SR* value is again clearly lower than the observed difference of 4 log units for the log *D* value.

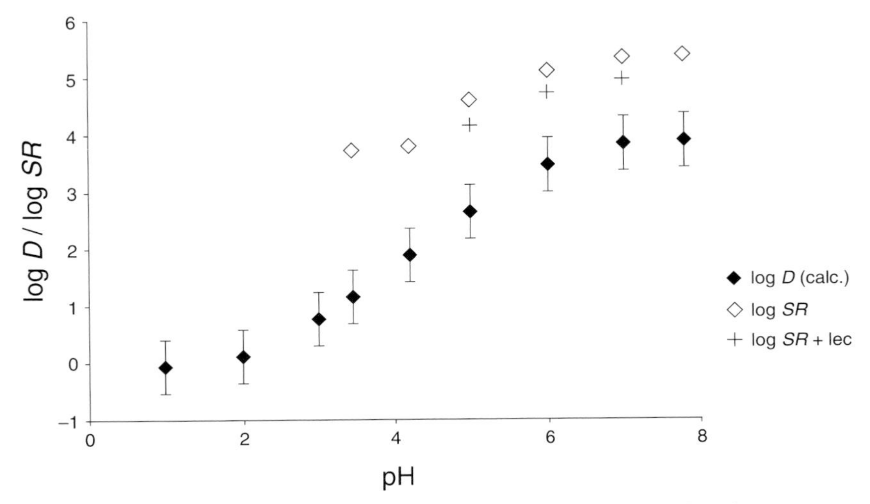

Fig. 3. *pH-Dependent log* D *and log* SR *profile of dipyridamole*

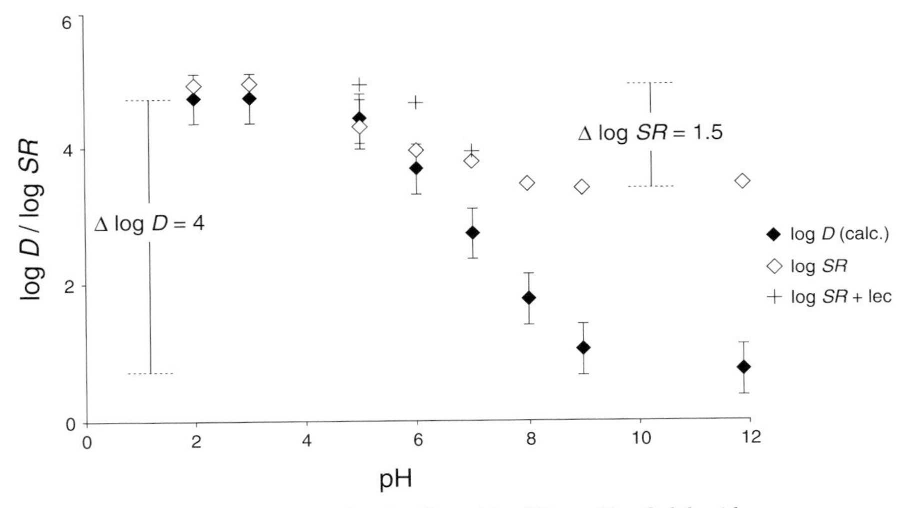

Fig. 4. *pH-Dependent log* D *and log* SR *profile of glyburide*

For mefenamic acid, the sigmoidal behavior was very similar to the pattern observed for dipyridamole and glyburide (*Fig. 5*). Comparison of *Figs. 4* and *5* shows one difference, however. In the nonionized state, *i.e.*, at acidic pH, the difference between log D and log SR for mefenamic acid is larger than for glyburide. In contrast, (log SR – log D) in the ionized state is virtually the same for both compounds. Consequently, a larger Δ log SR results between the limiting values of log SR for mefenamic acid.

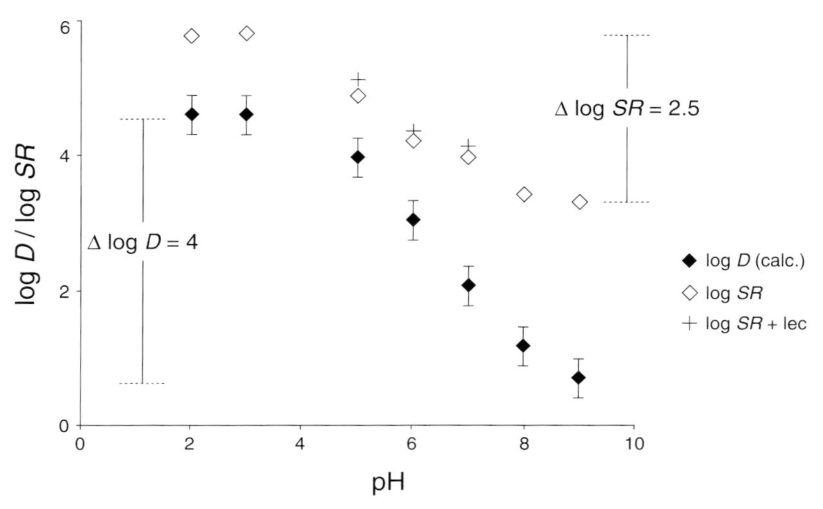

Fig. 5. *pH-Dependent log* D *and log* SR *profiles of mefenamic acid*

For mefenamic acid, the $\Delta \log SR$ is 2.5, for glyburide only 1.5. This difference is unexpected, because the compounds differ very little in their physicochemical characteristics (see *Tables* 5 and 6).

Why are the curves so different despite the similar pK_a and $\log P$ values? Some possible explanations are:

a) Solubility: The $\log P$ values and the melting points indicate that the solubilities of the two compounds lay in a similar range. This assumption was confirmed experimentally. Therefore, the aqueous solubility does not seem to be responsible for the observed differences.
b) Since the pK_a of mefenamic acid is 4.46 and that of glyburide 4.99, the solubility of mefenamic acid begins to rise at a slightly lower pH than that of glyburide.
c) The uncharged mefenamic acid seems to penetrate better into the lipophilic micelles than the uncharged glyburide. A possible reason could be the molecule size. Simple micelles of bile salts are very small with an aggregation number of lower than 10. Therefore, the smaller molecular weight of mefenamic acid (241.29 *vs.* 494.01) could be

Table 5. *Comparison of Physicochemical Properties of Mefenamic Acid and Glyburide*

Compound	M_r	M.p. [°]	pK_a	$\log P$	Intrinsic aqueous solubility ($\mu g\ ml^{-1}$)
Glyburide	494.0	173	4.99	4.73	0.07 ± 0.01
Mefenamic acid	241.3	230	4.46	4.58	0.06 ± 0.01

Table 6. *Solubility Ratio* (log *SR*) *and Octanol/Water log* D *of Acids at Different pH Values*

Compound	Lecithin[a])	Final pH	log *SR*	log *D*
Glyburide	0	2.0	4.98 ± 0.02	4.73 ± 0.36
		3.0	4.99 ± 0.09	4.73 ± 0.36
		5.0	4.36 ± 0.38	4.42 ± 0.36
		6.0	4.01 ± 0.12	3.68 ± 0.36
		7.0	3.82 ± 0.06	2.72 ± 0.36
		8.0	3.47 ± 0.03	1.76 ± 0.36
		9.0	3.41 ± 0.02	1.03 ± 0.36
		11.9	3.48 ± 0.01	0.73 ± 0.36
	w	5.0	4.95 ± 0.43	4.42 ± 0.36
		6.0	4.66 ± 0.12	3.68 ± 0.36
		7.0	3.95 ± 0.06	2.72 ± 0.36
Mefenamic acid	0	2.0	5.82 ± 0.07	4.58 ± 0.27
		3.0	5.84 ± 0.09	4.57 ± 0.27
		5.0	4.92 ± 0.06	3.93 ± 0.27
		6.0	4.25 ± 0.04	3.03 ± 0.27
		7.0	4.01 ± 0.02	2.05 ± 0.27
		8.0	3.47 ± 0.01	1.17 ± 0.27
		9.0	3.33 ± 0.01	0.69 ± 0.27
	w	5.0	5.11 ± 0.03	3.93 ± 0.27
		6.0	4.35 ± 0.02	3.03 ± 0.27
		7.0	4.12 ± 0.04	2.05 ± 0.27
Niclosamide	0	2.0	6.56 ± 0.12	4.67 ± 0.25
		3.0	6.6	4.67 ± 0.25
		4.1	6.49 ± 0.18	4.66 ± 0.25
		5.0	5.96 ± 0.12	4.60 ± 0.25
		6.0	5.33 ± 0.26	4.24 ± 0.25
		7.0	5.08 ± 0.04	3.42 ± 0.25
		8.0	5.07 ± 0.07	2.44 ± 0.25
		9.0	5.17	1.51 ± 0.25
	w	4.1	6.79 ± 0.18	4.66 ± 0.25
		5.0	6.33 ± 0.06	4.60 ± 0.25
		6.0	5.65 ± 0.24	4.24 ± 0.25
		7.0	4.93 ± 0.04	3.42 ± 0.25
		7.8	4.82 ± 0.10	2.64 ± 0.25

[a]) 0 = without lecithin (simple NaTC micelles), w = with lecithin (mixed micelles).

responsible for the fact that more mefenamic acid penetrates into micelles.

d) With rising basicity ($> pK_a$), the log *SR* values approach each other. A possible reason could be that the molecular volume of mefenamic acid changes with increasing ionization relative to the molecular volume of glyburide. The drugs are differently hydrated at their functional acidic groups. Mefenamic acid is a carboxylic acid, while glyburide is a sulfonylurea. The carboxylate ion of mefenamic acid is present in the molecule with no steric hindrance and can easily be hydrated, leading to an increased molecular volume. In contrast, the ionizable group of glyburide is well protected by the other functional groups and the molecule is less hydrated in the ionized state. As a result, the molecular volume of the ion is not much bigger than that of the nonionized compound.

Niclosamide, like glyburide, exhibits a Δlog *SR* value of 1.5 over the entire pH range. The water solubility of niclosamide (S_{pH3} = 0.005 μg ml^{-1}) is lower than that of the other two acids and this leads to higher log *SR* values (*Table 6*).

The assumed pH-dependent sigmoidal gradient of the solubility ratio for ionizable compounds was confirmed for both acids and bases. It is of note that, for the asymptotic values parallel to the *x* axis, there is a difference of only *ca.* 1–2.5 log units. This is an obvious difference to the pH profile of log *D*, for which a difference of 3.5–4.5 log units is commonly found in the literature. A correlation between log *SR* and log *D* values, which was possible for neutral compounds, could not be developed for ionizable drugs. On the one hand, the number of data generated was too low for the development of a predictive algorithm. On the other hand, the comparison of results for glyburide and mefenamic acid lead to the conclusion that additional factors (*e.g.*, molecular size, hydration) may play a role.

Nevertheless, an important conclusion can be drawn from the data set, namely the relatively small difference between the log *SR* value of the ionized and nonionized state. A reduction in the log *D* value of *ca.* 4 log units means that the completely ionized molecule prefers the aqueous phase about a factor of 10000 more than the nonionized molecule. This effect is clearly smaller in the case of the micellar solubilization ratio, indicating that the ionized drug is able to penetrate to a large extent into the micelle. The ionized molecules of glyburide and mefenamic acid, for example, penetrate micelles *ca.* 100 times better than they penetrate into octanol. For ionizable compounds, similar distributions have been observed in phosphatidylcholine bilayer dispersions. These are used to simulate the permeability of biological membranes which consist mainly of phospholipids. *Avdeef et al.* [16] examined the distribution coefficients of eight ionizable compounds in large unilamellar vesicle (LUV) suspensions which consisted of dioleylphosphatidylcholine. For this membrane, the difference in distribution coefficients between the neutral and ionizable state of a compound was determined. The difference was *ca.* 2 for the acids and *ca.* 1 for the bases. Similarly, *Miyazaki et al.* [17] observed a membrane difference of 2.2 for acids and 0.9 for bases in dimyristoylphosphatidylcholine bilayer dispersions.

7. Conclusions

The linear correlation for the solubility ratio of sodium taurocholate/ water and the distribution coefficient octanol/water set up by *Mithani et al.* [1] could be confirmed for a range of neutral compounds. The extent of the solubility increase depends primarily on the lipophilicity of the drug. A similar correlation has been established for the solubility ratio in mixed NaTC/lec (4:1) micelles. In comparison to simple NaTC micelles, the solubilization in mixed micelles is doubled. The obtained correlation for the mixed micelles is of more biorelevance, since it better reflects conditions in the small intestine.

This study could also be of interest for application of the BCS system to pharmaceutical development. Through the acquired correlation between water solubility, bile salt concentration, and log *D*, it is possible to calculate the *in vivo* solubilities of a neutral drug with just a measurement of aqueous solubility. Especially for lipophilic compounds of BCS class II, the solubility under gastrointestinal conditions is clearly higher than would be estimated from the solubility in a simple aqueous buffer. In some cases, the boost to the solubility may be sufficient to circumvent solubility limitations to absorption and they could be classified as relatively unproblematic for development of an oral dosage form. An example is felodipine, which is available on the market with 10 mg as maximum dose. With an intrinsic solubility of 0.86 $\mu g\ ml^{-1}$, a volume of 11.6 l would be necessary to dissolve the entire dose. The calculation of the dose/solubility ratio for the fasted state leads to a required volume of 0.17 l. To dissolve the entire dose in the fed state, however, only 40 ml of intestinal fluid are needed. In this case, limitations to absorption reside in a low dissolution rate, as opposed to insufficient volume of gastrointestinal fluids available, and this barrier can be overcome simply by micronizing the drug.

The bile salt/water solubility ratio was also determined for acids and bases over a pH range typically encountered in the small intestine. It was demonstrated that bile salts play a much more minor role for the solubility of acids and bases than for neutral drugs, and that a change in gastrointestinal pH affects the solubility far more than the solubilizing effect of bile salts. This is because solubility rises exponentially with pH but only linearly with the bile salt concentration. Accordingly, for the ionizable drugs examined, like dipyridamole, ketoconazole, glyburide, and mefenamic acid, none or only a very small food-induced increase in bioavailability has been described in the literature.

For ionizable compounds, no linear correlation between log *SR* and log *D* could be determined. This result was no surprise, since the octanol/ water distribution coefficient for ionizable substances shows no linearity over the pH range either. Just like the log *D* value, a sigmoidal curve for the pH-dependent log *SR* was observed. However, a very important difference was apparent. The tendency of an ionized drug to penetrate into the organic octanol phase is *ca.* 3 to 4 orders of magnitude lower than for the nonionized compound. This is called the 'rule of diff 3–4' in the literature [18]. The difference for the solubility ratio between mixed micelles and buffer is obviously smaller and is only 1–2 orders of magnitude. Since the values for log *D* and log *SR* are almost the same in the nonionized state, the difference is mainly caused by the ionized compound. It follows that the ionized molecule penetrates by a factor of *ca.* 2 log units better into the bile salt micelle than into octanol.

REFERENCES

[1] S. D. Mithani, V. Bakatselou, C. N. TenHoor, J. B. Dressman, *Pharm. Res.* **1996**, *13*, 163.
[2] W. F. Caspary, J. Stein, 'Darmkrankheiten. Klinik, Diagnostik und Therapie', Springer, Heidelberg, 1999.
[3] T. Scherstén, *Helv. Med. Acta* **1973**, *37(2)*, 161.
[4] A. Glomme, J. März, J. B. Dressman, *J. Pharm. Sci.* **2005**, 94, 1.
[5] E. Galia, E. Nicolaides, D. Hoerter, R. Löbenberg, C. Reppas, J. B. Dressman, *Pharm. Res.* **1998**, *15*, 698.
[6] V. Bakatselou, 'Dissolution of Steroidal Compounds at Physiological Bile Salt Concentrations', Ph.D. Thesis, University of Michigan, Ann Arbour, 1990.
[7] S. H. Yalkowsky, in 'Solubility and Solubilization in Aqueous Media', Ed. S. H. Yalkowsky, Am. Chem. Soc., Washington, DC, 1999, p. 49–80.
[8] G. L. Amidon, H. Lennernäs, V. P. Shah, J. R. Crison, *Pharm. Res.* **1995**, *12*, 413.
[9] J. B. Dressman, J. Butler, J. Hempenstall, C. Reppas, *Pharm. Tech.* **2001**, *25*, 68.
[10] V. H. Sunesen, R. Vedelsdal, H. G. Kristensen, L. Christrup, A. Müllertz, poster presentation, http://www.dfh.dk/forskningensdag/p15.doc, 2003.
[11] A. Scholz, B. Abrahamsson, S. M. Diebold, E. Kostewicz, B. I. Polentarutti, A. L. Ungell, J. B. Dressman, *Pharm. Res.* **2002**, *19*, 42.
[12] N. Aoyagi, H. Ogata, N. Kaniwa, A. Ejima, *J. Pharmacobiodyn.* **1982**, *5*, 120.
[13] J. A. Barone, J. G. Koh, R. H. Bierman, J. L. Colaizzi, K. A. Swanson, M. C. Gaffar, B. L. Moskovitz, W. Mechlinski, V. Van de Velde, *Antimicrob. Agents Chemother.* **1993**, *37*, 778.
[14] J. Cook, E. Randinitis, B. J. Wilder, *Neurology* **2001**, *57*, 698.
[15] D. Hoerter, 'Löslichkeit, Freisetzungsverhalten und galenische Präformulierung schwer wasserlöslicher Antimykotika unter Berücksichtigung der gastrointestinalen Physiologie', Ph.D. Thesis, University of Frankfurt, 1999.
[16] A. Avdeef, K. J. Box, J. E. Comer, C. Hibbert, K. Y. Tam, *Pharm. Res.* **1998**, *15*, 209.
[17] J. Miyazaki, K. Hideg, D. Marsh, *Biochim. Biophys. Acta* **1992**, *1103*, 62.
[18] A. Avdeef, *Curr. Top. Med. Chem.* **2001**, 1, 277.

Accelerated Stability Profiling in Drug Discovery

by **Edward H. Kerns*** and **Li Di**

Wyeth Research, Chemical Technologies, CN8000, Princeton, NJ 08543-8000, USA
(e-mail: kernse@wyeth.com)

Abbreviations
2D-NMR: Two-dimensional nuclear magnetic resonance spectroscopy; Caco-2: a human colon carcinoma cell line; CYP: cytochrome P450; DDI: drug–drug interaction; DTT: dithiothreitol; FDA: *United States Food and Drug Administration*; FIA: flow injection analysis; GI: gastrointestinal; HLM: human liver microsomes; HPLC: high-performance liquid chromatography; HTS: high-throughput screening; LC/MS: combined liquid chromatography/mass spectrometry; LC/MS/MS: combined liquid chromatography/tandem mass spectrometry; LT: low throughput; MS: mass spectrometry; MS/MS: tandem mass spectrometry; MT: moderate throughput; MUX: multiplexed MS interface; PAMPA: parallel artificial membrane permeability assay; PK: pharmacokinetics; QC: quality control; r.h.: relative humidity; S9: tissue preparation from homogenation and S9 sedimentation; SARs: structure–activity relationships; SPRs: structure–property relationships.

1. Introduction

Drug discovery is continually re-evaluated as a process. The goals, deliverables, quantitative performance, root causes of problems, solutions, and metrics are monitored to achieve improved product quality and productivity. Strategies are selected and implemented to improve the process. Drug discovery strives for novel, first-in-class/best-in-class therapies that improve patient life quality and longevity.

One of the problems of drug discovery was identified as the pharmaceutical quality of drug candidates. Not only are potency and selectivity important, but the properties of the molecule must be conducive to safe and effective delivery to the therapeutic target within the tissues and long-term storage. The terms 'drug-like molecule' and 'pharmaceutical properties' have taken greater importance in the process of drug discovery in recent years [1–3]. Thus, medicinal chemists work to enhance structure–property relationships (SPRs) [4] [5] *via* synthetic modification while they enhance traditional structure–activity relationships (SARs).

Pharmaceutical properties that can be improved include stability, solubility, permeability, and drug–drug interaction (DDI). Structural modifications are used to alter the underlying physicochemical characteristics. Physicochemically, a compound can be modified *via* changes in substructures, resulting in changes in lipophilicity/polarity, molecular weight, and pK_a. This is achieved through structural modifications such as cyclization, isostere substitution, and incorporation of substructures with different acidity or basicity.

Before the medicinal chemist can proceed, data is needed to quantitatively and qualitatively understand the performance of the lead molecule or series. We use the term pharmaceutical profiling for the process of obtaining these data [6] [7]. Assays for properties such as stability, solubility, permeability, and DDI have been widely implemented.

While activity and selectivity optimization may continue to be the highest priority for medicinal chemists, a comprehensive approach to drug discovery includes active selection and optimization of compounds for pharmaceutical properties. Reasons for attention to pharmaceutical properties during discovery have been discussed [8]. One major reason is that knowledge about the liabilities of a compound or series can assist with overcoming potential pitfalls in drug discovery. Property improvement can also increase delivery of a compound to the therapeutic target, which effectively increases its *in vivo* potency. Also, owing to the effect of properties on *in vitro* experiments, property information can provide better insight for planning and interpreting discovery experiments, thus increasing the quality and efficiency of discovery research. For these reasons, assays for pharmaceutical properties have been increasingly implemented to improve the discovery process. In this chapter, we will focus on stability profiling during discovery. An analytical approach is used, which examines the need for stability data, the limitations imposed by discovery resources, a tiered strategy that emphasizes appropriate methods at different discovery stages, analytical methodology overview, and case studies.

2. Selecting Stability Profiling Methods

2.1. *The Need for Stability Profiling in Drug Discovery*

Compounds encounter an array of barriers and environmental challenges in the lab during *in vitro* discovery testing, in storage, and *in vivo* [8]. Any of these can reduce the concentration of drug that reaches the therapeutic target tissue, generate toxic compounds, destabilize the compound prior to laboratory or clinical use, and compromise the results of discovery biological assays.

In the laboratory, compounds are exposed to reactions with light, oxygen, water, aqueous and mixed organic/aqueous solutions at various pHs. The solution phase can enhance reactivity, while long-term storage, elevated temperature, or humidity can result in compound degradation. Tests can be devised to assess whether compounds are susceptible to significant degradation under these physicochemical conditions.

Compounds are also exposed to degradation by *in vitro* bioassay test matrices that vary in pH and test matrix constituents. Solution equilibria can produce molecular forms (*e.g.*, adducts, aggregates, micelles, hydrates) that enhance or reduce bioactivity. Matrix constituents such as serum proteins can interact with test compounds. The resulting chemical modifications (*i.e.*, degradation) can confuse SAR conclusions of research teams and lead to dead ends and loss of valuable timeline for the drug discovery project.

Living systems have many stability barriers to xenobiotic compounds. The pH ranges *in vivo* in the gastrointestinal (GI) tract vary from pH 1 to 8. Hydrolytic enzymes are present in the GI, plasma, and tissues. Cytochrome P450 (CYP) enzymes oxidize most drug molecules. Other phase-I and -II metabolic reactions also occur. CYP Enzymes are encountered *in vivo* as early as the small intestine and continue to oxidize compounds in the liver and many other tissues. These enzymes reduce the concentration of dosed compound and produce potentially active or toxic metabolites. Understanding and reduction of compound metabolic degradation has been a long-standing goal of drug discovery.

The possibility of compound degradation in small or large percentage should be considered by all chemical, biological, and pharmaceutical scientists when conducting experiments in their area of study. Lack of knowledge of compound degradation and its effects on the experimental results can mislead research teams, confuse SAR, and lead to poorly informed decisions. The opportunity is that early knowledge of stability issues can aid data interpretation and trigger synthetic or formulation studies aimed at enhancing compound series stability.

2.2. *What Data Are Needed?*

Two types of stability data are of value to discovery research teams. First, quantitative data can be derived by measuring the concentration of compound in a test mixture before and after incubation. This data can help teams evaluate the overall stability of a compound or series, rank order analogs, and develop structure–stability relationships. Moreover, quantitative data can be used to calculate *kinetic parameters* for the degradation reaction (*e.g.*, half-life, clearance). Quantitative data are obtained from analyses in which the relative or absolute concentration of a compound is measured.

A second type of data is qualitative. Identification of the structures of degradants provides additional insight for discovery teams with: the *mechanism of degradation*, and the possibility of testing degradants for toxicity and activity. Mechanism information helps a team know how to modify a structure to improve stability by blocking labile sites. Isolation or synthesis of degradants can make material available for toxicity and activity testing. Toxicity data can help to avoid potential 'show-stoppers' in the further development of a compound. Activity data for a degradant can help to add value to the research project by discovering new SAR and leads or new substructures that enhance activity. Qualitative information is derived using analytical methods that provide structural data, such as mass spectrometry (MS) and nuclear magnetic resonance spectroscopy (NMR).

2.3. *The Requirements for Stability Profiling Methods in Drug Discovery*

Stability studies have traditionally been performed in the development phase. However, as discussed above, stability data are both valuable and necessary throughout the discovery phase. In the discovery phase, there are different constraints than are experienced during development. It is important to assure that methods used for discovery property profiling effectively address these requirements.

During much of discovery, only a *small quantity of compound* is available (*i.e.*, milligrams) for all biological and chemical studies. Methods must be designed to use microgram to milligram quantities of sample. The lack of material also suggests that methods for discovery profiling must be *sensitive*. Methods in this class often involve such analytical techniques as 96-well plates, plate readers, MS, and HPLC.

Methods used for stability profiling should be *specific*. Often, the test matrices for stability profiling contain compounds that can interfere with the detection and measurement of test compounds. For example, plasma, microsomes, and bioactivity test media all contain proteins, cofactors, lipids, *etc*. Thus, stability methods often include HPLC for separation and MS for selective detection and confirmation of the compound or degradant. By using MS, a specific signal is generated for the test compound, at its molecular ion (*e.g.*, $[M+H^+]$), even if it co-elutes with one or more other compounds of different molecular weight (M_r).

Discovery also requires rapid turn-around. Major decisions are often made in days to weeks. Stability data must be available at the same time as biological activity data, if it will have an impact on the discovery project.

Profiling methods must also be capable of handling a *large number of samples*, on the order of *moderate throughput* (MT), 100–1000 compounds per week per analytical chemist. This is because most pharmaceutical research organizations run multiple discovery projects simultaneously and each can generate many compounds for testing. The throughput does not need to be *high throughput* (10,000–100,000 compounds per week, HT), on the order of high-throughput screening (HTS). However, profiling must handle more than the traditional manual *low-throughput* analyses (1–10 compounds per week, LT) used in the development phase.

Profiling methods must also use organizational resources in an *efficient* manner. It is not efficient to use traditional LT methods, even if they produce higher levels of statistical confidence. An estimate of industrial research expense per compound is shown in *Table 1*. Clearly, with the number of samples from discovery for which profiling data are desired, as well as the need for multiple property assays, an organization would need 10–100 scientists for property profiling if LT methods are used. Inevitably, this requires compromises in the detail and statistical confidence of the data, but if acceptable levels of data quality can be achieved, there are great advantages in having data for more compounds. Another approach would be to only analyze selected compounds. However, discovery researchers constantly compare individual compounds to develop SAR and SPR, so data for selected series examples is not sufficient.

Table 1. *Approximate Expenses for Different Methods in Drug R&D*

Throughput	Samples/Week	Cost/Sample [a])
LT	10	$500
MT	1000	$5
HT	100000	$0.05

[a]) Based in industry benchmark of $250,000 per scientist.

2.4. *A Classification Strategy for Profiling Methods*

The limitation discussed above suggests a strategy for the efficient use of pharmaceutical profiling resources. This strategy is based on implementing methods at a particular discovery stage that are 'appropriate' to meet the needs of that stage. *Table 2* summarizes one such strategy.

In the early discovery lead selection stage, the questions are which compounds show activity and are nominally 'drug-like', have the potential to be improved to have drug-like properties, or have profound difficulties. At this stage, SPR can be used to rank drug-like compounds higher, because they have lower risk for the discovery process. The medicinal chemist has a multitude of issues to evaluate in selecting leads, and drug-like properties must be balanced against such factors as structural novelty, selectivity, and potency. Property information alerts research teams to potential liabilities of a compound series, from which decisions can be made to terminate seriously flawed series, or to plan structural modifications during the lead optimization phase to improve the properties. In the lead selection stage, appropriate methods have moderate throughput (MT) and rapidly provide the team with a profile of the major property issues for all compounds.

During the mid-discovery lead optimization stage, medicinal chemists make major modifications to the scaffold of the lead series. The primary focus is on improving potency and selectivity. Properties often take a lower, but important role. Increasingly, discovery researchers have recognized the effect of properties on *in vivo* pharmacokinetics (PK), toxicity, and pharmaceutical product quality. During the lead optimization stage it is important to continue to obtain the same MT property data on newly synthesized compounds (as above) to compare them to the baseline

Table 2. *Classification of Property Profiling Methods*

Discovery stage	Position on discovery timeline	Need	Appropriate method
Lead selection	Early	Broad overview of major properties of all compounds to assess property acceptability	Moderate-throughput profiling (MT)
Lead optimization	Middle	All compounds: SAR and SPR Selected compounds: focused biological and property studies	Moderate throughput profiling and diagnostic assays
Development selection	Late	Selected compounds: detailed review	In-depth profiling

values of their series. This provides insight on whether the series properties are improving or deteriorating.

In addition to MT profiles at the lead optimization stage, it becomes increasingly important to utilize *diagnostic assays*. These assays can range from MT to LT. They cover a broader array of property conditions than standardized MT assays. A toolbox of assays is predeveloped by the organization and held ready for use. Their purpose is to address specific questions for selected compounds or a series. This information is often of great value to research teams, because it helps them to resolve specific critical questions. The team can then make informed decisions on how to further modify structures, plan or interpret specific experiments, or rank a series at higher or lower priority.

At the stage of late discovery and development selection, detailed studies are needed to fully check all of the late-discovery and early-development property issues and fully understand the properties of the compounds under consideration for development advancement. At this stage, *in-depth profiling* methods are needed. These include a set of established tests that are conducted on a few selected compounds and are performed in low throughput (LT) to obtain data with high degrees of statistical confidence. In-depth methods require considerable resources and time compared to methods used for earlier stages.

In addition to considerations of appropriate methods, it is also important to consider appropriate experimental conditions. For example, in discovery, the experimental animal disease model usually uses mice or rats. Data generated using materials from humans (*e.g.*, human liver microsomes (HLM)) for metabolic stability is applicable for looking ahead to human clinical performance, but it may not provide an accurate understanding of the metabolic stability of the test compound in mice and rats. Having animal metabolism data gives researchers more information for interpreting animal model studies. If the compound is found to lack activity in the animal model, and it is known to have low stability, the research team can attempt to improve the stability and test the resulting analog for *in vivo* activity. Otherwise, the series might be reclassified to a lower priority without a holistic understanding of all the factors affecting *in vivo* activity. The study of metabolic degradation with human materials becomes very appropriate during development candidate selection. This same situation can occur for other profiling assays as well. Often, the end-user of the data is not in contact with the scientist that generates the property data, so it is important for the end-user to ensure that the conditions under which the data were generated are relevant to the question that is being asked.

2.5. *Stability Issues in Drug Discovery*

Before any methods are established for stability profiling, the critical stability issues and needs in drug discovery should be identified. It is useful to associate these with the three stages of discovery that were outlined above.

During lead selection, the key issue is whether compounds are metabolically stable. At this stage, metabolism is the most common risk for compound stability and success. Metabolic stability is a primary component of *in vivo* pharmacokinetics (PK) and bioavailability. Most organizations conduct a moderate-throughput assay for CYP oxidation for all compounds they evaluate as potential leads.

Diagnostic assays can be used during lead selection and throughout lead optimization. During lead selection, diagnostic assays are used to more-widely characterize selected compounds as part of the lead selection process. During lead optimization, they help to diagnose specific issues. For example, if the compound has lower than expected potency in an assay, this may be owing to instability in the pH, temperature, or components of the bioassay matrix. Or, if the PK is poor *in vivo*, this may be owing to instability in the pH range of the GI tract, enzymes in the GI, blood, or tissues, or phase-I or -II metabolism may be high. Diagnostic assays can be used to review a range of issues, or they can be used selectively to diagnose a specific potential cause. Conditions would include physicochemical ones: pH (1–8), light consisting of a range of wavelengths in the visible and UV spectral range that are common for indoor lighting, and elevated temperature that is often used to simulate a range of storage temperatures and to accelerate reactions compared to room temperature. Conditions can also include metabolism: phase-I liver microsome stability in different species, conjugations, and reactions occurring in plasma. Diagnostic studies can also include more-detailed time-course quantitation to obtain kinetic parameters. Diagnostic data can be used proactively to predict the behavior of a compound under a given set of conditions, or to retroactively diagnose the root cause(s) for an observed performance. For example, *Gan* and *Thakker* [9] described the use of several *in vitro* assays to retroactively diagnose the root causes of poor PK *in vivo* and then used selected assays, for the properties that were found to be poor, to monitor the improvement of properties in further compounds in the series.

Diagnostic studies also expand into the qualitative realm. The identification of degradant and metabolite structures allows the research team to understand mechanisms, block labile sites to increase stability, and identify potentially active or toxic analogs. Structural information requires more-sophisticated structure elucidation techniques, but these tools and experts are often available in pharmaceutical research organizations and the knowledge is very valuable.

During later portions of lead optimization and predevelopment, the focus is on assuring that the compound meets a wide array of criteria. Here, stability conditions include an array of simulated *in vivo* fluids, a wider array of physicochemical conditions (light at more wavelengths, heat, solution stability, buffers, pharmaceutical solvents, and excipients). Structure elucidation of degradants is expanded, because knowledge of these structures can accelerate early development studies to shorten the time to first-in-human phase-I studies.

3. Methods for Stability Profiling

3.1. *Common Analytical Tools for Stability Methods*

3.1.1. *General Considerations*

As with other profiling methods, the assay may, at first, seem uncomplicated and easy to implement. However, experience has shown that results from each assay can vary widely depending on the conditions and their day-to-day control. For example, metabolic stability results are greatly affected by DMSO as cosolvent, sample concentration, and microsomal preparation [10]. The data, even from seemingly simple *in vitro* methods can be misleading if the conditions of the assay are not well developed, validated, and performed properly.

3.1.2. *Automation*

Stability assays can be automated using standard laboratory robots. Benchtop robotic workstations are commercially available. These allow a flexible arrangement of assay tools, such as well plates, reagent reservoirs, and pipette tips. They are readily operated by an individual scientist and offer helpful graphical interfaces for robotic program development that allow a scientist to transfer a benchtop manual method into an automated method. Operations, such as liquid handling, dilutions, reagent additions, and timed events are readily carried out. Plates can be moved around the deck of the robot using an optional gripper arm.

The advantages of robotics are many. Automation reduces the needed human staff time, which is a major portion of any operating budget. The boredom of scientists also decreases as scientist's routine tasks are reduced, freeing them to perform more-sophisticated, high value activities. Robotics are also engineered to be very reproducible. Higher levels of data accuracy and precision are obtained, compared to manual human operations.

There is a current movement toward transferring benchtop robotic methods to large scale automation [11][12]. Such automation has previously been developed and used for HTS. Several methods are often integrated around a central articulated robotic arm that moves along the length of a bench. Scheduling software can time various events and one arm can support many assays. In this manner, time and attention from human staff is further reduced. There is, however, a considerable initial investment in equipment and system integration development.

3.1.3. *Well Plate Technology*

The 96- or 384-well plates have become the reaction vessel of choice for *in vitro* assays. Efficiency is obtained from batch parallel processing. Lab robots have built-in processes to manipulate 8–96 wells simultaneously. Scientists implementing stability assays should check to be sure that the assay reagents do not react with the well plate material, because compounds can be extracted and interfere with the assay.

3.1.4. *High Performance Liquid Chromatography/Mass Spectrometry* (LC/MS)

Stability assays place high demands on the analytical instrumentation used to quantitate the compound. They must have high sensitivity (often incubations are conducted at 1–5 μM, requiring the detection of nanogram quantities), selectivity (complex matrices that interfere with quantitation can be used), and high throughput (several 96-well plates need to be assayed each day). Generally, only MS-based methods can meet this challenge. The problem is that MS-based methods are typically slower than plate-reader-based methods. For example, at 5 min per analysis, an LC/MS method could require 480 min to analyze a 96-well plate, whereas a plate reader could analyze a plate in 5–20 min. To increase the throughput of MS-based methods, *Janizewski et al.* [13] developed a customized method using flow injection analysis tandem mass spectrometry (FIA/MS/MS). The throughput is *ca.* 40 min for a 96-well plate (> 120 samples/h). Another version of this design, using all commercially available subsystems and software, has also been described [14]. A diagram of

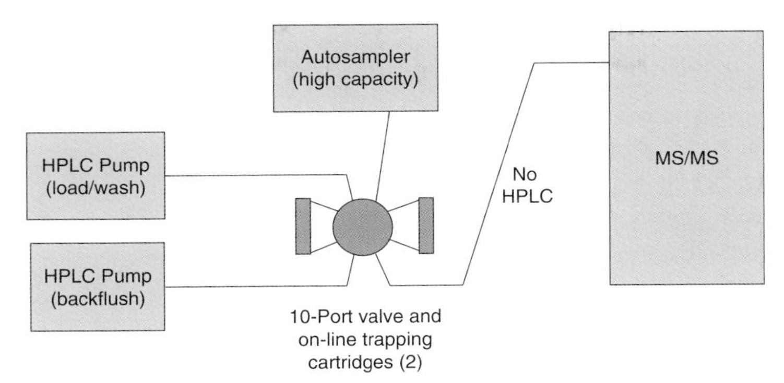

Fig. 1. *Diagram of a FIA/MS/MS system for HT quantitation*

the design is shown in *Fig. 1*. Incubated and prepared samples are placed in a 96-well plate in a high capacity autosampler and serially injected onto a trapping cartridge. The trap allows the aqueous wash to remove residual salts and proteins in the prepared sample that would interfere with quantitating the test compound or contaminate the mass spectrometer ion source. The test compound is then back flushed from the trap directly into the mass spectrometer ion source without analytical HPLC. Sensitive detection is achieved using the MS/MS system with parent product ion quantitation conditions that are selective for the test compound. MS/MS conditions are optimized for each test compound using an automated, unattended method development process. FIA/MS/MS systems are simple and inexpensive. A large number of samples can be processed under the sensitivity and selectivity requirements of the assay. FIA/MS/MS is growing in popularity for MT quantitation.

Multiplexed (MUX) LC/MS systems have also been developed for MT quantitation [15–17]. Four or eight HPLC columns are attached to one MS instrument *via* a special interface, as shown in *Fig. 2*. This interface switches HPLC effluent input to the MS in sequence among the parallel HPLC columns. Signals for each sample are obtained in a fraction of the time of serial LC/MS analyses of the same number of samples. This accelerates the analysis and reduces the number of MS systems that are needed. The maximum throughput is similar to the FIA/MS/MS system, because a 2–5-min HPLC separation is involved. MUX Methods are used in many labs. The disadvantages are that MUX instrumentation is initially more expensive, complicated to set up (plumbing), and requires more maintenance and analyst operation time.

HPLC with UV detection is often found to be sufficient for lower throughput methods in which individual compounds are subjected to an ensemble stability tests for in-depth analysis. Sensitivity and compound

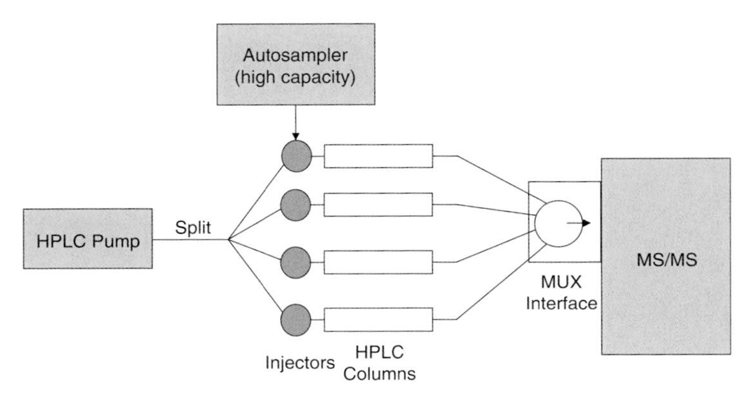

Fig. 2. *Diagram of a MUX/MS/MS system for HT quantitation*

amount are not often limiting and an HPLC method can be developed and applied to hundreds of analyses of a single compound.

3.2. *Overview of Methods*

3.2.1. *MT Metabolic Stability*

In vitro metabolic stability is one of the most-common stability assays. High levels of metabolic degradation are a severe limitation to the chemical series. Compounds undergo metabolism by CYP oxidation, resulting, for example, in hydroxylation and dealkylation [18]. The concentration of a dosed drug is reduced *in vivo* by metabolism, thus, affecting PK, bioavailability, and efficacy.

Various groups use liver microsomes [10] [19] [20] or liver S9 fraction [21] in MT 96-well format assays. MT Profiles often utilize materials from a single generic rodent species (*e.g.*, rat). For example, a microsomal stability assay typically involves incubation of the test compound with rat liver microsomes at a concentration of 1–20 μM, with the NADPH or NADPH enzymatic regenerating system. After a 5–60-min incubation, a quenching solution is added (*e.g.*, MeCN, acid) and the sample is prepared (*e.g.*, centrifugation, filtration) for LC/MS analysis. It is apparent that most organizations conduct the assay under slightly different conditions and no standard method exists.

Recent investigations have shown that the assay conditions can greatly affect results. For example, in the past, it has been common to incubate a test compound at concentration of 10–20 μM. However, it has been argued that this concentration is not physiologically relevant [10] and that the enzymes can be saturated and give false high estimates of stability. Furthermore, other assay conditions can greatly affect results. These include DMSO concentration, microsome source, and incubation time. It appears that a concentration of 1 μM is a physiologically more-relevant concentration and is less prone to saturate the enzyme capacity. DMSO Concentration should be kept below 0.2%. The activity of microsomes was found to vary greatly among vendors and same vendor lots. Thus, microsome activity should be monitored with QC standards. An automated method, with high precision and 250 compounds per day throughput is described [10].

3.2.2. *Diagnostic Metabolic Stability*

The MT assays only study initial range-finding issues. Many other metabolic stability issues can affect discovery compounds. Thus, diagnostic assays are often made available to answer such questions. Compounds are often modified by conjugation reactions (*e.g.*, glucuronidation, sulfation). Conjugations often occur for compounds that contain a hydroxy, carboxylic acid, or amine moiety. Many phase-I metabolites are further modified by conjugation. Glucuronidation can be addressed by adding UDP-glucuronic acid (UDPGA) to the microsomal reaction, because the enzyme (UDP-glucuronyl transferase) is present in microsomes. Another approach is to use cryopreserved hepatocytes [22] or liver slices that contain a broader ensemble of naturally occurring enzyme systems. Diagnostic assays can use alternate model species or human materials.

CYP Isozyme 'phenotyping' has also been used diagnostically. The test compound is incubated with pure CYP isozymes to see which isozyme(s) are responsible for the bulk of the compound's metabolism. Alternatively, compounds are incubated with microsomes and inhibitors for specific CYP isozymes are co-incubated with the compound to see which inhibitors reduce the compound degradation. These assays provide medicinal chemists with information from which SPR for metabolism can be developed, and this information can be combined with knowledge of the active site of the CYP isozyme to redesign the molecule to attempt stability improvement.

3.2.3. *In-Depth Metabolic Stability*

In-depth profiles of metabolic stability are performed in mid-to-late discovery phase. Multiple species are often assayed (*e.g.*, mouse, rat, dog, monkey, human). In addition, multiple time points are assayed to increase statistical confidence from multiple assays and to select time points that allow accurate calculation of half-life. Allometric scaling [23][24] is performed to predict a human half-life. In-depth metabolic stability assays are conducted in a similar manner as MT assays, but lower-throughput techniques for enhanced accuracy and precision are used. The ratio of compound evaluations for MT *vs.* in-depth assays is on the order of 10–1000-fold. Thus, each assay type has its appropriate application.

3.2.4. *Diagnostic Stability in Acidic and Neutral Conditions*

An assay for stability and solubility at acidic pH can yield a 'liberation ranking' in the stomach [25]. The *in vitro* stability of compounds at pH 2 for 75 min (the mean human stomach residence time) is measured. If pH 2 stability is >50% and solubility is >0.1 mg/ml, then the compound is considered to have a high liberation rate. This information is combined with absorption and metabolism data for a prediction of oral bioavailability classification (*i.e.*, high, intermediate, low).

Compounds may be incubated with HCl (pH 1) for 2 h in 96-well plate format (0.1 ml) to assess acidic stability [26]. The incubation is stopped by the addition of pH 7.4 phosphate buffer (0.2 ml) to adjust the pH. The samples are then analyzed using LC/MS. One problem with such an approach is that, while neutralization reduces the acidic stress on the compound, the sample is still sitting in buffer until it is introduced into the instrument. The compound may be susceptible to degradation at neutral pH. Furthermore, if the first sample is injected into the instrument at time zero, it will take 1–4 h until the final sample on the 96-well plate is injected, implying that samples are exposed to different conditions.

Our laboratory [27] has developed a methodology that eliminates these problems. The 96-well plate is placed in a HPLC autosampler. The autosampler and control software of the instrument is used to add the stability modifier (*e.g.*, acid, buffer) to different samples at a given time prior to injection. Thus, every sample is incubated for the same time, there is no need to perform the extra step of neutralizing the sample, and stability at the neutralized buffer pH is not an issue.

The above methods can be used at different pH values. This expands pH-stability tests beyond stomach pH to gastric pH (pH 3–8) and bioassay pH values, which can vary widely. It is possible that pH instability in the bioassay is the cause of poor activity. Knowledge of this instability can help a research team to better plan, interpret, and diagnose activity assay results. Medicinal chemists and experienced property profiling scientists can often recognize, in advance, substructures that are susceptible to pH-related instability and can select compounds for diagnostic evaluation.

3.2.5. *Diagnostic and In-Depth Stability*

It has long been recognized in development that physicochemical factors cause degradation of the active ingredient in the drug product. For this reason, a series of accelerated stability studies are performed during development to study the stability of development candidates under a wide array of conditions to uncover sources of degradation, quantitate kinetics, and identify degradants. This allows the opportunity to stabilize the candidates through physical storage conditions and excipients. The FDA has provided guidance to industry on the tests that should be performed (*Table 3*). The emphasis is on active ingredient and drug product storage stability and the environmental factors that enhance degradation: elevated temperature, humidity (water), photolysis, oxidation, and hydrolysis at various pH values. The structure identification of degradants is also required by the FDA (*Table 3*).

Table 3. *FDA Guidance for Industry on Stability Testing of New Drug Substances and Products* [28][29]

Effect	Stability test conditions[a])
Storage conditions (combined temperature and humidity)	12 months at 25°/60% r.h. 6 months at 30°/60% r.h. 6 months at 40°/75% r.h.
Temperature	10° increments
Humidity	75% r.h. or greater
Oxidation	High oxygen atmosphere
Photolysis	D65/ID65 standard (or cool white fluorescent lamp), 1.2 million lux h and UV (320–400 nm); overall illumination $\geq$200 W h/m^{-2}
Hydrolysis across a wide pH range	Conditions not specified
Examine degradation products	Identity and chemical structure for mechanism of formation

[a]) '*...should include testing those attributes of the drug substance that are susceptible to change during storage and are likely to influence quality, safety, and/or efficacy.*'

Upon the major causes of degradation that have been identified by the FDA, discovery researchers can decide which stability assays should be performed during discovery. The purpose of profiling stability during discovery is to detect unstable compounds early. Unstable compounds cause development failure, lead to more-time-consuming studies, more-expensive formulations, or more-restrictive storage conditions. These problems can be fixed during discovery by synthetic modification, if possible, to improve the drug product and save development time and the expense of sophisticated formulations. However, physicochemical stability is not often a high priority for discovery research teams. Exceptions to this include situations where medicinal chemists suspect compound instability based on their organic chemistry experience, when bioassays do not turn out the way that was expected and an explanation is being diagnosed, or when a change in compound solution is observed (*e.g.*, change in color of a compound solution). Thus, these assays are more of an interest for diagnosis (mid- to late-discovery) and in-depth predevelopment profiling (late discovery).

3.2.5.1. *Light Stability*

Certain classes of compounds are susceptible to degradation by light. The greatest concern here is indoor lighting in laboratories, pharmacies, clinics, and storage facilities. Thus, the test of light stability is performed under lights simulating the visible and UV light emitted by fluorescent and other indoor light sources. One light source that provides high-intensity light for accelerated stability assays during discovery is manufactured by *New England Ultraviolet* under the '*Rayonette*' brand name. The unit provides high intensity for indoor-like light and accommodates 96-well plates. Other manufacturers provide systems that meet the FDA guidelines for development studies and can also be used in discovery, although the smallest units often have much more capacity than are needed for discovery studies.

3.2.5.2. *Oxidation*

A common approach for susceptibility to oxidation is incubation in solution with 3% H_2O_2 [30]. This exposes moieties in the molecule that are susceptible to oxidation.

A recent methodology was introduced by *Lombardo* and *Campos* [31]. Test compounds are passed through a series of electrochemical cells (*ESA Inc.*) at increasing voltage. The cells are in-line with an HPLC, so that reduction of the concentration of compound by oxidation can be quantitated. Extensive method development and validation with a known set of compounds was performed to provide a rugged and reliable assay.

3.2.5.3. *Temperature*

For discovery purposes, test compounds can be incubated in an oven at elevated temperature to test the effect of temperature in accelerating degradation reactions. Such reactions are typically due to unimolecular decomposition, hydrolysis with water from the air, or oxidation with oxygen from the air. These conditions are not as rigorous as the FDA guidelines (*Table 3* and [28] [29]), and the humidity is low, but they can still reveal instability trends in compounds that are better to know early. Commercial incubators for development stress testing can be used, but they tend to be much larger size than is common in discovery labs.

3.2.5.4. *Diagnostic Bioassay Matrix Stability*

Another source of degradation is the bioassay matrix. These solutions are often complex, and degradation can occur from pH, hydrolytic enzymes, and reactive components (*e.g.*, DTT). Also, conditions can favor the formation of equilibrium forms of the molecule, which can have decreased or enhanced activity. The compound can be incubated with the assay matrix, or a subset of the solution components, to evaluate compound degradation.

3.2.5.5. *Diagnostic and In-Depth Stability in Plasma*

Hydrolytic enzymes can be responsible for compound degradation *in vivo*. Sometimes this is even an advantage for some purposes, such as prodrug release. In other cases, degradation in the plasma can cause a reduced *in vivo* half-life. Knowledge of stability in plasma is also important for pharmacokinetics studies, because plasma samples are collected following *in vivo* dosing and the test compound is in plasma for a period of time. Incubation of test compounds with animal plasma can reveal the potential for break-down in the blood stream. This assay requires a specific detection method, such as LC/MS, to selectively detect the test compound in the presence of the many other compounds found in plasma. A plasma stability method has been described by *Wang et al.* [32], who used a HPLC autosampler to incubate the compounds in plasma. A mixed function HPLC column (*Capcell MF C8*) coupled to MS/MS was used to rapidly analyze the incubated samples without sample preparation, because the sample protein material is unretained and is diverted to waste ahead of the MS/MS. *Di et al.* [33] have explored the various method conditions of the plasma incubation and analysis methodology.

3.2.5.6. *Diagnostic and In-Depth Degradant Structure Identification*

The structure elucidation of degradation products is more time consuming than quantitation and requires spectroscopic interpretation expertise. However, it can provide very useful information that leads to insights on degradation mechanism, active leads, and compounds requiring toxicity evaluation. In recent years, rapid identification of degradants has become more common through the application of LC/MS/MS techniques [34]. The application of a rapid method for the identification of trace level components using LC/MS/MS to degradation products has been discussed [30][35–39]. This approach uses the following procedure:

1) A standard of the test compound is run by full scan LC/MS to observe the molecule ion.
2) The standard is rerun and the MS/MS product ion spectrum of the molecule ion is obtained at the retention time of the test compound.
3) The incubated test compound is run by full scan LC/MS to observe the molecule ions of any HPLC peaks that are new compared to the nonincubated standard.
4) The incubated test compound is rerun and the MS/MS product ion spectra of the molecule ions of the new HPLC peaks are obtained at their respective retention times.
5) The MS/MS product ion spectrum of the test compound is interpreted, thus assigning substructures to specific MS/MS product ions to serve as a template.
6) This template is applied to the spectra of the new peaks in the incubated compound to assign the substructural modifications that have occurred in the degradation products.

The protocol is rapid and effective for proposing degradation product structures. Often, structure proposal is aided by discussions with the synthetic chemist for the series, by recognizing that polarity differences in the degradants from the parent compounds will be reflected in their retention times (increasing polarity shifts the retention time earlier in reversed phase HPLC), and the consistency of UV spectrum changes (from an in-line diode array UV detector) with structural changes (*e.g.*, extended conjugation shifts absorbance maxima to longer wavelengths).

If resources are available, a quantity of each important degradation product can be obtained by fraction collection and the sample can be analyzed using NMR [40]. MS is effective for rapid structural proposal for trace quantities of compound, but often two or more regioisomers are possible from the MS/MS data. ^{1}H-NMR using 10–100 μg of compound and extended data collection can often resolve the possible regioisomers to a specific isomer. Again, the NMR spectrum of the test compound serves as the template for assigning resonances. This template is used to examine the isolated degradant NMR spectrum to look for proton resonances that disappeared, thus indicating the position of substitution. 2D-NMR is sometimes needed to resolve the possible isomers.

Thus, a graduated scheme can be used for structure elucidation of degradation products:

a) Screen major degradants using LC/MS/MS and the test compound as the interpretation template for unknown degradants.
b) If greater structural detail is needed, isolate 10–100 μg of degradant using preparative HPLC and perform ^{1}H-NMR analysis using the test compound as the template to look for eliminated resonances (LC/NMR is an alternative if available).
c) Regioisomers are resolved using 2D-NMR with long data collection.

Data on degradation product structures that is obtained during discovery is very useful if the compound advances to development. Early development formulation, stability, and process scale-up all produce degradation products, whose identification during discovery can speed up early development.

3.2.5.7. *In-Depth Stability Profiling*

During the development selection stage, it is very valuable to subject selected compounds to in-depth physicochemical analysis. Useful stability assay conditions include the following:

- Aqueous buffers (37°, pH 1–12)
- Simulated intestinal fluid (37°, 1–24 h)
- Simulated gastric fluid (37°, 1–24 h)
- Simulated bile/lecithin mixture (37°, 1–24 h)
- Plasma (37°, 1–24 h)
- High-intensity light (room temperature, 1–7 days)
- Heat (30–75°, 1–7 days)

Tests are conducted over a prolonged time for extended exposure. An example of a time course study over 24 h that can be used for kinetic analysis is shown in *Fig. 3*. Samples are collected at multiple time points so that accurate kinetics can be calculated. The procedures are not as

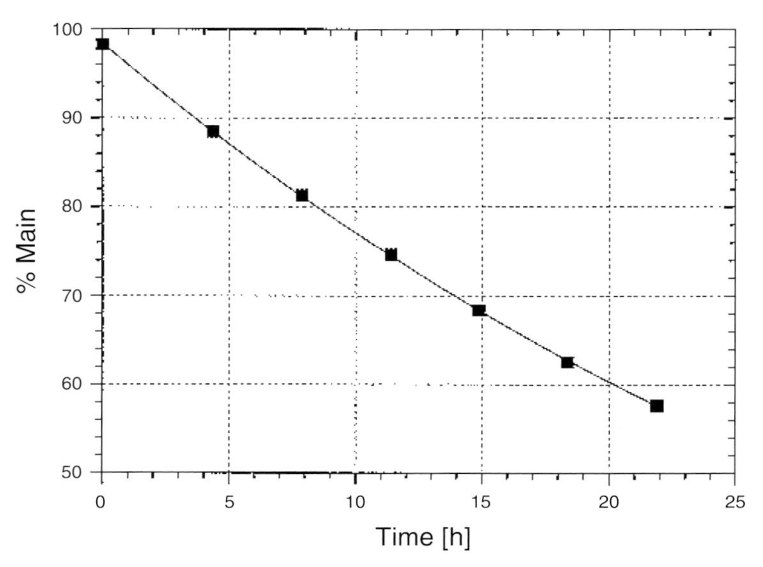

Fig. 3. *Kinetics plot of a development candidate in aqueous buffer at pH 9*

rigorous as those required for regulatory documentation. However, they reveal weaknesses before compounds are committed to expensive and time-consuming development activities. The information is also useful for providing development departments with an accelerated start to their work.

FDA Documents that define tests for regulatory documentation can be used as guides for these tests. A summary of FDA guidance on stability testing is provided in *Table 3*. FDA defines 'accelerated testing' as studies designed to increase the rate of chemical and physical degradation of a drug substance or drug product by using exaggerated storage conditions. The purpose is to determine kinetic parameters, to predict the tentative expiration dating period. The term 'accelerated testing' is often used synonymously with 'stress testing'. FDA also indicates that tests be conducted 'preferably in open containers, where applicable'.

4. Case Studies from the Literature

Following are examples of stability profiling methods and related technologies. These examples are intended to provide the reader with insights for developing, implementing, or improving methodology.

4.1. *Integrated Process for Physicochemical Properties Measurement*

Kibbey et al. [41] discussed an integrated process for measuring physicochemical properties and stability. They measure log *P*, pK_a, solubility, and chemical stability. Capillary electrophoresis was used to determine pK_a and log *P*. Solubility is determined using UV detection. Chemical stability is measured at *ca.* 100 μM by overnight incubation at pH 2, 7, and 12, and 3% H_2O_2. The solutions contain 50% MeCN to enhance dissolution. A *Gilson* robot is used for reagent additions and sample handling. Quantitation uses HPLC with a 5-min mobile-phase-gradient cycle and fresh compound for the control. Results are reported on a scale of 1–5. A maximum of 350 compounds per week can be profiled in this way for chemical stability.

4.2. *Rapid Structural Identification of Degradants*

The work of *Volk et al.* [30] on accelerated stability studies of paclitaxel provides a useful overview of stress conditions and LC/MS/MS methodology for degradation product production and identification for in-depth profiling. A sample of paclitaxel was treated with aqueous Na_2CO_3 at pH 8 for 10 min (basic stress), 0.7M HCl for 240 min (acidic stress), 23% H_2O_2 (oxidative stress), and bulk was exposed to 1000-foot-candle intensity light for 92 days (light). The stress conditions were selected based on experience with paclitaxel. Electrospray LC/MS/MS was sufficient to observe and identify five major degradants. These included the following compounds under the conditions noted in parentheses: side-chain methyl ester (base), baccatin III core (base), C(3)–C(11)-bridged paclitaxel isomer (light), oxetane-ring-opened paclitaxel, and 7-epipaclitaxel. LC/MS/MS was not sufficient to identify three minor degradants that had lost CO from paclitaxel. The analytical data for the stress-induced degradants (HPLC relative retention time, full-scan mass spectrum, product ion MS/MS spectrum, and structure) were tabulated in a 'library' for use in future studies. The data supports accelerated development work (*e.g.*, formulation, stability, analytical methods that are 'stability indicating'). The methodology provides structural identification for degradants 'in less than one day' of analysis.

4.3. *Rapid Automated System for Drug Stability Kinetics*

Shah et al. [42] described an integrated system for quantitative drug stability. The test compound is incubated in a thermostated reaction vessel and samples are obtained at regular intervals using a microdialysis probe inserted in the solution. Samples are injected automatically onto a HPLC with a fast isocratic analysis and detection using an UV detector. Experiments can be carried out at controlled temperature and pH to observe their effects on the degradation reactions. Repetitive samplings and analyses every 30 s or longer can be obtained to produce high-quality kinetics analysis with a half-life as low as 1 min. An example was given for hydrolysis.

4.4. *Electrochemical Stability*

Electrochemical cells can be used to mimic oxidative reactions. When they are coupled directly to a mass spectrometer, the structures of the degradation products can be rapidly determined. *Volk et al.* [43] demonstrated the detection of oxidation products in known biochemical pathways. *Jurva et al.* [44][45] adapted this methodology to produce phase-I metabolic oxidation products. Known one-electron metabolic oxidations (*e.g.*, tertiary amines) were observed, but oxidations known to proceed *via* direct H-atom abstraction (*e.g.*, *O*-dealkylation) were not observed. NMR would be needed to differentiate regioisomers. Regiospecific CYP oxidations are not differentiated by the method. The opportunity for this method is the rapid identification of oxidative reaction products, assessment of the sensitivity of a compound toward oxidation, and the likely position(s) of oxidation. The method is rapid, easy to use, and can be applied to a wide range of compounds to complement other metabolic stability methods.

4.5. *Single-Time-Point Quantitation for High-Throughput Methods*

In-depth studies typically include multiple-time-point samplings over an extended time course to obtain detailed data of enhanced statistical confidence. MT Methods, however, require rapid throughput. Sampling at a single time point is usually sufficient. The problem then becomes how to select the time point. *Di et al.* [46] examined the effect of three factors on single time point results: *a*) half-life, *b*) assay variability, and *c*) the nonlinear relation of half-life to percentage of compound remaining in the incubation for first-order kinetics. For example, short half-life (1 – 10 min) compounds have very little compound remaining in samples obtained after 30 – 120 min. Thus, calculations of half-life would be poor. Furthermore, since resolution of compounds is limited by the assay variability, there would be poor resolution of unstable compounds when sampled at a much later time point. Resolution is necessary for differentiating structure–metabolic-stability relations. Conversely, more-stable compounds (half-life > 60 min) would not be resolved by using samples obtained at 5 – 10 min (although a half-life is calculable and often is reported). The variability of the method establishes the ability to resolve calculated half-lives with statistical confidence. It was concluded that a single-point sampling time of 15 min allows differentiation of less-stable compounds while allowing statistically significant calculation of half-life out to *ca.* 30 min. Most MT methods are used early in discovery and the greatest needs in that stage are to screen for stability and differentiate less-stable compounds to determine whether structural modifications were successful in prolonging stability.

5. Concluding Remarks

Drug discovery is dedicated to achieving novel and effective therapies to improve a patient's quality and length of life. To meet this high goal, discovery continuously challenges itself to improve the effectiveness and productivity of its processes. Technologies and implementation strategies play a major role in successful processes. Over the past two decades, physicochemical and biological properties of compounds have emerged as a critical area of study in discovery, and novel technologies and strategies have developed to assist its effectiveness and productivity. Stability profiling, as a subset of this effort, provides information needed by discovery research teams to optimize the longevity and safety of leads as they interact with various physicochemical and biological barriers that chemically modify compounds. One strategy for stability profiling that addresses the needs and resource allocation issues in discovery with appropriate methods is to use a three tiered approach: MT assays for a broad profile of most compounds, diagnostic assays to answer key questions for a series or project, and in-depth assays to thoroughly examine the predevelopment issues (*Table 4*). In the complex process of discovery, profiling competes for resource allocation with all other technologies and strategies that benefit discovery. Thus, the future goals for profiling are the same as any discovery function: quality (better data), quantity (more data for less expense), quickness (faster access to data), and effective decision

Table 4. *Examples of Stability Assays*

Method type	Stability assays	Simulated exposure	Other assays used in a similar manner
Moderate throughput	• Rat liver microsomes	• Liver	• Integrity and purity
			• Solubility (pH 7.4)
			• Permeability (PAMPA)
			• CYP Inhibition
Diagnostic assays [26]	• Microsomes (other species)	• Liver	• Solubility (multiple pH)
	• Plasma	• Blood	• Permeability (PAMPA, multiple pH)
	• Aqueous acid (pH 1)	• Stomach	• Permeability (caco-2)
	• Aqueous acid (pH 3–7)	• Small intestine	• P-Glycoprotein efflux
	• Aqueous base (pH 8)	• Large intestine	• CYP Inhibition, LC/MS
	• Aqueous Neutral (pH 7.4)	• Bioactivity assay	
	• Phase II (glucuronidation, sulfation)	• Liver	
	• Structural screen of major degradants	• Stomach, intestine	
	• Simulated *in vivo* fluids	• Lab, pharmacy, clinical storage	
	• Light, heat	• Long-term storage and nonregulated temperature	
	• Oxidation (peroxide, electrochemical)	• Air/oxygen	
In-depth profile	• Aqueous buffers	• *In vivo* fluids	• Purity
	• Simulated *in vivo* fluids		• Crystallinity
	• Plasma		• pK_a
	• High-Intensity light	• Long-term storage	• Equilibrium solubility
	• Heat		• Plasma protein binding
	• Microsomes (other species)	• Liver	• Log *P*
	• Multiple time points		• CYP Induction

making. With regard to quality, it need not be assumed that moderate-to-high-throughput assays produce poor quality data. We should continue to strive for precise and accurate data and to implement methods that correlate well (by themselves or in combination with other data) to critical drug success factors (*e.g.*, half-life, storage stability). Quantity and quickness continue to improve with innovation in parallel and integrated analytical technologies (*e.g.*, LC/MS/MS) and should be supported. Effective decision making benefits from *1*) data that provides insights for important questions (*e.g.*, kinetics, mechanisms), *2*) effective communication of this information, *3*) application of the information by integration of profiling scientists into discovery teams, and *4*) knowledge by medicinal chemists and biologists about compound properties and their effects. A discovery process that effectively integrates properties ('property-based design', [4]) with activity ('structure-based design') to select and optimize compounds is an ongoing goal.

REFERENCES

[1] R. A. Prentis, Y. Lis, S. R. Walker, *Br. J. Clin. Pharmacol.* **1998**, *25*, 387.
[2] C. A. Lipinski, F. Lombardo, B. W. Dominy, P. J. Feeney, *Adv. Drug Delivery Rev.* **1997**, *23*, 3.
[3] C. A. Lipinski, *Curr. Drug Disc.* **2001**, April issue, 17.
[4] H. van de Waterbeemd, D. A. Smith, K. Beaumont, D. K. Walker, *J. Med. Chem.* **2001**, *44*, 1313.
[5] D. A. Smith, H. van de Waterbeemd, *Curr. Opin. Chem. Biol.* **1999**, *3*, 373.
[6] E. H. Kerns, *J. Pharm. Sci.* **2001**, *90*, 1838.
[7] E. H. Kerns, L. Di, *Curr. Topics Med. Chem.* **2002**, *2*, 87.
[8] E. H. Kerns, L. Di, *Drug Disc. Today* **2003**, *8*, 316.
[9] L. S. L. Gan, D. R. Thakker, *Adv. Drug Delivery Rev.* **1997**, *23*, 77.
[10] L. Di, E. H. Kerns, Y. Hong, T. Kleintop, O. J. McConnell, *J. Biomol. Screening* **2003**, *8*, 453.
[11] K. Dickinson, J. Herbst, J. Kolb, T. Zvyaga, *American Association of Pharmaceutical Scientists Conference*, Toronto, ON, November 13, 2002.
[12] C. Green, K. Saunders, D. Gibson, T. Letby, K. Hurst, H. van de Waterbeemd, B. Jones, *International Symposium on Laboratory Automation*, Boston, MA, October 20, 2003.
[13] J. S. Janiszewski, K. J. Rogers, K. M. Whalen, M. J. Cole, T. E. Liston, E. Duchoslav, H. G. Fouda, *Anal. Chem.* **2001**, *73*, 1495.
[14] E. H. Kerns, T. Kleintop, D. Little, T. Tobien, L. Mallis, L. Di, M. Hu, Y. Hong, O. J. McConnell, *J. Pharm. Biomed. Anal.* **2004**, *34*, 1.
[15] Y. Deng, J. T. Wu, T. L. Lloyd, C. L. Chi, T. V. Olah, S. E. Unger, *Rapid Commun. Mass Spectrom.* **2002**, *16*, 1116.
[16] D. Morrison, A. E. Davies, A. P. Watt, *Anal. Chem.* **2002**, *74*, 1896.
[17] L. Fang, J. Cournoyer, M. Demee, J. Zhao, D. Tokushige, B. Yan, *Rapid Commun. Mass Spectrom.* **2002**, *16*, 1440.
[18] Z. Yan, G. W. Caldwell, *Curr. Topics Med. Chem.* **2001**, *1*, 403.
[19] W. A. Korfmacher, C. A. Palmer, C. Nardo, K. Dunn-Meynell, D. Grotz, K. Cox, C. C. Lin, C. Elicone, C. Liu, E. Duchoslav, *Rapid Commun. Mass Spectrom.* **1999**, *13*, 901.
[20] R. Xu, C. Nemes, K. Jenkins, R. A. Rourick, D. B. Kassel, C. Z. C. Liu, *J. Soc. Mass Spectrom.* **2002**, *13*, 155.
[21] A. K. Mandagere, T. N. Thompson, K. K. Hwang, *J. Med. Chem.* **2002**, *45*, 304.
[22] A. P. Li, in 'Drugs and the Pharmaceutical Sciences, 114 (Handbook of Drug Screening)', Eds. R. Seethala, P. Fernandes, Dekker, New York, 2001, p. 383–402.
[23] R. S. Obach, J. G. Baxter, T. E. Liston, B. M. Silber, B. C. Jones, F. Macintyre, D. J. Rance, P. Wastall, *J. Pharmacol. Exp. Ther.* **1997**, *283*, 46.
[24] R. S. Obach, *Curr. Opin. Drug Disc. Dev.* **2001**, *4*, 36.
[25] G. W. Caldwell, *Curr. Opin. Drug Disc. Dev.* **2000**, *3*, 30.
[26] E. H. Kerns, *Symposium on Chemical and Pharmaceutical Structure Analysis*, Princeton, NJ, September 26–28, 2000.
[27] L. Di, unpublished data.
[28] U.S. Department of Health and Human Services, Food and Drug Administration, Center for Drug Evaluation and Research (CDER), Center for Biologics Evaluation and Research (CBER), ICH, August 2001, Revision 1, www.fda.gov/cder/guidance/4282fnl.htm, 'Guidance for Industry, Q1A Stability Testing of New Drug Substances and Products'.
[29] U.S. Department of Health and Human Services, Food and Drug Administration, Center for Drug Evaluation and Research (CDER), Center for Biologics Evaluation

and Research (CBER), ICH, November 1996, www.fda.gov/cder/guidance/1318.htm, 'Guidance for Industry, Q1B Photostability Testing of New Drug Substances and Products'.
[30] K. J. Volk, S. E. Hill, E. H. Kerns, M. S. Lee, *J. Chromatogr., B* **1997**, *696*, 99.
[31] F. Lombardo, G. Campos, *AAPS Workshop: Pharmaceutical Profiling in Drug Discovery for Lead Selection*, Whippany, NJ, May 19–21, 2003.
[32] G. Wang, Y. Hsieh, K.-C. Cheng, K. Ng, W. A. Korfmacher, *J. Chromatogr., B* **2002**, *780*, 451.
[33] L. Di, unpublished data.
[34] K. J. Volk, S. E. Klohr, R. A. Rourick, E. H. Kerns, M. S. Lee, *J. Pharm. Biomed. Anal.* **1996**, *14*, 1663.
[35] R. A. Rourick, K. J. Volk, S. E. Klohr, T. Spears, E. H. Kerns, M. S. Lee, *J. Pharm. Biomed. Anal.* **1996**, *14*, 1743.
[36] E. H. Kerns, R. A. Rourick, K. J. Volk, M. S. Lee, *J. Chromatogr., B* **1997**, *698*, 133.
[37] H. Kim, *J. Separat. Sci.* **2002**, *25*, 877.
[38] L. L. Lopez, X. Yu, D. Cui, M. R. Davis, *Rapid Commun. Mass Spectrom.* **1998**, *12*, 1756.
[39] N. J. Clarke, D. Rindgen, W. A. Korfmacher, K. A. Cox, *Anal. Chem.* **2001**, *73*, 430A.
[40] J. P. Shockor, S. E. Unger, P. Savina, J. K. Nicholson, J. C. Lindon, *J. Chromatogr., B* **2000**, *748*, 269.
[41] C. E. Kibbey, S. K. Poole, B. Robinson, D. L. Jackson, D. Durham, *J. Pharm. Sci.* **2001**, *90*, 1164.
[42] K. P. Shah, J. Zhou, R. Lee, R. L. Schowen, R. Elsbernd, J. M. Ault, J. F. Stobaugh, M. Slavik, C. M. Riley, *J. Pharm. Biomed. Anal.* **1994**, *12*, 993.
[43] K. J. Volk, R. A. Yost, A. Bajter-Toth, *Anal. Chem.* **1992**, *64*, 21A.
[44] U. Jurva, H. V. Vikström, A. P. Bruins, *Rapid Commun. Mass Spectrom.* **2000**, *14*, 529.
[45] U. Jurva, H. V. Vikström, L. Weidolf, A. P. Bruins, *Rapid Commun. Mass Spectrom.* **2003**, *17*, 800.
[46] L. Di, E. H. Kerns, N. Gao, S. Q. Li, Y. Huang, J. L. Bourassa, D. M. Huryn, G. T. Carter, *J. Pharm. Sci.* **2004**, *93*, 1537.

Physicochemical Characterization of the Solid State in Drug Development

by **Danielle Giron**

Chemical and Analytical Research and Development, *Novartis Pharma AG*, WKL-127.4.60, CH-4002 Basel
(e-mail: Danielle.Giron@novartis.com)

Abbreviations
DSC: Differential scanning calorimetry; ICH: International Conference on Harmonization; *IDR*: intrinsic dissolution rate; r.h.: relative humidity; SEM: scanning electron microscopy; TG: thermogravimetry.

1. Introduction

The intrinsic properties of drug substances are related to their chemical structure and include the pK_a value of ionizable groups, the lipophilicity (*e.g.*, log *P*), and the intrinsic stability in solutions (*e.g.*, *Fig. 1*).

The physicochemical properties of the solid state of a drug substance are related to the crystal structure of all possible solid phases associated with this drug substance. These physicochemical properties are critical

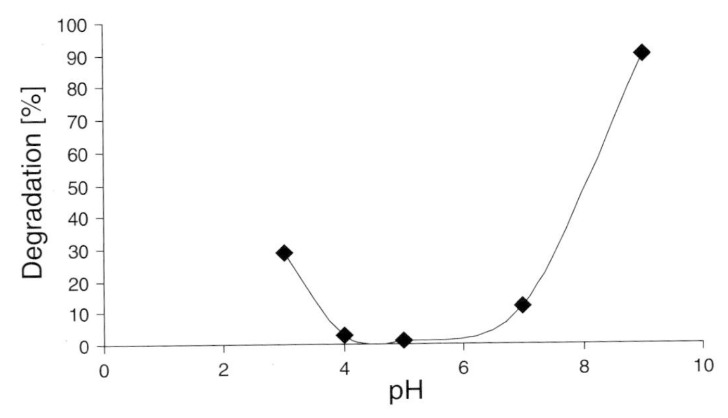

Fig. 1. *Example of a stability–pH profile of a drug candidate*

factors which may affect the therapeutic efficacy, toxicity, bioavailability, pharmaceutical processing, and stability of the drug product. For ionizable chemical entities, the choice of the salt form is the first step of preformulation. Comparison of the solid-state properties of different salt candidates may be quite complicated when the salt form exists as different solid phases, polymorphs, solvates, hydrates, or amorphous forms with different solubility profiles and different thermodynamic domains of stability.

Formulations are developed to improve solid-state properties and in certain cases affect the crystal properties of the drug substance in the drug product. The processing of the drug substances and drug products involve solvent(s), temperature and pressure changes as well as mechanical stress, and different solid phases may coexist in the drug product. Organic substances show supersaturation behavior, and unstable solid phases which should not exist at defined temperature, pressure, and humidity may behave like stable forms. These solid metastable phases obtained outside their domains of stability will convert to the thermodynamic stable forms at given temperatures, pressures, and relative humidities. These conversions driven by thermodynamics are also governed by kinetics and are influenced by impurities, particle size, crystal defects, and presence of seeds.

As a result of the above, solid-state properties such as solubility, dissolution, melting, density, morphology, hygroscopicity, density, crystal hardness, processability, stability, compatibility, and transformations during processing or storage have to be studied in the context of thermodynamic and kinetic viewpoints for each salt candidate [1][2] to start the development of a new entity with the appropriate salt and solid form which fulfills the needs of the targeted formulation.

For the registration of new entities, the International Conference for Harmonization (ICH) [3] requires a program for screening, characterization, and, if necessary, quantitation of polymorphs, solvates, hydrates, and amorphous forms in drug substances and drug products within routine analytical testing. For this challenging task, the application of systematic procedures with the use of high-throughput instrumentation and of combined sophisticated analytical techniques increase the probability to elucidate the relevant polymorph forms and their thermodynamic relationships very quickly. Computer modeling based on crystallographic data take a great part in the understanding and prediction of some properties and give basic information for quantitative determinations.

2. Definitions and Thermodynamic Aspects

Polymorphism is the tendency of any substance to form crystals of different crystalline states. The solid forms of the same compound are called polymorphs or crystalline modifications. Polymorphs have the same liquid or gaseous state but they behave as different substances in the solid state. If a solvent is part of the crystal lattice, a new compound called a solvate is formed. This behavior is called pseudo-polymorphism.

The amorphous state is produced by precipitation, milling, drying, melting, lyophilization, spray-drying, and crystallization in a nonordered, random system, related to the liquid state is characteristic.

The relationships between different phases are governed by *Gibbs* phase rule (*Eqn. 1*):

$$V = C + 2 - \Psi \tag{1}$$

where V is the variance, C is the number of constituents, and Ψ is the number of phases. In the case of polymorphism, C is 1 if two solid phases are present; if both pressure and temperature vary, the variance is unity. If the pressure is fixed, the variance is zero. Phase diagrams of pressure *vs.* temperature illustrate the different equilibrium curves for polymorphism (see *Fig. 2*).

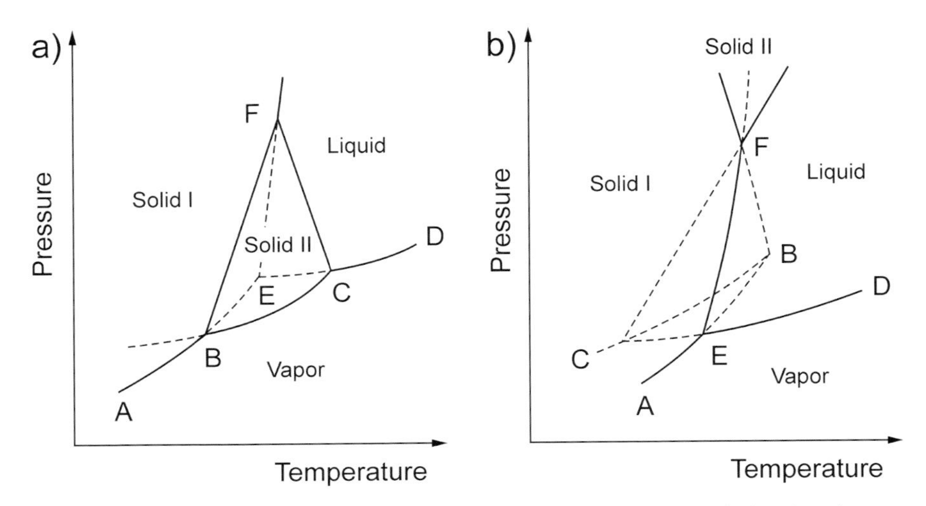

Fig. 2. *Phase diagram of pressure* vs. *temperature for a single compound, showing* a) *enantiotropy, and* b) *monotropy* (from S. C. Wallwork, D. J. W. Grant, 'Physical Chemistry for Students of Pharmacy and Biology', 3rd edn., Longman, London, 1977, p. 42).

For each solid form, there is a solid–liquid equilibrium curve and a solid–vapor equilibrium curve. The solid–vapor curves meet at a point (point B in *Fig. 2*). If the liquid–vapor equilibrium curve (CD) meets the two solid–vapor curves after this point of intersection, there will be a solid-I–solid-II equilibrium curve (curve BF in *Fig. 2,a*) and a reversible transition point I–II at a specific pressure. At the transition point, the free energy of the two forms is the same. This case where two forms convert reversibly is called 'enantiotropy' (*Fig. 2,a*). The term 'monotropy' applies in the case of an irreversible transition from a metastable to the thermodynamic stable form. The liquid–gas curve (curve CD in *Fig. 2,b*) crosses the solid–vapor curves for the two forms before their point of intersection B (*Fig. 2,b*).

In case of enantiotropy, the low-melting form is the thermodynamic stable form below the transition point, and, above this point, the high-melting form is the thermodynamic stable form. The transition point can be low, close to 40° in the case of tolbutamide or close to 100° in the case of propyphenazone or even above 200° [4]. *In case of monotropy, there is only the high-melting form which is the thermodynamic stable form whatever the temperature is.*

The phase diagrams of solvates and hydrates are more complex, since binary mixtures are implied with different compositions. *Fig. 3* illustrates the behavior of a new compound with a congruent and a noncongruent melting. A series of such binary phase diagrams have to be considered if several compounds are formed. *These diagrams are fundamental* for the understanding of crystallization and drying steps.

When a physical property of a crystalline substance is plotted against temperature, a sharp discontinuity occurs at the melting point. For

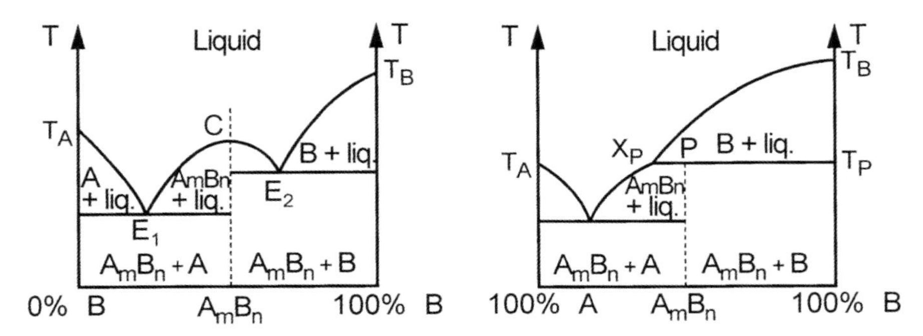

Fig. 3. *Solvates or salts: phase diagrams of binary mixtures of temperature* vs. *composition* (*e.g.*, mol fraction) *of two chemical compounds, A and B, showing the following behavior:* a) *formation of a compound with a congruent melting point at C;* b) *formation of a compound with an incongruent melting point at P*

amorphous substances, there is no melting point, and a change of slope occurs at the so-called glass transition temperature T_g. Below this temperature, the amorphous phase has certain properties of a crystalline solid (*e.g.*, plastic deformation) and is termed 'glassy'. Above this temperature, the substance retains some of the properties of a liquid, *e.g.*, molecular mobility, and is termed 'rubbery'. The increased molecular mobility facilitates then the spontaneous crystallization into the crystalline form with an exothermic enthalpy change after the glass transition. The glass transition temperature, T_g, is lowered by water or other additives, facilitating crystallization. The amorphous state is unstable, and the study of the glass transition with excipients under humidity is part of the preformulation.

All thermodynamically 'unstable' forms may behave like stable forms outside the phase diagrams for kinetic reasons. They are therefore called 'metastable' forms.

3. Basic Properties

The general properties and behavior of polymorphs and pseudo-polymorphs may be found in a number of references (*e.g.*, [4–9]).

3.1. *Appearance, Morphology*

The morphology of a crystal is relevant for its processability; therefore, samples are observed with microscopy or scanning electron microscopy (SEM). The intrinsic morphology (habitus) of a crystal is correlated with its crystal structure (see for example different morphologies of salts candidates in *Fig. 4*). However, a difference of morphology does not imply systematically a change of polymorph. Amorphous samples are easily detected by microscopy under polarized light or with SEM (see *Fig. 5*).

3.2. *Solubility*

The thermodynamic rules imply that, for an ionizable drug substance, the solubility in function of pH depends on its pK_a. The relationship of solubility and temperature obey *Gibbs* phase rule (*Eqn. 1*). In case of enantiotropy and solvate (or hydrate) formation, there are temperatures at which the enantiotrops or the different solvates have the same solubility (*Fig. 6*).

Fig. 4. *Scanning electron microscopy pictures showing different morphologies of several salt forms of the same compound. a*) Base, *b*) hydrochloride, *c*) hydrogen malonate, *d*) mesylate.

Fig. 5. *Scanning electron microscopy pictures showing an amorphous and a crystalline sample of the same drug substance*

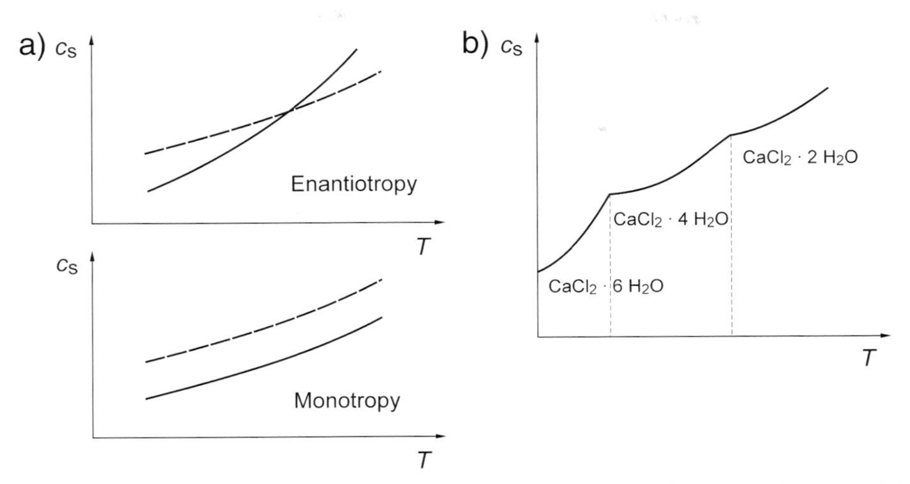

Fig. 6. *Solubility curves* c_S vs. *temperature* T: a) *for enantiotropic and monotropic systems;* b) *for solvates exemplified by the dihydrate, tetrahydrate, and hexahydrate of* $CaCl_2$. The solubility decreases as the number of bound H_2O molecules increases.

Therefore, the equilibrium solubility must be always determined knowing the different phases and the residual solid. In development, the solubilities in different media have to be determined very precisely, namely in water and different aqueous buffers, in the solvents to be used in the preparation of the drug and its medicines, and in the solvents used in analysis. Determinations are carried out according to the International Conference on Harmonization (ICH) [3] with the pure manufactured polymorphs and pseudo-polymorphs. The approximate solubility is first determined to avoid to use a too high amount of residual solid. The role of impurities and kinetic behavior are discussed in [1]. Some solvents (*e.g.*, MeCN, alcohols, THF, CH_2Cl_2, $CHCl_3$ [7]) have a tendency to give solvates, and the water activity in solvents is the driving force for the formation of hydrates [10][11]. The term 'thermodynamic solubility' is often used for equilibrium solubility in the presence of solid and liquid phase where dissolution and crystallization are in equilibrium. Because several thermodynamic stable species may exist depending on the solvent, pH, temperature, and relative humidity, it is more appropriate to define the 'equilibrium solubility' of the stable forms in the medium and at the temperature considered. *Table 1* gives an example for 'equilibrium solubility' of a drug, where the solubility is determined for the trihydrate, the monohydrate, and the anhydrous form [12]. In the case of salts, the exchange of the counter-ion with the buffers must be considered [1][2].

Table 1. *Equilibrium Solubility in Water of the Trihydrate, the Monohydrate, and the Anhydrate of a Drug Substance*

Residual solid	T [°]	Solubility [mg ml^{-1}]
Trihydrate	10	2.1
Trihydrate	25	2.8
Trihydrate	40	10.9
Monohydrate	60	25.3
Anhydrate	80	31.6

3.3. *Dissolution Rate and Intrinsic Dissolution Rate*

The dissolution rate of a drug is the measured amount dissolved as a function of time. It is generally measured by the flow-cell method. The profile depends on the solid form and on the particle size. The dissolution rate curves of formulations are required for bioequivalence studies. *Fig. 7* shows a drug for which the manufacturing of a metastable form seemed feasible. With scale-up, it was impossible to avoid the stable form. The slow dissolution rate of the stable form had a high impact on the behavior of the phase-I clinical formulation. It was necessary to develop a new formulation of the drug [4].

The dissolution rate per unit surface area, termed the intrinsic dissolution rate (*IDR*) [13], is independent of particle size. In the 'disc' method, the powder is compressed by a punch in a die to produce a compact disc or tablet. Only one face of the disc is exposed to the dissolution medium and the cumulative amount dissolved per unit surface area is determined by UV spectrophotometry until 10% of the solid is dissolved. The slope of the

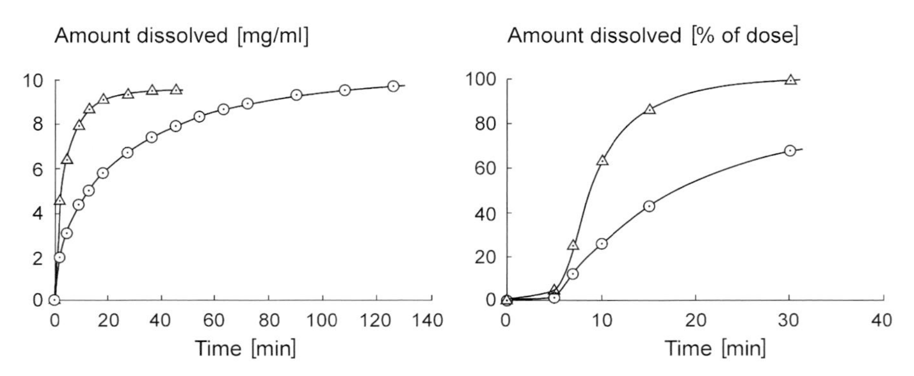

Fig. 7. a) *Dissolution rate curves of two crystalline modifications* (A and B) *of a drug candidate with the same particle-size distribution.* b) *The corresponding curves of the drug products consisting of capsules containing A and B.* △: Polymorh A; ⊙: polymorph B.

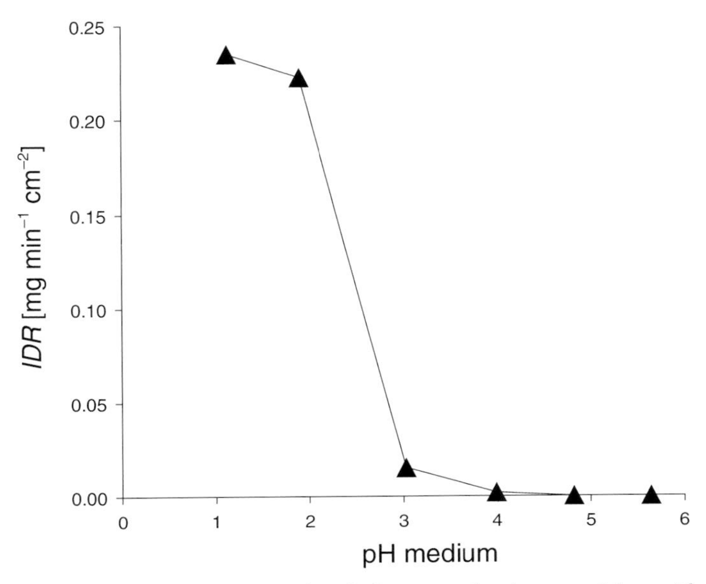

Fig. 8. *Influence of pH on the intrinsic dissolution rate of a drug candidate with the same counter-ion* (Cl^-)

plot of mass dissolved per unit surface area against time gives the intrinsic dissolution rate in appropriate units, *e.g.*, mg min^{-1} cm^{-2}. Very poorly soluble compounds have an $IDR < 0.1$. An $IDR > 1$ indicates that the compounds do not present a solubility problem.

The influence of the pH on the *IDR* of a drug candidate measured at 50 rpm with a *Vankel* instrument is shown in *Fig. 8* [2]. There are differences among salts resulting from their behavior in solution (*Fig. 9*). For a drug candidate with three ionizable functions, the succinate (counter-ion/drug 1 : 1) and the dimaleate (counter-ion/drug 2 : 1) were obtained and the solubility as well as the *IDR* behaved differently at pH 1 or at pH 3 (*Table 2*).

Table 2. *Influence of the Counter-Ion on the Solubility Behavior of a Drug Substance*

Counter-ion	Solubility [%]		*IDR* [mg min^{-1} cm^{-2}]		pH of 1% suspension
	in 0.1N HCl soln.	in tartrate buffer (pH 3)	in 0.1N HCl soln.	in tartrate buffer (pH 3)	
Base	2.4	0.28	3.9	0.06	6.2
Succinate 1:1	2.5	0.38	4.2	0.07	5.8
Maleate 1:2	1.6	0.86	1.57	0.25	3.9

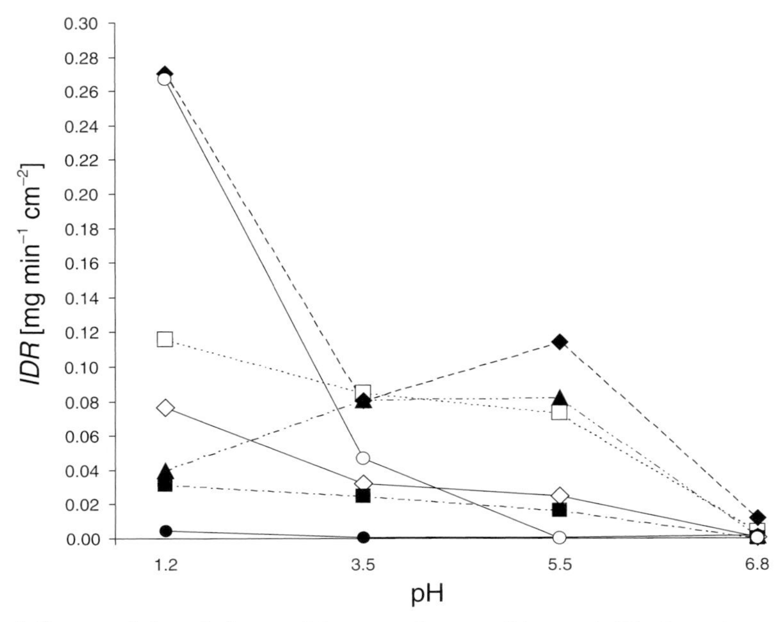

Fig. 9. *Influence of the salt forms of the same drug candidate as in* Fig. 8 *on the intrinsic dissolution rate in different buffers.* ▲: Hydrochloride; □: malonate; ◇: maleate; ◆: malate; ■: oxalate; ●: pamoate; ○: base.

Table 3. *Impact of Polymorphism on the Intrinsic Dissolution Rate* (*IDR*) [mg min^{-1} cm^{-2}]

Drug substance as base		Drug substance neutral	
Polymorph	*IDR* in water with 0.2% LDAO[a])	Polymorph	*IDR* in water
Amorphous form	0.048	Amorphous form	0.269
Form B	0.035	Form A	0.117
Form D	0.011	Form B	0.085

[a]) LDAO: Lauryldimethylamine oxide (= *N,N*-dimethyldodecylamine *N*-oxide).

Polymorphism has also relevance for poorly soluble substances as demonstrated in *Table 3*. The amorphous form improves the behavior, but the factor is <5. When different salt forms are possible with the same counter-ion, their *IDR* is also relevant as shown in *Table 4*. The mono-sodium salt is very soluble. The monohydrate is less soluble, but the *IDR* of the hemi-salt decreases by a factor of *ca.* 10 [2].

Table 4. *Impact of Salt and of Hydrate Formation on the Intrinsic Dissolution Rate (IDR)* [mg min^{-1} cm^{-2}]

Salt form	*IDR*	
	in water	in buffer (pH 6.8)
Na-Monosalt	43.6	22.6
Na-Monosalt-monohydrate	17.6	16.5
Hemi-salt	0.40	0.35

3.4. *Bioavailability*

The influence of particle size on bioavailability is exemplified by proquazone [14]. Several examples dealing with the influence of particle size and polymorphism on dissolution rate and bioavailability can be found in [15]. *Table 5* gives some examples of marketed drugs for which sufficient published data demonstrate problems of bioavailability and bioequivalence.

For poorly soluble drugs, formulations are developed to stabilize the amorphous state and improve bioavailability. The characterization and properties of the amorphous state have been discussed [25]. *Table 6* gives some examples of solubility improvement due to the amorphous state.

Table 5. *Examples of Bioavailability or Toxicity Differences Correlated with Polymorphism*

Drugs	Refs.
Ampicilline anhydrous and trihydrate	[16][17]
Carbamazepine, anhydrous, hydrate	[18][19]
Chloramphenicol palmitate	[20]
Griseofulvine	[21]
Mebendazole	[22]
Novobiocine	[23]
Ritonavir	[24]

Table 6. *Published Examples of Drugs Exhibiting a Relevant Solubility Ratio between Amorphous and Crystalline Forms*

Drugs	Forms	Solubility ratio	Ref.
Glibenclamide	Amorphous/crystalline	14 (23°, buffer)	[26]
Griseofulvin	Amorphous/crystalline	1.4 (21°, water)	[27]
Indomethacin	Amorphous/gamma	4.5 (5° and 25°, water)	[28]
Iopanoic acid	Amorphous/crystalline I	3.7 (37°, phosphate buffer)	[29]
MK-0591 sodium	Amorphous/crystalline	*ca.* 1000	[30]

3.5. *Hygroscopicity*

Water vapor is an omnipresent component of the atmosphere. Most excipients contain water. For solid dosage forms, granulation under humid conditions are generally used. Therefore the study of the behavior of compounds in water vapor atmospheres is a prerequisite in the studies for the choice of the salt form. Some salts are deliquescent and cannot survive high humidity. At a given temperature, the ratio of actual water vapor pressure over saturated vapor pressure at that temperature is called the relative humidity (r.h.) given as percentage of saturation. The environmental humidity depends of the climatic zone.

Sorption–desorption isotherms are measured as the mass change observed during the change in r.h. A hysteresis in desorption is generally an indication of hydrate formation. But reversible desorption may also be observed for hydrates. X-Ray diffraction during such studies is very fruitful. *Fig. 10* shows the complexity of sorption–desorption of several salts of a drug candidate. The base and the hydrogen maleate salt were not hygroscopic while the hydrochloride transformed into a hydrate and the hydrogen malonate took up to 22% water. The hydrogen tartrate was slightly hygroscopic. In such studies, the polymorphic form as well as the amorphous content of the samples used for comparison among salts is very important.

Fig. 11 shows the different behavior of two polymorphs of a hydrochloride [1][4] having an enantiotropic relationship. The high-melting

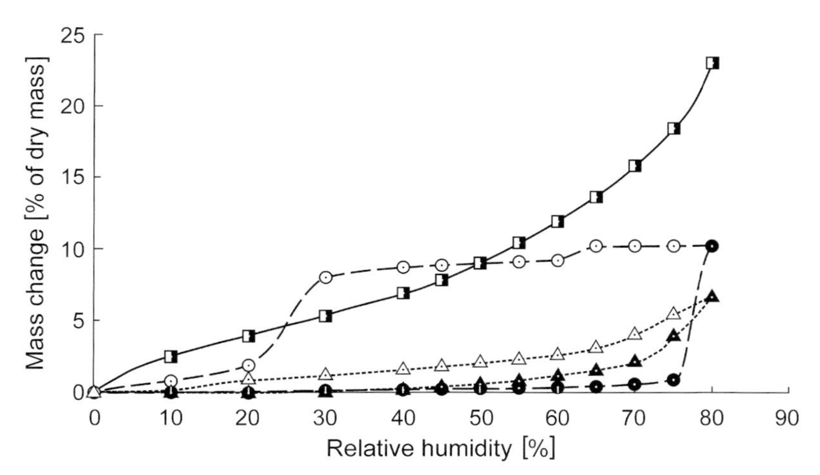

Fig. 10. *Water vapor sorption isotherms of several salts of the same basic investigational compound at 25°*. ⊙: Hydrochloride; △: hydrogen tartrate; ⊡: hydrogen malonate (sorption and desorption without hysteresis); filled symbols: adsorption; open symbols: desorption.

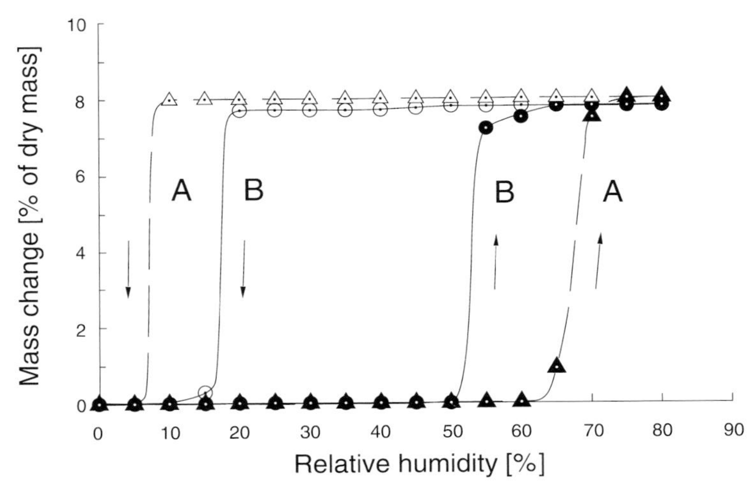

Fig. 11. *Hygroscopicity and polymorphism: Examples of water sorption–desorption isotherms of two enantiotropic forms A and B at 25°*. The two polymorphs A and B transform into the same hydrated form. The higher-melting form B, which is metastable at ambient temperature, takes up water at lower relative humidity (r.h.) than the stable form A. The hydrate form loses water at r.h. values <20%.

form was metastable at ambient temperature and adsorbed water at lower r.h., whereas the hydrate form lost water at r.h. values <20%. A second polymorph of the hydrated form was also obtained in crystallization studies in aqueous media, and the anhydrous low-melting enantiotrop was chosen for further development.

Fig. 12 shows the hygroscopicity behavior of the amorphous form, of the stable anhydrous form, and of the metastable anhydrous form, which transformed reversibly into the monohydrate. The transformation was followed by X-ray diffraction in cell with variable humidity and also in a heating cell.

Very often, the loosely bound solvent of solvated forms is replaced by water and the critical water activity expressed as relative humidity can be determined.

3.6. *Stability*

Chemical reactivity in the solid state is correlated with the nature of the crystalline modifications. Thus, the two crystalline modifications of fenretinide were found to behave quite differently [31]. After 4 weeks at 25°, the stable form showed no detectable degradation, whereas the unstable form showed 8% degradation.

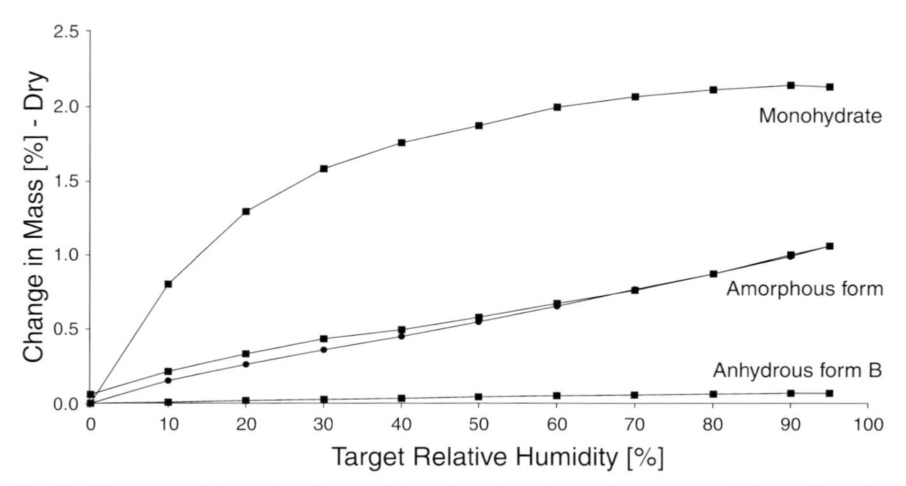

Fig. 12. *Hygroscopicity and polymorphism: Sorption–desorption curves of the stable anhydrous form, of the amorphous form, and of the hydrate.* For this hydrate, a reversible hydrate–metastable anhydrous form transformation occurs during desorption as demonstrated by X-ray diffraction.

The amorphous state is very reactive. *Table 7* illustrates the difference in reactivity between the two crystalline forms and the amorphous form of a drug candidate. The crystalline form A is more stable than both form B and the amorphous form. Larger differences are observed for the two polymorphs of a dihydrate. *Example 3* deals with a peptide drug candidate. In the amorphous state, both the base and its hydrochloride were very unstable. However, the base could be obtained as a crystalline material with a substantial gain in stability. *Examples 2* and *3* also show that the

Table 7. *Influence of Polymorphism on the Stability Behavior in the Solid State*

	Degradation [%] (HPLC)				
	1 month at 80° (oxygen/water)	2 weeks at 50°	1 week at 70°	Exposition 1200 klux h	Exposition 300 klux h
Example 1					
Crystalline form A	0				
Crystalline form B	0.5–1.50				
Amorphous form	2–3.5				
Example 2					
Monohydrate A		0		10	
Monohydrate B		12		23	
Example 3					
Crystalline form			10		2
Amorphous form			80		38

stability under light exposure can be very different. In the case of ethoxycinnamic acid, different photolytic degradation products were obtained for each form. The α-form gave one degradation product, the β-form a second degradation product, the γ-form did not decompose under light exposure [32]. As a rule, stability decreases as particle size decreases [33].

4. Phase Transformations, Kinetic Aspects

4.1. *Solvent-Mediated Transformations*

Kinetic factors are responsible for the existence of solid phases outside thermodynamic phase diagrams. Transformations may be accelerated by the presence of a solvent such as water. *Solvent-mediated transformations* occur by a continuous dissolution–crystallization process. This type of transformation may occur during crystallization or granulation. Hydrates may be formed by this process in mixtures involving moisture. Solvent-mediated transformations may occur when measuring solubilities, so that, given sufficient time, recrystallization of a more-stable state may be complete. *Fig. 13* shows the solubility in water *vs.* time of two anhydrous forms of a drug candidate. A quick transformation of the two anhydrous

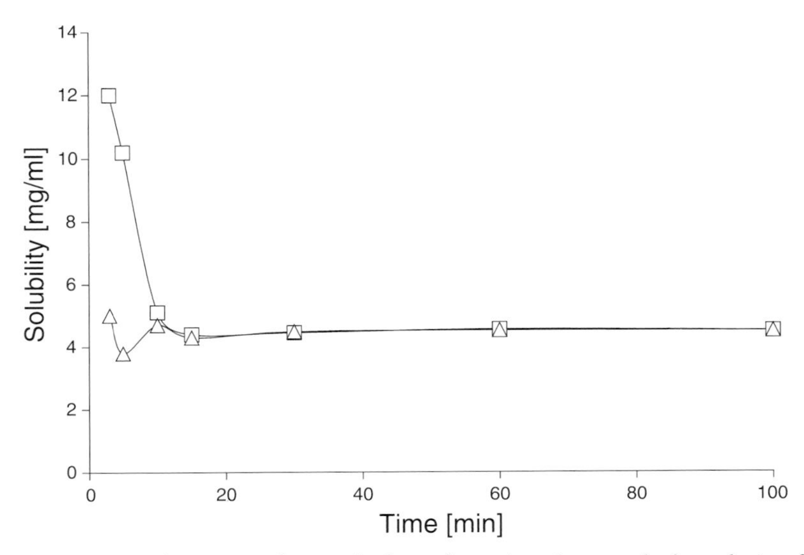

Fig. 13. *Rapid transformation of two anhydrous forms into the monohydrate during dissolution in water.* △: Form 1; □: form 2.

forms into the hydrated form was observed. A strong decrease of solubility was observed for the metastable anhydrous form 2.

4.2. *Dissociation of Salt Form or Exchange of Counter-Ion*

Dissociation of the salt forms is often observed. For example, a methanesulfonate dihydrate was selected for a parenteral formulation. Upon storage, a precipitate was observed in the formulation. The anhydrous monomethanesulfonate was less soluble and more stable in the formulation. In another case, the salt was a hydrochloride and the base was undissolved. In solubility experiments, the amount of substance in solution increased with the amount of solid added resulting in a strong decrease in pH, but the remaining base was undissolved and the solubility results completely erroneous [2].

Most critical are the transformations during measurements, especially when determining intrinsic dissolution rates and solubilities. *Fig. 14* shows the *IDR* curves of a hydrochloride in 0.1N aqueous HCl solution and acetate buffer. In acetate buffer, an abrupt change occurred due to the formation of the acetate salt as confirmed by X-ray diffraction.

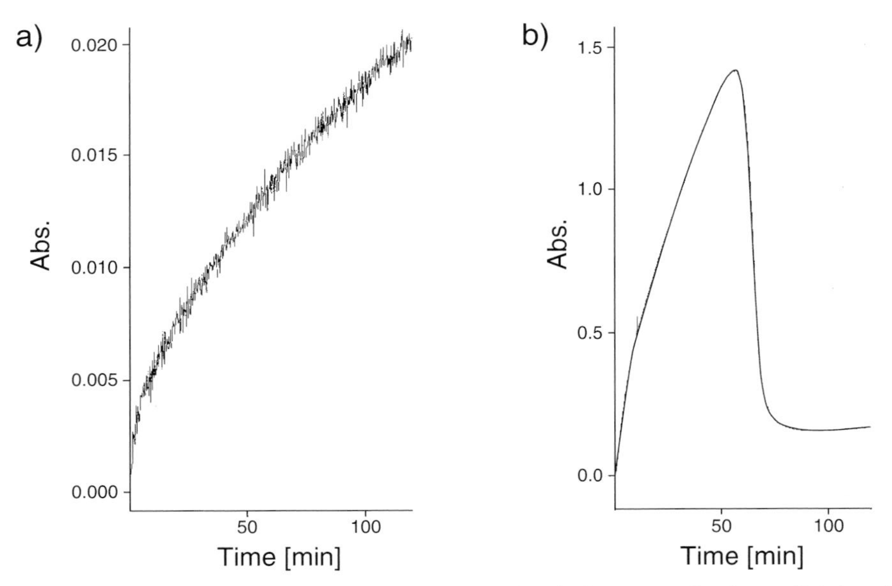

Fig. 14. *Intrinsic dissolution rate curves of a hydrochloride* a) *in aq. HCl solution* (0.1N) *and* b) *in acetate buffer* (pH 4.6). In acetate buffer, an abrupt change occurs due to the formation of the acetate salt.

4.3. *Slow Transformations after Storage of Drug Products*

Solid-state characterization would be easier if the stable form would not be hidden by the metastable forms which do not transform due to kinetic factors. This was observed for ritonavir [34]. In 1998, many batches failed the dissolution test due to precipitation from the final (semisolid) formulation. A new thermodynamically polymorphic form had emerged. In ethanol/water 90:10, form I has a solubility of 234 mg ml^{-1}, and the solubility of the new form is 60 mg ml^{-1}. In ethanol/water 75:25, the solubility of form I is 170 mg ml^{-1} and the new form has a solubility of 30 mg ml^{-1}.

Fig. 15 shows the microscopic picture of the orthorhombic crystal appeared in a stability sample of soft gelatin capsules containing a liquid formulation. The new form is extremely insoluble. It was manufactured and characterized with its single-crystal structure, thermal analysis, and spectroscopy. Stability and solubility studies were performed. The formulation was newly optimized using this new form for the solubility measurements in pharmaceutical liquid excipients. In parallel, the solubility of the soluble form was determined. *Fig. 16* exemplifies the acceleration of the solvent-mediated transition of the metastable form A into the stable form B by the temperature of measurement. The data were obtained after equilibration for 24 h at different temperatures in soybean oil.

Fig. 15. *Formation of a new, very insoluble crystalline modification during storage of a liquid formulation*

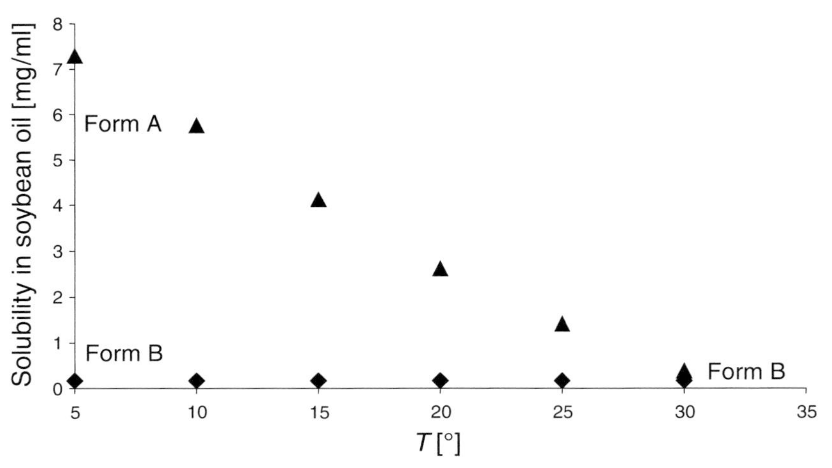

Fig. 16. *Quantitative monitoring of the example in* Fig. 15: *comparison of the equilibrium solubility measured with the two forms after 24 h equilibration at different temperatures.* The solvent-mediated transformation of the soluble form A into the insoluble form B is accelerated as the temperature of measurement increases.

To prevent the transformation of an amorphous form into crystalline forms in solid dispersion is a challenging objective for an enhanced bioavailability of insoluble drugs.

4.4. *Influence of Seeds*

Fig. 17 shows a substance for which a metastable form A was selected and two batches were subjected to stability tests under tropical conditions. The content of the stable form B in the two batches of the metastable form A was measured by X-ray diffraction and the results confirmed by IR spectroscopy. Seeds of form B initiated the transformation in batch 2 while no change was observed in batch 1 [33]. This example demonstrated the need of a very sensitive analytical method for rapid detection of such effects.

Seeds of stable forms can affect solubility as demonstrated in *Fig. 18*. Here, the solubility of four forms of MKS492 was examined in water at 20° and 40°. Forms A, C, and D were monotrops of form B. When B was added, solvent-mediated transformation occurred immediately [35].

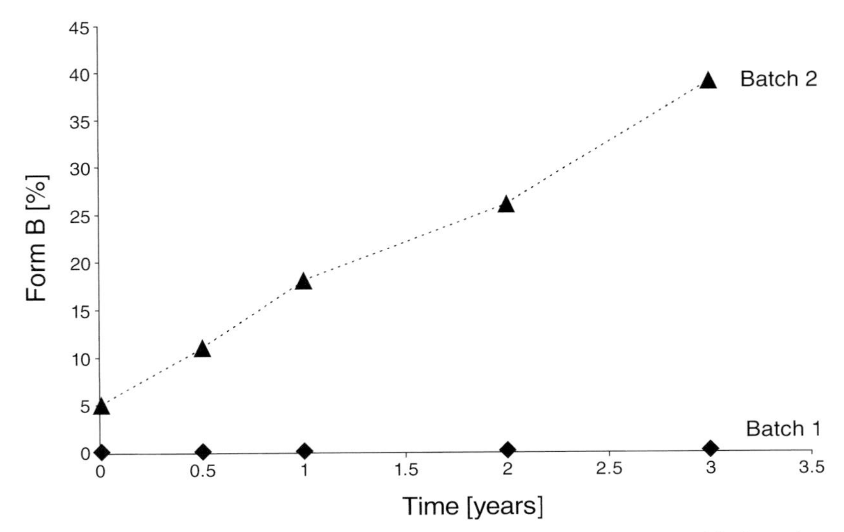

Fig. 17. *Content of the stable form B in samples of two batches of the metastable form A stored under tropical conditions measured by X-ray diffraction.* Seeds of form B initiate the transformation in batch 2 while no change is observed in batch 1.

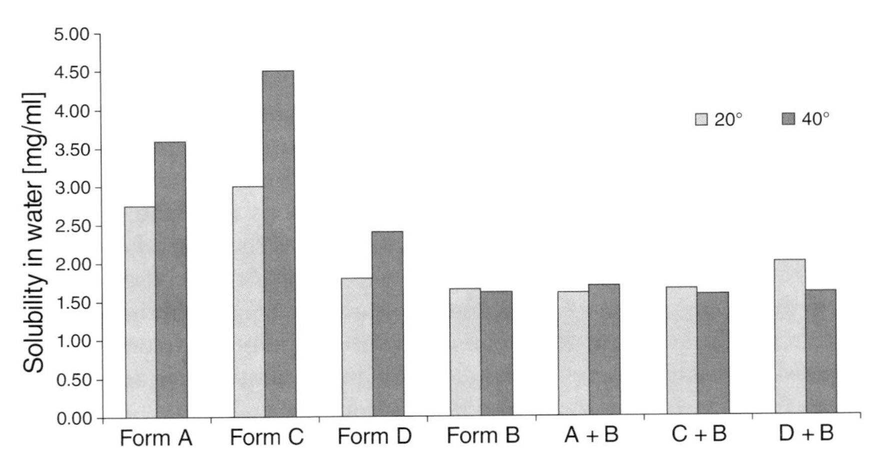

Fig. 18. *Influence of seeds on solubility: solubility of four forms of MKS492 in water at 20°* (light) *and 40°* (dark). Forms A, C, and D are monotrops of form B. When form B is added, the solvent-mediated transformation occurs immediately.

5. Analytical Technologies

5.1. *Polymorphic Screening*

For the screening, crystallizations, precipitations at different temperatures and equilibrations in slurry by solvent-mediated transformation should allow detection of both metastable and stable forms. Heating and cooling in differential scanning calorimetry (DSC) permit the identification of forms not obtainable from solvents. Water atmospheres as well as solvent atmospheres allow to study hydrates and solvates. Detailed discussions have been published [36] [37].

A variety of analytical techniques are available for the detection and identification of each solid phase. Each technique has its advantages and disadvantages. X-Ray diffraction is the most useful technique since it is directly related to the crystal structure, which is the characteristic of polymorphs. Commercial high-throughput instruments are available.

5.2. *Identification of Phases and Thermodynamic Relations*

Once differences in the X-ray pattern or in DSC are observed, further steps are the identification of phases, but the interpretation of data must refer to thermodynamics. Energy diagrams of free energy and enthalpy *vs.* temperature at a given pressure reflect the transition observed between solid phases, and between solid and liquid phases. Differential scanning calorimetry delivers melting temperature and melting enthalpy and allows to distinguish between enantiotropy and monotropy according to *Burger* [4] [8]. Microcalorimetry with the measurement of the heat of dissolution also allows to correlate thermodynamics and to calculate the transition energy between phases [4]. Thermogravimetry (TG) is the most appropriate technique for the detection of solvates. Today's instruments offer the possibility to measure accurately a few milligrams or less and they are able to measure automatically up to 50 samples.

Combined techniques are needed to understand complex situations. Temperature X-ray resolved diffraction, thermomicroscopy, IR or *Raman* spectroscopy with heating cells, and thermogravimetry coupled with mass spectrometry (TG/MS) or with IR spectroscopy (TG/IR) are very efficient tools in such studies [38].

5.3. *Quantitative Analysis*

Quantitative methods generally require samples of polymorphs for routine analysis, a difficult condition for metastable solid phases. X-Ray diffraction is attractive in all steps of development from detection to quantitation, especially when combined with thermal analysis. Computational X-ray diffraction can demonstrate the purity of the different forms for their further characterization. Isothermal microcalorimetry is a growing and successful technique for the determination of very low levels of the amorphous phases [36] [37].

5.4. *Characterization*

Physicochemical properties are analyzed by DSC (melting point, melting energy), solubility (equilibrium solubility, HPLC), *IDR*, and dynamic vapor absorption instruments for the study of hygroscopicity. Surface property is analyzed by scanning electron microscopy, atom force microscopy, and surface measurements. Density, wettability, electrostaticity, flowability are typical properties relevant for the processability of drug products.

Fig. 19 decomposes the tasks in development. For a monosodium salt of a drug substance, one monohydrate, different solvates, and different salt forms were isolated and manufactured. They were then characterized by their single-crystal structure, *IDR*, solubility, stability, and the relationships determined. Quantitative methods were also developed to ensure the purity of the selected form [39].

6. Conclusions

Improvement of the physicochemical properties of a drug substance for targeted formulation is possible thanks to a good knowledge of all possible salt forms and polymorphs thereof. The physicochemical characterization of the solid state of new drug substances in development requires first the isolation of all forms to be considered. The most-important task is to recognize very rapidly the thermodynamically stable form. Selection of this form is generally preferred since it allows a robust manufacturing process and delivers a constant quality for the manufacture of the drug product. However, the stabilization of metastable forms and especially the amorphous state increase the challenging task of development.

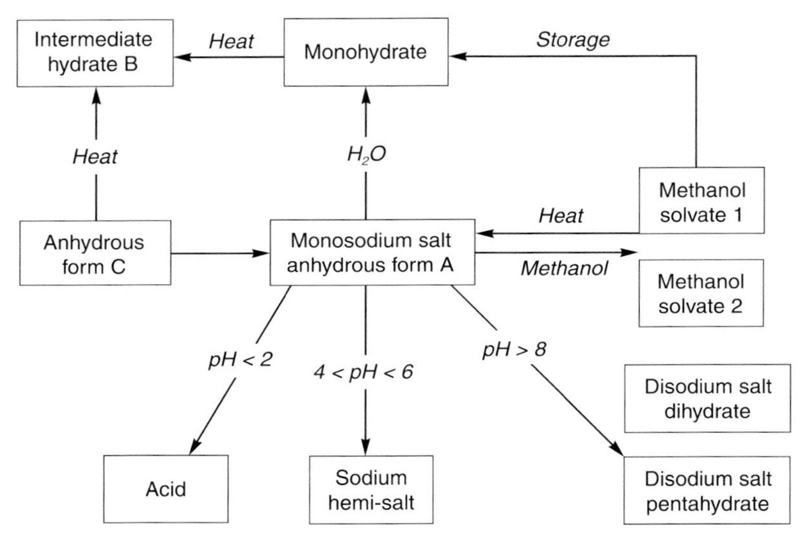

Fig. 19. *Examples of solid phases to consider for a monosodium salt of a drug substance*

The solid properties of the salt forms and polymorphs have to be studied in a thermodynamic and kinetic context. The quality of samples influences the quality of the results. Highly sophisticated automatic combined analysis techniques and modeling tools need to be used for a proper design of the drug substance in development.

REFERENCES

[1] D. Giron, D. J. W. Grant, in 'Handbook of Pharmaceutical Salts. Properties, Selection, and Use', Eds. P. H. Stahl, C. G. Wermuth, Verlag Helvetica Chimica Acta, Zürich, 2002, p. 42–81.
[2] D. Giron, *J. Therm. Anal. Calorim.* **2003**, *73*, 441.
[3] International Conference on Harmonization (ICH), 1999, Guideline Specification Q6A.
[4] D. Giron, *Thermochim. Acta* **1995**, *248*, 1.
[5] J. Haleblian, *J. Pharm. Sci.* **1975**, *64*, 1269.
[6] D. Giron, *Labo-Pharma, Probl. Tech.* **1981**, *307*, 151.
[7] S. R. Byrn, R. R. Pfeiffer, J. G. Stowell, 'Solid State Chemistry of Drugs', SSCI, West Lafayette, Indiana, 2003.
[8] A. Burger, R. Ramberger, *Mikrochim. Acta* **1979**, *2*, 259; A. Burger, R. Ramberger, *Mikrochim. Acta* **1979**, *2*, 273.
[9] H. G. Brittain, 'Polymorphism in Pharmaceutical Solids', Dekker, New York, 1999.
[10] R. Khankari, D. J. W. Grant, *Thermochim. Acta* **1995**, *248*, 61.
[11] M. D. Ticehurst, R. A. Storey, C. Watt, *Int. J. Pharm.* **2002**, *247*, 1.
[12] D. Giron, C. Goldbronn, M. Mutz, S. Pfeffer, P. Piechon, P. Schwab, *J. Therm. Anal. Calorim.* **2002**, *68*, 453.

[13] 'The United States Pharmacopeia 24', United States Pharmacopeial Convention, Rockville, 2000, Suppl. 1, p. 2706–2707; 'The United States Pharmacopeia 26', United States Pharmacopeial Convention, Rockville, 2003.
[14] F. Nimmerfall, J. Rosenthaler, *J. Pharm. Sci.* **1980**, *69*, 605.
[15] D. Giron, *S.T.P. Pharma* **1988**, *4*, 330.
[16] J. W. Poole, C. K. Bahal, *J. Pharm. Sci.* **1968**, *57*, 1945.
[17] A. A. Ali, A. Farouk, *Int. J. Pharm.* **1981**, *9*, 239.
[18] P. Kahela, R. Aaltonen, E. Lewing, *Int. J. Pharm.* **1983**, *14*, 103.
[19] Y. Kobayashi, S. Ito, S. Itai, K. Yamamoto, *Int. J. Pharm.* **2000**, *193*, 137.
[20] A. J. Aguiar, J. Krc , A. W. Kinkel, J. C. Samyn, *J. Pharm. Sci.* **1967**, *56*, 847.
[21] M. Krieger, Dissertation, University Erlangen-Nürnberg, 1980.
[22] F. Rodriguez-Caabeiro, A. Criado-Fornelio, A. Jimenez-Gonzales, L. Guzman, A. Igual, A. Perez, M. Pujol, *Chemotherapy* **1987**, *33*, 266.
[23] J. D. Mullins, T. J. Macek, *J. Am. Pharm. Assoc. Sci. Ed.* **1960**, *49*, 245.
[24] D. Law, E. A. Schmitt, C. Marsh, E. A. Everitt, W. Wang, J. J. Fort, Y. Qiu, *J. Pharm. Sci.* **2004**, *93*, 563.
[25] B. C. Hancock, G. Zografi, *J. Pharm. Sci.* **1987**, *86*, 1.
[26] M. Mosharraf, C. Nystrom, *Pharm. Sci.* **1998**, *1*, S268.
[27] A. A. Flamin, C. Ahlneck, G. Alderbon, C. Nystrom, *Int. J. Pharm.* **1994**, *111*, 159.
[28] B. C. Hancock, M. Parks, *Pharm. Res.* **2000**, *17*, 397.
[29] W. C. Stagner, J. K. Guillory, *J. Pharm. Sci.* **1979**, *68*, 1005.
[30] S. D. Clas, R. Faizer, R. E. O'Connor, E. B. Vadas, *Int. J. Pharm.* **1995**, *121*, 73.
[31] W. D. Walkling, W. R. Sisco, M. P. Newton, B. Y. Fegely, H. N. Plampin, F. A. Chrzanowski, *Acta Pharm. Technol.* **1986**, *32*, 10.
[32] M. D. Cohen, B. S. Green, *Chem. Br.* **1973**, *9*, 490.
[33] D. Giron, B. Edel, P. Piechon, *Mol. Cryst. Liq. Cryst.* **1990**, *187*, 297.
[34] S. R. Chemburkar, *Org. Proc. Res. Dev.* **2000**, *4*, 413.
[35] D. Giron, C. Goldbronn, P. Piechon, S. Pfeffer, *J. Therm. Anal. Calorim.* **1999**, *57*, 61.
[36] D. Giron, J. Therm. *Anal. Calorim.* **2001**, *64*, 37.
[37] D. Giron, *Eng. Life Sci.* **2003**, *3*,103.
[38] D. Giron, J. Therm. *Anal. Calorim.* **2001**, *64*, 37.
[39] S. Pfeffer, P. Piechon, C. Goldbronn, D. Giron, unpublished.

Part IV. Computational Strategies

Calculation of Lipophilicity: A Classification of Methods

by **Raimund Mannhold**

Department of Lasermedicine, Molecular Drug Research Group,
Heinrich-Heine-Universität Düsseldorf, Universitätsstraße 1, D-40225 Düsseldorf

Abbreviations
ADME: Absorption, distribution, metabolism, excretion; HTS: high-throughput screening; log *D*: octanol/water distribution coefficient; log *P*: octanol/water partition coefficient; MLP: molecular lipophilicity potential.

1. Introduction

Drug disposition and bioactivity are guided by lipophilicity in a comprehensive way [1–3]. Increased lipophilicity was shown to correlate with poorer aqueous solubility, increased plasma protein binding, increased storage in tissues, more-rapid metabolism and elimination, increased biological activity, and faster rate of onset of action, to mention a few.

An increasing impact of lipophilicity is given in modern drug discovery [4][5]. Late-stage failures in classical drug development were often due to poor pharmacokinetics and toxicity. Thus, nowadays it is common sense in pharmaceutical research to address these areas as soon as possible within the drug discovery process. High-throughput screening (HTS) facilities and combinatorial chemistry have dramatically increased the rate at which biological screening data are available for large-size databases. With the increased capacity for biological screening and chemical synthesis, demands for large amounts of information on ADME (absorption, distribution, metabolism, excretion) properties of drugs increased as well. Beyond the development of HTS measurements of ADME properties, there is increasing interest in adequate tools for predicting these properties.

Among the physicochemical descriptors of ADME properties, log *P* is of key importance. The extent of existing experimental log *P* data is negligible compared to the exponentially increasing number of compounds for which log P data are highly desired. Correspondingly, there is great interest in methods which allow to derive log P from molecular structure.

The first method of calculating log P was the π-system, developed by *Hansch* and *Fujita* [6]. Shortcomings in the π-system led *Rekker* to develop the first fragmental contribution approach [7–10]. Calculation systems based on atomic contributions were first described *by Broto et al.* [11]. Whereas these methods are based on substructure definitions, more-recent approaches inspect the molecule as a whole and use different molecular descriptors to quantify log P. Up to date, most calculation methods have been focused on log P, considering the molecule in its neutral state. Only very recently, attempts have been made to estimate log D, taking into account the ionization of molecules.

In the first part of this chapter, representative log P calculation approaches are described and their advantages and shortcomings are discussed. A main goal is to classify log P programs according to their methodology. Aspects of log D calculation are treated in the second part of this overview.

2. Classification of Methodological Approaches

Methods for calculating log P can be divided into substructure and whole molecule approaches (*Table 1*). *Substructure approaches* cut molecules into fragments (fragmental methods) or down to the single-atom level (atom contribution methods); summing the single-atom or fragmental contributions results in the final log P. Molecules, however, are not mere collections of fragments or atoms. Thus, fragmental methods apply correction rules coupled with molecular connectivity. With one exception, atom contribution methods work without correction factors:

$$\textit{Fragmental Methods:}\ \log P = \sum_{i=1}^{n} a_i \cdot f_i + \sum_{j=1}^{m} b_j \cdot F_j \qquad (1)$$

f: fragmental constant; a: number of fragments; F: correction factor; b_j: frequency of F_j.

$$\textit{Atom Contribution Methods:}\ \log P = \Sigma\, n_i \cdot a_i \qquad (2)$$

N_i: number of atoms of type i; a_i: contribution of an atom of type i.

Table 1. *Classification Scheme for log* P *Calculation Programs*

Substructure Approaches[a])	
Fragmental methods	*Atom contribution methods*
CLOGP	MOLCAD (*Crippen*, original)
Σf-SYBYL (*Rekker*, revised)	TSAR (*Crippen*, original)
SANALOGP_ER (*Rekker*, revised)	PrologP (*Crippen*, original)
KLOGP	ALOGP98 (*Crippen*, revised)
LOGKOW	XLOGP
ACD/LogP	
AB/LogP	

Whole Molecule Approaches[b])		
MLP-related	*Topological indices*	*Molecular properties*
CLIP	MLOGP	BLOGP
HINT	VLOGP	QLOGP
VEGA	T-LOGP	
	AUTOLOGP	
	CSLogP	

[a]) Substructure approaches have in common that molecules are cut into groups (fragmental methods) or atoms (atom contribution methods). Most fragmental methods use correction rules. Atom contribution methods work without correction factors except XLOGP. The high quality of XLOGP underlines the importance of using correction rules. [b]) Whole molecule approaches inspect the entire molecule; they use either molecular lipophilicity potentials (MLP), topological indices, or molecular properties to quantify log *P*. The quite large number of attempts to model log *P* with molecular descriptors sharply contrasts to the limited number of elaborated programs available. This is particularly true for whole molecule approaches using molecular properties, while almost all approaches using topological descriptors are commercially available.

Fragmentation of a molecule can be somewhat arbitrary and any fragmentation approach has advantages and disadvantages. Fragments larger than a single atom can be defined, so that significant electronic interactions are comprised within one fragment, which represents a main advantage of using fragments. An advantage of atom contribution methods is that ambiguities are avoided; shortcomings are the huge number of atom types needed for describing a diverse set of molecules and the failure to deal with long-range interactions as found, *e.g.*, in *p*-nitrophenol. Substructure approaches fail to calculate structural isomers and do not consider conformational flexibility. Atom contribution [12] and fragmental methods [13][14] have been extended to cope with these shortcomings.

Whole molecule approaches utilize descriptions of the entire molecule to calculate log *P*. These models attempt to circumvent shortcomings of fragmental approaches such as the simplification of steric effects, the failure to calculate log *P* for structures with missing fragments or the lacking differentiation between structural isomers. Whole molecule approaches use *a*) molecular lipophilicity potentials (MLP), *b*) topological indices, or *c*) molecular properties such as charge densities, surface area, volume, and electrostatic potential to quantify log *P*.

a) The MLP defines the influence of all lipophilic fragmental contributions of a molecule on its environment and offers a quantitative three-dimensional description of lipophilicity. At a given point in space, the MLP value represents the results of the intermolecular interactions between all fragments and the solvent system at that point. Two components are necessary to calculate the MLP: a substructure system and a distance function. The group of *Dubost* [15] was the first to introduce the MLP approach, using the Σf-system as fragmental method and a hyperbolic distance function. *Fauchère et al.* [16] applied the fragmental systems of *Rekker* and/or *Hansch* and *Leo* and an exponential distance function. Others use the atom contribution method of *Ghose–Crippen* and a hyperbolic distance function [17].

b) Topology concerns properties and spatial relations unaffected by continuous change of shape or size. In relation to molecules, connectivity deals with which atoms are connected to which other atoms. Since this is one unambiguous feature of well-defined molecules, molecular connectivity indices may be deduced directly from molecular structure and used to predict log *P*.

c) Increasing speed and accuracy of molecular orbital calculations allow the derivation of models based on quantum-chemical approaches. Accordingly, a variety of quantum-chemical descriptors such as ionization potentials, dipole moments, electrostatic potentials, charge densities, charge-transfer energies, as well as HOMO and LUMO energies are used to model log *P*.

In the following sections, selected substructure and whole molecule approaches are individually described. Selection is primarily based on common use and software availability. Instead of a complete overview, the main scope is to highlight the characteristics of these log *P* models and to discuss their advantages and disadvantages.

3. Fragmental Methods

Fragmental methods (*Table 2*) have in common to cut molecules down into fragments and to apply correction rules. Main representatives are CLOGP, Σf, KLOGP, LOGKOW, ACD/LogP, and AB/LogP. CLOGP is based on the principle of constructionism, whereas the remaining approaches are reductionistic. More-recently developed fragmental methods such as LOGKOW, ACD/LogP, and AB/LogP comprise much larger numbers of correction rules than first generation methods.

3.1. *CLOGP*

CLOGP [18–22] is based on the principles of '*constructionism*'. The basic fragmental values were derived from measured log P data of simple molecules such as H_2 and CH_4, then the remaining fragment set was constructed. A main concept within fragmentation rules was the definition of isolating C-atoms (sp^3 C-atoms with at least two bonds linked to other C-atoms). In contrast to the Σf-system, a correction for branching is applied. In [19], 200 fragment values and 25 correction factors were given. The method was first adapted for computational use by *Chou* and *Jurs* [23]. CLOGP is the most-frequently used log P calculation program. Recent versions include the FRAGCALC algorithm [24], which was devised to calculate fragment values from scratch. It is based on a test set of 600 dependably measured fragments having only aliphatic or aromatic bonds.

3.2. Σf-*System*

Rekker's group developed the first fragmental method [7–10]. Experimental log P values of simple organic compounds were used to derive fragmental values by *Free–Wilson* analyses; hence, this approach has been labeled '*reductionistic*'. The development of the Σf-system comprised three main phases.

The first period resulted in a valuable system for log P calculation based on 126 fragment values. Fragmentation leaves functional groups with direct resonance interaction intact. Fragments range from atoms over substituents to complicated, in particular heterocyclic rings; fragments are differentiated according to aliphatic or aromatic attachment. Regression analyses revealed systematic differences between measured log P and log P calculations based on the mere summation of fragment values. Attribution of these differences to chemical characteristics of the molecules allowed the definition of correction rules. Correction values were found to represent multiples of a constant value of 0.289. This approach is known as the *original* Σf-*system* [7–9].

In the second phase, thorough revision of the original system resulted in a better fit of log P_{oct} for simple halo-alkanes and aliphatic hydrocarbons

Table 2. *Substructure Approaches*

Program	Fragmentation and correction rules	Provider	Internet access
Fragmental methods			
CLOGP	200 fragment values, 25 correction rules[a])	*Daylight* *Biobyte*	www.daylight.com www.biobyte.com
Σf, manual	126 fragment values, 10 correction rules		
Σf-SYBYL	169 fragment values, 13 correction rules	*Tripos*	www.tripos.com
SANALOGP_ER	302 fragment values, 13 correction rules		
KLOGP	68 contribution values, 30 correction factors	*MULTICASE*	www.multicase.com
LOGKOW	144 group contributions, 235 correction factors	*Syracuse Res. Corp.*	esc.syrres.com/interkow/kowdemo.htm
		US EPA	www.epa.gov/oppt/exposure/docs/episuitedl.htm
ACD/LogP	537 group contributions, 2206 correction factors	*Advanced Chemistry Development*	www.acdlabs.com
AB/LogP	473 group contributions, 1076 clusters of correction factors	*Advanced Pharma Algorithms*	www.ap-algorithms.com
Atom contribution methods			
TSAR	120 atom contributions, no correction rules	*Accelrys*	www.accelrys.com
MOLCAD	120 atom contributions, no correction rules	*Tripos*	www.tripos.com
PrologP	120 atom contributions, no correction rules	*Compudrug*	www.compudrug.com, www.compudrug.hu
ALOGP98 within Cerius^2	68 atom contribution, no correction rules	*Accelrys*	www.accelrys.com
XLOGP 2.0	90 atom contributions, 10 correction rules	*Luhua Lai*	lai@ipc.pku.edu.cn

[a]) According to [19], present data not known.

with Σf-data and a refinement of the correction factor to a value of 0.219. Details of the *revised* Σf-*system* are given in [25].

Complex multi-halogenation in aliphatic hydrocarbons was treated in a third phase. The most-recent version of the Σf-system is given in [26]; it lists 13 correction rules and 169 fragment values, including 14 new heterocyclic fragments as well as doubly and triply halogenated methyl groups.

The Σf-system is the only fragment method allowing manual log P calculation. Computerized versions such as Σf-SYBYL, based on the revised Σf-system and SANALOGP_ER, based on an extended, revised

Σf with 302 fragmental constants [27], allow the calculation of larger databases.

3.3. *KLOGP*

An 'extended group contribution' approach was developed by *Klopman et al.* [28]:

$$\log P = b_0 + \Sigma b_i \cdot N_i + \Sigma F_j \cdot N_j \tag{3}$$

N_i is the occurrence of the ith atom-centred fragment and N_j are the occurrences of particular fragments accounting for the interaction between groups, whose influence on final log P is described by calculated correction factors F_j. KLOGP was derived *via* regression analysis ($r^2 = 0.93$; $s = 0.38$) from a database containing 1663 diverse organic compounds. Basic atom-centred groups and correction factors were automatically identified by the artificial intelligence system CASE (computer automated structure evaluation). Ninety-eight contribution values and correction factors are given; corrections reflect tautomerization effects, zwitterion effects, proximity effects, and conjugated multi-heteroatomic effects. A revised, unpublished version of KLOGP is available in the software distributed by MULTICASE.

3.4. *KOWWIN*

This 'atom/fragment contribution method' was introduced by *Meylan* and *Howard* [29]. Like the Σf-system, it is a reductionistic approach derived by regression analysis; the model is defined as:

$$\log P = 0.229 + \Sigma f_k \cdot N_k + \Sigma F_j \cdot N_j \tag{4}$$

N_k is the occurrence of the kth fragment or atom type and N_j is the occurrence of the jth correction factor. The hydrophobic constants were evaluated by a first regression analysis of 1120 compounds without considering correction factors. The latter were then derived *via* linear regression of additional 1231 compounds correlating the differences between experimental log P and the log P estimated by the first regression model. The software is continuously upgraded; version 1.54 contained 144 atom/fragment values and 235 correction factors. The most-recent version of the LOGKOW program is KOWWIN v1.67. The estimation program can be downloaded for free from the corresponding websites; for details see *Table 2*.

3.5. *ACD/LogP*

A further example for pure fragmental methods is the ACD/LogP approach [30][31]. Fragmentation rules in ACD/LogP are based on the *Hansch–Leo* approach, but differ in several respects from the definition in CLOGP. H-Atoms, for instance, are never detached from ICs, eliminating the need for several structural correction factors.

The ACD/LogP algorithm uses 532 group contributions, f, and 2206 intramolecular correction factors, F_{ij}. A three-step procedure is underlying ACD/LogP calculations: *a*) structure fragmentation and assignment of f constants; missing fragments are estimated by atomic increments similar, *e.g.*, to *Ghose–Crippen*; *b*) assignment of implemented F_{ij} constants; missing interfragmental interactions are calculated by a polylinear expression similar to the *Hammett–Palm* equation; and *c*) summation of the implemented and estimated f and F_{ij} constants. ACD/LogP uses the following equation:

$$\log P = \Sigma f_j + (\Sigma Q_j) + \Sigma\text{aliph-}F_{ijk} + \Sigma\text{vinyl-}F_{ijk} + \Sigma\text{arom-}F_{ijk} \quad (5)$$

where f_i: fragmental increments, Q_j: increments of 'superfragments' (shown in parentheses due to very occasional use), F_{ijk}: increments of interactions between any two (ith and jth) groups separated by k-number of aliphatic, vinylic, or aromatic atoms.

The main weakness of ACD/LogP is its use of large numbers of increments for aromatic interactions. A remarkable advantage is that it correctly addresses tautomeric forms.

3.6. *AB/LogP*

Fragmentation rules in AB/LogP [32] are based on the *Hansch–Leo* approach as well. AB/LogP uses *Eqn. 5* without 'superfragments'. All interactions were generalized before optimizing the increments. Generalization was done by hierarchical clustering analysis based on similarity of H-bonding and electronic interactions. A similarity key was derived from analysis of 10000 *Abraham*'s beta (H-accepting) parameters. The latter are known to play a dominant role in determining log P values. The obtained similarity key is implemented as a standard function in AB/LogP.

Hierarchical clustering analysis produced clusters of fragments and interactions, which were assigned with generalized increments ($\{F_{ijk}\}_{\text{Clusters}}$) by optimizing the following equation:

$$f(X) = \Sigma f_j + \Sigma\{F_{ijk}\}_{\text{Clusters}} \quad (6)$$

In ACD/LogP, increments were first optimized and then generalized; in AB/LogP, they were first generalized and then optimized, which reduced the number of increments, avoided many single point determinations and manual errors in generalized increments and increased statistical significance and predictive power.

4. Atom Contribution Methods

Atom contribution methods (*Table 2*) cut molecules down to the single-atom level and commonly work without correction rules. MOLCAD, TSAR, and ALOGP98 are based on the *Ghose–Crippen* approach. XLOGP is the only atom-additive method applying corrections.

4.1. *The* Ghose–Crippen *Approach*

The group of *Crippen* [33–36] has described the development of a purely atom-based procedure, which exclusively applies atom contributions and avoids correction factors:

$$\log P = \Sigma a_k \cdot N_k \quad (7)$$

N_k is the occurrenced of the kth atom type. C-, H-, O-, N-, S-, and halogen-atoms are classified into 110 atom types; after several revisions, the number of atom classifications has increased to 120 [33] obtained from a training set of 893 structures ($r^2 = 0.86, s = 0.50$). H- and halogen-atoms are classified by the hybridization and oxidation state of the C-atom they are bonded to; C-atoms are classified by their hybridization state and the chemical nature of their neighboring atoms. The complexity of classification is attested by a total of 44 C-atom types alone. The original *Ghose–Crippen* approach underlies the MOLCAD and TSAR software.

The PrologP software is also based on the atomic fragment collection of *Ghose* and *Crippen*. In its current version 7.0, PrologP uses as prediction method a feedforward neural network model that is able to recognize hidden and nonlinear relationships between chemical structure and log *P*.

Model building within PrologP was based on nearly 13000 experimental log *P* values.

ALOGP98 [37][38] is a refinement of the original *Ghose–Crippen* approach aimed at considering earlier criticisms, in particular the chemical sense of atomic contributions. The new version comprises 68 atomic definitions obtained *via* SMARTS from *Daylight*. The chemical interpretation of the atomic definitions is improved by constraining several C-atom types to have positive contributions to log *P* in the fitting process. The training set was expanded to the 9000 structures in the POMONA database, a standard deviation of 0.67 is reported.

4.2. *XLOGP*

Wang et al. [39] published a further atom-additive method classifying atoms by their hybridization states and their neighboring atoms. Seventy-six basic atom types and 4 pseudoatom types for functional groups (CN, SCN, NO, and NO_2) gave a total of 80 descriptors in the atom classification.

In contrast to pure atom-based methods, correction rules are defined to account for intramolecular interactions: *1*) the number of 'hydrophobic C-atoms' (= sp^3 and sp^2 C-atoms without any attached heteroatom); *2*) an indicator variable of amino acids; *3*) presence of intramolecular H-bonds, and *4*) two corrections for 'poly-halogenation' (two or more halogen-atoms are attached to the same atom); the latter corrections differ depending on the presence or absence of F-atoms.

In the newest version, XLOGP 2.0, the number of atom types for C-, N-, O-, S-, P-, and halogen-atoms is increased to 90 [40]. Ten correction factors are derived to correctly handle hydrophobic C-atoms, internal H-bond, halogen-atom 1–3 pair, aromatic N-atom 1–4 pair, *ortho* sp^3 O-atom pair, *para* donor pair, sp^2 O-atom 1–5 pair, α-amino acid, salicylic acid, and *p*-amino sulfonic acid. The training set to derive XLOGP 2.0 comprised 1853 compounds. The correlation coefficient for fitting this set to experimental log *P* is 0.93 and the standard deviation 0.349.

The preferential use of atomic contributions classifies XLOGP as an atom-based approach, but the use of correction factors is characteristic of fragmental methods. The use of correction factors might explain the rather good performance of XLOGP *vs.* pure atom contribution methods.

5. Approaches Based on Molecular Lipohilicity Potentials (MLP) and Related Methods

Calculation approaches on the basis of MLPs and related methods are summarized in *Table 3*. CLIP, HINT, and MOLFESD are exemplary described in this section.

Table 3. *Whole Molecule Approaches*

Program	Descriptors and validation	[a])	Provider	Internet access
Approaches based on MLP and related methods				
CLIP	molecular lipophilicity potential	MLR	*P.-A. Carrupt*	www-ict.unil.ch/ict/clip/docs/clip.html
HINT	hydrophobic atom constant, central and frontier atoms	MLR	*eduSoft*	www.eslc.vabiotech.com/hint
VEGA	molecular lipophilicity potential, polar surface area, dipole moment, lipole		*A. Pedretti, G. Vistoli*	www.ddl.unimi.it
Approaches based on topological indices				
MLOGP	Σ lipophil. and Σ hydrophil. atoms, unsat. bonds, amphot. properties, proximity effects	MLR		
VLOGP in TOPKAT	electrotopological state (E values), size-corrected E values, topological shape descriptors	MLR	*Accelrys Network Science*	www.accelrys.com www.netsci.org
T-LOGP	uniform-length descriptors	PLS	*Upstream Solutions*	www.upstream.ch
AUTO-LOGP	4 autocorrelation vectors encoding hydrophobicity, molar refractivity, H-bond acceptor and donor ability	NN	*J. Devillers*	j.devillers@ctis.fr
CSLogP	topological descriptors including *Kier–Hall* descriptors	NN	*ChemSilico*	www.chemsilico.com
Approaches based on molecular properties				
BLOGP	μ, indicator for alkanes, charges on N and O, M_r, surface, ovality, number of C-atoms	MLR	*N. Bodor*	www.otl.ufl.edu/ufrf/
QLOGP	size, *N* (relat. to H-bonding), correction for alkanes	MLR	*N. Bodor*	www.otl.ufl.edu/ufrf/

[a]) Model validation was performed within the different approaches *via* multiple linear regression (MLR), partial least squares (PLS) analysis, or neural nets (NN).

5.1. *CLIP*

The group of *Testa* [41] developed an MLP approach based on the atom contribution method of *Broto et al.* [11] and a modification of the distance function from *Fauchère et al.* [16]. This MLP-based log *P* calculation procedure is available in the software package CLIP (computed lipophilicity properties). CLIP exhibits the following capabilities: computation and representation of the MLP on the solvent-accessible surface of molecules and macromolecules; calculation of log *P* from the MLP on the solvent-accessible surface; calculation of virtual log *P* values for individual conformers, and exploration of the lipophilicity range accessible to a compound; computation and representation of the MLP in a given region of space around molecules and macromolecules; computation and incorporation of the MLP into comparative molecular field analysis (CoMFA).

5.2. *HINT*

The program HINT [42–46] is another approach to reflect three-dimensionality by combining substructure contributions and conformational effects. The key parameter is the hydrophobic atom constant a_i, derived from *Leo*'s fragment constants. HINT calculates hydrophobic atom constants using the following criteria: *1*) the sum of atom constants within a fragment equals the fragment constant value, *2*) bond, branching, or vicinal-halogen factors are applied to all eligible atoms, while polar proximity factors are applied to the central atom of fragments, and *3*) superficial atoms are considered to be more important than central atoms.

6. Approaches Based on Topological Indices

A particularly broad class of calculation approaches is based on topological indices (*Table 3*) including MLOGP, VLOGP, T-LOGP, AUTO-LOGP, and CSLogP.

6.1. *MLOGP*

Multiple regression analysis of a set of 1230 organic compounds including general aliphatic, aromatic, and heterocyclic compounds together with complex drugs and agrochemicals was used by *Moriguchi et al.* [47] to derive their calculation method. The final regression equation involved 13 parameters, including summation of hydrophobic atoms, summation of hydrophilic atoms, proximity effects, unsaturated bonds, amphoteric properties and specific functionalities such as the presence of a quaternary N-atom or the number of NO_2 groups. An acceptable correlation coefficient of 0.952 was obtained; log *P* of the 1230 compounds was calculated with a standard deviation of ± 0.411. Examples for shortcomings are given, which, however, also occurred in comparative CLOGP calculations.

6.2. *VLOGP*

Gombar and *Enslein* [48] [49] introduced the VLOGP approach, which employs electrotopological state values (*E* values), size-corrected *E* values and topological shape descriptors. For 6675 compounds from the Pomona Starlist, a 363-variable model was developed. Explained variance amounted to 98.5%, and a standard error of estimate of 0.201 was calculated. A helpful feature of VLOGP is the definition of an optimum prediction space; it allows an *a priori* identification of compounds for which the model should not be applied. VLOGP is available within TOPKAT 3.0.

6.3. *T-LOGP*

Junghans and *Pretsch* [50] describe the estimation of log *P* from a reference database using local predictive models. Structures are represented by uniform length vectors generated from 3D-structures and substructures. A global model is built from all entries in the database using partial least squares, and, in addition, individual local models are derived for each structure cluster by complete linkage clustering. The authors use a dataset of 245 structures from [51]; 123 compounds for the training set and the remaining 122 for testing. The quality of prediction depends on the presence of a chemically similar compound in the training set, enabling the use of the appropriate local model; otherwise, a less accurate prediction is possible with the global model.

6.4. *AUTOLOGP*

Devillers et al. [52] use an autocorrelation method combining a topological and physicochemical description. Molecules in their database are described by means of four different autocorrelation vectors encoding hydrophobicity, molar refractivity, H-bonding acceptor ability and H-bonding donor ability. The model is developed from a remarkable training set of 7200 compounds with 35 descriptors; the composite neural network has 1185 weights; a standard deviation of 0.37 for the training set and 0.39 for the test set of 519 compounds indicates good statistical performance.

6.5. *CSLogP*

CSLogP was developed by *Joe Votano* and *Lowell Hall* and is built around artificial neural net analysis based on newly developed topological descriptors and the common *Kier–Hall* descriptors. A dataset of *ca.* 12 800 compounds was used for training and testing. In external validation with 2890 compounds – not used in model development – 70% were found to exhibit mean absolute errors < 0.5 log units and 90.1% within 1 log unit. The approach is not yet published.

7. Approaches Based on Molecular Properties

Increasing speed and accuracy of molecular orbital calculations allow the derivation of models based on quantum-chemical approaches. Accordingly, a variety of quantum-chemical descriptors such as charge densities, surface area, volume, and electrostatic potential are used to model log *P*. In contrast to the large number of published models, only a few software packages are available (*Table 3*).

7.1. *BLOGP*

Klopman and *Iroff* [53] presented a charge density method using MINDO/3 and *Hückel*-type calculations; for 61 simple organic compounds, the results were favorably compared with data obtained from fragment analysis. *Bodor et al.* [51][54] started from this method with an attempt to bypass some of its shortcomings. *Klopman*'s method was applicable only to compounds containing C-, H-, N-, and O-atoms, and only standard molecular geometry was used for MINDO/3 calculations. In addition, the calculated charge distribution alone seemed to be insufficient to characterize compound solubility. Thus, *Bodor et al.* [54] also examined the contribution of volume, surface, shape, and dipole moment. The dataset of *Klopman* was enlarged to 118 molecules for which fully optimized geometries were obtained from AM1 calculations. *Bodor* and *Huang* [51] published an extended version of the AM1 method, in which the number of C-atoms, the sum of the absolute values of atomic charges on each atom, and the fourth power of the ovality were introduced as additional parameters. For a dataset of 302 molecules, a regression equation with 17 parameters and $r=0.978$ was derived.

7.2. *QLOGP*

A 'Molecular-Size-Based Approach' was developed by *Bodor* and *Buchwald* [55]; they used an algorithm combining analytical and numerical techniques to compute *van der Waals* volume and surface area. Only one additional parameter was necessary to adequately describe log P in a diverse test set of 320 molecules. This parameter, labelled N, is a positive integer increased in an additive manner by each functional group within a test molecule. The authors hypothesize that N could be related to the H-bonds formed at the acceptor sites of the solute molecule when it is transferred from octanol to water. By adding a correction, which reflects the peculiar partitioning behavior of alkanes, *Bodor* and *Buchwald* derived with these three parameters a final model for the training set with $r=0.989$ and $s=0.214$. The predictive power of their model was tested using a validation set of 438 molecules comprising such diverse structures as H_2 and prednisolone. An r value of 0.975 and a standard deviation of 0.365 indicate the validity of the model.

8. Octanol/Water Distribution Coefficients (log *D*)

Up to date, most calculation methods have been focused on log P, considering the molecule in its neutral state. Only very recently, attempts have been made to estimate log D, taking into account the ionization of molecules.

8.1. *General Aspects*

The great majority of currently available drugs are ionized at physiological pH to varying degrees. Due to their positive or negative charges, ions exhibit a much more pronounced polarity as compared to their neutral counterparts. Consequently, the degree of ionization and protonation has a strong impact on the lipophilicity of ionizable compounds. To differentiate between the lipophilicity of neutral and ionizable compounds, partition coefficients (log P) and distribution coefficients (log D) are used as the corresponding descriptors.

The *partition coefficient* (log *P*) expresses the ratio of neutral solute concentrations in an organic and an aqueous phase of a two-component system under equilibrium conditions. Thus, log *P* describes the intrinsic lipophilicity of a compound in the absence of dissociation or ionization. In contrast, the overall ratio of a compound – including ionized and unionized fractions – between the two phases is defined as the *distribution coefficient* (log *D*). This term is used to describe the effective or net lipophilicity of a compound at a given pH taking into account both its intrinsic lipophilicity and its degree of ionization. Partition and distribution coefficients are interrelated for, *e.g.*, monoprotic organic acids *via Eqn. 8* and for, *e.g.*, monoprotic organic bases *via Eqn. 9*:

$$\log D = \log P - \log(1 + 10^{\mathrm{pH}-\mathrm{p}K_\mathrm{a}}) \quad (8)$$

$$\log D = \log P - \log(1 + 10^{\mathrm{p}K_\mathrm{a}-\mathrm{pH}}) \quad (9)$$

Methods for measuring distribution coefficients have been comprehensively reviewed [56][57]. The classical measurement of log *D* is *via* shake-flask experiments. Another source of high-quality measurement of log *D* is the pH-metric method, in which log *D* is calculated from the difference between the apparent $\mathrm{p}K_\mathrm{a}$, measured in a dual-phase system, the volume ratio of the phases, and the aqueous $\mathrm{p}K_\mathrm{a}$. Some chromatographical approaches allow higher-throughput measurement of log *D*.

8.2. *Log* D *Calculation Programs*

Currently, only a rather limited number of software programs for calculating log *D* are available. Some representative examples are collected in *Table 4* and in brief described below.

Table 4. *Calculation Software for Distribution Coefficients log* D

Program	Provider	Internet access
AB/LogD within ADME boxes	*Advanced Pharma Algorithms*	www.ap-algorithms.com
ACD/LogD	*Advanced Chemistry Development*	www.acdlabs.com
BioPrint/LogD	*CEREP*	www.cerep.com
CSLogD	*ChemSilico*	www.chemsilico.com
PrologD 2.0 within PALLAS	*Compudrug*	www.compudrug.com
SLIPPER	*MOLPRO PROJECT*	www.ibmh.msk.su/molpro/

Within ADME boxes from *Pharma Algorithms*, the AB/LogD tool is available that calculates log D values at physiologically relevant pH values and displays log D–pH plots. 2D-Structures or SMILES strings are used as input. Also the ACD/LogD software allows the calculation of octanol/water partition coefficients for partially dissociated compounds at any set pH from 0 to 14. BioPrint/LogD offers the possibility to calculate log D at pH 7.4 and at pH 6.5. A combination of linear or neural net equation with neighborhood behavior is used to predict properties. The data used for the construction of the model are coming from the BioPrint database. CSLogD is a module from *ChemSilico* allowing log D calculations. PrologD 3.1 is based on the pK_a and log P prediction of the neutral form of a solute and on the calculation of its micro- and macro-dissociation constants. PrologD 3.1 is a module within the PALLAS software from *CompuDrug*. The methodology of the program SLIPPER-2003 is based on a combination of similarity and physicochemical, in particular volume-related and H-bond descriptors [58] [59]. It also offers complete log D–pH plots.

9. Conclusions

The aim of this chapter was to describe currently available lipophilicity calculation software and to give a classification scheme on the basis of their methodological background.

Main classification refers to substructure and whole molecule approaches characterized by application of fragmentation rules in the former or by inspection of the intact molecule in the latter case. Considering precision of calculation as well as information content, substructure approaches seem to be more precise, whereas whole molecule approaches allow better theoretical interpretation.

Among substructure approaches, fragmental methods perform better than atom-additive methods which is presumably due to the advantageous application of correction rules. The additional use of correction factors presumably underlies the rather good performance of the atom-additive XLOGP approach *vs.* pure atom contribution methods.

More-recent approaches inspect the molecule as a whole; they use molecular lipophilicity potentials, topological indices, or molecular properties to quantify log P and some reflect the impact of 3D structure. Within the first subgroup, CLIP and HINT are most commonly used. CLIP exhibits the advantage of calculating virtual log P values for individual conformers.

Amongst whole molecule approaches, software developers seem to favor approaches based on topological indices. *Table 1* lists six commercially available programs based on these descriptors.

Increasing speed and accuracy of molecular orbital calculations allow the derivation of models based on molecular properties, a lot of which were used to model log *P*. In contrast to the large number of published models, the only available software package represents QLOGP. Despite the rather large number of available calculation programs, validity checks *vs.* experimental log *P* indicate a rather limited reliability [60–62]. Thus, the use of a diverse 'program set' instead of a single log *P* program is highly recommended. An adequate selection – covering the entire spectrum of methodologies – could comprise the following programs: CLOGP, AB/LogP, ALOGP98, XLOGP 2.0, CLIP, AUTOLOGP, and QLOGP.

Application of such program sets would facilitate to judge the quality of the obtained results. In this context it is interesting to note that *Tetko* [63] reviewed in detail the availability of WEB tools for calculating molecular descriptors. For lipophilicity descriptors, the website www.vcclab.org allows calculations of CLOGP, KOWWIN, IA_LogP, XLOGP, and ALOGPS 2.1. A special feature of the latter is the possibility to add user's molecules *via* a LIBRARY mode without a need to retrain the neural networks or to generate new molecular indices. This approach can significantly increase the predictivity of the method for the user's molecules [64]. Histograms of the obtained results enable a quick check of calculation reliability *via* dispersion of the calculated data. Further WEB tools for calculating lipophilicity are collected in *Table 5*.

Table 5. *WEB Tools for Calculating Lipophilicity*

Program	Internet access	Program	Internet access
ABSOLV	www.sirius-analytical.com	QMPRPlus	www.simulations-plus.com
SPARC	ibmlc2.chem.uga.edu/sparc	QikProp	www.schrodinger.com
PROPRED	www.capek.kt.dtu.dk	IA	www.logp.com

For partition processes in the body, distribution coefficients log *D* – for which an aqueous buffer of pH 7.4 (blood) or 6.5 (intestine) is applied in experimental determination – often provides a more-meaningful description of lipophilicity, considering the large amount of ionizable drugs. However, software packages that validly predict log *D* are rather scarce at present.

Taken together, the development of reliable lipophilicity calculation software remains a challenging task. Whole molecule approaches split log *P* into components and then calculate these components; *i.e.*, they put forward correct log *P* interpretation (meaningful components) and give minor priority to preciseness of calculation. Substructure approaches put forward preciseness on the basis of meaningless fragmentation and thereby give minor priority to interpretation. Thus, finding correct combinations of these two aspects – interpretation and accuracy – might represent a key to future developments of log *P* calculation approaches.

A crucial point is the availability of sufficiently large as well as diverse training sets of experimental log *P* data. Access to high-throughput technologies for log *P* and log *D* measurement might serve to fill this gap.

REFERENCES

[1] V. Pliska, B. Testa, H. van de Waterbeemd, 'Lipophilicity in Drug Action and Toxicology', VCH, Weinheim, 1996.
[2] P.-A. Carrupt, B. Testa, P. Gaillard, in 'Reviews in Computational Chemistry', Eds. K. B. Lipkowitz, D. B. Boyd, Wiley-VCH, Weinheim, 1997, Vol. 11, p. 241–315.
[3] C. A. Lipinski, F. Lombardo, B. W. Dominy, P. J. Feeney, *Adv. Drug Delivery Rev.* **1997**, *23*, 3.
[4] H. van de Waterbeemd, E. Gifford, *Nat. Drug Discov. Rev.* **2003**, *2*, 192.
[5] F. Lombardo, E. Gifford, M. Y. Shalaeva, *Mini-Rev. Med. Chem.* **2003**, *3*, 861.
[6] T. Fujita, J. Iwasa, J. C. Hansch, *J. Am. Chem. Soc.* **1964**, *86*, 5175.
[7] G. G. Nys, R. F. Rekker, *Chim. Ther.* **1973**, *8*, 521.
[8] G. G. Nys, R. F. Rekker, *Chim. Ther.* **1974**, *9*, 361.
[9] R. F. Rekker, 'The Hydrophobic Fragmental Constant. Its Derivation and Application', Elsevier, Amsterdam, 1977.
[10] R. F. Rekker, H. M. de Kort, *Eur. J. Med. Chem.* **1979**, *14*, 479.
[11] P. Broto, G. Moreau, C. Vandycke, *Eur. J. Med. Chem.* **1984**, *19*, 71.
[12] T. Masuda, T. Jikihara, K. Nakamura, A. Kimura, T. Takagi, H. Fujiwara, *J. Pharm. Sci.* **1997**, *86*, 57.
[13] N. G. J. Richards, P. B. Williams, M. S. Tute, *Int. J. Quantum Chem., Quantum Biol. Symp.* **1991**, *40*(S18), 299.
[14] N. G. J. Richards, P. B. Williams, M. S. Tute, *Int. J. Quantum Chem.* **1992**, *44*, 219.
[15] E. Audry, J.-P. Dubost, J.-C. Colleter, P. Dallet, *Eur. J. Med. Chem.* **1986**, *21*, 71.
[16] J.-L. Fauchère, P. Quarendon, L. Kaetterer, *J. Mol. Graphics* **1988**, *6*, 203.
[17] P. Furet, A. Sele, N. C. Cohen, *J. Mol. Graphics* **1988**, *6*, 182.
[18] A. J. Leo, P. Y. C. Jow, C. Silipo, C. Hansch, *J. Med. Chem.* **1975**, *18*, 865.
[19] C. Hansch, A. J. Leo, 'Substituent Constants for Correlation Analysis in Chemistry and Biology', Wiley, New York, 1979.
[20] A. J. Leo, *J. Pharm. Sci.* **1987**, *76*, 166.
[21] A. J. Leo, *Methods Enzymol.* **1991**, *202*, 544.
[22] A. J. Leo, *Chem. Rev.* **1993**, *93*, 1281.
[23] J. T. Chou, P. C. Jurs, *J. Chem. Inf. Comput. Sci.* **1979**, *19*, 172.
[24] A. J. Leo, D. Hoekman, *Perspect. Drug Discov. Des.* **2000**, *18*, 19.
[25] R. F. Rekker, R. Mannhold, 'Calculation of Drug Lipophilicity', VCH, Weinheim, 1992.
[26] R. Mannhold, R. F. Rekker, K. Dross, G. Bijloo, G. de Vries, *Quant. Struct.-Act. Relat.* **1998**, *17*, 517.
[27] D. E. Petelin, N. A. Arslanov, V. A. Palyulin, N. S. Zefirov, in 'Proceedings of the 10th European Symposium on Structure-Activity Relationships', Prous Science Publishers, Barcelona, 1995, abstract B263.
[28] G. Klopman, J. W. Li, S. Wang, M. Dimayuga, *J. Chem. Inf. Comput. Sci.* **1994**, *34*, 752.

[29] W. M. Meylan, P. H. Howard, *J. Pharm. Sci.* **1995**, *84*, 83.
[30] A. A. Petrauskas, E. A. Kolovanov, *Perspect. Drug Discov. Des.* **2000**, *19*, 99.
[31] A. A. Petrauskas, E. A. Kolovanov, *13th European Symposium on Quantitative Structure–Activity Relationships*, Abstract Book P.4, Düsseldorf, 2000.
[32] P. Japertas, R. Didziapetris, A. A. Petrauskas, *Quant. Struct.-Act. Relat.* **2002**, *21*, 23.
[33] A. K. Ghose, G. M. Crippen, *J. Comput. Chem.* **1986**, *7*, 565.
[34] A. K. Ghose, G. M. Crippen, *J. Chem. Inf. Comput. Sci.* **1987**, *27*, 21.
[35] A. K. Ghose, A. Pritchett, G. M. Crippen, *J. Comput. Chem.* **1988**, *9*, 80.
[36] V. N. Viswanadhan, A. K. Ghose, G. R. Revankar, R. K. Robins, *J. Chem. Inf. Comput. Sci.* **1989**, *29*, 163.
[37] A. K. Ghose, V. N. Viswanadhan, J. J. Wendoloski, *J. Phys. Chem. A* **1998**, *102*, 3762.
[38] S. A. Wildman, G. M. Crippen, *J. Chem. Inf. Comput. Sci.* **1999**, *39*, 868.
[39] R. Wang, Y. Fu, L. Lai, *J. Chem. Inf. Comput. Sci.* **1999**, *39*, 868.
[40] R. Wang, Y. Gao, L. Lai, *Perspect. Drug Discov. Des.* **2000**, *19*, 47.
[41] P. Gaillard, P.-A. Carrupt, B. Testa, A. Boudon, *J. Comput.-Aided Mol. Des.* **1994**, *8*, 83.
[42] G. E. Kellogg, S. F. Semus, D. J. Abraham, *J. Comput.-Aided Mol. Des.* **1991**, *5*, 545.
[43] G. E. Kellogg, G. J. Joshi, D. J. Abraham, *Med. Chem. Res.* **1992**, *1*, 444.
[44] D. J. Abraham, G. E. Kellogg, *J. Comput.-Aided Mol. Des.* **1994**, *8*, 41.
[45] G. E. Kellogg, D. J. Abraham, *Analusis* **1999**, *2*, 19.
[46] G. E. Kellogg, D. J. Abraham, *Eur. J. Med. Chem.* **2000**, *35*, 651.
[47] I. Moriguchi, S. Hirono, Q. Liu, I. Nakagome, Y. Matsushita, *Chem. Pharm. Bull.* **1992**, *40*, 127.
[48] V. K. Gombar, K. Enslein, *J. Chem. Inf. Comput. Sci.* **1996**, *36*, 1127.
[49] V. K. Gombar, *SAR QSAR Environ. Res.* **1999**, *10*, 371.
[50] M. Junghans, E. Pretsch, *Fresenius J. Anal. Chem.* **1997**, *359*, 88.
[51] N. Bodor, M. J. Huang, *J. Pharm. Sci.* **1992**, *81*, 272.
[52] J. Devillers, D. Domine, C. Guillon, W. Karcher, *J. Pharm. Sci.* **1998**, *87*, 1086.
[53] G. Klopman, L. D. Iroff, *J. Comput. Chem.* **1981**, *2*, 157.
[54] N. Bodor, Z. Gabanyi, C. K. Wong, *J. Am. Chem. Soc.* **1989**, *111*, 3783.
[55] N. Bodor, P. Buchwald, *J. Phys. Chem. B* **1997**, *101*, 3404.
[56] A. Avdeef, in 'Lipophilicity in Drug Action and Toxicology', Eds. V. Pliska, B. Testa, H. van de Waterbeemd, VCH, Weinheim, 1996, p. 109–139.
[57] J. Comer, in 'Drug Bioavailability', Eds. H. van de Waterbeemd, H. Lennernäs, P. Artursson, Wiley-VCH, Weinheim, 2003, p. 21–45.
[58] O. A. Raevsky, *SAR QSAR Environ. Res.* **2001**, *12*, 367.
[59] O. A. Raevsky, S. V. Trepalin, H. P. Trepalina, V. A. Gerasimenko, O. E. Raevskaja, *J. Chem. Inf. Comput. Sci.* **2002**, *42*, 540.
[60] H. van de Waterbeemd, R. Mannhold, in 'Lipophilicity in Drug Action and Toxicology', Eds. V. Pliska, B. Testa, H. van de Waterbeemd, VCH, Weinheim, 1996, p. 401–418.
[61] R. Mannhold, H. van de Waterbeemd, *J. Comput.-Aided Mol. Des.* **2001**, *15*, 337.
[62] P. Buchwald, N. Bodor, *Curr. Med. Chem.* **1998**, *5*, 353.
[63] I. V. Tetko, *Mini-Rev. Med. Chem.* **2003**, *3*, 809.
[64] I. V. Tetko, V. Y. Tanchuk, *LogP2004 Symposium*, ETH Zürich, Abstract Book C-11, 2004.

The Concept of Property Space: The Case of Acetylcholine

by **Giulio Vistoli*[a]**, **Alessandro Pedretti[a]**, **Luigi Villa[a]**, and **Bernard Testa[b]**

[a]) Istituto di Chimica Farmaceutica, Facoltà di Farmacia, Università di Milano, Viale Abruzzi 42, I-20131, Milano (e-mail: giulio.vistoli@unimi.it)
[b]) Pharmacy Deptartment, University Hospital Centre (CHUV), Rue du Bugnon, CH-1011 Lausanne

Abbreviations
MD: Molecular dynamics; POPC: 1-palmitoyl-2-oleoyl-*sn*-glycero-3-phosphocholine; PSA: polar surface area; QSPRs: quantitative structure–property relationships; SAS: solvent-accessible surface.

1. Introduction

Helmholtz repeatedly stated that there exists no intrinsic property, but that a property is always the ability of a subject to exert some effects on an object, even if the latter cannot be defined or remains implicit in the reasoning [1]. Such an implicit object is well illustrated by the environment or medium surrounding the subject. This model underlines the interactive nature of a property and implies a dynamic profile, since a property requires a process by which it can emerge and a discrete duration for its manifestation [2].

Such a dynamic vision of properties, which finds an obvious embodiment in chemistry, thus requires time as the fourth dimension for its analysis. In other words, a molecule cannot be considered as a static object, but is an animated subject, the structural changes of which significantly affect any property profile [3]. This concept is also reflected on the social nature of living organisms, which are currently seen as dynamic complex systems that develop and evolve in relation to other organisms and complex environments [4][5].

The growing computational power available to researchers proves an invaluable tool to investigate the time profile of molecules and more complex systems. Molecular dynamics (MD) simulations have, thus, become a pivotal technique to explore the dynamic dimension of physicochemical properties. For example, molecular lipophilicity represents a noteworthy application for this dynamic analysis. Indeed, the computational methods to predict lipophilicity allow the elaboration of a 3D-based log *P* value for each conformation (virtual log *P*) [6], assuming that the experimentally measured log *P* value will be a weighted average of the virtual log *P* values of all existing conformers. The recent studies of *Kraszni et al.* [7], who used ^{1}H-NMR vicinal coupling constants to determine conformer specific log *P* values, show that the calculation of virtual log *P* values is not an unrealistic operation, but that experimental evidence now exists to validate these computational results.

The dynamic nature of a property can be monitored during MD simulations by observing its fluctuations and their sensitivity to the molecular environment (*e.g.*, the solvent). It thus becomes possible to analyze in detail how the accessible geometric properties, for example torsion angles or interatomic distances, affect physicochemical properties and how the latter are interrelated.

The implications of such a dynamic and interactive vision is that it is no longer possible to describe a property by a single value, but that a set of values (basically one per conformer) must be considered. This set of values defines a property space, the range and distribution of which will depend on other molecular properties and on the environment.

Up to now, quantitative structure–property relationships (QSPRs) have always considered sets of different compounds, correlating their structural (*i.e.*, geometric) properties with physicochemical properties. By means of the dynamic approach outlined above, we can consider each recognizable conformer of a flexible molecule as a discrete entity and correlate its geometric properties with the corresponding physicochemical properties calculated by a 3D method (*e.g.*, virtual log *P*, dipole moment, polar surface area, grid energies, *etc.*).

These correlations highlight if and how two molecular properties change coherently. Indeed, there will be molecules, the property profile of which varies markedly with small geometric changes, and others showing properties constant even upon significant geometric fluctuations. It is evident that such a molecular sensibility can affect biological activity, as the latter is a dynamic property in itself, the emergence of which will depend on the ability of the molecule to fit into the target cavity.

The present study takes acetylcholine as the object of study and analyzes correlations between its geometric properties and its lipophilicity profile, other 3D physicochemical properties being also considered. Acetylcholine was chosen for its interesting structure, major biological role, and the abundant data available on its conformational behavior [8–16]. The present paper lays the ground for such a dynamic approach, focusing the reader's attention on the behavior of both geometric and physicochemical properties and on the nature and meaning of their interrelations. Finally, some insights on solvent effects are offered, confirming the role of the molecular environment on the lipophilicity space of acetylcholine.

2. The Property Profiles of Acetylcholine

As illustrated in *Fig. 1*, acetylcholine has four dihedral angles. Preliminary calculations and literature data indicate that τ_1 and τ_4 vary in a narrow range ($\tau_1 = 60 \pm 20$, $\tau_4 = 0 \pm 20°$) due to the symmetry of a triple rotor (τ_1) and the rigidity of the ester group (τ_4). We, therefore, focus here on the dihedral angles τ_2 and τ_3, which describe most of the conformational behavior of the molecule.

Fig. 2 plots the τ_2 *vs.* τ_3 values obtained from a 30-ns MD simulation *in vacuo*. Seven low-energy conformational classes are apparent and organized in a symmetrical distribution around the central extended conformation (τ_2 = **t**; τ_3 = **t**; **tt**). It can be seen that both torsion angles can assume all three low-energy geometries (**g+**, **t**, **g−**) and that acetylcholine can assume all possible combinations of τ_2 and τ_3 except **g−g+** and **g+g−**. Furthermore, the conformers can be clustered into three enantiomeric pairs of conformers, which show quite similar geometric properties (**g+g+** ≅ **g−g−**, **g+t** ≅ **g−t**, and **tg+** ≅ **tg−**), plus the fully relaxed **tt** conformer, which represents the symmetry center. *In vacuo* intermolecular attraction between the ammonium head and the ester group plays a major role, making the folded conformers the most abundant ones.

To gain a better insight into the conformational behavior of acetylcholine, we monitored the intermolecular distance between the N-atom and the methyl C(8) atom, since preliminary tests showed it to be the best indicator of folding. It is interesting to observe that this geometric descriptor has a discontinuous behavior, as depicted by a distribution plot.

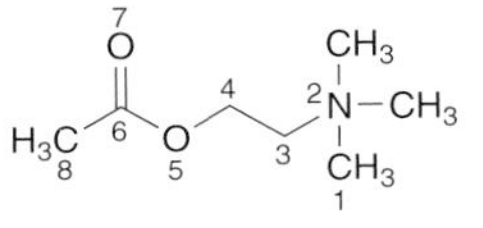

τ_1 = C(1)–N(2)–C(3)–C(4)
τ_2 = N(2)–C(3)–C(4)–O(5)
τ_3 = C(3)–C(4)–O(5)–C(6)
τ_4 = C(4)–O(5)–C(6)–O(7)
distance = N(2)···C(8)

Fig. 1. *Dihedral angles in acetylcholine.* Their values are defined according to *Klyne* and *Prelog* [17].

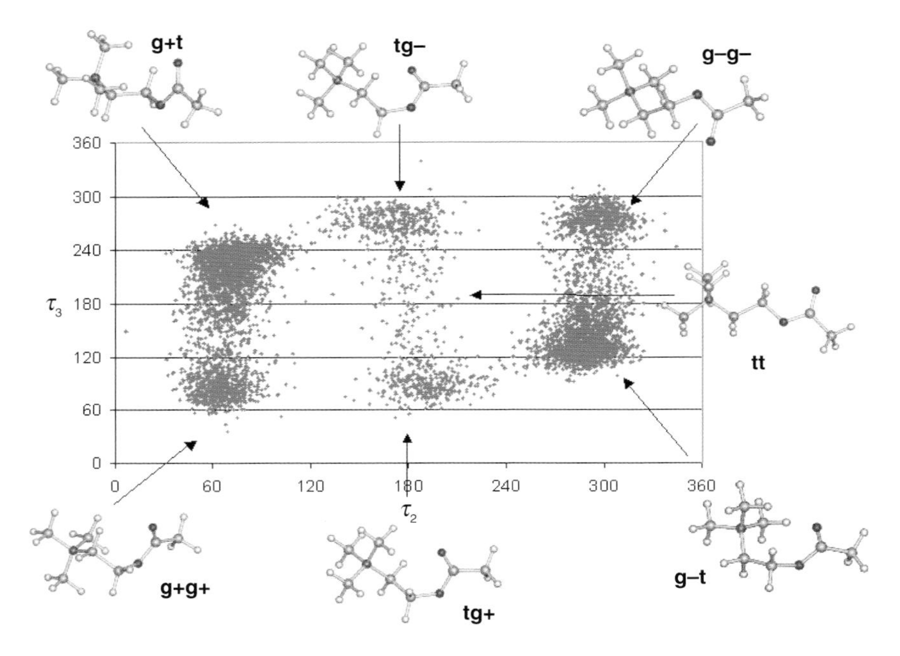

Fig. 2. *The conformational behavior of acetylcholine* (τ_2 *vs.* τ_3 plot) in vacuo (30-ns MD simulation at 300 K, conformations stored each 5 ps) *showing the seven low-energy conformers*

Fig. 3,a clearly shows that the values are clustered in a bimodal distribution, the first peak being centered at 5.1 Å and the second at 5.9 Å. The average distances calculated per conformation (data not shown) indicate that the first peak is composed of conformers with τ_2=gauche, while the second peak includes conformers with τ_3=antiperiplanar, highlighting thus the major influence of τ_2 on the geometry of acetylcholine. This multiple distribution indicates that the conformers obtained can be neatly divided into folded and extended ones with little if any intermediate conformations. This behavior reflects the discontinuous profile of the τ_2 angle, which can never assume anticlinal geometries.

The physicochemical properties monitored (*i.e.*, lipophilicity [18], dipole moment, polar surface area (PSA) [19], and solvent-accessible surface (SAS)) do not appear to feel the effects of the geometric discontinuity, since their values are distributed in a single Gaussian curve (*Fig. 3,b* for the distribution of log *P*), the average values of which will be discussed later. When the averages and the span of ranges of physicochemical properties are examined for each conformational class (*Table 1*), it appears that the above-mentioned enantiomeric pairs of

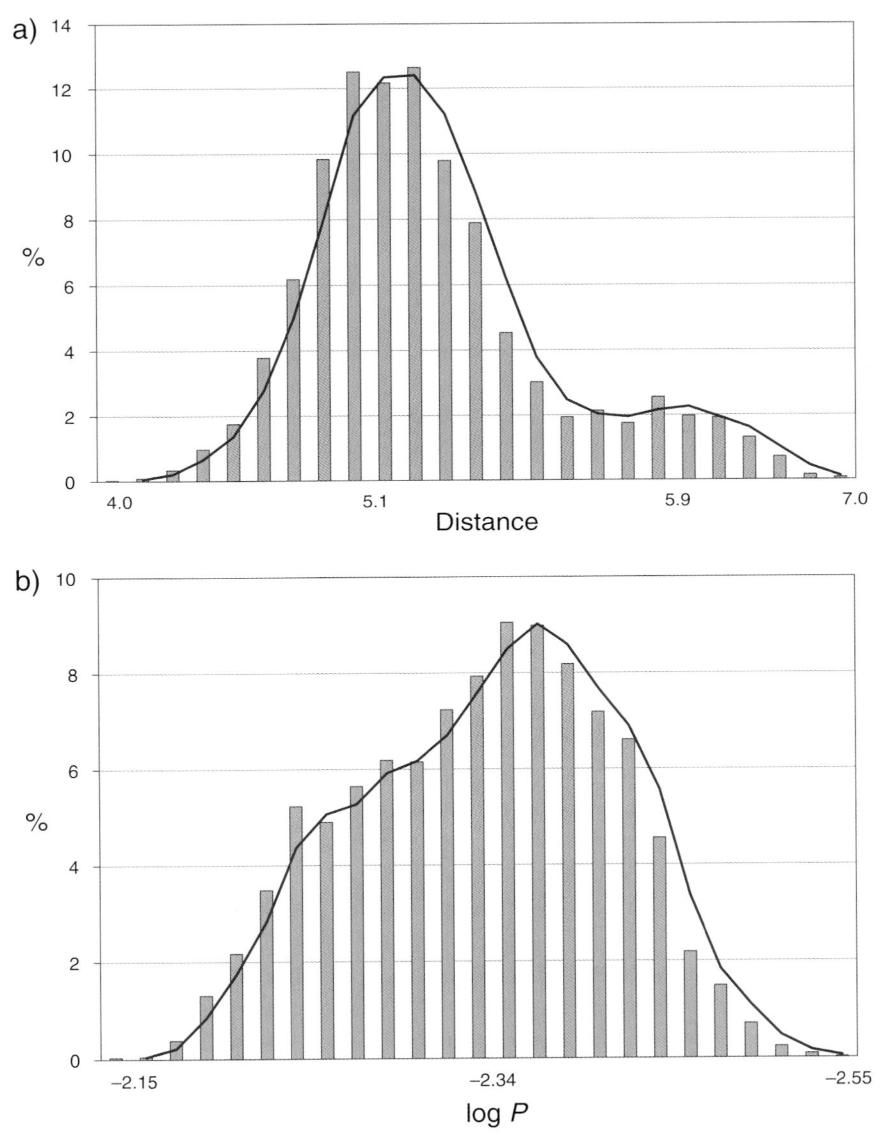

Fig. 3. *Distribution plots of some properties monitored* in vacuo. *a*) Distribution of distances. *b*) Distribution of virtual log *P* values.

conformers show quite similar values for all properties examined, while the **tt** conformers represent a unique case with significantly different average values. The ranges seem to be proportional to the relative abundance of each conformer, and, hence, the **tt** conformers again show a much

Table 1. *Average Values of Physicochemical Properties in Each Conformational Class, Calculated from MD* in vacuo

Conformational class	log P		Dipole moment		PSA[a])		SAS[b])	
	mean	span of range	mean	span of range	mean	span of range	mean	span of range
g+g+	−2.36	0.28	7.861	4.13	34.83	13.96	357.0	26.59
g−g−	−2.36	0.30	7.81	4.24	34.68	12.22	356.9	26.03
g+t	−2.34	0.32	7.727	3.91	34.63	16.90	356.9	25.45
g−t	−2.33	0.37	7.706	4.30	34.61	17.79	356.8	28.82
tg+	−2.33	0.17	7.624	2.26	35.99	12.14	362.8	17.16
tg−	−2.33	0.18	7.629	2.25	35.21	11.11	363.3	14.36
tt	−2.47	0.09	9.277	1.48	40.66	5.40	372.9	9.81

[a]) Polar surface area. [b]) Solvent-accessible surface.

smaller range than the other structures. The property profiles of the **tt** conformer demonstrate that this fully relaxed structure is an upper limit for acetylcholine and that its narrow property ranges can be explained by its poor flexibility.

3. Correlations between Properties Monitored *in vacuo*

As described in the *Introduction*, the dynamic nature of a property can be seen as its ability to span a possible range and simultaneously influence the behavior of other properties. This implies that the profile of a property may be fully understood only by considering its variations in relation to other properties. From a mathematical viewpoint, this analysis can be carried out by considering the regression coefficients obtained by correlating pairs of properties. A good coefficient implies that the two properties change coherently, while poor coefficients reveal a lack of interdependence.

We first examine the relations between geometric and physicochemical properties and focus our attention on lipophilicity *vs.* torsion angles or intermolecular distance. Thus, it is interesting to observe that τ_2, the major determinant of the conformational profile of acetylcholine, does not influence log P (*Fig. 4,a*). Indeed, the same log P range is covered irrespective whether τ_2 is **g** or **t**. In contrast, log P shows an intriguing relation with τ_3, seemingly a multiple parabolic distribution, although its average is similar for each conformational class except the **tt** conformers (see *Table 1*). This distribution can be split into three parabolas, corresponding to the conformers having τ_2=**g+** (*Fig. 4,b*, r^2=0.79), τ_2=**g−** (*Fig. 4,c*, r^2=0.79), and τ_2=**t** (*Fig. 4,d*, r^2=0.63). This behavior may mean that the ammonium head, which has a strong influence on the hydrophilicity of acetylcholine, is always accessible irrespective of τ_2. In the **tt** conformers however, the fully exposed ammonium group appears to make a particularly strong contribution to the molecule's hydrophilicity, as seen in the average values per conformational class (*Table 1*, log *P* averages ranging from −2.33 for **g+g+** to −2.47 for **tt**). In contrast, the accessibility of the ester group appears to depend mainly on τ_3 and, even if its influence on lipophilicity is minor, it varies coherently with τ_3. Thus, the ester group is associated with higher log *P* values when τ_3=**g** (*i.e.*, when it is partly masked by the ammonium group) and lower log *P* values when τ_3=**t** (*i.e.*, when it is fully exposed).

The discontinuous character of the distance parameter is reflected in all its correlations with other properties. For example, no linear correlation exists (r^2=0.13) in the log *P vs.* distance relation. A closer examination of this plot (*Fig. 5*) reveals three clusters, namely *a*) the cluster of the **gg** and **gt** conformers (log *P* ranging from −2.51 to −2.20, no detectable subclustering), *b*) the cluster of the **tg** conformers (log *P* ranging from −2.46 to −2.26), and *c*) the cluster of the **tt** conformers (log *P* ranging from −2.54 to −2.41). The same conclusion emerges when examining plots of dipole moment *vs.* distance (not shown).

Due to the mutual position of the two polar groups, the dipole moment also shows a composite parabolic relation with τ_3, with *trans* geometries showing maximal dipole moment values.

Other correlations can be searched between two properties of the same type. With three exceptions, no pair of parameters in *Table 2* shows a significant linear correlation (*i.e.*, r^2>0.5). The first exception is the correlation between two geometric parameters, namely distance and SAS, with an r^2 value equal to 0.72 (*in vacuo*). This correlation is expected and understandable, since folded conformers have a decreased SAS compared to the extended forms. The second correlation is between the PSA and dipole moment (r^2=0.58) and depends on a reasonable relationship between molecular polarity and accessibility of its polar groups.

The third, and most-noteworthy correlation is between dipole moment and log *P*, the r^2 value of which is equal to 0.77 *in vacuo*. Clearly, a higher dipole moment implies a lower lipophilicity, and the fact that the two parameters correlate despite their different nature can be seen as a mutual validation of the respective algorithms used to calculate them.

These correlations underline the fact that the information content of some properties overlaps to some extent (*e.g.*, log *P*, dipole moment, and polar area). The fact that these correlations are only partial indicates that

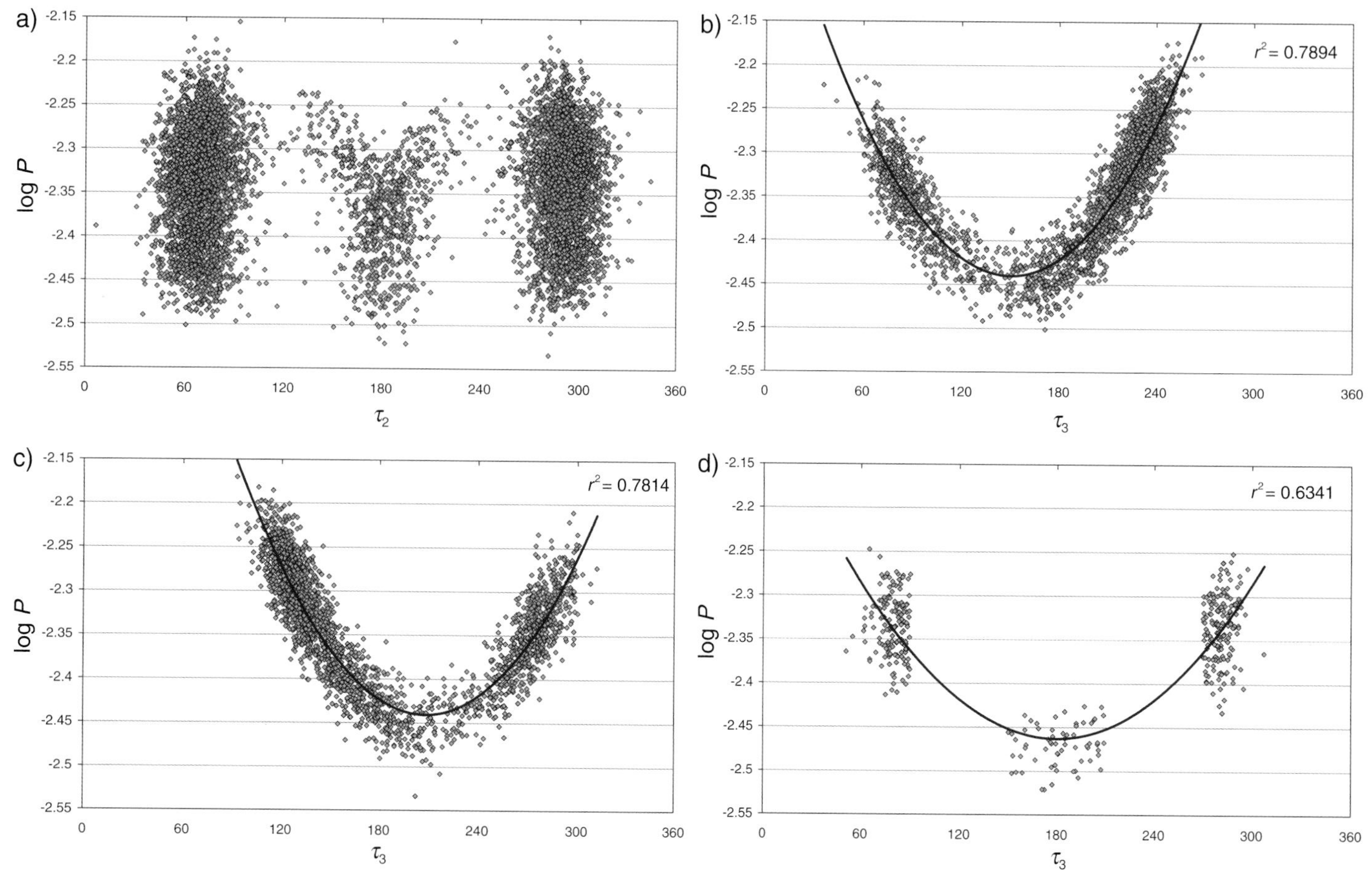

Fig. 4. *Relations of log* P *with τ_2 and τ_3 for the structures obtained during the MD simulation* in vacuo. *a*) Relation of log *P* with τ_2 for all conformers. *b*) Correlation of log *P vs.* τ_3 for the conformers having τ_2 = **g+**. *c*) Correlation of log *P vs.* τ_3 for the conformers having τ_2 = **g−**. *d*) Correlation of log *P vs.* τ_3 for the conformers having τ_2 = **t**.

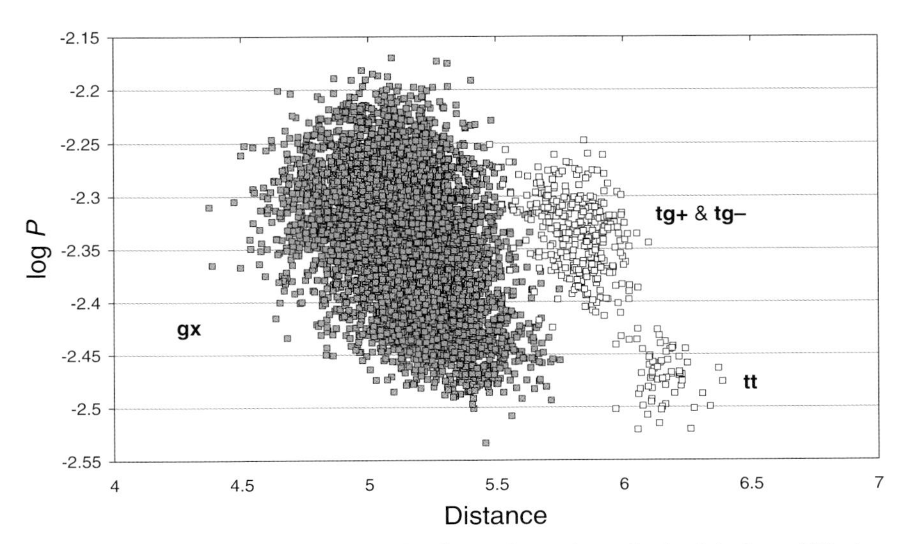

Fig. 5. *Correlation of log* P vs. *distance for the conformations obtained during a MD simulation* in vacuo (the conformers with $\tau_2 = \mathbf{g}\pm$ are indicated in gray, those with $\tau_2 = \mathbf{t}$ are white)

Table 2. *Correlation Matrix* (expressed as r^2) *between Some Average Properties of the Conformers of Acetylcholine*

Properties	log *P*	Dipole moment	PSA	SAS	Distance
log *P*	–				
Dipole moment	0.77	–			
PSA	0.48	0.58	–		
SAS	0.39	0.40	0.35	–	
Distance	0.13	0.08	0.14	0.72	–

acetylcholine despite its apparent simplicity behaves in complex ways. This might be due to its overall structure, with two polar ends separated by a hydrophobic linker.

4. Influence of Solvent on Property Space

Similar MD simulations were performed with acetylcholine in four different solvent clusters. The solvents were water and $CHCl_3$ to simulate isotropic systems of different polarity [20], hydrated octanol, and a membrane model (hydrated 1-palmitoyl-2-oleoyl-*sn*-glycero-3-phosphocholine, POPC) to simulate environments of different complexity and capability to interact with the solute [21].

Table 3. *Solvent Effects on the Conformational Profiles for Acetylcholine as Monitored from the Five Performed MD Simulations*

Conformational class	Populations [%] of conformational classes in various media				
	Vacuum	Water	$CHCl_3$	Octanol	Membrane
g+g+	12.1	12.4	9.3	9.0	15.7
g−g−	12.1	13.4	12.8	10.2	9.4
g+t	29.9	15.4	15.1	10.1	50.3
g−t	27.0	16.1	17.8	10.0	21.4
tg+	2.5	3.6	8.4	6.7	0
tg−	2.5	3.8	7.7	7.4	0
tt	1.1	9.3	4.1	8.8	0
Transitional conformers	12.8	26.0	24.8	37.8	3.2

The solvent effects on the conformational profile of acetylcholine (*Table 3*) show that the solvents act as filters, selecting conformers with the properties most similar to theirs. The solvent effects can be summarized as follows:

a) *Polarity*: The solute assumes conformers best able to mimic the properties of the solvent. This can be seen with the significant increase of fully extended (**tt**) conformers in water compared to the vacuum (the percentage varies from 1.1% *in vacuo* to 9.3% in water). This effect is twofold, namely intrinsic polarity and a solute–solvent interaction, which selects the conformers able to maximize interactions with the solvent. For example, the **tt** conformers are simultaneously the most-polar ones and the best able to form H-bonds with H_2O molecules.

b) *Friction*: The solvent slows down all molecular movements irrespective of its physicochemical properties, an effect mainly proportional to solvent molecular weight, as seen with the percentage of anticlinal structures that dramatically increase from the vacuum (12.8%) to octanol (37.8%). Indeed, the anticlinal forms can be considered transitional conformations, the percentage of which increases in relation to the slowing down of the molecular movements. This slowing down is maximal in the membrane, where acetylcholine remains blocked in folded conformations (as seen also when starting with an extended conformer, results not shown).

c) *Shape*: When the size of the solvent molecules becomes comparable to that of the solute, the latter minimizes steric hindrance by mimicking the shape of the solvent. This effect is seen in octanol, where acetylcholine shows a significant percentage of antiperiplanar geometries, which mimic the preferred zig-zag conformation of octanol.

Finally, we observe that the solvent can select solute conformers by two mechanisms. When solvent–solute and solute–solute interactions are of comparable strength, the solvent will favor some conformers over others. But when solvent–solute interactions are stronger than solute–solute interactions, the solvent will not only favor some conformers, but also exclude others. This behavior is evident in the membrane, where acetylcholine exists in folded geometries only, which allow it to form strong electrostatic interactions with phospholipid polar heads.

Table 4 reports the average values of properties in the different solvents, showing that the values of log *P* and dipole moment are coherent with solvent properties. Indeed, all solvents, $CHCl_3$ included, interact with acetylcholine and compete with intramolecular interactions, thus increasing its polarity (log *P* −2.36 to −2.41). In the vacuum, intramolecular interactions are not hampered and result in a slightly higher average log *P* value.

The shape effect becomes evident for octanol when one considers the average distance and PSA. Indeed, both properties reach their highest values in octanol, in agreement with the high percentage of extended conformers. It is intriguing to observe that acetylcholine can modulate its properties in such a manner that it can assume rather extended geometries while retaining a intermediate polarity (log *P* −2.39).

In contrast, in a membrane model, acetylcholine shows the lowest average values for distance and SAS due to the exclusive presence of folded conformers, which maximize electrostatic interactions with phospholipids. Nevertheless, acetylcholine in the membrane possesses a log *P* and dipole moment quite similar to those in octanol. This suggests that a molecule can retain the same average log *P* and dipole moment while exhibiting different conformational profiles.

The spans of ranges (shown in brackets in *Table 4*) decrease with increasing structural complexity of the solvent, an effect particularly

Table 4. *Average Values* (and span of ranges) *for the Physicochemical Properties Monitored During the MD Simulations*

Property	Medium				
	Vacuum	Water	$CHCl_3$	Octanol	Membrane
log *P*	−2.33 (0.37)	−2.41 (0.35)	−2.36 (0.33)	−2.39 (0.31)	−2.39 (0.28)
Distance	5.25 (2.01)	5.43 (1.99)	5.40 (2.01)	5.57 (2.01)	5.01 (1.89)
PSA	35.0 (19.8)	37.8 (20.4)	40.1 (21.9)	42.8 (21.9)	40.9 (19.2)
SAS	358 (34.3)	361 (36.4)	356 (39.1)	359 (39.1)	353 (34.3)
Dipole moment	7.78 (4.61)	8.88 (3.91)	8.40 (4.10)	8.75 (3.90)	8.67 (3.83)

Table 5. *Average log* P *Values from the MD Simulations*

Medium	Conformational class						
	g+g+	**g−g−**	**g+t**	**g−t**	**tg+**	**tg−**	**tt**
Vacuum	−2.357	−2.356	−2.336	−2.334	−2.334	−2.334	−2.471
Water	−2.396	−2.398	−2.392	−2.388	−2.365	−2.363	−2.481
$CHCl_3$	−2.363	−2.366	−2.349	−2.352	−2.344	−2.347	−2.467
Octanol	−2.375	−2.361	−2.389	−2.385	−2.362	−2.354	−2.447
Membrane	−2.377	−2.376	−2.381	−2.39	−	−	−

evident for log *P.* This trend suggests that the greater the structural complexity of the solvent, the more constrained the property spaces of the solute.

Table 5 shows the average log *P* values calculated in different solvents. It is evident that they are all lower in water than *in vacuo*, suggesting that the higher hydrophilicity of acetylcholine in polar solvents is due to both a higher percentage of extended conformers and a lower lipophilicity in all conformational classes. Thus, acetylcholine can adjust its lipophilicity behavior and assume the most-suitable conformers within each conformational class. This behavior becomes evident when considering the average log *P* value of the most-hydrophilic conformers, namely the **tt** conformers, which is somewhat higher in octanol than in other solvents. It, thus, appears that acetylcholine can assume extended conformers to mimic the shape of octanol while avoiding its highest polarity.

5. Conclusions

This chapter presents a brief overview of the property space, focusing on the lipophilicity space of acetylcholine and on its relations with other physicochemical and geometric properties. Solvent effects on property space are also analyzed, considering four solvent systems of different polarity and structural complexity. This study examines molecular flexibility in terms of both geometric features (*e.g.*, interatomic distances and torsion angles) and physicochemical properties of great pharmacological and biological relevance, given that some of these properties (*i.e.*, hydrophobicity, H-bond donor and acceptor capacity, and lipophilicity) are major recognition forces.

As such, the study contributes to a better understanding of molecular structure and it invites efforts to transform the concept of property space from a speculative to a practical one, deriving statistical descriptors to pursue dynamic-dependent QSAR analyses.

REFERENCES

[1] H. Helmholtz, *Gesammelte Abhandlungen* **1887**, *3*, 356.
[2] J. Kim, 'Events as Property Exemplification in Action Theory', Reidel, Dordrecht, 1976, p. 196–215.
[3] B. Testa, L. B. Kier, A. J. Bojarski, *SEED* **2002**, *2*, 84; www.library.utoronto.ca/see/pages/SEED_Journal.html.
[4] B. Testa, L. B. Kier, *Entropy* **2000**, *2*, 1.
[5] G. Van de Vijver, L. Van Speybroeck, W. Vandevyvere, *Acta Biotheor.* **2003**, *51*, 101.
[6] P. A. Carrupt, B. Testa, P. Gaillard, *Rev. Comput. Chem.* **1997**, *11*, 241.
[7] M. Kraszni, I. Bányai, B. Noszál, *J. Med. Chem.* **2003**, *46*, 2241.
[8] C. Chothia, P. J. Pauling, *Nature* **1969**, *223*, 919.
[9] R. W. Behling, T. Yamane, G. Navon, L. W. Jelinsky, *Proc. Natl. Acad. Sci. USA* **1988**, *85*, 6721.
[10] P. Partington, J. Feeney, A. S. V. Burgen, *Mol. Pharmacol.* **1972**, *8*, 269.
[11] D. L. Beveridge, M. M. Kelly, R. J. Radna, *J. Am. Chem. Soc.* **1974**, *96*, 3769.
[12] J. Langlet, P. Claverie, B. Pullman, D. Piazzola, J. P. Daudey, *Theor. Chim. Acta* **1977**, *46*, 105.
[13] C. Margheritis, G. Corongiu, *J. Comput. Chem.* **1988**, *9*, 1.
[14] Y. J. Kim, S. C. Kim, Y. K. Kang, *J. Mol. Struct.* **1992**, *269*, 231.
[15] M. D. Segall, M. C. Payne, R. N. Boyes, *Mol. Phys.* **1998**, *93*, 365.
[16] G. Vistoli, A. Pedretti, L. Villa, B. Testa, *J. Am. Chem. Soc.* **2002**, *124*, 7472.
[17] W. Klyne, V. Prelog, *Experientia* **1960**, *17*, 521.
[18] P. Gaillard, P. A. Carrupt, B. Testa, A. Boudon, *J. Comput.-Aided Mol. Des.* **1994**, *8*, 83.
[19] D. F. Veber, S. R. Johnson, H. Y. Cheng, B. R. Smith, K. W. Ward, K. D. Kopple, *J. Med. Chem.* **2002**, *45*, 2615.
[20] G. Vistoli, A. Pedretti, L. Villa, B. Testa, *J. Med. Chem.* **2005**, *48*, 1759.
[21] G. Vistoli, A. Pedretti, L. Villa, B. Testa, *J. Med. Chem.* **2005**, *48*, 6926.

Prediction of Site of Metabolism in Humans: Case Studies of Cytochromes P450 2C9, 2D6, and 3A4

by **Gabriele Cruciani***[a]), **Riccardo Vianello**[b]), and **Ismael Zamora**[c])

[a]) Laboratory for Chemometrics and Cheminformatics, Chemistry Department, University of Perugia, Via Elce di Sotto, 10, I-06123 Perugia (e-mail: gabri@chemiome.chm.unipg.it)
[b]) *Molecular Discovery Ltd.*, 215 Marsh Road, Pinner, Middlesex HA5 5NE, UK
[c]) *Lead Molecular Design S.L.*, Fransesc Cabanes i Alibau, 1-3, 2-1, E-08190 Sant Cugat del Valles, Barcelona

Abbreviations
CYP: Cytochrome P450; GRIND: grid-independent descriptors; MEP: molecular electrostatic potential; MetaSite: site-of-metabolism prediction; MIF: molecular interaction field.

1. Introduction

The superfamily of cytochrome P450 (CYP) enzymes provides one of the most sophisticated catalysts of drug metabolism. These enzymes catalyze a wide variety of oxidative and reductive reactions and has activity towards chemically diverse substrates [1][2]. Since only a small subset of the CYP enzymes is responsible for the majority of drug-metabolizing events, it is unavoidable that different drugs will compete for the active site of a given CYP. Several aspects of these enzymes, such as the rate and position of their metabolic attack, their inhibition and induction, and the specificity of the various isoforms, must be taken into account in the lead optimization process during the development of new therapeutic agents.

To this end, the pharmaceutical industry needs computational predictive methods to identify the major CYP enzymes responsible for the metabolism of a given drug, and to be able to avoid potential drug–drug interactions. Despite the large amount of information available on the functional role of these enzymes, the knowledge of their three-dimensional structure is still incomplete. At the time of writing, the only X-ray structure publicly available is that of human 2C9.

Although several papers report the development of computational models to predict cytochrome P450 inhibition [3–6], rate and position of metabolism [7–14], and selective interactions [15], this remains a major challenge due to the many different isoforms that can be involved in the metabolism of a single compound, the number of possible positions of metabolism for each isoform, and, as stated above, the lack of structural information about most human CYPs.

While *in vitro* screens for inhibition and metabolic stability can provide some of the above information, the experimental elucidation of the position (*i.e.*, site) of metabolism is usually a high resource-demanding task which requires several experimental techniques and consumes large amounts of the compounds so investigated. Nevertheless, a recognition of the site(s) of metabolism could be of great help in designing new compounds with better pharmacokinetic profile and in avoiding the presence of toxic metabolites by chemically protecting the metabolic labile moieties. Another interest in predicting site(s) of metabolism is in prodrug design, where the compound needs to be metabolized to become active.

The aim of the present chapter is to describe a recent, fast, easy, and computationally inexpensive method to predict CYP2C9, 2D6, and 3A4 regioselective metabolism using *ad hoc* developed 3D homology enzyme models and the 3D structure of their potential substrates. The method requires only the 3D structure of the potential substrates and automatically determines the interaction of the virtual compounds with the enzymes using GRID flexible molecular interaction fields, providing the site(s) of metabolism in graphical output.

The computational procedure is fully automated and fast. The method thus appears as a valuable new tool in virtual screening and in early ADME/Toxicity evaluation, where potential drug–drug interactions and metabolic stability must be evaluated to facilitate drug design.

2. Description of the Method

The proposed methodology involves the calculation of different sets of descriptors, one for the CYP enzymes and one for the potential substrates. The set of descriptors used to characterize CYP enzymes is based on the GRID flexible molecular interaction field (MIF). The 3D structure of the CYP enzymes is required to derive the GRID-MIF interaction.

2.1. *Structure of CYPs*

The structure of a rabbit CYP2C5 [16] was used as a template in homology modeling of the CYP2C9 enzyme. In fact, this enzyme shows a high degree of similarity (>82%) and identity (>77%) with the human CYP2C family [3][15]. More recently, the crystal structure of human CYP2C9 was resolved and deposited in the *Protein Data Bank* [17]. However, this structure appears biased by the cocrystallized substrate. The above-mentioned 3D structure from homology model was therefore used in our work.

The initial 3D structures of CYP2D6 and 3A4 were kindly provided by *DeRienzo et al.* [18]. The 3D models were built by restraint-based comparative modeling using the X-ray crystallographic structure of bacterial cytochromes P450 BM3, CAM, TERP, and ERYF as templates (PDB entries 2bmh, 3cpp, 1cpt, and 1oxa, resp.). Secondary structure predictions were obtained by the method of *Rost* and *Sunders* [19]. The heme molecule, with its Fe-atom in the ferric (Fe^{III}) oxidation state, was extracted from the structure of P450-BM3 and fitted into the active site of each of the three cytochromes. The starting structures were submitted to dynamic runs, without any ligand, to select an average bioconformation for all the isoenzymes.

2.2. *3D Structure of Substrates*

The majority of CYP substrates contain flexible moieties. Since the conformation of the substrates had a sensible impact on the outcome of the method, two different protocols were used and tested to build their 3D structure. In the first method, a conformational search followed by energy minimization was performed for each substrate using the CONFORT program [20]. The runs were constrained to obtain a population of diverse low-energy minimum conformations.

In the second method, which was implemented only later, each substrate was submitted to a conformational search followed by energy minimization by means of an in-house software. The runs were constrained to obtain a population of conformers with a 3D structure induced by the shape and the *interaction fields* of the CYP active site.

2.3. *CYP Active Site Requirements*

It is known that CYP2C9 binds compounds with large dipoles or negative charges. Thus, oxygen-rich compounds such as carboxylic acids, sulfonamides, and alcohols are substrates for CYP2C9. Site-directed mutagenesis experiments have demonstrated that lipophilic interactions are extremely important for binding in the enzyme cavity. The GRID force field applied to CYP2C9 shows that its side chains have a great flexibility and that the binding site can accomodate a variety of substrates. Side-chain flexibility, in turn, modifies the physicochemical enviroment of the cavity. H_2O Molecules were introduced or removed according to the relative position of side chains. GRID shows that the CYP2C9 cavity is *ca.* 600 $Å^3$ wide, with hydrophobic interactions filling 20–40% of the cavity volume depending on the 3D rearrangment of the side chains induced by ligand binding.

CYP2D6 binds compounds with a basic N-atom and/or a positive charge. Thus, nitrogen-rich compounds such as arylalkylamines are potential substrates for CYP2D6. It is known that the 3D pharmacophore of CYP2D6 substrates needs one site of oxidation and at least one basic N-atom at 5–7 Å from the oxidation site [21]. However, several substrates do exist which show a greater distance between the oxidation site and the basic N-atom, *e.g.*, tamoxifen ($>$10 Å). This example demonstrates the important role played by CYP flexibility in substrate recognition.

CYP3A4 tends to exhibit a broad substrate specificity. It binds low-molecular- and high-molecular-weight compounds and shows no pharmacophoric preferences or special structural constraints for its substrates, due to its large cavity. Since its substrates probably adopt more than one orientation in the active site, CYP3A4 attacks ligand positions mainly in function of their chemical reactivity.

In the resting state, the central Fe-atom of CYPs is hexacoordinated with a serine OH group. Furthermore, the exclusion of H_2O molecules from a functional water channel is essential for effective enzymatic function. Upon substrate binding, H_2O must be displaced from the active site to prevent electron uncoupling [22]. It has been postulated that the water channel is located in the proximity of the thiolate side of the heme. Methods able to predict H_2O movement due to side-chain flexibility or substrate binding are essential for a correct prediction of CYP–substrate interactions.

2.4. *CYP–Substrate Interactions*

The molecular interaction fields (MIFs) in the binding site of CYP2C9, 2D6, and 3A4 are calculated by the GRID force field [23]. Five MIFs are generated in this analysis: the DRY molecular interaction field simulating hydrophobic interactions, the N1-amide nitrogen probe simulating H-bond donor interactions, the O-carbonyl oxygen probe simulating H-bond acceptor interactions, and two charged probes (one positive, the other negative) simulating charge–charge electrostatic interactions. All MIFs are obtained using a 0.5 Å grid and a self-accommodating dielectrical constant [19]. The grid box size for the three isoforms is carefully placed around the active-site cavities. The active-site cavities range from *ca.* 600 Å^3 for 2C9 to 1100 Å^3 for 3A4. The MIFs are generated using the flexible mode in GRID (directive MOVE = 1) [24]. With this option, some of the residue side chains can automatically move reflecting attractive or repulsive interactions with the probe. The side-chain flexibility in GRID can mimic this movement to accommodate different substrates depending on size, shape, and interaction pattern.

When a ligand approaches the side chains of residues, their movements are always influenced by neighbors and by the ligand. For example, the CH_2 groups in the side chain of lysine will tend to move toward the hydrophobic moiety of an interacting ligand. However, if the ligand contains a positively charged group, the charged N-atom of lysine will tend to move away from it. What actually happens depends on the overall balance between attractive and repulsive effects, and GRID is calibrated to simulate these movements. *Fig. 1* shows these phenomena in the cavity of CYP2D6. However, we point out that the flexible GRID map cannot take the large movements of the protein backbone into account.

2.5. *Transformation of the Molecular Interaction Fields*

The MIFs are subsequently transformed and simplified as shown in *Fig. 2*. In a first step, the regions close to the binding site but not accessible to the substrates are removed from the analysis. Then, the selected interaction points are used to calculate a new set of descriptors using the GRIND technology [25]. For each CYP-probe interaction, this approach transforms the interaction energies at a certain spatial position (MIF descriptors) into a number of histograms that capture the 3D pharmacophoric interactions of the flexible protein (correlograms).

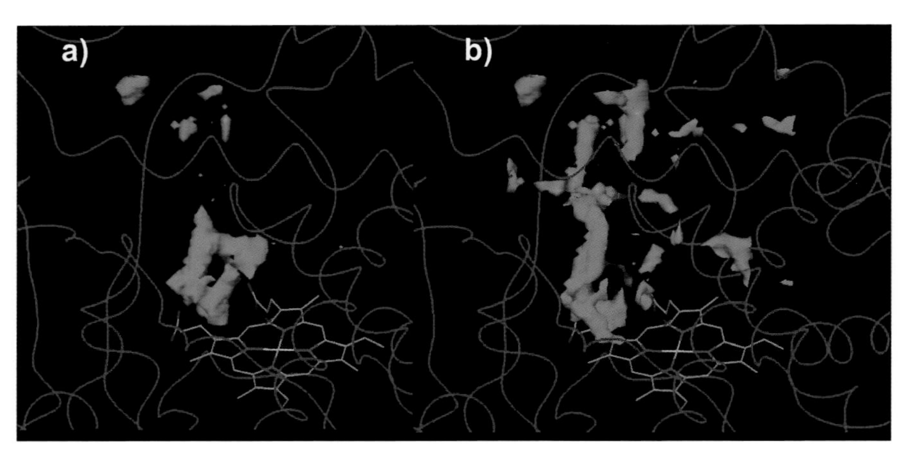

Fig. 1. a) *Rigid and* b) *flexible GRID maps with the hydrophobic DRY probe at* -0.5 *kcal/mol in the active-site cavity of CYP2D6.* It is important to note that, due to the starting 3D structure, some of the hydrophobic potential regions shown in *b*) are not present in *a*). This demonstrates the important role plaid by MIF flexibility calculations in CYP–substrate recognition.

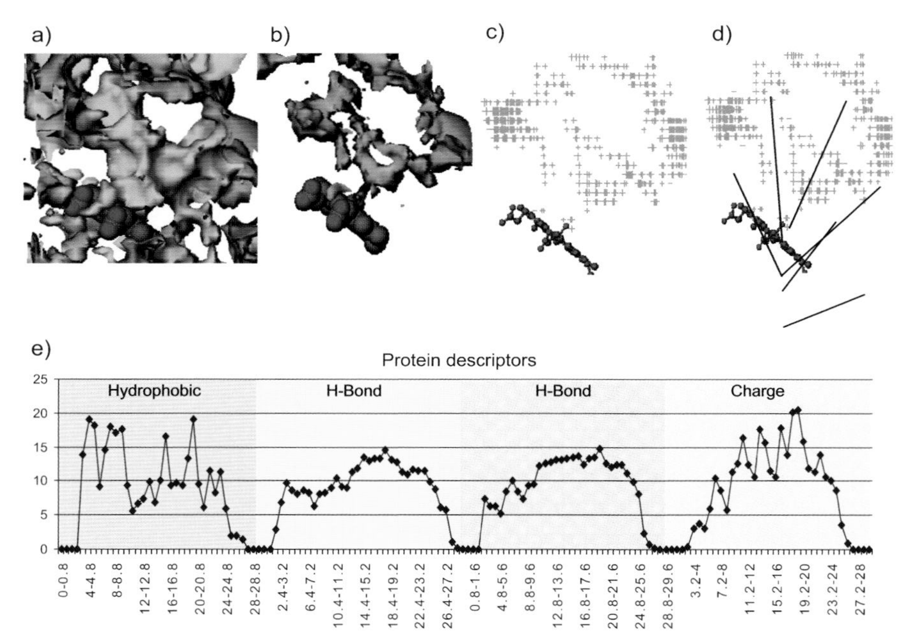

Fig. 2. *Protein treatment:* a) *molecular interaction field for CYP2C9 model using the N1 probe;* b) *MIF filtered by considering only the negative interactions with the N1 probe;* c) *the selected points for the distance-energy criteria;* d) *descriptors calculated for the different interaction fields;* e) *representation of the new descriptors in the different bin distances.* The overall process from *a*) to *e*) is fully automated.

2.6. *Substrate Treatment*

The descriptors developed to characterize the substrate chemotypes are obtained from a combination of molecular orbitals calculations and GRID probe–pharmacophore recognition. Molecular orbital calculations are first performed to compute the substrate's electron-density distribution. All atom charges are determined using the AM1 Hamiltonian. The computed charges are used to derive a 3D pharmacophore based on the molecular electrostatic potential (MEP) around the substrate.

Moreover, all atoms in the substrate are classified into GRID probe categories depending on their hydrophobic and H-bond donor and acceptor capabilities. Their intramolecular distances are then binned and transformed into clustered distances (see *Fig. 3*). One set of descriptors is computed for each atom-type category: hydrophobic, H-bond acceptor, and H-bond donor, yielding a fingerprint for each atom in the molecule. The distances between the different atomic positions classified using the previous criteria are then transformed into binned distances. In this case, the distances between the different atoms are calculated and a value of one or zero is assigned to each bin distance indicating the presence or the absence of such distance in the substrate.

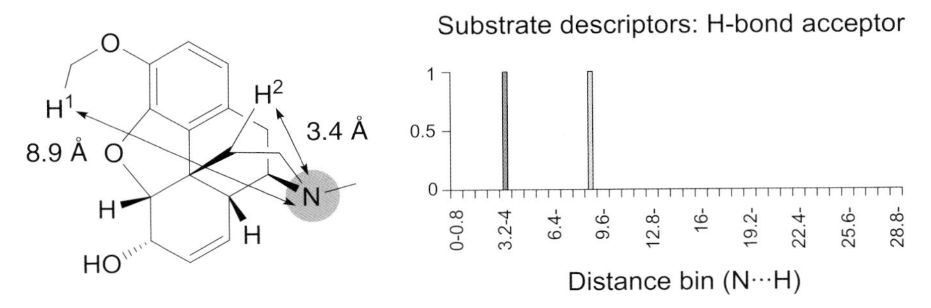

Fig. 3. *Ligand code treatment using codeine as an example.* Only two H-atoms are highlighted for clarity: H^1 and H^2.

2.7. *Ligand–CYP Protein Comparison: The Recognition Component*

Once the protein interaction pattern is translated from Cartesian coordinates to distances in the receptor and the structure of the ligand is described with similar fingerprints, both sets of descriptors can be compared (see *Fig. 4*). The complementarity of hydropobicities, charges,

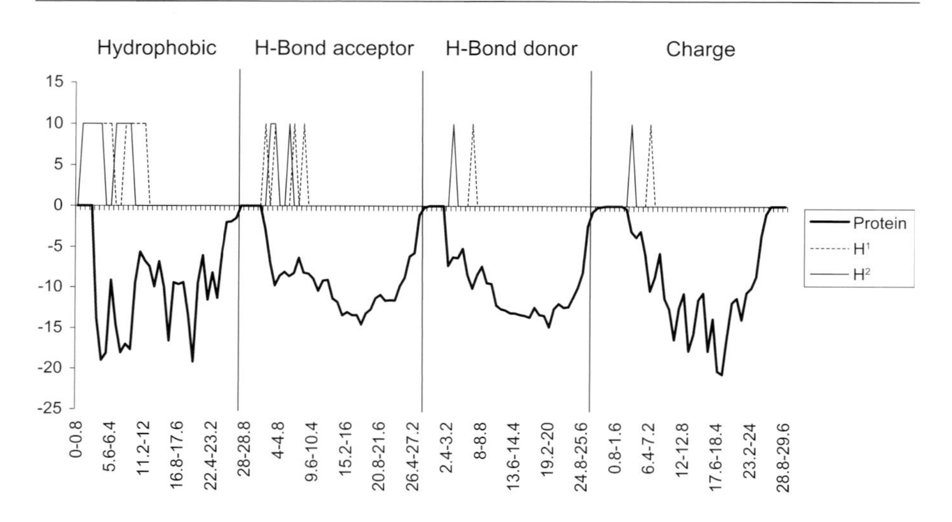

Fig. 4. *Comparison between the molecule fingerprint and the descriptors generated using the CYP2D6 homology model and the GRID molecular interaction field*

and H-bonds between the protein and the substrate are then computed using *Carbó* similarity indices [26].

Finally, the different atoms in the substrate are ranked according to the computed total similarity index.

2.8. *The Reactivity Component*

Cytochromes P450 catalyze oxidative and reductive reactions. Oxidative biotransformations are more frequent and include aromatic and side-chain hydroxylation, deamination, *N*-, *O*-, *S*-dealkylation, *N*-oxidation, sulfoxidation, dehalogenation, and desulfuration. The majority of these reactions can be rationalized considering a FeO^{3+} intermediate and a one-electron abstraction-rebound mechanism. This high-valent complex (FeO^{3+}, sometime written as $Fe^{IV}=O$) is electron-poor and abstracts either a H-atom or an electron from the substrate, generating an intermediate species. A subsequent collapse of the intermediate generates the product.

Although many different reactions are possible, we have addressed only the most-common metabolic reactions. Less-frequent or exotic reactions will be probably addressed later. The major P450 reaction groups considered are carbon hydroxylation, homolytic heteroatom oxygenation, heteroatom release (dealkylation), heteroatom oxygenation (*N*-oxydation and *S*-oxydation).

Carbon hydroxylation is probably the most-common reaction. The mechanism requires the formation of a radical species. This basic metabolism can be extended to the oxidation of alcohols to carbonyl compounds and of hydrated aldehydes to carboxylic acids. In any case, for all the above reactions, the site of metabolism can be described by a probability function P_{SM} which is correlated to, and can be considered as, the free energy of the overall process:

$$P_{SM_i} = E_i \cdot R_i$$

where P is the probability of an atom i to be the site of metabolism, E is the accessibility of atom i to the heme, R is the reactivity of atom i in the actual mechanism of reaction. E_i is, thus, the recognition score between the CYP protein and the ligand when the latter is positioned in the CYP protein and exposes its atom i towards the heme. E_i depends on the 3D structure of the ligand, on its conformation and chirality, and on the 3D structure of the enzyme. The E_i score is proportional to the exposure of the atom i to the heme group.

Similarly, R_i is the reactivity of atom i in the appropriate reaction mechanism, and represents the activation energy needed to produce the reactive intermediate. It depends on the ligand 3D structure and on the reaction mechanism.

For the same ligand, the P_{SM} function assumes different values for different atoms according to the E_i and R_i components. The flowchart of the procedure is reported in *Fig. 5*.

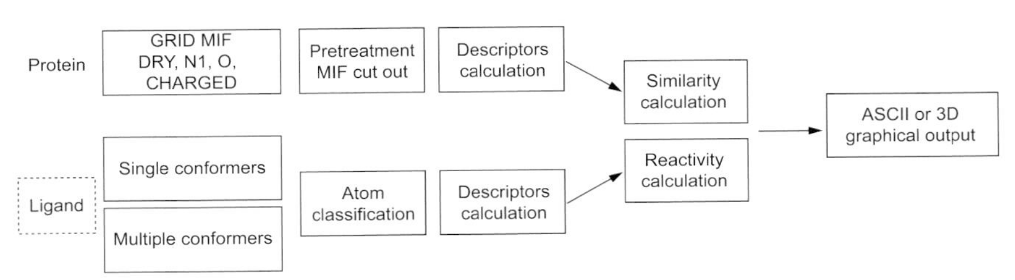

Fig. 5. *The MetaSite computation flowchart.* The user needs only to introduce the 3D structure of the ligand (dotted rectangle). All remaining processes are automated.

2.9. *Software Package*

The procedure is called MetaSite (site-of-metabolism prediction). The MetaSite procedure is fully automated and does not require any user assistance. All the work can be handled and submitted in batch queue. The molecular interaction field for CYPs obtained from the GRID package are precomputed and stored inside the software. The semiempirical calculations, pharmacophoric recognition, descriptor handling, similarity computation and reactivity computation, are made automatically once the 3D structure of the compounds is provided. The complete calculation is performed in a few seconds in IRIX SGI machines and is even faster in the Linux or Windows environment. For example, processing a database of 100 compounds, starting from 3D molecular structures, takes *ca.* 3 min at full resolution with a R14000 Silicon Graphics 500 MHz CPU, less than 1 min in a Windows Pentium machine and *ca.* 30 s in a Linux Pentium machine. MetaSite is a free software for nonprofit institutions, available at www.moldiscovery.com.

3. An Overview of Major Results

To validate the methodology in the three human enzymes, 120 metabolic reactions catalyzed by CYP2C9, 130 metabolic reactions catalyzed by CYP2D6, and 230 metabolic reactions catalyzed by CYP3A4, together with information concerning their sites of oxidation, were used.

A short selection of the reactions used as tests is reported in *Fig. 6*. The compounds show different metabolic pathways. Some are metabolized only at a single site, others present two sites of metabolism, and very seldom three. Moreover, substrates show a large structural diversity including rigid compounds (*e.g.*, steroids) and very flexible ones with more than ten rotatable bonds, and a wide range of molecular weight and lipophilicity.

In more than 68% of CYP2C9 reactions, the first option selected by the methodology matches the experimental one. Moreover, in more than 13% and 5% of cases, the second and third atoms, respectively, are the ones that fit the experimental one. Therefore, in considering the overall ranking list for the multiple sites of metabolism, in *ca.* 86% of the reactions, the methodology predicts the site of metabolism for CYP2C9 within the first three atoms selected, independently of the conformer used.

In more than 68% of CYP2D6 substrates, the first option suggested by the methodology was in agreement with the reported site of metabolism. When the first three options suggested by the method were considered as

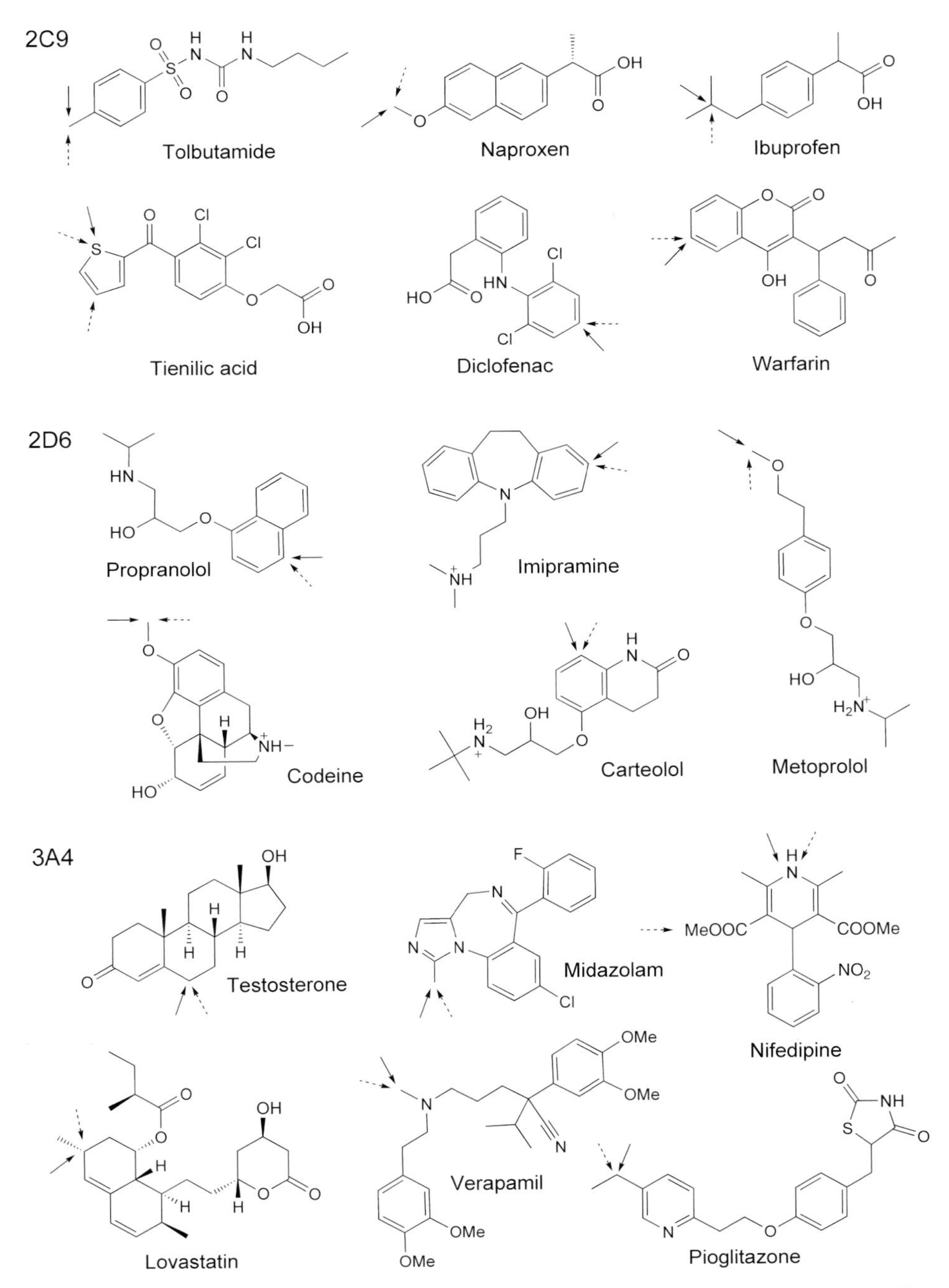

Fig. 6. *Some of the metabolic reactions used to* validate *the site of metabolism for CYP2C9, 2D6, and 3A4 with the experimental site of metabolism* (plain arrows) *and the predicted site(s) of metabolism* (dashed arrows)

the potential site of metabolism, 85% of the substrates were well predicted.

In more than 50% of CYP3A4 substrates, the first option suggested by the methodology was in agreement with the reported site of metabolism. When the first three options suggested by the method were considered as the potential site of metabolism, 80% of the substrates were well predicted.

4. Conclusions

A methodology has been developed to predict the site of metabolism for substrates of CYP2C9, 2D6, and 3A4. On average, for *ca.* 84% of cases, the method predicted the correct site of metabolism within the first three atoms in the ranking list.

The method is based on flexible molecular interaction fields generated by the GRID force field on CYP homology modeling structures that were treated and filtered to extract the most-relevant information. The methodology is very fast. To predict a site of metabolism for drug-like substrates, the method requires a few seconds per molecule. It is important to note that the method does not use any training set or statistical model or supervised technique, and it has proven to be predictive for extensively diverse validation sets examined in different pharmaceutical companies.

The 3D structure of the substrate to be analyzed (the starting conformation) has an impact on the outcome of the method. Satisfactory results were obtained using an in-house conformer generation which takes the MIFs and the flexible shape of the active site of the enzymes into account. The latter procedure is automatically performed when a molecule or a set of molecules are provided in 3D coordinates.

The methodology can be used either to suggest new positions that should be protected to avoid metabolic attack, or to check the suitability of a prodrug. Moreover, this procedure can be used to determine potential interactions of virtual compounds for early toxicity filtering.

REFERENCES

[1] B. Testa, 'The Metabolism of Drugs and Other Xenobiotics – Biochemistry of Redox Ractions', Academic Press, London, 1995.

[2] B. Testa, W. Soine, in 'Burger's Medicinal Chemistry and Drug Discovery', 6th edn., Ed. D. J. Abraham, Wiley-Interscience, Hoboken, NJ, 2003, Vol. 2, p. 431–498.

[3] L. Afzelius, I. Zamora, M. Ridderström, A. Kalén, T. B. Andersson, C. Masimirembwa, *Mol. Pharmacol.* **2001**, *59*, 909.

[4] J. P. Jones, M. He, W. F. Trager, A. E. Rettie, *Drug Metab. Dispos.* **1996**, *24*, 1.

[5] S. Ekins, G. Bravi, S. Binkley, J. S. Gillespie, B. Ring, J. Wikel, S. Wrighton, *Drug Metab. Dispos.* **2000**, *28*, 994.
[6] S. Rao, R. Aoyama, M. Schrag, W. F. Trager, A. Rettie, P. J. Jones. *J. Med. Chem.* **2000**, *43*, 2789.
[7] G. Cruciani, M. Pastor, S. Clementi, S. Clementi, in 'Rational Approaches to Drug Design: 13th European Symposium on Quantitative Structure–Activity Relationships', Eds. H. D. Höltje, W. Sippl, Prous Science, Barcelona, 2001, p. 251–261.
[8] K. R. Korzekwa, J. Grogan, S. DeVito, J. P. Jones, *Adv. Exp. Med. Biol.* **1996**, *38*, 361.
[9] D. F. Lewis, M. Dickins, P. J. Eddershaw, M. H. Tarbit, P. S. Goldfarb, *Drug Metab. Drug Interact.* **1999**, *15*, 1.
[10] M. J. De Groot, M. Ackland, V. Horne, A. Alexander, J. Barry, *J. Med. Chem.* **1999**, *42*, 4062.
[11] B. C. Jones, G. Hawksworth, V. A. Horne, A. Newlands, J. Morsman, M. S. Tute, D. A. Smith, *Drug Metab. Dispos.* **1996**, *24*, 260.
[12] A. Mancy, P. Broto, S. Dijols, P. M. Dansette, D. Mansuy, *Biochemistry* **1995**, *34*, 10365.
[13] S. Singh, L. Shen, M. Walker, R. Sheridan, *J. Med. Chem.* **2003**, *46*, 1330.
[14] I. Zamora, L. Afzelious, G. Cruciani, *J. Med. Chem.* **2003**, *46*, 2313.
[15] M. Riddestrom, I. Zamora, O. Fjäström, T. B. Andersson, *J. Med. Chem.* **2001**, *44*, 4072.
[16] P. A. Williams, J. Cosme, V. Sridhar, E. F. Johnson, D. E. McRee, *Mol. Cell* **2000**, *5*, 121.
[17] P. A. Williams, J. Cosme, A. Ward, H. C. Angove, D. Matak Vinkovic, H. Jhoti, *Nature* **2003**, *424*, 464.
[18] F. DeRienzo, F. Fanelli, M. C. Menziani, P. G. De Benedetti, *J. Comput.-Aided Mol. Des.* **2000**, *14*, 93.
[19] C. Saunders, B. Rost, *Proteins* **1994**, *19*, 55.
[20] R. S. Perlman, R. Balducci, 'Confort: A Novel Algorithm for Conformational Analysis', *National Meeting of the American Chemical Society*, New Orleans, 1998.
[21] M. J. de Groot, M. J. Ackland, V. A. Horne, A. A. Alex, B. C. Jones, *J. Med. Chem.* **1999**, *42*, 4062.
[22] T. I. Oprea, G. Hummer, A. E. Garcia, *Biochemistry* **1997**, *94*, 2133.
[23] P. J. Goodford, *J. Med. Chem.* **1985**, *28*, 849.
[24] P. J. Goodford, in 'Rational Molecular Design in Drug Research, Alfred Benzon Symposium 42', Eds. T. Liljefors, F. S. Jorgensen, P. Krogsgaard-Larsen, Munkgaard, Copenhagen, 1998, p. 215–230; GRID V.21, *Molecular Discovery Ltd.*, 2003, http:\\www.moldiscovery.com.
[25] M. Pastor, G. Cruciani, I. McLay, S. Pickett, S. Clementi, *J. Med. Chem.* **2000**, *43*, 3233.
[26] R. Carbó-Dorca, L. Amat, *J. Comput. Chem.* **1999**, *20*, 911.

Use of Pharmacophores in Predictive ADME

by **Omoshile O. Clement***[1]) and **Osman F. Güner**

Accelrys Inc., 9685 Scranton Road, San Diego, CA 92121, USA

Abbreviations
CoMFA: Comparative molecular field analysis; CYP: cytochrome P450; K_i: kinetic rate constant for enzyme inhibition; MS-WHIM: molecular surface weighted holistic invariant molecular modeling technique; PLS: partial least squares.

1. Introduction

The earliest proposed definition of the term 'pharmacophore' was made by *Ehrlich* [1a] (also quoted by *Ariëns* [1b]), who defined it as the molecular framework that carries (*phoros*) the essential features responsible for a drug's (*pharmacon*) biological activity. More-recent definitions of what a pharmacophore entails have been proposed by various researchers; many of these newer definitions can be found in the first published book on pharmacophores [2]. Today, elucidation of the pharmacophore is considered as one of the most-important first step towards understanding ligand–target interaction, and hence pharmacophore-modeling methodology is widely used in drug discovery research as an aid in the early identification of new chemical entities [3].

The pharmacophore concept derives its coinage from an intuitive perception of which chemical functional entities are important in ligand–target recognition, especially where no information about the structure of the native or bound receptor or protein target is available to researchers. Until recently, the pharmacophore concept has been mostly applied to ligand-based approaches in drug discovery research. Today, increasing advancements in high-throughput crystallography, NMR imaging techniques, and protein-modeling tools have significantly increased the number of protein structures available to researchers, concomitantly

[1]) Current address: *Bio-Rad Laboratories*, Informatics Division, 3316 Spring Garden St., Philadelphia, PA 19104, USA (e-mail: omoshile_clement@bio-rad.com).

elevating application of the pharmacophore concept to protein-guided drug discovery research. A review of applications of pharmacophores in structure-based drug designs (as of mid-2001) was recently published by *Clement* and *Mehl* [4].

As mentioned earlier, pharmacophore modeling has been mostly applied to the discovery of new lead candidates in the absence of target structural information. Much less use of this approach has been applied to candidate evaluation for clinical viability, *i.e.*, use of pharmacophore models as predictive tools for understanding the ADME properties and potential toxicities of new lead candidates. The earliest pioneering studies of pharmacophore modeling of cytochrome P450 enzyme inhibition and/or substrate specificity were described by *Wolff et al.* [5a], *Meyer et al.* [5b], *Islam et al.* [5c], *Koymans et al.* [5d], *Jones et al.* [6a–c], and *Mancy et al.* [7]. The models derived in these earlier studies were qualitative with no predictive ability, yet provided significant insights into substrate specificity or ligand inhibition for cytochrome P450 mediated metabolism.

Predictive pharmacophore-based 3D-QSAR models of substrates and inhibitors of some cytochromes P450 have been reported by *Mancy et al.* [8], *Ekins et al.* [9–13], and *de Groot et al.* [14][15]. These studies applied pharmacophore-modeling methodology to investigate drug–drug interactions mediated by several cytochrome P450 metabolism enzymes, *e.g.*, CYP2D6, CYP1A2, CYP3A4, and CYP2C9.

2. Predictive ADME Models Derived by Pharmacophore Modeling

2.1. *CYP Enzymes*

As a rule, metabolism increases the aqueous solubility of a drug such that it is easily excreted from the body. Such metabolism is mediated by a host of CYP enzymes, notably CYP2D6, CYP3A4, CYP1A2, and CYP2C9. A list of drugs metabolized by P450 enzymes can be found at http://medicine.iupui.edu/flockhart/table.htm. Out of the *ca.* 20% of all relevant CYPs known to date, the 3A4 subsystem accounts for *ca.* 50% of CYP metabolism, 2D6 accounts for *ca.* 30%, while 1A2, 2C9, 2C10, 2C19, and 2E1 account for the remaining 20% [16]. Since eukaryotic CYPs are membrane-bound proteins, crystal-structure information has been difficult to obtain. The first reported structure of a CYP enzyme was reported for CYP51 from *Mycobacterium tuberculosis* (MTCYP51) recently appeared in the literature [17]. Recently, *Williams et al.* [18] reported a crystal-structure determination of the human CYP2C9 enzyme with bound warfarin. Prior to these two reports, a number of homology models of CYP-dependent enzymes were generated [19][20] to aid research efforts dedicated to the study of drug–drug interactions and drug metabolism involving the CYP enzyme subtypes.

There are vast amounts of *in vitro* and *in vivo* CYP data available in the public domains that are suitable for use in molecular-modeling studies. Data exists for 1A1 [21], 1A2 [21–23], 1B1 [21], 2A6 [24], 2B6 [9], 2C9 [7][12][25–27], 2C19 [28][29], 2D6 [13][30][31], and 3A4 [10][12][32–34]. With the large amount of data generated for substrate specificity or inhibition of these enzymes, studies involving *in silico* model development for predicting drug–drug interactions and drug metabolism have become very important. Applications of pharmacophore modeling techniques for evaluating ligand requirements for specificity and/or inhibition of the CYP enzymes have recently been summarized [35]. The studies described below expand on this previous review and includes recent work, which employed the pharmacophore concept to elucidate difficult information on predicting potential drug–drug interactions and drug metabolism.

2.1.1. *CYP2B6*

The CYP2B6 enzyme represents $<0.2\%$ of total human hepatic CYPs [36], yet it plays a very prominent role in the metabolism of many xenobiotics [37]. There is no crystal-structure information for this enzyme due to its membrane-bound nature, but the residues believed to be responsible for enzyme interaction were deduced from a homology model of a rat CYP2B1 aligned with P450-BM3 [20b]. This homology model allowed researchers to deduce that substrate specificity for the CYP2B6 enzyme may involve π-stacking interactions among the side chains of Phe^{181} and/or Phe^{263} [8][20b].

The earliest study of pharmacophore modeling of ligands with potential specificity or inhibition of the CYP2B6 enzyme was reported by *Ekins et al.* [9]. The study involved two different statistical methods using a pharmacophore-based 3D-QSAR modeling package available within the Catalyst software [38], and a partial-least-squares (PLS)-based approach termed MS-WHIM, or molecular surface weighted holistic invariant molecular modeling technique [39]. For the pharmacophore-based study, 16 compounds were selected as training set for model generation (see, *e.g.*, *Fig. 1*). Conformations of each compound were sampled within 20 kcal/mol energy threshold, with a maximum number set to 255. Biological activity data in the form of K_i, the kinetic rate constant for enzyme inhibition, was used as input data for the QSAR model generation.

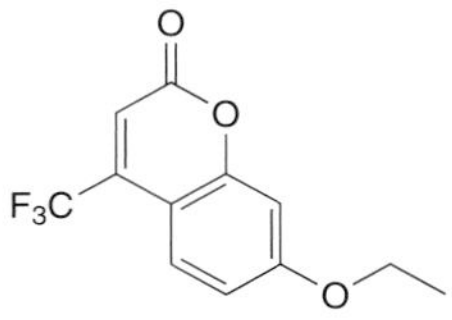

7-Ethoxy-4-(trifluoromethyl)coumarin

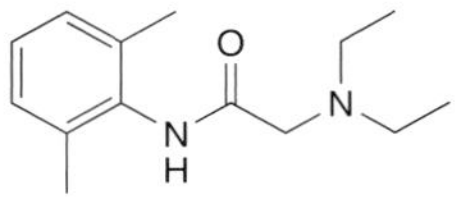

Lidocaine

Fig. 1. *Examples of CYP2B6 substrates used in the pharmacophore model generation study: 7-ethoxy-4-trifluoromethylcoumarin and lidocaine* [10]

For substrate binding to the CYP2B6 enzyme, the pharmacophore hypothesis identified four geometric binding features – three hydrophobic features and one H-bond acceptor at distances of 5.3, 3.1, and 4.6 Å. The pharmacophore model was shown to correlate the variations in the biological activities of training-set compounds as a function of the geometric features in the model. The model was used to predict the activities of four test compounds, and the estimated activities were within 1 log unit with $r = 0.85$. In general, some agreement was found between the Catalyst pharmacophore model for substrate binding to CYP2B6, and the active-site characterization of a homology model of CYP2B6 reported by *Lewis* and *Lake* [20b].

2.1.2. *CYP2C9*

Pharmacophore perception of substrate specificity and active-site characterization of the CYP2C9 enzyme is the second-most investigated of the CYP enzymes. The first study involving a pharmacophore-based elucidation of requirements for substrate specificity for a CYP metabolism enzyme was conducted by *Jones et al.* [6a]. Using a manual pharmacophore mapping of the active site of a homology model of CYP2C9, these workers suggested that the active site is dominated by H-bond interaction features. A follow-up study [6b] involving alignment of eight substrates and one inhibitor implicated a H-bond donor heteroatom situated at a distance of *ca.* 7 Å from the catalytic site. *Mancy et al.* [7] identified a similar inter-feature distance (*ca.* 7.8 Å) between an anionic site and a hydrophilic site for tight substrate binding to CYP2C9 based on manual superposition of 20 substrates of the enzyme. This study corroborated earlier findings reported by *Jones et al.* [6a] with regard to the geometric features required for substrate binding to CYP2C9, as well as the location constraints between these features.

An extension of this work to sulfaphenazoles confirmed findings that strong ligand–CYP2C9 binding involves hydrophobic interactions as well as a cationic functional group, which can be represented as a positive ionizable N-atom. The studies conducted by *Jones et al.* [6a–b] and *Mancy et al.* [7] provided the first glimpse into geometric features responsible for substrate binding to the CYP2C9 enzyme. However, the models generated from these studies were qualitative with no predictive ability.

The first predictive pharmacophore model for estimating binding affinities for the CYP2C9 enzyme was reported by *Ekins et al.* [12], using the Catalyst/Hypogen program [38]. The dataset (see, *e.g.*, *Fig. 2*) was partitioned into three training sets to build models that would identify the requisite geometric functional groups responsible for the substrate specificity of the enzyme. Each training set was subjected to a pharmacophore-modeling run, and results were compared against each other. Each model differed from the other and contained different combinations of requisite binding features. All models implicate at least one hydrophobic group and one H-bond acceptor function as an important feature for binding to the enzyme, with the location of a H-bond donor at 3.4–5.7 Å from a neighboring H-bond acceptor [12].

The pharmacophore models for inhibition of the CYP2C9 enzyme [12] derived using Catalyst/Hypogen was found to be in good agreement with a CoMFA analysis [15c] of a collection of 27 CYP2C9 compounds, which identified two cationic sites, an aromatic group, and a steric region as important to substrate recognition by the CYP2C9 enzyme. Additionally, *Ekins et al.* [12] employed a PLS-optimized MS-WHIM technique [39] against the same dataset, and the sets of QSAR models obtained from this approach were found to contain chemical descriptors, which were internally consistent with chemical functional groups identified in the Catalyst/Hypogen pharmacophore model. As mentioned earlier, a new crystal structure of human CYP2C9 has been reported [18], and it is hoped that

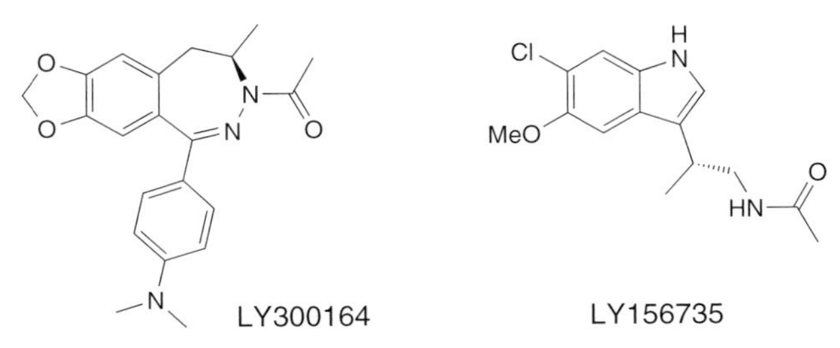

Fig. 2. *Examples of compounds used in the pharmacophore modeling of inhibitors of CYP2C9* [12]

this will spur additional structure-based investigation of inhibitors and substrates for this enzyme.

2.1.3. *CYP2D6*

Earliest reports [5a,b] of small molecule based computational models for inhibition or substrate specificity for this enzyme were based on manual alignments of a variety of substrates or inhibitors of this enzyme. These models were found to be inconsistent in rationalizing substrate specificity requirements for this enzyme. Later studies conducted by *Islam et al.* [5c] using information from related crystal structure of CYP101 [40] and by *Koymans et al.* [5d] provided a more-detailed picture of substrate specificity for this enzyme.

A pharmacophore model for inhibition of CYP2D6 was derived using a template of six strongly reversible inhibitors of the enzyme [31]. The two most-important geometric features in the pharmacophore model were a positive ionizable N-center (protonated at physiological conditions), and an aromatic hydrophobic group. The presence of a H-bond interaction was also linked to enhanced inhibitory potency [31]. This inhibitor-based model consisting of a tertiary N-center, an aromatic hydrophobic group, where enhanced inhibitory potency is observed with inclusion of H-bond interactions, and another aromatic hydrophobic site, which plays no role in inhibitory potency, compared favorably with the substrate specificity models reported by others [5c,d][14]. It is of note that the model for substrate specificity derived by these workers [5d][14] was used to successfully design a novel and selective CYP2D6 substrate [41] and to investigate the hydroxylation of debrisoquine [42].

More recently, *de Groot et. al.* [14][15] reported a combined pharmacophore/homology modeling study of the CYP2D6 enzyme. The study identified two pharmacophore models: one for *O*-dealkylation and oxidation reactions, and another for CYP2D6-catalyzed *N*-dealkylation reactions. The latter model correctly predicted the metabolism of a wide range of compounds.

Predictive 3D/4D-QSAR pharmacophore-based models for competitive inhibition of CYP2D6 were also reported by *Ekins et al.* [12]. The first model was generated from an in-house (*Eli Lilly Corp.*) dataset of 20 inhibitors of bufuralol-1′-hydroxylation (see, *e.g.*, *Fig. 3*). The correlation coefficient (r) for estimated activities (K_i) of the training set compounds was 0.75. A second model was derived using 31 compounds from the literature. The observed correlation coefficient for estimated activities

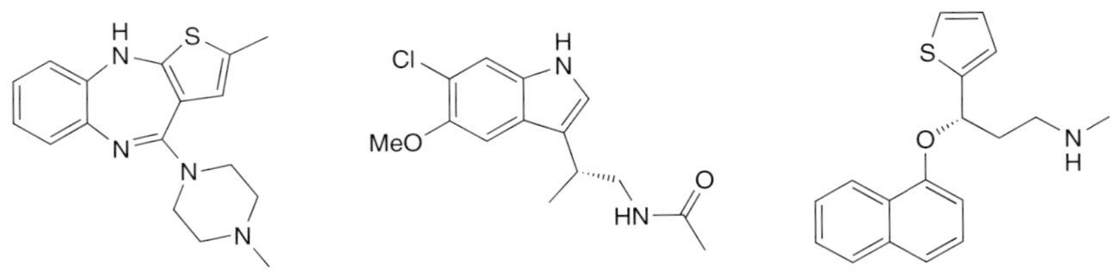

LY170053 LY156735 LY24868

Fig. 3. *Examples of inhibitors of CYP2D6 used to generate the 3D pharmacophore model for enzyme inhibition* [13]

(K_i) was 0.91 [13]. Both models correctly estimated activities (K_i) of 9–10 of 15 test compounds [13].

Fukushima et al. [43] also employed the Catalyst/Hypogen program to generate predictive models for ligand inhibition of CYP2D6. In the study, four pharmacophore models were generated from training and test sets derived from either published data or in-house data. Predictive models were built with input of four chemical features defined for the Hypogen program – hydrophobe, H-bond donor, H-bond acceptor, and a basic amine. The latter feature was used in place of a positive ionizable feature, since most CYP2D6 inhibitors are basic compounds, and it is not clear what the functional form of such feature is. Interestingly, all four pharmacophore models generated contained the basic amine function as a required feature for enzyme inhibition, along with one or more hydrophobic groups and a H-bond donor group. The four models correctly estimate activities of *ca.* 60% of a test set of 41 compounds. Poor estimated activities were obtained for *a*) compounds with aromatic nitrogen groups, which can act as a heme chelator, *b*) compounds containing basic functional groups other than amines, and *c*) compounds with poor fits to any of the four pharmacophores generated in the study.

2.1.4. *CYP3A4*

This enzyme is important in the metabolism of many classes of drugs and can be described as the most-significant human CYP enzyme involved in drug metabolism. A homology model of the CYP3A4 enzyme derived from soluble bacterial CYP structures as templates found that the active site of this enzyme is likely dominated by hydrophilic groups, but includes H-bond donor/acceptor features. In addition, the homology model suggested a conformationally flexible active site, which is believed to

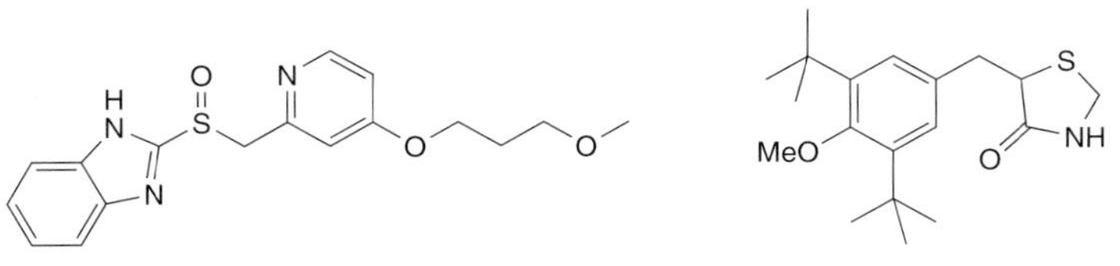

LY307640 LY213829

Fig. 4. *Examples of competitive inhibitors of CYP3A4 used for generating a pharmacophore model for the enzyme* [10]

account for the wide range of structurally diverse ligands that interact with this enzyme [44].

Three sets of pharmacophore-modeling experiments of ligand requirements for CYP3A4 inhibition were performed by *Ekins et al.* [10]. The first run consisted of inhibitors of CYP3A4-mediated midazolam-1′-hydroxylase (see, *e.g.*, *Fig. 4*). The pharmacophore model derived for this class of ligands contained four chemical functional requirements – three hydrophobic centers and a H-bond acceptor, located 5.2–8.8 Å apart. The model had a correlation coefficient of 0.91 between observed and estimated activities (K_i). The second pharmacophore experiment was conducted on ligands that competitively inhibit CYP3A4-mediated cyclosporin A metabolism [34][35]. The model derived for this class of compounds contained five chemical functional requirements – three hydrophilic centers and two H-bond acceptor groups. As a predictive tool, the model had a correlation coefficient of 0.77, between observed and estimated activities (K_i). The third pharmacophore model experiment was conducted on data from inhibition (IC_{50}) of CYP3A4-mediated quinine hydroxylation [33]. The derived model contained four feature types – one hydrophobic center and three H-bond acceptors [10]. The correlation coefficient between observed and estimated activities was 0.92.

Comparing the two pharmacophore models generated using K_i data showed strong similarities in location and type of chemical features. The merged model identified the critical requirements for CYP3A4 inhibition: two hydrophobic regions separated by H-bond acceptor groups [10]. Another study to evaluate requirements for substrate specificity for CYP3A4 enzyme was conducted using a dataset of 38 known substrates of the enzyme [11]. The pharmacophore model so derived contained four geometric features – two hydrophobic centers, one H-bond donor, and one H-bond acceptor group. Although the correlation coefficient of the difference between observed and estimated K_m was poor ($r=0.67$), predicted activities for 12 test compounds were well within an order of magnitude of their observed activities [11].

Both the inhibition model and the substrate-binding model had a level of commonality with presence of at least one H-bond-acceptor feature. This finding supports information derived from the homology model of CYP3A4 [20] which implicated interaction with Asn^{74} in both substrate binding and ligand inhibition processes.

2.1.5. *CYP51*

The enzyme 14α-lanosterol demethylase (CYP51) is widely distributed in various biological species as the major sterol 14-demethylase [45]. This enzyme catalyzes the removal of the 14α-methyl group of lanosterol in the biosynthesis of ergosterol. Inhibition of CYP51 across many species is believed to lead to undesirable side-effects, and hence a need for very selective drugs against this enzyme is desired.

Several reports have shown that azole-based antifungal agents inhibit CYP51. A number of these antifungal agents block ergosterol biosynthesis depleting its availability while causing excessive accumulation of lanosterol and other 14-methylsterols [45]. A recent pharmacophore-based investigation of derivatives of 1-[(aryl)[4-aryl-1*H*-pyrrol-3-yl}-methyl]-1*H*-imidazole by *Tafi et al.* [46] identified the following geometrical features required for CYP51 inhibition: *a*) an aromatic N-atom with an accessible lone pair, *b*) a di(arylmethyl) moiety at the azole N(1) position, *c*) a second aryl group containing two coplanar aromatic rings. *Fig. 5* shows the alignment of these geometrical features with the template

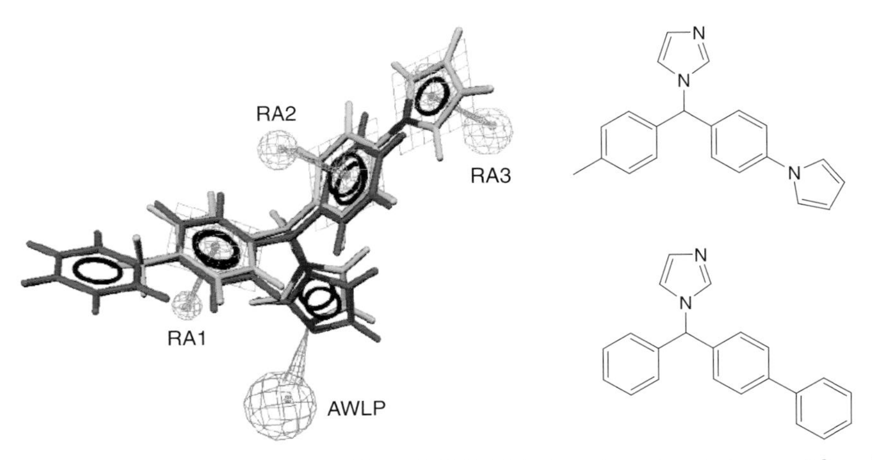

Fig. 5. *Catalyst pharmacophore model for CYP51 inhibition, aligned against two antifungal inhibitors of CYP51* [47]. RA1, RA2, RA3: Aromatic rings; AWLP; aromatic N-atom with lone pair [46].

antifungal agent and the most-active ligand in the dataset used for the study. Similar pharmacophoric features were identified in the 4D-QSAR study reported by *Hopfinger* and co-workers [47].

3. Conclusions

Although current understanding of the influence of CYP enzymes in drug metabolism is well known and characterized, there remain areas that research has yet to shed light on. Questions such as *a*) to what extent a given drug will induce or inhibit a specific CYP enzyme, or *b*) to what extent are specific CYP enzymes responsible for drug metabolism of specific drugs remain unanswered [16]. Thankfully, research to shed light into these areas is ongoing. Recent reports indicate that clinical efficacy of a drug as a function of ADME-Tox properties has undergone some dramatic change in the past decade. Failures associated with poor ADME profiles of lead drugs have reduced from *ca.* 50% of all clinical candidates to *ca.* 10%. Conversely, challenges of drug selectivity and toxicities have increased from *ca.* 10% of all drug failures to *ca.* 20%, becoming one of the most-problematic areas requiring solution today.

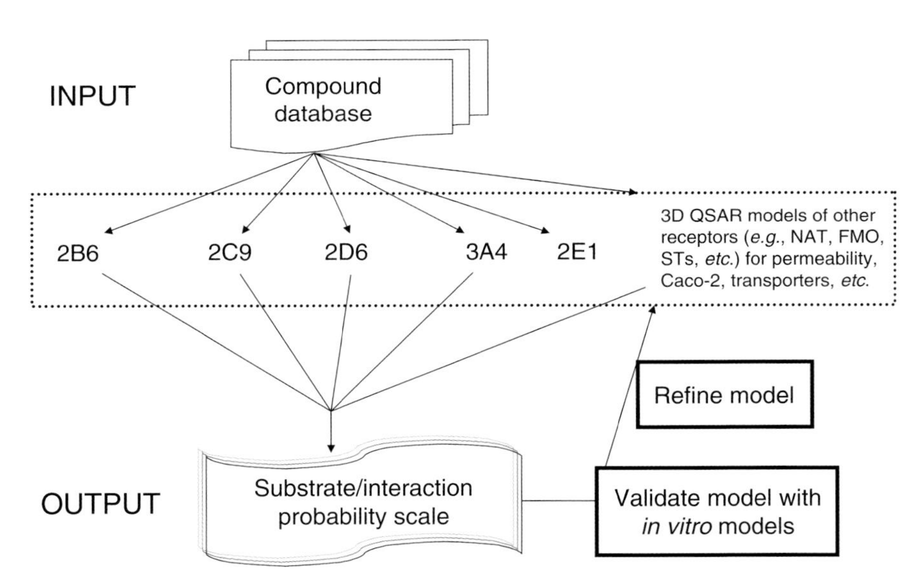

Fig. 6. *A scheme of an* in silico *parallel-screening strategy for drug-metabolizing enzymes and transporters 3D-QSAR, and the steps necessary for model validation and iterative 3D-QSAR design* [48]

This chapter has attempted to provide an overview of the new paradigm, which leverages recent advances in computational algorithms and increased experimental data, to address these problems. *Ekins et al.* [48] have proposed a workflow (see *Fig. 6*), where computational chemistry plays a central role in building virtual filters that can be applied to large virtual or real chemical libraries to eliminate ligand candidates that may potentially act as substrates or inhibitors for many of the drug metabolism enzymes. The pruned outputs from these *in silico* filters can be used to further refine these models to make them even better in clinical candidate evaluation.

In closing, clinical efficacy of lead candidates can also be improved if *a*) the candidate can be metabolized by more than one CYP enzyme, *b*) its metabolism is not dependent on a CYP enzyme with a significant genetic polymorphism, and *c*) the candidate does not significantly inhibit any of the CYP enzymes that mediate oxidative pathways of drug metabolism, *i.e.*, 3A3/4, 2D6, 1A2, 2C19, and 2C9/10 [16]. Drug candidates meeting all of these conditions have a greater chance of making it through the clinic and ultimately into commercial viability.

REFERENCES

[1] a) P. Ehrlich, *Chem. Ber.* **1909**, *42*, 17; b) E. J. Ariëns, *Prog. Drug Res.* **1966**, *10*, 429.

[2] 'Pharmacophore Perception, Development, and Use in Drug Design', Ed. O. F. Güner, IUL Biotechnology Series, La Jolla, CA, 2000.

[3] P. S. Charifson, in 'Practical Applications of Computer-Aided Drug Design', Marcel Dekker, New York, 1997.

[4] O. O. Clement, A. T. Mehl, in 'Protein Structure: Determination, Analysis, and Applications for Drug Discovery', Ed. D. Chasman, Marcel Dekker, New York, 2003, p. 453–462.

[5] a) T. Wolff, L. M. Distelath, M. T. Worthington, J. D. Groopman, G. J. Hammons, F. F. Kadlubar, R. A. Prough, M. V. Martin, F. P. Guengerich, *Cancer Res.* **1985**, *45*, 2116; b) U. A. Meyer, J. Gut, T. Kronbach, C. Skoda, U. T. Meier, T. Catin, *Xenobiotica* **1986**, *16*, 449; c) S. A. Islam, C. R. Wolf, M. S. Lennard, J. E. Sternberg, *Carcinogenesis* **1991**, *12*, d) L. Koymans, N. P. E. Vermeulen, S. A. B. E. van Acker, J. M. te Keppele, J. J. P. Heykants, K. Lavrijsen, W. Meuldermans, G. M. Donne-Op den Kelder, *Chem. Res. Toxicol.* **1992**, *5*, 211.

[6] a) B. C. Jones, G. Hawksworth, V. Horne, A. Newlands, M. Tute, D. A. Smith, *Br. J. Clin. Pharmacol.* **1993**, *34*, 143P; b) J. P. Jones, M. He, W. F. Trager, A. E. Rettie, *Drug Metab. Dispos.* **1996**, *24*, 1; c) B. C. Jones, G. Hawksworth, V. Horne, A. Newlands, J. Morsman, M. S. Tute, D. A. Smith, *Drug Metab. Dispos.* **1996**, *24*, 260.

[7] A. Mancy, P. Broto, S. Dijols, P. M. Dansette, D. Mansuy, *Biochemistry* **1995**, *34*, 10365.

[8] A. Mancy, S. Dijols, S. Poli, P. Guengerich, D. Mansuy, *Biochemistry* **1996**, *35*, 16205.

[9] S. Ekins, G. Bravi, B. J. Ring, T. A. Gillespie, J. S. Gillespie, M. Vandenbranden, S. A. Wrighton, J. H. Wikel, *J. Pharmacol. Exp. Ther.* **1999**, *288*, 21.

[10] S. Ekins, G. Bravi, S. Binkley, J. S. Gillespie, B. J. Ring, J. H. Wikel, S. A. Wrighton, *J. Pharmacol. Exp. Ther.* **1999**, *290*, 429.

[11] S. Ekins, G. Bravi, J. H. Wikel, S. A. Wrighton, *J. Pharmacol. Exp. Ther.* **1999**, *291*, 424.

[12] S. Ekins, G. Bravi, S. Binkley, J. S. Gillespie, B. J. Ring, J. Wikel, S. A. Wrighton, *Drug Metab. Dispos.* **2000**, *28*, 994.
[13] S. Ekins, G. Bravi, S. Binkley, J. S. Gillespie, B. J. Ring, J. H. Wikel, S. A. Wrighton, *Pharmacogenetics* **1999**, *9*, 477.
[14] M. J. de Groot, M. J. Ackland, V. A. Horne, A. A. Alex, B. C. Jones, *J. Med. Chem.* **1999**, *42*, 1515.
[15] a) M. J. de Groot, G. J. Bijloo, K. T. Hansen, N. P. E. Vermeulen, *Drug Metab. Dispos.* **1995**, *23*, 667; b) M. J. de Groot, N. P. E. Vermeulen, *Drug Metab. Dispos.* **1997**, *29*, 747; c) M. J. de Groot, M. J. Ackland, V. A. Horne, A. A. Alex, B. C. Jones, *J. Med. Chem.* **1999**, *42*, 4062.
[16] S. H. Preskorn, A. T. Harvey, http://www.acnp.org/g4/GN401000086/CH085.html.
[17] L. M. Podust, T. L. Poulos, M. R. Waterman, *Proc. Natl. Acad. Sci. U.S.A.* **2001**, *98*, 3068.
[18] P. A. Williams, J. Cosme, A. Ward, H. C. Angove, D. M. Vinković, H. Jhotl, *Nature* **2004**, *424*, 464.
[19] T. L. Poulos, B. C. Finzel, A. J. Howard, *J. Mol. Biol.* **1987**, *195*, 687; C. A. Hasemann, K. G. Ravichandran, J. A. Peterson, J. Diesenhofer, *J. Mol. Biol.* **1994**, *236*, 1169; J. R. Cupp-Vickery, T. L. Poulos, *Nat. Struct. Biol.* **1995**, *2*, 144.
[20] a) D. F. V. Lewis, *Drug Metab. Rev.* **1997,** *29*, 589; b) D. F. V. Lewis, B. G. Lake, *Toxicology* **1998**, *125*, 31.
[21] T. Shimada, H. Yamazaki, M. Foroozesh, N. E. Hopkins, W. L. Alworth, F. P. Guengerich, *Chem. Res. Toxicol.* **1998**, *11*, 1048.
[22] K. Kobayashi, M. Nakajima, K. Chiba, T. Yamamoto, M. Tani, T. Ishizaki, Y. Kuroiwa, *Br. J. Clin. Pharmacol.* **1998**, *45*, 361.
[23] U. Fuhr, G. Strobl, F. Manaut, E.-M. Anders, F. Sorgel, E. Lopez-de-Brinas, D. T. W. Chu, A. G. Pernet, G. Mahr, F. Sanz, A. H. Staib, *Mol. Pharmacol.* **1993**, *43*, 191.
[24] A. L. Draper, A. Madan, A. Parkinson, *Arch. Biochem. Biophys.* **1997**, *341*, 47.
[25] J. P. Jones, M. He, W. F. Trager, A. E. Rettie, *Drug Metab. Dispos.* **1996**, *24*, 1.
[26] J. M. Morsman, D. A. Smith, B. C. Jones, G. M. Hawksworth, *ISSX Proceedings* **1995**, *8*, 259.
[27] D. J. Back, J. F. Tjia, J. Karbwang, J. Colbert, *Br. J. Clin. Pharmacol.* **1988**, *26*, 23.
[28] R. E. Lock, B. C. Jones, D. A. Smith, G. M. Hawksworth, *Hum. Exp. Toxicol.* **1998**, *17*, 514.
[29] R. E. Lock, B. C. Jones, D. A. Smith, G. M. Hawksworth, *Br. J. Clin. Pharmacol.* **1998**, *45*, 511P.
[30] G. R. Strobl, S. von Kruedener, J. Stockigt, F. P. Guengerich, T. Wolff, *J. Med. Chem.* **1993**, *36*, 1136.
[31] D. Wu, S. V. Otton, T. Inaba, W. Kalo, E. M. Sellers, *Biochem. Pharmacol.* **1997**, *53*, 1605.
[32] X-J. Zhao, T. Ishizaki, *Br. J. Clin. Pharmacol.* **1997**, *44*, 505.
[33] L. Pichard, I. Fabre, G. Fabre, J. Domergue, B. S. Aubert, G. Mourad, P. Maurel, *Drug Metab. Dispos.* **1990**, *18*, 595.
[34] L. Pichard, J. Domerergue, G. Fourtainer, P. Koch, H. F. Schran, P. Maurel, *Biochem. Pharmacol.* **1996**, *51*, 591.
[35] S. Ekins, M. J. De Groot, J. P. Jones, *Drug Metab. Dispos.* **2001**, *29*, 936.
[36] T. Shimada, H. Yamazaki, M. Mimura, Y. Inui, F. P. Guengerich, *J. Pharmacol. Exp. Ther.* **1994**, *270*, 414.
[37] S. Ekins, M. VandenBrunden, B. J. Ring, J. S. Gillespie, S. A. Wrighton, *Pharmacogenetics* **1997**, *7*, 165.
[38] Catalyst Software, *Accelrys Inc.*, San Diego, CA, http://www.accelrys.com.
[39] G. Bravi, E. Gancia, P. Mascani, M. Pegna, R. Todeschini, A. Zaliani, *J. Comput.-Aided Mol. Des.* **1997**, *11*, 79.
[40] T. L. Poulos, B. C. Finzel, A. J. Howard, *Biochemistry* **1986**, *25*, 5314.
[41] R. C. Onderwater, J. Venhorst, J. N. M. Commandeur, N. P. E. Vermeulen, *Chem. Res. Toxicol.* **1999**, *212*, 555.

[42] T. Lightfoot, S. W. Ellis, J. Mahling, M. J. Ackland, F. E. Blaney, G. J. Bijloo, M. J. de Groot, N. P. E. Vermeulen, *Xenobiotica* **2000**, *30*, 219.
[43] C. Fukushima, T. Harada, T. Kume, K. Fukuda, *16th Annual Meeting of the Japanese Society for the Study of Xenobiotics*, 2001.
[44] D. A. Smith, M. J. Ackland, B. C. Jones, *Drug Discov. Today* **1997**, *2*, 406;D. A. Smith, M. J. Ackland, B. C. Jones, *Drug Discov. Today* **1997**, *2*, 479.
[45] Y. Yoshida, Y. Aoyama, M. Noshiro, O. Gotoh, *Biochem. Biophys. Res. Commun.* **2000**, *273*, 799.
[46] A. Tafi, R. Costi, M. Botha, R. Di Snato, F. Corelli, S. Massa, A. Ciacci, F. Manetti, M. Artico, *J. Med. Chem.* **2002**, *45*, 2720.
[47] J. Liu, D. Pan, Y. Tseng, A. J. Hopfinger, *J. Chem. Inf. Comput. Sci.* **2003**, *43*, 2170.
[48] S. Ekins, B. J. Ring, G. Bravi, J. H. Wikel, S. A. Wrighton, in 'Pharmacophore Perception, Development, and Use in Drug Design', Ed. O. F. Güner, IUL Biotechnology Series, La Jolla, CA, 2000, p. 269–299.

The *BioPrint*® Approach for the Evaluation of ADMET Properties: Application to the Prediction of Cytochrome P450 2D6 Inhibition

by **Rafael Gozalbes***, **Frédérique Barbosa***, **Nicolas Froloff***, and **Dragos Horvath**[1])

Cerep, Molecular Modeling Department, 19 avenue du Québec, Courtaboeuf 1, Villebon sur Yvette, F-91951 Courtaboeuf Cedex
(phone and fax: 0033-1 60 92 60 00; e-mail: n.froloff@cerep.fr)

Abbreviations
ADMET: Absorption, distribution, metabolism, excretion, toxicity; CM-PNB: complete metric predictive neighborhood behavior models; CYP: cytochrome P450; DS-PNB: descriptor selection based predictive neighborhood behavior models; EFO: extended field overlap; FBPA: fuzzy bipolar pharmacophore autocorrelogram; GA: generic algorithm; LS: learning set; PK: pharmacokinetics; PTA: pharmacophore type areas; QSAR: quantitative structure–activity relationships; RMS: root mean square; VS: validation set.

---

1. Introduction

A good drug candidate requires a balance of potency, safety, and good pharmacokinetic (PK) properties. Traditionally, researchers from pharmaceutical companies concentrated their efforts on maximizing potency against biological targets, and the PK and toxicity issues were addressed later. Nevertheless, some of the main reasons for failure in drug discovery and development are poor PK and toxicity properties, allegedly responsible for more than 40% of all failures [1]. In consequence, PK and toxicity testing is now carried out at a much earlier stage of drug discovery, frequently in parallel with activity and selectivity studies, thus resulting in significant money and time savings.

In silico tools are particularly appreciated in this field, since they have several advantages with respect to *in vitro* and/or *in vivo* assessments: *a*) the possibility to screen a large number of compounds in a reduced time;

---

[1]) Current address: Institut de Biologie de Lille, CNRS-UMR 8525, Campus Institut Pasteur, 1 rue Calmette, F-59800 Lille.

b) the limitation of chemical synthesis efforts by working with virtual compounds, generated by computer; *c*) the possibility to evaluate simultaneously different properties for a given compound; and *d*) a better understanding of the relationship between ADMET (absorption, distribution, metabolism, excretion, toxicity) properties and molecular structure. The ADMET properties for which *in silico* prediction is desirable include properties such as solubility, permeability, metabolic stability, protein binding, central nervous system (CNS) penetration, oral bioavailability, cytochrome P450 interaction and inhibition, mutagenicity, or carcinogenicity.

There are different computational approaches for the prediction of ADMET properties, which are based on the development of mathematical models linking structure and activity [2]. These models vary in descriptors and statistical tools, as well as in prediction accuracy. The descriptors commonly applied are physicochemical descriptors (such as H-bond donors and acceptors, size, calculated or experimental log *P*, *etc.*), topological descriptors based on the application of the 'Graph Theory' to chemistry [3], or descriptors based on the recognition of molecular features (fragments, groups, or sites) [4]. Specifically tailored descriptors for ADMET predictions have also been described, such as those of the VolSurf program [5].

Concerning metabolism issues, several commercial products are designed to predict metabolites of organic molecules, most of them based on rule-based expert systems, such as META (*MultiCASE*) [6], METEOR (*Lhasa Ltd.*) [7] and MetabolExpert (*CompuDrug International Inc.*) [8]. Particular interest is devoted to the cytochromes P450 (CYPs), a superfamily of monooxygenases present in both eukaryotes and prokaryotes, and which are involved in a majority of redox metabolic pathways [9]. In humans, several CYPs are present in the liver, each with a spectrum of specific substrates, which can be overlapping [10]. Relevant drug–drug interactions are often associated with the inhibition or the induction of specific CYP enzymes.

The modeling of CYPs has been limited until now by the fact that crystallographic structures of CYPs were only available for bacterial and some mammalian (unfortunately not human) CYPs [11]. Molecular models of various mammalian CYPs have been constructed by homology modeling techniques, particularly for CYP2C9, 2C19, 2D6 [12][13], and 3A4 [14]. Homology models for human CYPs from bacterial enzyme sequences have been reviewed by *Lewis* [15] and cover human CYP1A2, 2C5, 2B6, 2C8, 2C9, 2C19, 2D6, and 2E1 from the crystal structure of CYP2C5, as well as a model of CYP3A4 based on CYP1A2. Recently, the first human CYP2C9 crystallographic structure has been described [16][17].

Various modeling techniques have proved to be useful in the understanding and prediction of CYP substrate recognition and metabolism, such as a combination of homology modeling and site-directed mutagenesis [18], pharmacophore and 3D-QSAR analysis [19], and protein and pharmacophore models for CYP 2D6 explaining the involvement of this enzyme in hydroxylation, *O*-demethylation, and *N*-dealkylation reactions [20]. Recently, a conformer and alignment independent CYP2C9 inhibition 3D-model has been reported, based on molecular interaction fields [21]. A method for predicting the likely sites at which molecules will be metabolized by CYP2C9 has been reported by the same authors [22]. This method relies on an original docking procedure. The ligand is placed into the CYP binding site to determine a ranking list of all the H-atoms of the ligand that can be metabolized by CYP2C9. *Ab initio* quantum-modeling techniques have also been employed to predict whether catalytic reactions will proceed [23] or to predict reaction mechanism based on activation energies [24].

Nowadays, higher-throughput experimental screening tests using human CYPs are available [25], and rapid microtiter plate assays to conduct CYP inhibition studies have been developed [26][27]. These improvements enhance the acquisition of biological activity data, which is of particular importance for the development of accurate predictive models. Nevertheless, one of the main issues in ADMET prediction remains the need of large databases on marketed drugs and unsuccessful drug candidates with unfavorable ADMET properties (as counter-examples) to elaborate robust models [28]. Several ADMET datasets have been published, but their relevance in terms of size and data quality remains questionable. In this context, we present here a QSAR model for CYP2D6 based on *BioPrint*® data [29].

The *BioPrint*® database is a large and homogeneous set of experimental data generated at *Cerep*, which includes more than 2200 'drug-like' compounds (including commercialized drugs, failed drugs, or reference compounds) tested internally on more than 170 assays (receptors, enzymes, ion channels, cellular functional tests and *in vitro* ADMET assays). In-house QSAR models based on neighborhood behavior, linear equations, and neural networks are systematically built wherever there is a good balance between the number of actives and inactives in a given assay [29]. Up to now, more than 20 QSAR models have been built at *Cerep*, some of which are listed in *Table 1*.

Table 1. *List of* Cerep *Models for Predicting ADME Properties*

Solubility	
Permeability	Apical to basolateral
	Basolateral to apical
	Passive
Log *D*	At pH 7.4
	At pH 6.5
CYP2D6	Inhibition
Serum protein binding	Percentage of binding
	Logarithm of the affinity constant for serum protein binding
	[log $f_{bound}/(1-f_{bound})$], where f_{bound} symbolizes the bound fraction.

2. *BioPrint*-Based QSAR Methodology

2.1. *Descriptors*

The elaboration of a QSAR model requires the previous encoding of compound structures as molecular descriptors, which capture information regarding the structural features responsible for the activity under a numerical form. Different descriptors encode molecular information of different nature [4] [30], and the ones that best suit for the construction of a given model have to be selected in one way or another (by practical limitations such as software availability, by the know-how of the modeler, by automated procedures that search for the relevant ones out of an extended pool of available choices, or usually by a combination of these three). There follows here a short overview of *Cerep* proprietary descriptors which consist of various 3D fingerprints carrying information about the nature and the spatial distribution of the various pharmacophoric groups in a molecule.

2.1.1. *Fuzzy Bipolar Pharmacophore Autocorrelograms* (FBPAs)

The FBPAs represent a pharmacophore fingerprint of the various conformers of a molecule. To generate the FBPA [31] of a compound, its atoms are first classified according to their pharmacophoric features (hydrophobicity, aromaticity, H-bond donor or acceptor propensity, positive or negative charge). Any atom may possess one or more such features, detected by a feature assignment routine, according to empirical rules [31]. The 21 pairs of these 6 features are defined (hydrophobic–hydrophobic, hydrophobic–aromatic, *etc.*). All the atom pairs occurring in a molecule are first assigned to one of these 21 pharmacophore pair categories and furthermore broken down into 12 interatomic distance related bins of 1 Å width, going from 3 to 15 Å. This defines a total of 252 classes to which an atom pair may belong. For example, a pair featuring an aromatic 'Ar' and a positive charge 'PC' at 6.5 Å apart belongs to the class labeled Ar-PC6 (see *Fig. 1* for a schematic representation). The fingerprint is thus a 252-dimensional vector in which every component represents the number of atom pairs associated to the given category, averaged over a diverse sample of conformers. Fuzzy logics is used to define the classification of a pair in a specific distance bin, avoiding artifacts for pairs for which their classification in the lower or upper bin could be decided by small fluctuations setting the actual distance slightly above or below the threshold [31].

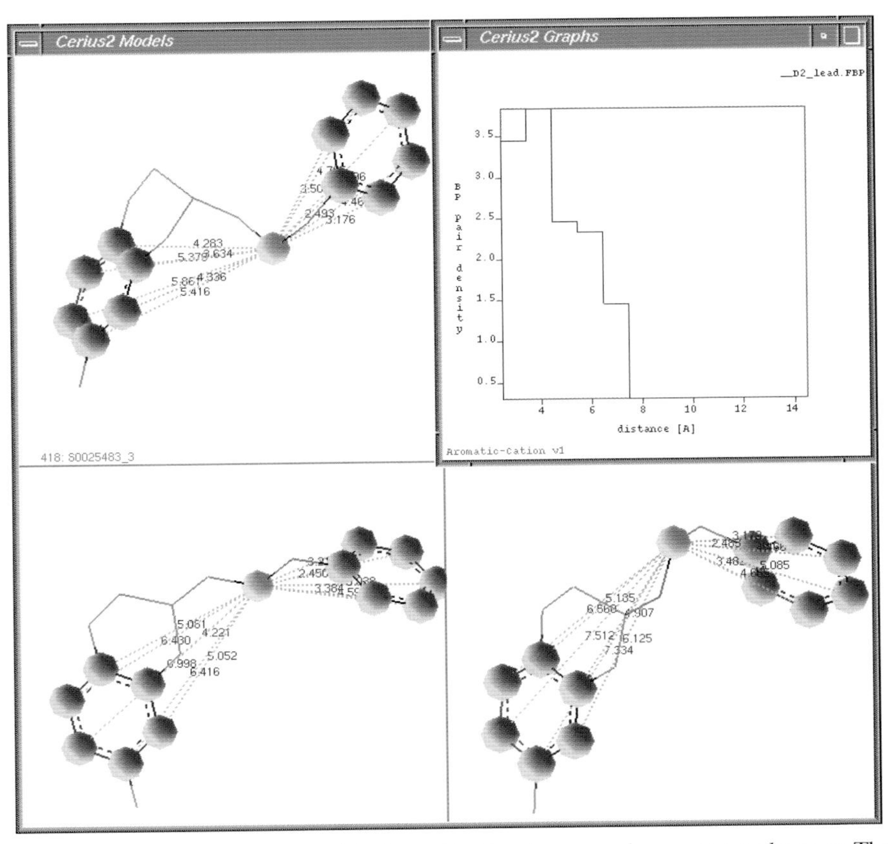

Fig. 1. *Graphical representation of the fuzzy bipolar pharmacophore autocorrelograms.* The distances between a positively charged atom (central, yellow) and the aromatic atoms are shown on the upper-left and the windows on the bottom for three different conformations. On the upper-right window, the average number of all the positive charge–aromatic pairs is represented.

2.1.2. *Pharmacophore Type Areas* (PTA)

The PTA report the molecular areas corresponding to each type of pharmacophoric features (aliphatic, aromatic, H-bond acceptor and donor, cationic and anionic areas).

2.1.3. *Extended Field Overlap* (EFO) *Descriptors*

The EFO descriptors offer a synthetic characterization of the spatial distribution of pharmacophoric features in the molecule: these terms are volume integrals of the pairwise products of local field intensities associated to each of the possible combinations of the pharmacophore types. For these descriptors, a more-detailed ('extended') definition of the pharmacophore types is used, featuring ten rather than six explicit pharmacophore types. For example, the H-bond donor class is further split into 'aromatic H-bond donors' (*e.g.*, indole >NH), 'H-bond donors and acceptors' (alcohol –OH), or 'H-bond donors only' (amide –CONH).

2.1.4. *Accounting for Nonlinear Transformations of the Descriptors*

Nonlinear transformations of the above-mentioned descriptors are under certain conditions used as novel, independent terms to enrich the initial descriptor pool, in quest for a suitable way to account for potentially nonlinear activity descriptor dependences in a (multi)linear regression model. For each descriptor D, sigmoid ($1/[1+e^{-z(D)}]$) or Gaussian ($\exp[-z(D)^2]$) transformations will be added as independent descriptors to the initial list if they are not redundant (correlated to already existing descriptors). In the equations above, $z(D)$ stands for the average/variance normalization function of D, placing D in context of its average and variance over the set of current drugs. This allows for a simple interpretation of the nonlinear transformations of descriptors: for example, the sigmoid transformation will tend to 0 whenever the D value is by several variance units larger than the average of D. Such a molecule 'stands out' from the set of current drugs due to its atypically large D value. On the contrary, Gaussian functions are maximal for the molecules characterized by descriptor values as close as possible to the average.

2.2. *QSAR Model Building*

After the filtering out of constant or strongly intercorrelated descriptor columns, an activity descriptor matrix relating experimental activity to the candidate molecular descriptors is built, and various descriptor selection algorithms integrated in our QSAR internal builder tool are applied to find the optimal structure–activity relationships. Synergy models based on two different approaches, linear regression or neural network on one hand, and predictive neighborhood behavior on the other, are employed [29].

The learning and validation sets are defined to proceed with the construction of the models. The validation set is picked randomly from the complete set. However, the selection procedure is designed such that the distribution of actives/inactives is as similar as possible for both the learning and validation sets.

2.3. *Observable Weighing*

Unlike in the classical medicinal chemistry series of analogues, the *BioPrint* collection is highly diverse and regroups compounds from many therapeutic areas. Therefore, only a minority of compounds will 'hit' any given target, and systematic prediction errors of the activities of the rare actives might therefore have little impact on the root-mean-square (RMS) prediction error used to assess the quality of the model. In other words, a null model assuming all compounds to be inactive may appear to perform 'quite well' in the sense that its predictions are almost always accurate – for all the compounds but the few percents of actives! To avoid such artifacts and compensate for the different population densities at the extremes of the activity range, an observable weighing procedure is applied to the learning set. The activity range is split into three user-defined domains of low, medium, and high activities, and each molecule is assigned a weighing factor that is inversely proportional to the population size of its category. Molecules belonging to an under-populated category are heavily weighed and will therefore contribute to the RMS as much as the ubiquitous but under-weighed inactives. R^2, Q^2, and RMS for the 'weighed learning set' are the default quality criteria for the statistical models reported here.

2.4. *Linear Models*

A genetic algorithm is used to select, from the activity descriptor matrix, several combinations of descriptor columns that yield near optimal multilinear regression models of the experimental activity in function of selected descriptors. The optimality (fitness) criterion used to determine which equation is best, corresponds to the cross-validated correlation coefficient Q^2 minus a penalty that is linearly related to the number of variables entering the equation. The pool of reasonably predictive linear equations is further subjected to a diversity analysis procedure to discard redundant equations based on roughly similar descriptor choices. This procedure assesses whether some of the different descriptors used by different equations are intercorrelated, and therefore interchangeable [32]. The remaining diverse QSAR equations are further classified by 'size' (number of descriptors they include). The best equations of each encountered size are kept for final validation on hand of the validation set.

2.5. *Neural Networks*

The QSAR builder supports neural net fitting as a generalization of the best linear models selected at the previous step. For each selected equation including less than 45 descriptors, the tool allows the user to proceed with fitting of three layer fully interconnected neural net, where each input neuron is associated to one of the descriptors of the linear source model, the hidden layer includes either one or three hidden neurons and the signal of the output neuron is related to experimental activity values by means of a linear regression equation. Each neuron is characterized by a discrete list of possible 'configurations' fully describing the state of the neuron, *e.g.*, specifying both the functional form of the transformation function to be used as well as the parameters associated to this function. A configuration is a triplet including an integer to code for the functional form (1: sigmoid, 2: Gaussian, 3: piecewise linear, 4: sinus) and two real parameters, the interpretation of which will depend on the current transformation function (for sigmoids, these will be the x coordinate of the inflexion point and the slope at the inflexion point, while, for Gaussians, they represent the coordinate of the peak and the peak width at half-height). Prior to neural net fitting, a list of 'meaningful' and 'diverse' configurations is assembled for each neuron, in function of the nature of input 'seen' by it. Meaningful configurations exclude the use of transformation functions set up such as to yield a constant output signal throughout the LS of compounds presented to the net. The predefined configurations have to be 'diverse' in the sense that the output firing pattern of the neuron over the entire LS of compounds must be significantly different from all other output patterns achieved with already accepted configurations. For example, if a neuron would be fed in a binary descriptor $D = \{0,1\}$, using a Gaussian transformer function $f(D) = \exp[-(D - D^0)^2]$ with the peak at $D^0 = 0.5$ would return a same value $f(0) = f(1)$ for any molecule, which is of no use. Also, the use of a sigmoid transformer returning $f(0) \approx 0$ and $f(1) \approx 1$ would be redundant, as the same firing pattern can be, in this case, achieved with the simpler linear 'transformer' $f(x) = x$. Once a list of potentially interesting configurations is built for each of the neurons, a genetic algorithm will sample the space of all possible combinations of these configurations, as well as the space of the synapse weighing factors. A conjugate gradient optimization is used to locally adjust the latter, at fixed neuron configurations, whenever a new 'fittest' individual has been evolved. This procedure accelerates the convergence towards a family of robust and predictive neural net models.

2.6. *Neighborhood Behavior Models*

Neighborhood behavior models – including the popular K-NN (K nearest neighbors) approaches [33], tend to extrapolate the property of a novel compound as an average of properties of reference molecules that are shown to be structurally similar – according to a well-defined computed similarity score. The success of such an approach based on the 'neighborhood behavior' principle (similar structures → similar properties) heavily depends on the exact definition of the similarity score used to select the nearest neighbors that serve for property prediction, as the set of N nearest neighbors $N_1, N_2, \ldots, N_n$, is defined as the compounds of minimal dissimilarity $S(C,N)$ with respect to the candidate compound C. The neighbor list will therefore depend on the nature of the similarity scoring scheme or similarity metric $S(C,N)$ – an empirical function returning a similarity score based on a comparison of the molecular descriptor values for molecules N and C.

2.6.1. *Complete Metric Predictive Neighborhood Behavior Models* (CM-PNB)

The similarity metric used in CM-PNB is a tunable empirical similarity score based on the fuzzy bipolar pharmacophore autocorrelogram (FBPA) metric [34][35], possibly combined with other scoring schemes using either EFO or PTA descriptors (*Eqn. 1*):

$$S(C,N) = S^{\mathrm{FBPA}}(C,N) + w^{\mathrm{PTA}} S^{\mathrm{PTA}}(C,N) + w^{\mathrm{EFO}} S^{\mathrm{EFO}}(C,N) \tag{1}$$

Here, S^{FBPA} is a fuzzy scoring scheme specially designed for use with FBPA descriptors, while S^{PTA} and S^{EFO} are average variance normalized Dice scores based on the totality of descriptors $D_i(C)$ and $D_i(N)$ from the PTA and EFO sets, respectively (*Eqn. 2*):

$$S^D(C,N) = \frac{2\sum D_i(C)D_i(N)}{\sum D_i^2(C) + \sum D_i^2(N)} \tag{2}$$

The FBPA similarity score includes fittable parameters for the relative weighs of the six considered pharmacophore feature types and the degree of fuzziness [31]. By contrast, PTA and EFO metrics are not tunable – their only degree of freedom is their weigh w^{EFO} or w^{PTA} in the consensus similarity score used for neighbor picking. The other two fittable parameters of CM-PNB models are:

- The dissimilarity cutoff value $s^{\max}$ sets an upper threshold for the dissimilarity of reference neighbors that may still contribute to the calculation of the average property returned as predicted property value for the candidate molecule.
- The dissimilarity distrust factor ω controls how steeply the relative contribution of remoter neighbors decreases (exponentially) in function of dissimilarity, when calculating the weighed average of the properties of the selected neighbors N (*Eqn. 3*):

$$P_{\mathrm{pred}}(C) = \frac{\sum_N P_{\mathrm{expt}}(N) \mathrm{e}^{-\omega S(C,N)}}{\sum_N \mathrm{e}^{-\omega S(C,N)}} \tag{3}$$

where the averaging concerns only the experimental properties $P_{\mathrm{expt}}(N)$ of the neighbors N that are close enough to C to be relevant ($S(C,N) < s^{\max}$).

The CM-PNB model is fitted by a Monte Carlo procedure exploring the 11-dimensional parameter space composed of the 7 FBPA metric parameters, w^{PTA}, w^{EFO}, $s^{\max}$, and ω, in search of the parameter combination triggering, for each compound C in the LS, the selection of neighbors N (different from the compound itself!) such that the above calculated $P_{\mathrm{pred}}(C)$ gets as close as possible to the actual value $P_{\mathrm{expt}}(C)$.

In addition to the calculated property, the model also returns, for each compound,

- a density criterion $\rho(C) = \Sigma e^{-\omega S(C,N)}$, expressing how well the current compound C is surrounded by relevant neighbors, and
- a homogeneity criterion $\eta^2(C)$ measuring the (weighed) variance of the property P within the set of selected neighbors N with respect to the overall variance of P throughout the LS. The term η^2 can be assimilated to a correlation coefficient (R^2), equaling 1 whenever all neighbors N have identical property values (variance within neighbor subset being 0) and reaching a minimum of 0 if the properties of selected neighbors differ with respect to each other as much as – or more than – the properties of any randomly selected compounds in the LS.

High density and homogeneity scores suggest that the prediction for a given compound is reliable, as C is shown to have many neighbors of roughly identical properties and is therefore likely to display a similar behavior as well. In CM-PNB, a predicted value is returned for *every* compound, even when $\rho(C)$ and $\eta^2(C)$ are low – however, an expected prediction uncertainty is estimated on hand of these quality scores and returned to the user. In the extreme, if a compound C possesses no neighbors within the dissimilarity radius s^{max}, the value $P(N)$ of its *nearest* neighbor will be returned as a prediction, no matter how far this neighbor actually is, but both density and homogeneity scores will be set to 0, thus signaling a high uncertainty of the output value.

2.6.2. *Descriptor Selection Based Predictive Neighborhood Behavior Models* (DS-PNB)

Unlike in CM-PNB, this methodology relies on the selection, out of the set of initially available terms from the activity descriptor matrix, of a subspace of relevant molecular descriptors to be used for neighbor selection. In this case, the metric based on the FBPA is not used since all 252 FBPA descriptors are necessary for the calculation of the distance between two neighbors.

The appropriate descriptors and metric are selected using a generic algorithm (GA)-based approach. Like in the procedure of GA-driven multilinear regression used for linear model fitting, a 'chromosome' is used to encode the status (used/discarded) of each candidate descriptor present in the input activity descriptor matrix. Each chromosome encodes thus a particular structural subspace, within which the molecular dissimilarity scores $S(C,N)$ are functions of selected descriptors only.

Besides its role in descriptor selection, the chromosome also accounts for DS-PNB specific parameters and further encodes a series of fittable parameters that need to be optimized. The optimal functional form of $S(C,N)$ is one of these: a specific locus in the chromosome possesses an integer code that stands for various possible choices: 1: *Euclidean* distance, 2: *Dice* index, 3: *Tanimoto* index, 4: cosine function [35]. Like in CM-PNB, the dissimilarity cutoff value s^{max} and the dissimilarity distrust factor ω also figure on the fittable parameter list. Unlike in CM-PNB, however, this methodology includes two more degrees of freedom: the minimal density (ρ_{min}) and the minimal homogeneity (η^2_{min}) criteria. Compounds with $\rho(C) < \rho_{min}$ or $\eta^2(C) \geq \eta^2_{min}$ will be considered 'unpredictable' as the neighborhood of C offers no guarantees for a meaningful application of the neighborhood behavior principle. The quality of the fit $P_{pred}(C)$ *vs.* $P_{expt}(C)$ will, of course, be evaluated only with respect to 'predictable' molecules with $\rho(C) \geq \rho_{min}$ or $\eta^2(C) \geq \eta^2_{min}$, and the objective function to be maximized by the genetic algorithm is chosen such as to ensure a simultaneous maximization of both the correlation between calculated and predicted properties and the fraction of predictable molecules. The GA will thus both find optimal 'recipes' to define a relevant structural subspace and the acceptance criteria of the PNB prediction.

2.7. *Synergy Models*

Linear models and neural nets on one hand and neighborhood behavior models on the other are two conceptually distinct and fully independent approaches to the modeling of a given property. The availability of two independent 'opinions' on the question of what the property of a novel compound should be is certainly of great practical value. If both predictions agree, then this can be interpreted as a supplementary indicator of confidence in model output. If, however, they do not, then it is expectable to find the 'truth' somewhere in between these two extreme values P_{lin} and P_{PNB} (here, the index 'lin' may also stand for the neural net derived on the basis of that equation). Synergy models assess the optimal way to calculate a most-representative weighed average of P_{lin} and P_{PNB}, depending on the confidence indices of the PNB prediction. At high density and homogeneity scores, it is assumed that the PNB prediction, based on concrete experimental data of doubtlessly related compounds, should be preferred to the linear/neural net estimation. At low density and/or homogeneity – and, in the extreme, for compounds that are 'unpredictable' according to the DS-PNB method – the output of the linear/neural net model would be a better, or the only available, estimation (although high dissimilarity with respect to LS examples is *per se* an indicator of high uncertainty of linear models as well). Without going into technical details, the calibration of a synergy model consists in finding an empirical function $f_{lin}(C)=f[\rho(C),\eta^2(C)]$ that returns the optimal participation of the linear/neural net prediction *vs.* the PNB prediction in establishing the optimally weighed average $P_{syn}(C)=f_{lin}(C)\,P_{lin}(C)+f_{PNB}(C)\,P_{PNB}(C)$ supposed to represent the 'synergy' prediction for a compound C, where the weighing of the neighborhood contribution is complementary to the one of the linear term ($f_{PNB}=1-f_{lin}$).

2.8. *Validation*

The fitness criterion employed by the GA-driven search for linear models is a *cross-validated* correlation coefficient, the chances of fortuitous correlations being picked up at this stage is already diminished. Cross-validation is performed in 10 steps: the learning set is divided into 10 subsets, each one being iteratively removed while a regression equation is being fitted on the basis of the remaining 9 others and then used to predict the activities of the compounds currently kept out. The herein predicted activities are then compared to the actual value to estimate the reported cross-validated RMS and correlation coefficient. A bad cross-validation suggests that the linear model has too many variables, or that some variables were included to specifically minimize the prediction error of a few examples of atypical structures.

Before confronting the obtained models with the VS compounds that have not been used for fitting, a randomization test is run on the LS compounds to assess whether the sheer number of initially available descriptors to select from would not allow for by-chance activity descriptor correlations to emerge. If so, correlations of similar quality are expected to arise when scrambling the activity data with respect to the descriptors: two independent GA-driven searches, in any respect identical to the actual mining for the actual linear models, are carried out after randomly associating activity values to other molecule's descriptors. Significantly higher degrees of correlation in the cases when activities are associated to the proper molecular descriptors are thus a necessary condition for a robust QSAR.

Eventually, the created model is used to predict the activity of the VS structures that were withheld during fitting. The comparison of the predicted and experimental activity on new structures gives indication on the generality of the model. The large number of data present in *BioPrint* gives the possibility of keeping part of the molecules for the validation set. This validation is a test for the ability of the constructed QSAR model to predict the activity of new molecules.

3. Predicting the Inhibition of CYP2D6 Using a Synergy Model

3.1. *Introduction*

The prediction of CYP2D6 inhibition is of great interest, since this isoform is one of the most-important drug-metabolizing CYPs, particularly for numerous psychotropic compounds. We have built QSAR models for predicting CYP2D6 inhibition propensity of small drug-like molecules on the basis of *BioPrint* data. Using our automatic QSAR builder, several linear, neural network, CM-PNB, and DS-PNB models have been generated and combined to yield optimal synergy models.

3.2. *Data Set*

The *BioPrint* data set is a diverse collection featuring most of the compounds that are currently marketed as drugs, withdrawn drugs, and 'failed' drug candidates, as well as drug-like candidates that are representative of typical 'leads' likely to be encountered in drug discovery. A distribution of this collection in terms of different physicochemical properties is presented in *Fig. 2*.

For the construction of the CYP2D6 inhibition QSAR model, 1995 molecules with valid inhibition data were selected. These molecules were split into a learning set (LS) and a validation set (VS) (1595 and 400 molecules, resp.). The CYP2D6 inhibition is expressed as $\mathrm{p}IC_{50}$ ($-\log(IC_{50})$) values. The distribution of LS and VS compounds with respect to their CYP2D6 inhibition propensity can be found in *Table 2*.

3.3. *Linear Regression*

The QSAR builder generated 32 linear models, containing from 4 to 35 descriptors. The weighed learning set RMS error varied from 1.27 to 0.80 ($\mathrm{p}IC_{50}$ units), making up for squared correlation coefficients R^2 ranging from 0.52 to 0.81 ($R^2 = 1 - \mathrm{RMS}^2/\sigma^2$). In absence of observable weighing, the correlation coefficient of the best model is $R^2 = 0.30$ (see *Table 3*). This can be first explained by the fact that the property variance within the

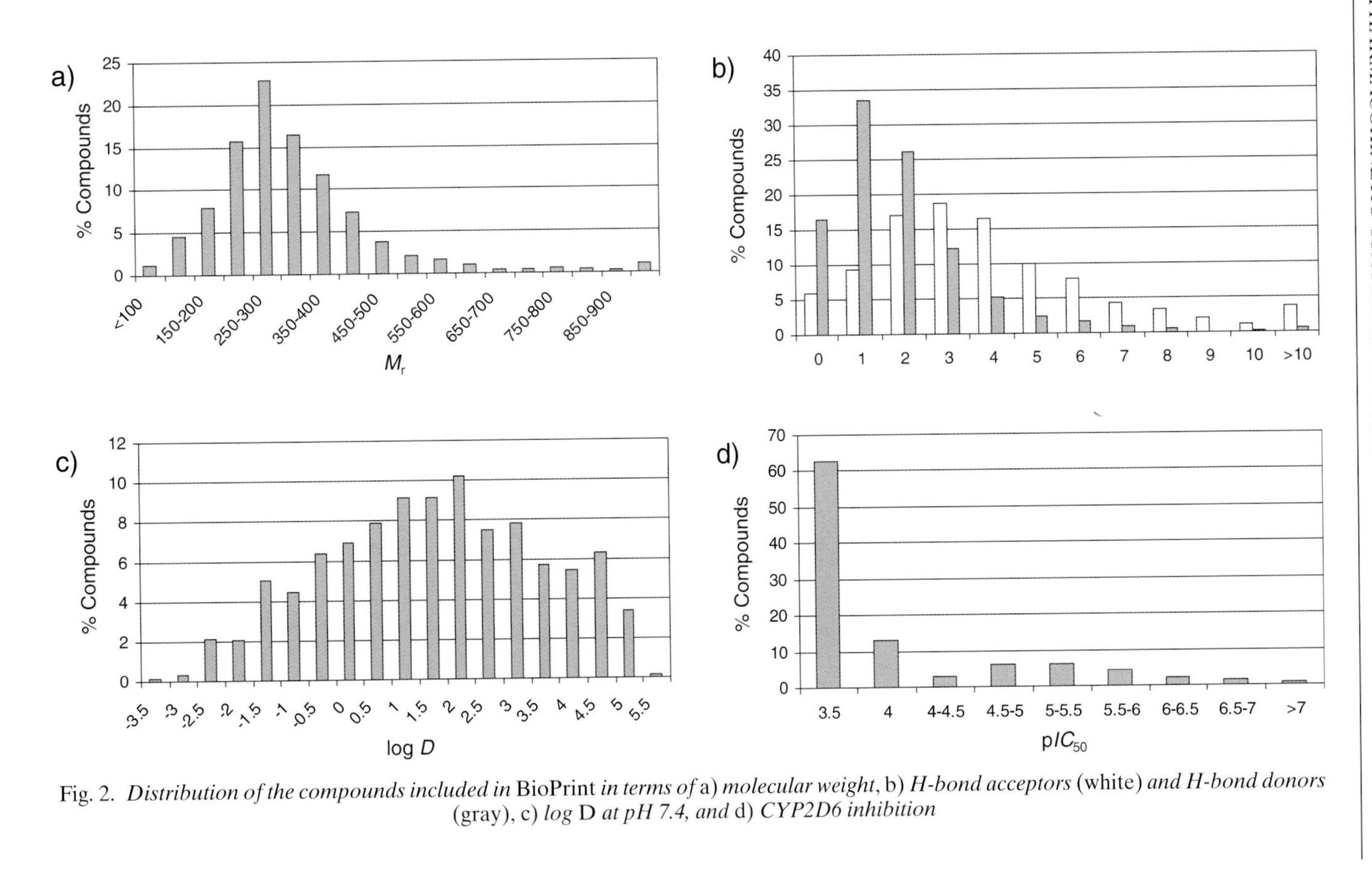

Fig. 2. *Distribution of the compounds included in* BioPrint *in terms of* a) *molecular weight*, b) *H-bond acceptors* (white) *and H-bond donors* (gray), c) *log* D *at pH 7.4, and* d) *CYP2D6 inhibition*

Table 2. *Distribution of* pIC_{50} *Values of the Inhibition of CYP 2D6 for the Learning Set and Validation Set*

	pIC_{50}		
	≤ 5.0	between 5.0 and 7.0	≥ 7.0
Learning set	1299	275	21
Validation set	326	69	5

Table 3. *Comparison of the* R^2 *and RMS Values for the Best Linear Model, the Neural Network, and the Best DS-PNB Model.* For the DS-PNB model, the percentage of predicted compounds is also presented.

		R^2	RMS	% predicted
Linear	Weighed learning set	0.81	0.80	
	Unweighed learning set	0.30	0.81	
	Validation set	0.31	0.81	
Neural network	Learning set	0.48	0.70	
	Validation set	0.44	0.73	
DS-PNB	Weighed learning set	0.60	0.79	84
	Unweighed learning set	0.58	0.51	84
	Validation set	0.57	0.58	85
Synergy model	Learning set	0.49	0.60	
	Validation set	0.53	0.63	

weighed LS σ^2 is much lower than for the unweighed LS, as, in the unweighed LS, the few strong actives have little impact on the overall property variance. The RMS error values calculated with and without observable weighing are almost identical (0.80 and 0.81, resp., for the best model), showing that the model did not accumulate systematic errors at either extreme of the activity range.

Furthermore, the prediction accuracy did not significantly decrease as the models were used to estimate VS compound affinities: validation RMS errors (in absence of weighing) varied from 1.05 to 0.81 pIC_{50} units. The validation R^2 values are low (up to 0.31) but comparable to the ones obtained for the unweighed LS.

As expected, models with more parameters systematically achieve lower fitting RMS errors with respect to the learning set. However, allowing more and more descriptors to enter the model will eventually lead to overfitting artifacts: higher-order equations are meaningful only if their validation performance is also more convincing than the one of models with fewer variables.

One of the descriptors with a major contribution is the number of aromatic positive charge pairs within the distance bin 5–6 Å. This is in accordance with published small molecule models for CYP2D6 [36–38] which state the importance of having a charged N-atom within 5–7 Å of an abstractable H-atom.

3.4. *Neural Networks*

A family of neural networks, with a unique hidden layer of either one or three hidden neurons, were fitted on the basis of the 35 descriptors selected in the equation giving the best prediction. For all the neural network models, the RMS and R^2 values for the LS are *ca.* 0.70 and 0.47 resp. (see *Table 3*). These values should be compared to those obtained for the linear model with the unweighed learning set. The improvement for the RMS in the overall predictive power of neural networks with respect to their parent linear models is *ca.* 15%. The VS is also well predicted with a RMS of 0.73 and a correlation coefficient R^2 of 0.44 (see *Table 3*).

3.5. *Predictive Neighborhood Model*

For the prediction of CYP2D6 inhibition, the DS-PNB approach proved more efficient than the CM-PNB. The best DS-PNB model led to a strong correlation of calculated *vs.* experimental values for the weighed LS (RMS = 0.79; R^2 = 0.60, see *Table 3*). The correlation (RMS = 0.51; R^2 = 0.58, see *Table 3*) within the unweighed LS is as expected slightly less significant. The DS-PNB applies to 84% of the learning set compounds, *i.e.* those which have a minimal number of effective neighbors fulfilling the required homogeneity criterion.

An RMS of 0.58 for the validation set is quite acceptable as compared to the one for the unweighed learning set (0.51). Of the VS, 85% are predictable according to the best DS-PNB model (see *Table 3*).

3.6. *Synergy Model*

The combination of prediction using neural network and DS-PNB leads to 20 models. The RMS were *ca.* 0.60 and 0.64 for the LS and VS, respectively, for all combinations (see *Table 3*). The RMS largely decreased with respect to the values obtained with the linear or neural network models. There is a slight increase of the RMS when compared to the DS-PNB predictions, which, however, included only 85% of the molecules found to be well surrounded by neighbors in structure space.

Table 4. *Statistics of the CYP2D6 Inhibition Synergy Model for the Learning Set.* The neural net and PNB performances are given in parentheses and brackets, resp. The PNB statistics apply only to the PNB-predictable compound subset. The 'expt' indices refer to the experimental values measured at *Cerep* and available in *BioPrint*. The 'pred' indices refer to the predicted values. The L, M, and H refer to low, medium, and high pIC_{50} categories with cutoff values of 5.0 and 7.0, resp. The '% Pure' represents the percentages of correctly predicted compounds within one category. The '% Found' represents the percentages of correctly found compounds within one category. The overall percentages of correctly predicted compound are found in the lower right box.

	L^{expt}	M^{expt}	H^{expt}	% Pure
L^{pred}	78.0 (77.2) [72.3]	4.1 (3.7) [3.1]	0.0 (0.0) [0.1]	94.94 (95.38) [95.78]
M^{pred}	4.8 (5.4) [2.7]	10.5 (10.4) [5.1]	0.9 (0.7) [0.4]	64.85 (62.71) [62.27]
H^{pred}	0.2 (0.4) [0.0]	0.6 (1.1) [0.0]	0.9 (1.0) [0.2]	52.40 (39.84) [97.56]
% Found	93.97 (92.99) [96.43]	68.86 (68.35) [62.39]	49.58 (56.04) [22.58]	89.38 (88.60) [92.49]

Table 5. *Statistic of the CYP2D6 Inhibition Synergy Model for the Validation Set.* See *Table 4* for explanation.

	L^{expt}	M^{expt}	H^{expt}	% Pure
L^{pred}	77.5 (76.4) [72.0]	5.0 (4.3) [3.4]	0.0 (0.1) [0.0]	93.94 (94.58) [95.47]
M^{pred}	5.3 (6.4) [3.2]	9.7 (8.9) [5.3]	1.9 (1.5) [1.2]	57.37 (52.89) [54.64]
H^{pred}	0.2 (0.2) [0.0]	0.4 (1.9) [0.1]	0.0 (0.3) [0.0]	0.00 (10.85) [0.00]
% Found	93.30 (92.03) [95.72]	64.13 (58.93) [60.04]	0.00 (13.89) [0.00]	87.16 (85.59) [90.66]

The confusion matrices for the LS and VS are shown in *Tables 4* and *5*, respectively.

Figs. 3 and *4* are experimental *vs.* predicted pIC_{50} plots for the LS and the VS, respectively, outlining the quality of the overall model.

4. Conclusions

We have briefly reviewed herein the computational approaches that are used in the field of ADMET prediction, and we presented a QSAR model that we developed for the prediction of the cytochrome P450 2D6 inhibition propensity of small organic molecules.

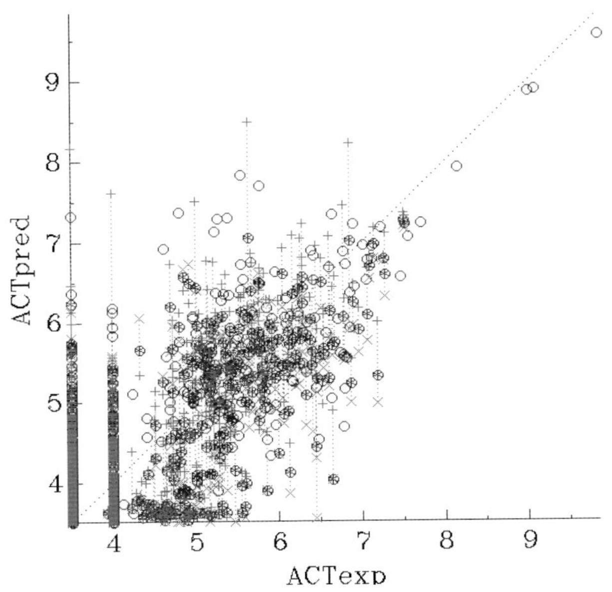

Fig. 3. *Experimental* vs. *predicted values for the inhibition of CYP2D6 of the learning set.* Blue circles mark the position of the synergy model predictions: when both neural and PNB predictions are available, the corresponding marker is spiked. By contrast, hollow blue circles mark the predicted values for the molecules failing to be predicted by the PNB approach – in the case when the synergy prediction equals the neural net based prediction. When both neural net and PNB values exist, the dotted red bars span the range between the predicted values of the neural (red +) and PNB models (red ×), resp. As synergy predictions are a weighed average of the two extreme outputs of the linear and PNB models, each blue spiked dot will fall somewhere inside the range spanned by the red bar, though not necessarily at its center (the better the PNB confidence criteria, the closer it will be to the × marker of the PNB extreme). Although the bars are not error bars in the classical sense of the term, coincidence of PNB and linear predictions, *e.g.*, short bars can be interpreted as an indication of reliability of prediction.

The originality and strength of our approach relies, first, on the high quality and completeness of the *BioPrint* data that were used for building the model, second, on the original descriptors that capture the essential pharmacophoric features of the modeled ligands, and third, on the combination (synergy) of linear/neural net models and neighborhood behavior models which are independent ways of identifying correlations between molecular description and experimental activity.

This has led to the development of a robust and validated QSAR model for CYP2D6 inhibition to be used as a virtual screening tool in our 'toolbox' of QSAR mass prediction models derived on hand of the *BioPrint* database. The ultimate goal is to integrate all essential ADMET

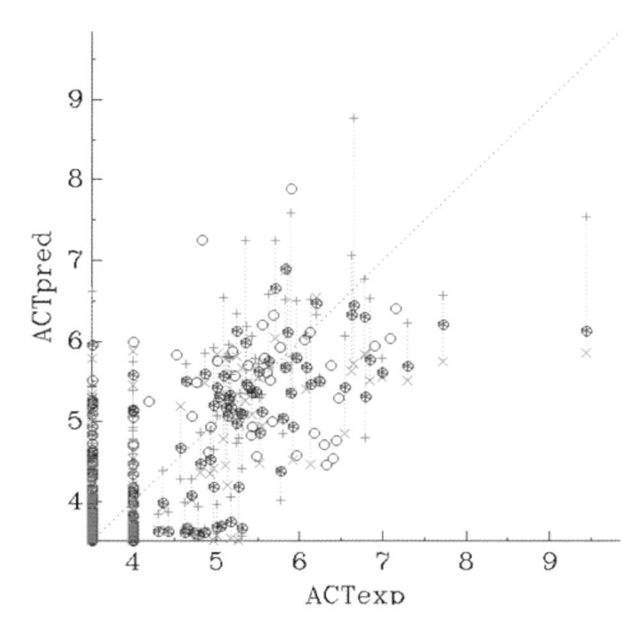

Fig. 4. *Experimental* vs. *predicted values for the inhibition of CYP2D6 of the validation set.* For explanations, see *Fig. 3*.

characteristics as early as possible in the design of new chemical libraries for accelerated drug discovery and development.

We acknowledge the dedication of all the people at *Cerep* involved in the *BioPrint* project and in the data acquisition. The *in vitro* ADMET data were created in the laboratory headed by *Cheryl Wu*.

REFERENCES

[1] R. A. Prentis, Y. Lis, S. R. Walker, *Br. J. Clin. Pharmacol.* **1998**, *25*, 387.
[2] H. van de Waterbeemd, *Curr. Opin. Drug Disc. Dev.* **2002**, *5*, 33.
[3] G. W. Milne, *J. Chem. Inf. Comput. Sci.* **1997**, *37*, 639.
[4] 'Handbook of Molecular Descriptors', Eds. R. Todeschini, V. Consonni, Wiley-VCH, Weinheim, 2003.
[5] G. Cruciani, P. Crivori, P. A. Carrupt, B. Testa, *THEOCHEM* **2000**, *503*, 17.
[6] G. Klopman, M. Dimayuga, J. Talafous, *J. Chem. Inf. Comput. Sci.* **1994**, *34*, 1320.
[7] J. Langowski, A. Long, *Adv. Drug Delivery Rev.* **2002**, *54*, 407.
[8] D. Butina, M. D. Segall, K. Frankcombe, *Drug Disc. Today* **2002**, *7*, S83.
[9] D. R. Nelson, L. Koymans, T. Kamataki, J. J. Stegeman, R. Feyereisen, D. J. Waxman, M. R. Waterman, O. Gotoh, M. J. Coon, R. W. Estabrook, I. C. Gunsalus, D. W. Nebert, *Pharmacogenetics* **1996**, *6*, 1.
[10] S. Cholerton, A. K. Daly, J. R. Idle, *Trends Pharmacol. Sci.* **1992**, *13*, 434.
[11] S. B. Kirton, C. A. Baxter, M. J. Sutcliffe, *Adv. Drug Delivery Rev.* **2002**, *4*, 385.
[12] D. F. Lewis, P. J. Eddershaw, P. S. Goldfarb, M. H. Tarbit, *Xenobiotica* **1997**, *27*, 319.

[13] D. F. Lewis, M. Dickins, R. J. Weaver, P. J. Eddershaw, P. S. Goldfarb, M. H. Tarbit, *Xenobiotica* **1998**, *28*, 235.
[14] G. D. Szklarz, J. R. Halpert, *J. Comput.-Aided Mol. Des.* **1997**, *11*, 265.
[15] D. F. Lewis, *J.* Inorg. *Biochem.* **2002**, *91*, 502.
[16] L. Wes, *BioCentury* **2001**, *9*, A11.
[17] H. Jhoti, P. Williams, A. Ward, J. Cosme, *Drug Metab. Rev.* **2002**, *34*, 10.
[18] T. L. Domanski, J. R. Halpert, *Curr. Drug Metab.* **2001**, *2*, 117.
[19] S. Ekins, M. J. de Groot, J. P. Jones, *Drug Metab. Dispos.* **2001**, *29*, 936.
[20] M. J. de Groot, M. J. Ackland, V. A. Home, A. A. Alex, B. C. Jones, *J. Med. Chem.* **1999**, *42*, 1515.
[21] L. Afzelius, I. Zamora, C. M. Masimirembwa, A. Karlen, T. B. Andersson, S. Mecucci, M. Baroni, G. Cruciani, *J. Med. Chem.* **2004**, *47*, 907.
[22] I. Zamora, L. Afzelius, G. Cruciani, *J. Med. Chem.* **2003**, *46*, 2313.
[23] M. D. Segall, M. C. Payne, S. W. Ellis, G. T. Tucker, R. N. Boyes, *Eur. J. Drug Metab. Pharmacokinet.* **1997**, *22*, 283.
[24] J. P. Jones, M. Mysinger, K. R. Korzekwa, *Drug Metab. Dispos.* **2002**, *30*, 7.
[25] C. L. Crespi, *Curr. Opin. Drug Disc. Dev.* **1999**, *2*, 15.
[26] P. J. Eddershaw, M. Dickins, *Pharm. Sci. Technol. Today* **1999**, *2*, 13.
[27] C. L. Crespi, D. M. Stresser, *J. Pharmacol. Toxicol. Methods* **2000**, *44*, 325.
[28] A. D. Rodrigues, G. A. Winchell, M. R. Dobrinska, *J. Clin. Pharmacol.* **2001**, *41*, 368.
[29] C. M. Krejsa, D. Horvath, S. L. Rogalski, J. E. Penzotti, B. Mao, F. Barbosa, J. C. Migeon, *Curr. Opin. Drug Disc. Dev.* **2003**, *6*, 470.
[30] D. J. Livingstone, *J. Chem. Inf. Comput. Sci.* **2000**, *40*, 195.
[31] D. Horvath, in 'Combinatorial Library Design and Evaluation: Principles, Software Tools and Applications', Eds. A. Ghose, V. Viswanadhan, Dekker, New York, 2001, p. 429–472.
[32] R. Todeschini, *Anal. Chim. Acta* **1997**, *348*, 419.
[33] A. Tropsha, W. Zheng, *Curr. Pharm. Design* **2001**, *7*, 599.
[34] D. Horvath, C. Jeandenans, *J. Chem. Inf. Comput. Sci.* **2003**, *43*, 680.
[35] D. Horvath, C. Jeandenans, *J. Chem. Inf. Comput. Sci.* **2003**, *43*, 691.
[36] T. Wolff, L. M. Distelrath, M. T. Worthington, J. D. Groopman, G. J. Hammons, F. F. Kadlubar, R. A. Prough, M. V. Martin, F. P. Guengerich, *Cancer Res.* **1985**, *45*, 2116.
[37] U. A. Meyer, J. Gut, T. Kronbach, C. Skoda, U. T. Meier, T. Catin, *Xenobiotica* **1986**, *16*, 449.
[38] S. A. Islam, C. R. Wolf, M. S. Lennard, J. E. Sternberg, *Carcinogenesis* **1991**, *12*, 2211.

Using Computer Reasoning about Qualitative and Quantitative Information to Predict Metabolism and Toxicity

by **Philip Judson**

LHASA Ltd., 22–23 Blenheim Terrace, Woodhouse Lane, Leeds LS2 9HD, UK
(e-mail: philip.judson@lhasalimited.org)

Abbreviations
DfW: DEREK for Windows; GC/MS: gas chromatography and mass spectrometry; LA: logic of argumentation.

---

1. Introduction

Scientists developing new chemicals need to answer questions such as 'will this compound be toxic in humans?' and 'I have found something by GC/MS in a sample from an animal study but is it a metabolite of my test compound and, if so, what is it and how was it formed?'. In the first case, the compound may not exist at the time when the question is asked, so that prediction has to be based on theory or speculation. In the second case, it is necessary to construct reasonable, putative metabolic trees leading to products with the formulae determined by MS.

Broadly, three approaches have been taken to trying to deal with problems of this kind: quantitative prediction using adaptations of standard methods for multivariate data analysis [1–5], probabilistic methods based on frequency of occurrence of molecular fragments in molecules [5] [6], and a heuristic approach in which human experts make predictions from judgments based on experience [6–10] (the classification proposed here is a simplification and the references provide illustrations rather than falling strictly into these classes).

The first approach can be helpful in lead optimization, where structures in the set under study are fairly similar and the researcher has some control over the variables that influence activity; its weakness is that the values it predicts are unreliable when a new variable is introduced, or one that had not been perceived to be relevant turns out to be, and it is not always easy to judge when these circumstances might apply. A further weakness is that input data must be in forms suitable for numerical analysis and so qualitative observations must either be ignored or represented by numbers, which may be of dubious validity. The output from this approach is numerical; humans tend to attach high credence to numerical values, which is good when the predictions are correct but potentially very damaging when they are not.

The second approach can be effective in discovering features in molecules that appear to influence, or even be responsible for, activity or reactivity. Analyses do not depend upon the input of numerical values for properties and so this approach can operate where conventional statistical methods are unsatisfactory. A weakness, shared with the conventional statistical methods, is that the resultant models may have no mechanistic basis and, if they do, that information is not explicit. Where systems apply probability theory in a simplistic way, a problem arises in that probability theory is about chance events and biologically controlled chemistry may not obey the rules of chance.

The third approach attempts to solve the problem of making predictions when only qualitative observations are available. Human experts are moderately successful at predicting toxicological hazard, or suggesting likely metabolic transformations, for novel compounds by looking at their structures and drawing on previous experience. Computer systems based on a heuristic approach seek to mimic this method. Knowledge is compiled by human experts and stored in a knowledge base to which the program can refer when it is presented with a query about a novel structure. So, to take a trivial example, a human expert might record in the knowledge base for a metabolism prediction system that many *N*-methyl compounds undergo metabolic demethylation. Given any specific query compound containing an *N*-methyl substituent, the system will be able to predict potential demethylation. In practice, there will be potential for competition between different metabolic reactions, and a useful system needs also to contain knowledge about which reactions dominate and under what circumstances.

A human expert with access to tools of these kinds would probably make use of information from all of them. The expert might, for example, note that a molecule of interest contained no structural features normally associated with toxicity but that it did contain a feature that could be converted by metabolism to one causing concern. A quantitative model appropriate for the putative metabolite might predict high activity. On the other hand, calculation of the physical properties of the original molecule, together with a consideration of its intended use, might suggest that it was unlikely to reach a site of metabolism in a mammalian system. Asked by a colleague for advice, the expert would consider all these factors to make a judgment about how much concern to express about the compound, and would qualify his/her advice by describing the salient points about competing indications.

This paper describes the use of reasoning in computer systems intended to give the kind of help that a human adviser might give: DEREK for Windows (DfW) advises on the potential toxicity and METEOR on the potential metabolic fate of chemicals.

2. DEREK for Windows and METEOR

These two programs are incorporated into a single application. A user wishing to get advice about a compound draws the structure of the compound using a computer mouse or similar device, or imports the structure from a Molfile or SDFile[1]).

DfW maps toxicological alerts – substructural features associated with toxicity and stored in its knowledge base – against the query molecule. If it finds a mapping it reports the hazard to the user. The alert is highlighted in color in the display of the query molecule and the toxicological end point associated with the alert is reported. The reasoning engine takes into account information about the likelihood that toxicity will be expressed, which can vary according to the species for which the prediction is being made, the physicochemical properties of the query molecule, *etc.* If the exact structure of the query is in the database of examples together with biological data, this influences the program's assessment of likelihood (the degree of influence will depend on whether activity in the query species is included in the data, for example). The likelihood of activity is included in the report to the user, and is expressed using a set of linguistic terms: certain, probable, plausible, equivocal, doubted, improbable, impossible, open, and contradicted. The terms are formally defined within the system – for example, 'probable' means that there are arguments in favor of the prediction and no arguments against it. Justification for the rules and alerts in the system is provided in the form of notes about them by their creators, literature references supporting the generalizations on which the rules or alerts are based, and specific examples from the literature of compounds that support the generalizations.

[1]) Molfiles and SDFiles are regarded as *de facto* standards for the communication of chemical structural information between computer programs. Their formats were developed and published by *MDL Information Systems*.

METEOR maps substructural fragments that key biotransformations in its knowledge base onto the query compound. It takes account of factors that make it more or less likely that biotransformations will be seen in practice and displays those that are expected. The program goes on to process the first level metabolites similarly and will continue to deeper levels until it finds no further likely reactions under the constraints set for the search. Where reactions are in competition, the program considers which are more likely to dominate, where knowledge is available. The user can set constraints such as a block on processing of metabolites with partition properties that would make excretion more likely than further metabolism, or a restriction to displaying and further processing only the products of reactions at or above a particular level of likelihood.

Compounds entered into DfW can be transferred to METEOR, and metabolites generated by METEOR can be transferred to DfW, for processing.

3. Human Reasoning

A human expert reasons about information from diverse sources. In a favorable situation, the answer to a question may have been determined experimentally and recorded in a database. On finding the entry, the expert will make judgments about the usefulness of the information. It may be necessary to explore how it was determined, whether there is corroboration from an independent source, and so on; but if the right validation criteria are met, the answer is frequently regarded as reliable enough to be accepted and used.

If the answer to a question has not been determined experimentally, and it is not practical or desirable to carry out an experiment, the expert seeks to predict the answer; even if the answer has been determined experimentally, the expert may wish to compare it with predictions as a cross-check. There may be a quantitative structure–activity relationship that is valid for the query compound; there may be general principles of chemical reactivity or metabolic chemistry that an expert believes to apply to the compound; the compound might fit a well-tried model for substrates to a particular enzyme; the physical properties and the uses of the compound might be relevant, or even decisive.

When dealing with complicated questions, human experts try to consider all the information available, weighing arguments for and against particular conclusions. This process of reasoning may be conscious or even formalized – for example, in the way that cases are presented and considered in a court of law – or more intuitive. Although it is human nature to ask for a categorical answer, preferably a reliable, numerical one, this is frequently not possible, and experts often express qualified opinions rather than firm answers. A person receiving such advice from an expert finds it easier to make decisions if the expert explains how the conclusion was reached. Being told, for example, that 'travel by road today may be hazardous' is less useful than being told 'travel by road today may be hazardous because there may be icy patches on high ground'.

So, to mimic the human expert, a computer needs to draw on a wide variety of types of information, some quantitative, some qualitative, and perhaps some speculative. In favorable cases, there may be sufficient reliable information to reach a numerical conclusion or to attach a numerical probability to a prediction. In other cases, predictions will be less clear. For predictions to be useful, the computer must be able to explain how they were reached.

4. Using the Logic of Argumentation

The logic of argumentation (LA) [11] [12] seeks to formalize reasoning. Arguments for and against a proposition are aggregated and weighed against each other to reach a view about how likely it is that the proposition will be true (or false). The assessment of likelihood may be qualitative or quantitative depending upon the reliability of the grounds of the arguments and upon the reliability of the arguments themselves, but the complexity of our fields of interest – xenobiotic metabolism and chemical toxicity – means that likelihood is currently always expressed qualitatively in METEOR and DfW.

As originally formulated and implemented in computer systems, LA was designed to work with the likelihood of individual propositions. For example, how likely it is that a query compound will have a particular toxicity. In metabolism, it may not be most useful to ask how likely a reaction is, but rather whether the reaction is more likely than others. This may seem to be of no great importance as far as LA is concerned – if one reaction is very likely and another reaction is distinctly unlikely then clearly the first is more likely than the second. Conversely, given some scale of likelihood, if the likelihood of one reaction is known and another reaction is more likely than that, then it must be at a place higher on the scale. It is hypothetically possible to construct complete descriptions of a domain of knowledge based entirely on *absolute* or entirely on *relative* reasoning (where a statement such as 'A is probable' is classed as 'absolute', and one such as 'B is more likely than C' is classed as 'relative'). In practice, there are big gaps in our knowledge, and so neither absolute nor relative reasoning can describe a domain completely.

These issues are discussed in more detail elsewhere, where refinements to LA [13] and the basic use of reasoning in DfW [14] and METEOR [15] are described. For the purposes of this chapter, it is sufficient to state that we use implementations of LA in both programs and that METEOR applies absolute and relative reasoning in parallel. The user can explore the line of reasoning leading DfW or METEOR to a conclusion.

Our aim in using LA is to make it possible for predictions to take account of diverse information. It is unnecessary, and probably near impossible, to build a program incorporating all the methods used to support prediction of metabolism or toxicity; METEOR and DfW get information from external packages to support their reasoning. So, for example, if a rule says that the likelihood of a conclusion depends upon log *P* for the query compound, a value can be sought automatically in a database or requested from a program such as ClogP. At this stage in their development, the programs make limited use of calls to other programs and databases, but we are adding links to more of them.

5. Illustrations of Advice

The following are idealized illustrations of advice a computer system needs to give if it is to mimic a human expert. They are intended to show the kind of knowledge that should be communicated. DfW and METEOR do not generate reports in the form of sentences, but they can provide the same information content.

5.1. *Illustration 1: Neurotoxicity Influenced by Physical Properties*

The following prediction can be generated by reasoning from computer-generated information that the structure of a query substance contains a particular substructure, alert_X, and that the log *P* of the substance is estimated to be -1.5, given these rules:

- if acetylcholinesterase_inhibition is certain, then neural_toxicity is probable;
- if substance_reaches_synapses is impossible, then acetylcholinesterase_inhibition is improbable;
- if log $P < -1.0$ is certain, then substance_reaches_synapses is improbable;

- if alert_X_present is certain, then acetylcholinesterase_inhibition is plausible.

'It is doubted that your substance is a neurotoxin because, although it contains alert_X, which is associated with acetylcholinesterase inhibition, it has a log *P* of -1.5 and it is therefore improbable that it will reach nerve synapses.'

5.2. *Illustration 2: Toxic Metabolites*

Having a link between a toxicity prediction system, a metabolism prediction system, and a database system makes predictions of the following kind possible:

'It is plausible that your substance will be of high acute toxicity, because, although it contains no alerts for toxicity, its expected primary metabolite is a known toxin with acute oral $LD_{50} = 0.1$ mg/kg.'

Although DfW and METEOR would be technically capable of supporting this kind of prediction, they do not currently do so.

6. Examples

These examples show how DfW and METEOR currently use reasoning to qualify their predictions. The terms used to express likelihood have defined meanings in the programs [14], but, for the purposes of this paper, it is sufficient to state that strength of belief in the truth of a proposition is ranked, in decreasing confidence, in the order CERTAIN > PROBABLE > PLAUSIBLE, and in the falseness of a proposition in the order IMPOSSIBLE > IMPROBABLE > DOUBTED. EQUIVOCAL indicates an equal degree of support for truth and falseness, OPEN means that there is no pertinent information on which to make a judgment, and CONTRADICTED means that there is apparently evidence for both the certainty and the impossibility of a proposition.

6.1. *DfW – Skin Sensitization Potential of 2,3,4-Trihydroxybutanal*

The skin sensitization prediction for 2,3,4-trihydroxybutanal shown in *Fig. 1* illustrates the use of information about the presence of a structural alert (toxicophore) in a query compound and the influence of a physicochemical property on activity. The figure shows the window in which an

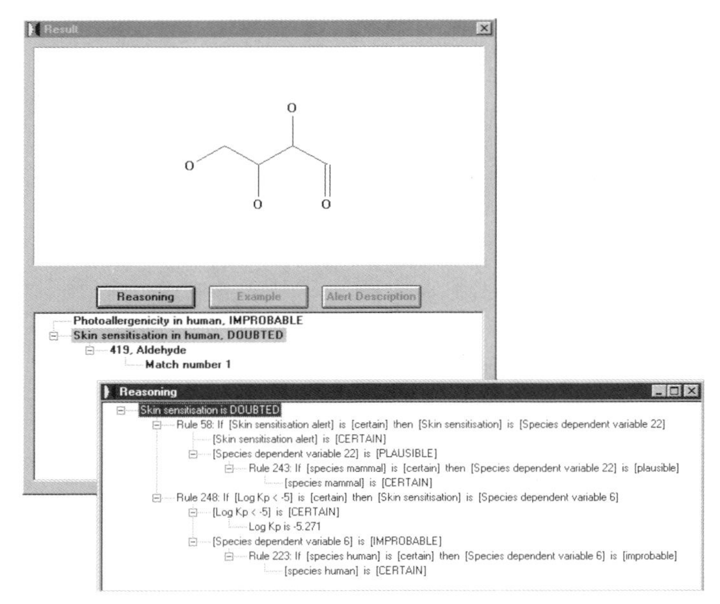

Fig. 1. *Result and reasoning displays for the skin sensitization prediction of 2,3,4-trihydroxybutanal in humans in DEREK for Windows*

overall assessment is presented and the window in which the reasoning process can be explored. The level of belief associated with this prediction is DOUBTED and the reasoning window shows that the conclusion is the result of two arguments.

Aldehyde alert 419 is present in the query structure, and, according to rule 58, if there is a skin sensitization alert then the likelihood of activity depends upon the species of interest. Rule 243 states that the value should be PLAUSIBLE if the user is asking about activity in mammals and the substatement appended to it reports that to be the case ('[species_ mammal] is CERTAIN'). So the likelihood of skin sensitization predicted by rule 58 is PLAUSIBLE.

Rule 248 relates activity to skin permeability (log K_p), which is believed to influence the potential for chemicals to cause skin sensitization in practice [16], and the values used for its supporting rules cause it to argue against activity: the chosen species is human and the value of log K_p, estimated outside the reasoning system according to the *Potts–Guy* equation [17][2]) and returned to it, is less than -5, and so rule 248 considers

[2]) This equation requires an estimate of log P which DfW currently gets from the ClogP plug-in, which is produced and supplied by *BioByte Corp.* and is also available from *LHASA Ltd.*

skin sensitization to be IMPROBABLE. Balancing the terms PLAUSIBLE from rule 58 and IMPROBABLE from rule 248 according to the resolution matrix currently used in DfW [14] leads to the conclusion that it is DOUBTED that 2,3,4-trihydroxybutanal will cause skin sensitization in humans.

6.2. *DfW – Peroxisome Proliferation*

There is believed to be a relationship between peroxisome proliferation and the observation of tumors of the liver in certain rodent species [18]. In the DfW knowledge base, this relationship is expressed in a rule:

'If [Peroxisome proliferation] is [certain] then [Carcinogenicity] is [probable]'.

Fig. 2 shows the result of processing a compound containing an alert, number 255 in the knowledge base, for peroxisome proliferation in the rat. The presence of the alert leads to peroxisome proliferation being predicted as PLAUSIBLE. Because of the rule above, this leads to the additional prediction that carcinogenicity in the rat is PLAUSIBLE. The conclusion from the argument is PLAUSIBLE and not PROBABLE because peroxisome proliferation is only PLAUSIBLE. Our LA model deals automatically with cases where the level of belief in the grounds of an argument falls short of the threshold (CERTAIN in this rule).

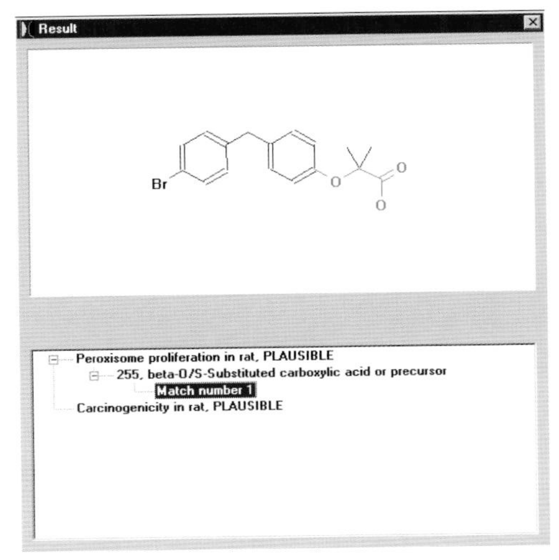

Fig. 2. *Peroxisome proliferation and carcinogenicity predictions in the rat for 2-{4-[(4-bromophenyl)methyl]phenoxy}-2-methylpropanoic acid in DEREK for Windows*

Processing the same compound for the human species would cause the level of belief in peroxisome proliferation to become IMPROBABLE, since humans are known to be much less susceptible to this effect [18]. In this case, the rule relating peroxisome proliferation to carcinogenicity would generate no conclusion about carcinogenicity because it only describes the relationship between the *presence* of peroxisome proliferation and carcinogenicity. In LA, the failure of an argument for a proposition does not imply that there is an argument against it, and *vice versa.*

6.3. *METEOR – Limiting and Ranking Predictions*

In *Fig. 3*, the user has chosen to view one reaction in a metabolic tree generated by METEOR, the conjugation of a carboxylic acid with glycine. The lower window contains a representation of the metabolic tree. Biotransformations are selected and grouped according to absolute reasoning rules about their likelihood. The likelihood that a given biotransformation will take place is dynamic, being modified by rules about the influences of physical properties of potential substrates for

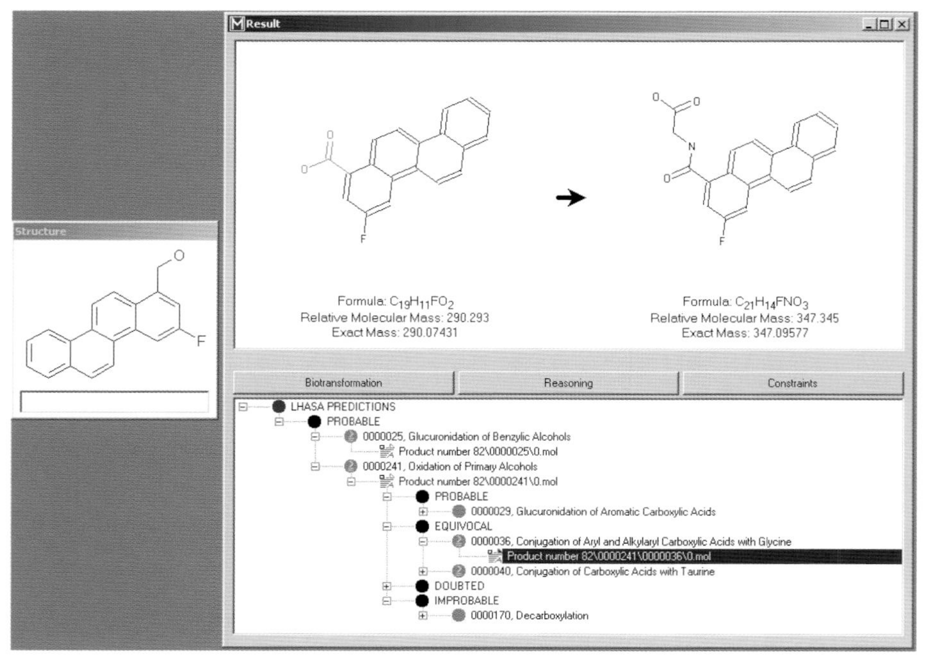

Fig. 3. *Example showing restriction and ranking of predictions in METEOR*

example. Rules about relative reasoning each specify which is the more likely to predominate in pairs of competing biotransformations. The reasoning system constructs ordered sets of biotransformations from these pairs. The most likely biotransformation in a list is designated a level-1 biotransformation, the next most likely level-2, and so on.

Whether biotransformations of a given likelihood are displayed and/or considered for further processing takes account of constraints set by the user. In this example, the constraints were set to allow all biotransformations not classed as IMPOSSIBLE and at relative reasoning level-1 to be displayed. If the user had chosen a cutoff for relative reasoning of 2 rather than 1, for example, some of the groups of biotransformations in the metabolism tree would have included additional less-dominant biotransformations.

METEOR considers both glucuronidation and oxidation of the $HOCH_2$ group in the query molecule (shown on the left in *Fig. 3*) to be PROBABLE, and proposes no other first metabolic steps. The user chose a constraint in METEOR which prevents further processing of phase-II products, and so the tree grew no further from the glucuronidation product.

Several potential reactions of the carboxylic acid resulting from oxidation of the $HOCH_2$ group are reported in addition to conjugation with glycine. Glucuronidation of the carboxylic acid is listed as PROBABLE whereas conjugation with glycine or taurine are only considered EQUIVOCAL. There is a biotransformation included on the tree which is DOUBTED, but the user has not expanded this part of the tree in the display. The user has expanded the tree to find out about the biotransformation listed as IMPROBABLE, which turns out to be decarboxylation.

Beside both of the biotransformations listed as EQUIVOCAL, there is a symbol in a grey circle formed from a '>' sign above an '=' sign. This indicates to the user that the biotransformations feature in relative reasoning rules. The user can click on the symbol to see more information – for example, in this case, the program would report that conjugation with glycine or taurine would be equally likely but that both are more likely than conjugation with glutamine.

6.4 *METEOR – Using Results of Calculations in Reasoning*

Fig. 4 shows the results of processing 3-fluorobenzyl alcohol in METEOR with the same constraints as those used for the substituted benzophenanthrene of *Example 6.3* shown in *Fig. 3*. In *Fig. 4*, biotrans-

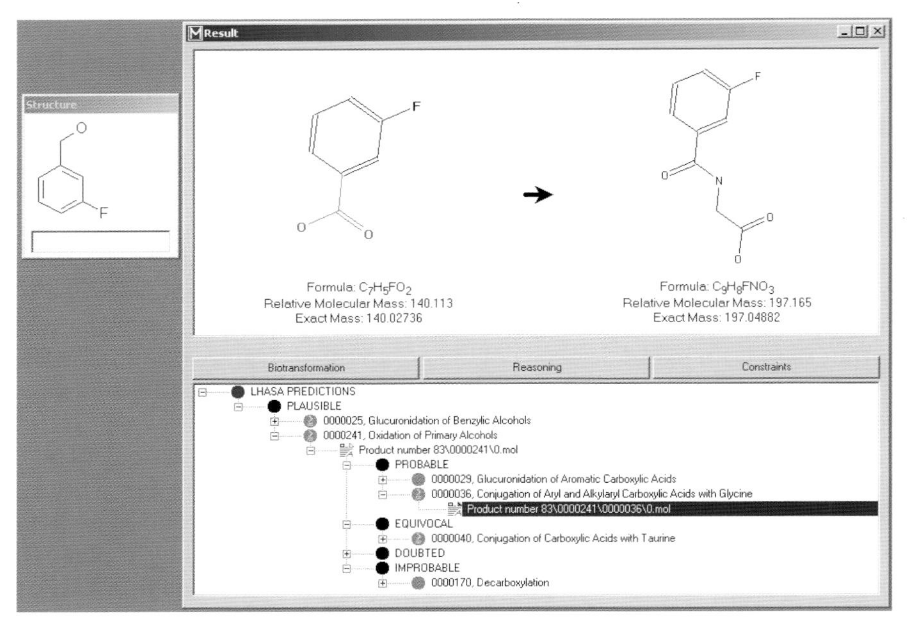

Fig. 4. *Example showing use of molecular weight to influence a prediction in METEOR*

formation 36, conjugation of a carboxylic acid with glycine, is ranked as PROBABLE whereas it is EQUIVOCAL in *Fig. 3*. This is because, as a generalization, acids of lower molecular weight are more frequently conjugated with glycine than acids of higher molecular weight, and there is an absolute reasoning rule in METEOR to that effect.

7. Conclusions

There is concern in many fields about the need for better ways to communicate about risk under uncertainty. Reasoning based on the logic of argumentation (LA) has the potential to make predictive computer systems more effective as sources of advice to decision makers. Its application in DEREK for Windows and METEOR is at an early stage of development but already offers benefits.

I thank *Carol Marchant* and *Anthony Long* for providing the examples used in this chapter, and *Jonathan Vessey* and *William Button* for their ideas about reasoning under uncertainty, which influenced the content of this chapter.

REFERENCES

[1] D. J. Livingstone, *Pestic. Sci.* **1989**, *27*, 287.
[2] L. Eriksson, E. Johansson, N. Kettaneh-Wold, S. Wold, 'Multi- and Megavariate Data Analysis: Principles and Applications', Umetrics Academy, Umeå, 2001.
[3] C. Hansch, *Acc. Chem. Res.* **1969**, *2*, 232.
[4] K. Enslein, V. K. Gombar, B. W. Blake, *Mutat. Res.* **1994**, *305*, 47.
[5] G. Klopman, H. S. Rosenkranz, *Mutat. Res.* **1994**, *305*, 33.
[6] M. P. Smithing, F. Darvas, in 'Food Safety Assessment', Eds. J. W. Finley, S. F. Robinson, D. J. Armstrong, American Chemical Society, Washington DC, 1992, p. 191–200.
[7] J. J. Kaufman, *Int. J. Quant. Chem. Quant. Biol. Symp.* **1981**, *8*, 419.
[8] F. Darvas, in 'QSAR in Environmental Toxicology', Ed. K. Kaiser, Riedel, Dordrecht, 1987, p. 71–81.
[9] G. Klopman, M. Dimayuga, J. Talafous, *J. Chem. Inf. Comput. Sci.* **1994**, *34*, 1320.
[10] N. Greene, P. N. Judson, J. J. Langowski, C. A. Marchant, *SAR QSAR Environ. Res.* **1999**, *10*, 299.
[11] J. Fox, P. J. Krause, S. A. Ambler, in 'Proceedings of ECAI '92', John Wiley and Sons, Chichester, 1992, p. 623–627.
[12] M. Elvang-Gøransson, P. J. Krause, J. Fox, in 'Uncertainty in Artificial Intelligence: Proceedings of the Ninth Conference', Eds. D. Heckerman, A. Mamdani, Morgan Kaufmann, San Francisco, 1993, p. 114–121.
[13] P. N. Judson, J. D. Vessey, *J. Chem. Inf. Comput. Sci.* **2003**, *43*, 1356.
[14] P. N. Judson, C. A. Marchant, J. D. Vessey, *J. Chem. Inf. Comput. Sci.* **2003**, *43*, 1364.
[15] W. G. Button, P. N. Judson, A. Long, J. D. Vessey, *J. Chem. Inf. Comput. Sci.* **2003**, *43*, 1371.
[16] D. A. Basketter, E. W. Scholes, M. Chamberlain, M. D. Barratt, *Food Chem. Toxicol.* **1995**, *33*, 1051.
[17] R. O. Potts, R. H. Guy, *Pharm. Res.* **1992**, *9*, 663.
[18] J. C. Corton, P. J. Lapinskas, F. J. Gonzalez, *Mutat. Res.* **2000**, *448*, 139.

Physiologically Based Pharmacokinetic Models

by **Thierry Lavé***, **Hannah Jones**, **Nicolas Paquerau**, **Patrick Poulin**, **Peter Theil**, and **Neil Parrott**

F. Hoffmann-La Roche AG, Pharmaceuticals Division, CH-4070 Basel
(e-mail: thierry.lave@roche.com)

Abbreviations
ADME: Absorption, distribution, metabolism, excretion, AUC: area under the curve; GIT: gastrointestinal tract; PBPK: physiologically based pharmacokinetic(s); PD: pharmacodynamics; PK: pharmacokinetics.

1. Introduction

Although technical advances in drug discovery are identifying an increasing number of biologically active compounds, many of these are still eliminated during the selection and development phases. Historically, a high proportion of these failures have been due to poor pharmacokinetic properties. To reduce this failure rate, candidate compounds are now being screened for ADME properties (absorption, distribution, metabolism, excretion) and the derived parameters are then being used to predict their human pharmacokinetic profiles. These predicted profiles not only help to select the best candidates for development but can also provide a starting dose for the first clinical studies. Such predictions can, therefore, drastically reduce the time and expense of drug research and development [1]. Furthermore, because ADME issues are considered during the selection process, fewer compounds are now dropping out of development because of pharmacokinetic reasons.

The approaches used to predict human pharmacokinetics tend to fall into two categories: empirical interspecies scaling and physiologically based pharmacokinetic (PBPK) modeling [2–12]. With the recent developments of *in silico* and *in vitro* tools together with a marked increase in computing power, PBPK modeling is rapidly becoming a powerful tool for predicting human pharmacokinetics.

The present chapter reviews recent developments in physiologically based methods for predicting human pharmacokinetics. First, some general background information is given on these methods. A number of applications in drug research are also described, together with some strategic issues that may be considered when applying such simulation tools.

2. Methodological Overview

2.1. *Disposition Models*

Physiologically based pharmacokinetic models divide the body into compartments [13–17], including the eliminating organs, *e.g.*, kidney and liver, and noneliminating tissue compartments, *e.g.*, fat, muscle, and brain, which are connected by the circulatory system (*Fig. 1*). The models use physiological and species-specific parameters such as blood flow rates and tissue volumes to describe the pharmacokinetic processes. These physiological parameters are coupled with physicochemical, biochemical and compound specific parameters (*e.g.*, tissue/blood partition coefficients and metabolic clearance) to predict the plasma and tissue concentration *vs.* time profiles of a compound in an *in vivo* animal or human system.

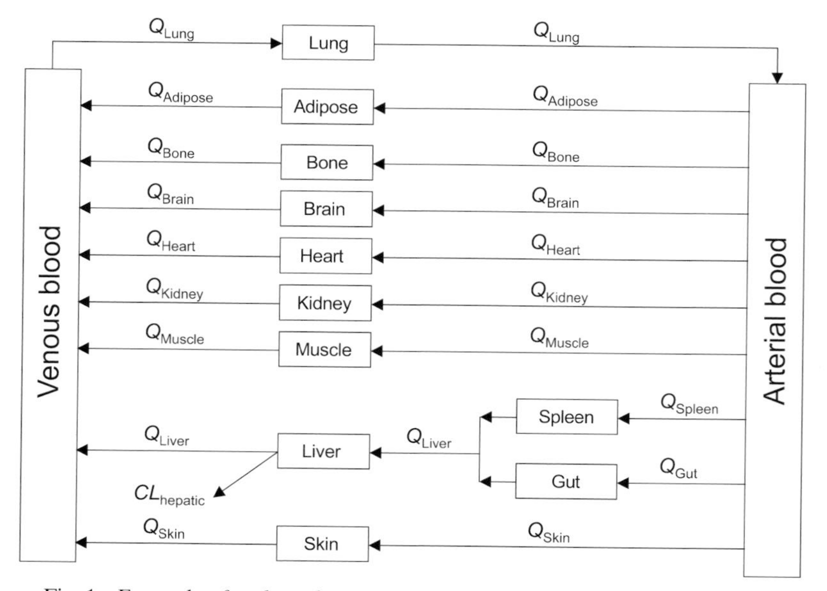

Fig. 1. *Example of a physiologically based pharmacokinetic* (PBPK) *model*

Once a model has been developed, the concentrations in the various tissues can be determined by using the following mass balance (*Eqn. 1*):

drug concentration in tissue =

rate of drug distribution into tissue

– rate of drug distribution out

– rate of drug elimination within the tissue (1)

Depending on the drug and tissue, the distribution can be perfusion rate or diffusion rate limited. Perfusion rate limited kinetics tend to occur with relatively low-molecular-weight, hydrophobic drugs which have no problem crossing the lipid barrier of the cell wall. In this case, the process limiting the penetration of the drug into the cells is the rate at which it is delivered to the tissue, *i.e.*, blood flow is the limiting process. By contrast diffusion rate limited kinetics occur with more-polar and/or larger drugs that do not freely dissolve in the lipid of the cell membrane and, therefore, have difficulty in penetrating into the cell. In this case, the diffusion of drug across the membrane, which is independent of blood flow, becomes the limiting process.

For perfusion limitation, the rate of change of drug concentration in a tissue, where no elimination occurs, can be described in *Eqn. 2*:

$$V\frac{dC}{dt} = Q\left(C_{\mathrm{in}} - \frac{C_{\mathrm{in}}}{K_{\mathrm{p}}}\right) \quad (2)$$

where V is the physical volume of the tissue, C the drug concentration in tissue, Q the blood flow to the tissue, C_{in} the drug concentration entering the tissue, and K_{p} the partition of the drug between tissue and blood.

When diffusion rate limitation occurs, diffusion to and from the extracellular space must be taken into account (*Eqn. 3*):

$$V_{\mathrm{e}}\frac{dC_{\mathrm{e}}}{dt} = Q(C_{\mathrm{in}} - C_{\mathrm{e}}) - P(C_{\mathrm{e}} - C_{\mathrm{i}}) \quad (3)$$

where P is the membrane permeability coefficient, C_{e} the free extracellular drug concentration, C_{i} the free intracellular drug concentration, and V_{e} the anatomical extracellular volume.

With organs such as liver, where elimination can occur, the rate of elimination must be included in the mass balance equation. The rate of elimination is described in *Eqn. 4*:

$$\text{Rate of elimination} = CL_{\text{int}} C_{\text{t}} \tag{4}$$

where CL_{int} is the intrinsic clearance of the drug from that organ, and C_{t} the concentration in tissue. In some instances, C_{t} may also be replaced by C_{vt} (concentration in the venous blood leaving the tissue equivalent to $C_{\text{in}}/K_{\text{p}}$). A comparative validation is still needed to find out which of these terms (C_{t} or C_{vt}) is most relevant for the *in vivo* situation. This term can then be inserted into the mass balance equation, such as that for the perfusion limitation (*Eqn. 5*):

$$V\frac{dC}{dt} = Q\left(C_{\text{in}} - \frac{C_{\text{in}}}{K_{\text{p}}}\right) - CL_{\text{int}} C_{\text{t}} \tag{5}$$

As the above equations indicate, a considerable amount of information is required to construct physiologically based pharmacokinetic models. Thus, estimates of tissue/blood partitioning and intrinsic clearance are required for each drug. Recently, tissue composition models have been developed which allow the tissue/blood partition coefficients to be estimated either *in silico* or to be measured experimentally as physicochemical descriptors. These approaches have dramatically reduced the amount of experimentation needed to support the use of flow models, considerably extending their utility [18]. Such approaches to predict tissue distribution are discussed elsewhere [19].

2.2. *Absorption Models*

Oral absorption is determined by complex mechanisms, which are governed by physiology and biochemical processes (*e.g.*, pH in the gastrointestinal tract (GIT), gastric emptying, intestinal transit, active transport, and gastrointestinal metabolism), drug-specific properties (*e.g.*, lipophilicity, pK_{a}, solubility, particle size, permeability, metabolic stability) and formulation factors (*e.g.*, release kinetics, dissolution kinetics). These are some of the main determinants which could be important parts of an absorption model. The interplay of parameters describing these processes determines the rate and extent of absorption.

The available simulation tools to predict oral absorption in animals and humans have been recently reviewed [20][21]. Different absorption models have been developed and in part described in the literature [20]. These GIT models are developed to a degree that they are commercially available as software tools (GASTROPLUS® from *Simulations Plus Inc.*). In brief, these models are physiologically based transit models segmenting the GIT into different compartments, where the kinetics of transit, dissolution, and uptake are described by sets of differential equations. The simulation models for oral absorption use a variety of measured or calculated *in vitro* input data such as permeability, solubility, pK_a, and dose.

2.3. *Utility of Physiologically Based PK Models*

A number of publications illustrate the potential of this approach, both for predicting human pharmacokinetics and for mechanistic purposes. The utility of physiologically based pharmacokinetic models to drug development and rational drug discovery candidate selection was also reviewed recently [19]. The following advantages of the mechanistic PBPK modeling framework can be considered: *a*) They predict plasma (blood) and tissue PK of drug candidates prior to *in vivo* experiments; *b*) They support a better mechanistic understanding of PK properties and help the development of more-rationale PK–PD relationships from tissue kinetic data predicted, thus facilitating a more-rational decision during clinical candidate selection. And *c*) they allow the extrapolation across species, routes of administration, and dose levels.

Two of these applications are discussed below.

3. Applications of Physiologically Based PK Models

3.1. *Generic Simulations During Drug Discovery*

During drug discovery, considerable resources are required to assess the pharmacokinetic properties of potential clinical candidates *in vivo* in animals, and there is interest in optimizing the use of such testing by applying simulation techniques.

Physiologically based models have the potential to do this by predicting pharmacokinetics based on *in vitro* and *in silico* input data. Such approaches may help to rank compounds based on their predicted profiles, provide the project teams with a balanced view on the properties of potential drug candidates, and help to select the optimal molecules for further *in vivo* experiments. However, before such tools are routinely used and accepted, there is a need for extensive validation.

Our in-house developed PBPK model has been applied in a generic mode to predict plasma profiles after i.v. and p.o. dosing to the rat in a number of discovery projects. The predictions were made on the basis of a minimum of measured data, namely calculated log P, calculated pK_a values, calculated protein binding, and intrinsic clearance determined in hepatocytes. The results support the use of the generic PBPK approach at the early stages of drug discovery. Even when based on a minimum of data, the models are able to give reasonable initial estimates of the expected pharmacokinetics of novel compounds.

The generic PBPK approach based upon minimal input data is better able to rank the PK properties of compounds across different chemical classes than within a close series. In addition, a good prediction of *in vivo* solubility still represents a major challenge for the prediction of the oral absorption profile of low-soluble compounds. At this time, it is recommended that generic PBPK models should only be applied for lead optimization after verification of the simulations with *in vivo* PK for a few compounds of a given chemical class. Such verification will help to identify invalid model assumptions or important missing processes, where additional data is needed, and will allow an assessment of the prediction error expected.

3.2. *Prediction of Pharmacokinetics in Humans*

To reduce failures related to pharmacokinetic (PK) issues in the drug development process and to find out about the suitability of compounds for an intended dosing regimen, it is important to predict human PK as early as possible. Empirical methods (*e.g.*, allometric scaling) have been traditionally used for this purpose. Although in some cases, these methods give good predictions, their physiological basis and predictive value is questionable. Recently, mechanistic physiologically based PK (PBPK) models have been developed. These models are mathematically more complex, and until recently their use in drug development has been limited.

Recently, empirical and PBPK approaches were compared for the prediction of human PK using 19 *Roche* compounds having reached clinical development, covering a broad range of physicochemical properties. Predicted values (PK parameters, plasma concentrations) were compared to observed values to assess the accuracy of the prediction methods.

The PBPK approach gave more-realistic predictions than the classical empirical methods for all 19 compounds (*Fig. 2*). A greater proportion of the predicted parameters (*e.g.*, C_{max}, AUC, $t_{1/2}$) and plasma concentrations were within twofold error of the observed values. For example, 76% and

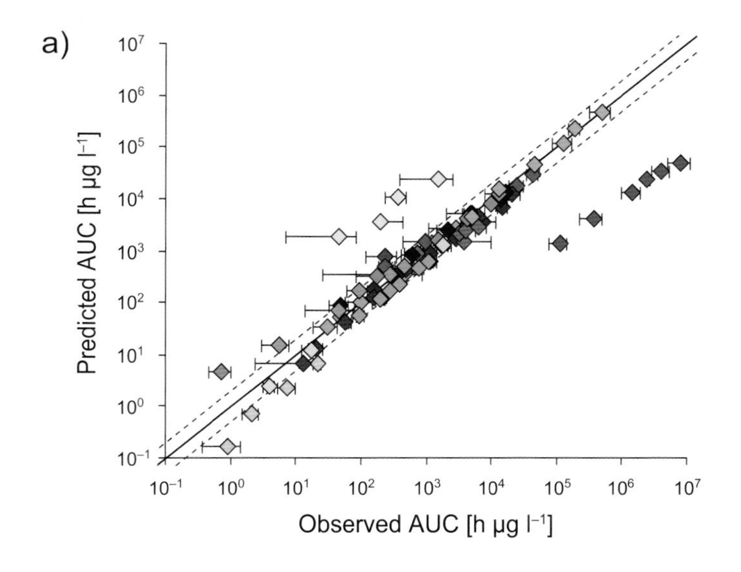

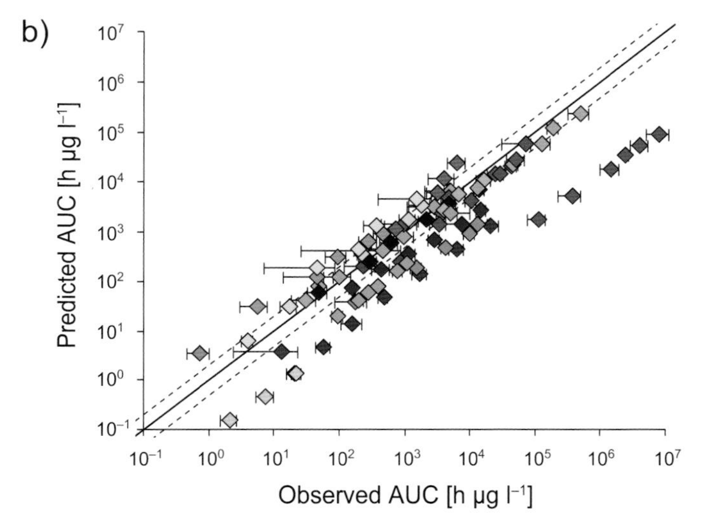

Fig. 2. *Prediction of the area under the curve* (AUC) *for in-house compounds. a*) Prediction based on a PBPK approach; *b*) prediction based on the *Dedrick* method.

42% of the compounds had a predicted AUC within twofold error of the observed value using PBPK and the *Dedrick* approach, respectively. Any poor prediction was generally a result of biliary elimination and/or enterohepatic recirculation processes that were not incorporated into the model.

In addition to improved prediction accuracy, PBPK approaches offer more potential in the early stages of the drug development process, including a reduction in the quantity of animal *in vivo* input data required, as well as greater insights into the mechanistic understanding of the compound characteristics, ultimately resulting in an improved selection process.

4. Conclusions

The use of both empirical methods and physiologically based models to predict human pharmacokinetic profiles can help to select the best candidates for drug development. They can also help to select doses for the first clinical studies.

Allometric scaling explores the mathematical relationships between pharmacokinetic parameters from various animal species, and these can then be used to predict the corresponding values in other species, including human. Such methods are also built to some extent on physiological principles. They are relatively easy to apply but resource demanding for the collection of *in vivo* data in animals. Nevertheless, their application has led to useful predictions of individual pharmacokinetic parameters (*e.g.*, clearance, fraction absorbed, volume of distribution).

Physiologically based models (PBPK) can be used to explore, and help to explain, the mechanisms that lie behind species differences in pharmacokinetics and drug metabolism. Such models can provide a rational basis for interspecies scaling of individual parameters which can then be integrated to provide quantitative and time-dependent estimates of both the plasma and tissue concentrations in humans. Furthermore, being mechanistically based, they can be used diagnostically to generate information on new compounds and to understand the sensitivity of the *in vivo* profile to compound properties. Despite this great potential, the use of PBPK models in drug discovery and development has been relatively limited. However, recent developments of *in silico* and *in vitro* models, which can provide estimates of these input parameters, have dramatically reduced the amount of experimental work required [19]. Another reason for the limited use of PBPK models relates to their mathematical complexity, so that a high level of expertise is needed to develop such models. Again, recent improvements in the availability of well-validated and user-friendly software packages should remove this barrier. In the very near future, therefore, the use of physiologically based models is likely to increase dramatically in the prediction of concentration–time profiles and as diagnostic tools to better understand potential development compounds.

REFERENCES

[1] D. A. Norris, G. D. Leesman, P. J. Sinko, G. M. Grass, *J. Controlled Release* **2000**, *65*, 55.
[2] H. Boxenbaum, M. Battle, *J. Clin. Pharmacol.* **1995**, *35*, 763.
[3] B. Boxenbaum, R. W. D'Souza, in 'Advances in Drug Research', Vol. 19, Ed. B. Testa, Academic Press, London, 1990, p. 139–196.
[4] D. B. Campbell, *Drug Inf. J.* **1994**, *28*, 235.
[5] J. Mordenti, *J. Pharm. Sci.* **1986**, *75*, 1028.
[6] R. M. Ings, *Xenobiotica* **1990**, *20*, 1201.
[7] J. H. Lin, *Drug Metab. Dispos.* **1995**, *23*, 1008.
[8] J. H. Lin, A. Y. Lu, *Pharmacol. Rev.* **1997**, *49*, 403.
[9] T. Lavé, O. Luttringer, J. Zuegge, G. Schneider, P. Coassolo, F. P. Theil, *Ernst Schering Research Foundation Workshop* **2002**, *37*, 81.
[10] T. Lavé, P. Coassolo, B. Reigner, *Clin. Pharmacokinet.* **1999**, *36*, 211.
[11] P. J. McNamara, in 'Pharmaceutical Bioequivalence', P. G. Welling, F. L. S. Tse, S. V. Dighe, Eds., Dekker, New York, 1991, p. 267–300.
[12] J. Zuegge, G. Schneider, P. Coassolo, T. Lavé, *Clin. Pharmacokinet.* **2001**, *40*, 553.
[13] L. E. Gerlowski, R. K. Jain, *J. Pharm. Sci.* **1983**, *72*, 1103.
[14] M. E. Andersen, *Toxicol. Lett.* **1995**, *79*, 35.
[15] M. E. Andersen, H. J. Clewell, M. L. Gargas, F. A. Smith, R. H. Reitz, *Toxicol. Appl. Pharmacol.* **1987**, *87*, 185.
[16] J. C. Ramsey, M. E. Andersen, *Toxicology* **1984**, *73*, 159.
[17] R. C. Ward, C. C. Travis, D. M. Hetrick, M. E. Andersen, M. L. Gargas, *Toxicol. Appl. Pharmacol.* **1988**, *93*, 108.
[18] P. Poulin, F. P. Theil, *J. Pharm. Sci.* **2000**, *89*, 16.
[19] F. P. Theil, T. W. Guentert, S. Haddad, P. Poulin, *Toxicol. Lett.*, in press.
[20] B. Agoram, W. S. Woltosz, M. B. Bolger, *Adv. Drug Delivery Rev.* **2001**, *50*(1), S41.
[21] M. G. Grass, J. P. Sinko, *Adv. Drug Delivery Rev.* **2002**, *54*, 433.

Processing of Biopharmaceutical Profiling Data in Drug Discovery

by **Kiyohiko Sugano***[1]), **Kouki Obata**, **Ryoichi Saitoh**, **Atsuko Higashida**, and **Hirokazu Hamada**

Preclinical Research Deptartment I, *Chugai Pharmaceutical Co. Ltd.*, 1–135 Komakado, Gotemba, Shizuoka 412-8513, Japan

Abbreviations
BCS: Biopharmaceutics classification system; CAT: compartmental absorption transit; *Do*: dose number (= Dose/(solubility × 250 ml)); log D_{oct}: log of octanol/buffer distribution coefficient; *Fa*: fraction of a dose absorbed in humans; IVIVC: *in vitro–in vivo* correlations; P_{am}: artificial membrane permeability in PAMPA [cm s^{-1}]; PAMPA: parallel artificial membrane permeation assay; P_{app}: apparent membrane permeability in Caco-2 [cm s^{-1}]; P_{eff}: effective intestinal membrane permeability in humans [cm s^{-1}]; PK: pharmacokinetics; log P_{oct}: log octanol/water partition coefficient; P_{para}: paracellular permeability [cm s^{-1}]; P_{tot}: total membrane permeability [cm s^{-1}]; P_{trans}: transcellular permeability [cm s^{-1}]; P_{UWL}: unstirred water layer permeability [cm s^{-1}]; SPRs: structure–property relationship; TPAM: theoretical passive absorption model; UWL: unstirred water layer.

1. Introduction

Since the 1990s and even earlier, Absorption, Distribution, Metabolism, and Excretion (ADME) have been recognized as crucial properties for a successful drug development, because *ca.* 40% of development withdrawal has been reported to be derived from poor ADME properties [1]. Therefore, ADME assays and physicochemical assays which are relevant for ADME have been incorporated into the lead optimization process [2–4]. Various medium-to-high-throughput assays have been developed, *e.g.*, octanol/buffer partition coefficients [5][6], pK_a [7], solubility [8–10], permeability [11], metabolic stability [12], drug–drug interactions [13], protein binding [14], *in vivo* pharmacokinetics (PK) studies (cassette dosing) [15], *etc.* Furthermore, *in silico* physicochemical

[1]) Present address: *Pfizer Inc.*, Global Research & Development, Nagoya Laboratories, Pharmaceutical R&D, 5-2 Taketoyo, Aichi 470-2393, Japan (e-mail: Kiyohiko.Sugano@pfizer.com).

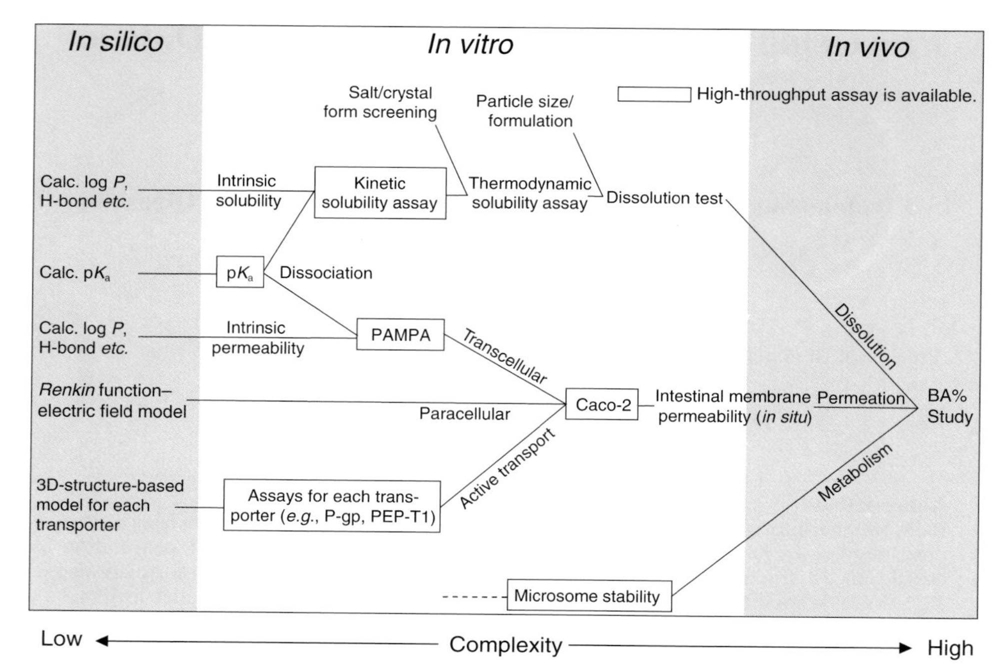

Fig. 1. *Reducing complex* in vivo *oral absorption assays to single-process* in vitro *assays and* in silico *models.* The physiological complexity of the assays differs at each level. For example, PAMPA is a single-process *in vitro* assay for transcellular pathways, and Caco-2 is a multiple-process *in vitro* assay consisting of the transcellular pathway, the paracellular pathway, and active transport systems. Usually, *in silico* predictions and single-process *in vitro* assays are frequently employed in the early stages of drug discovery, and *in vivo* PK studies tend to be employed at later stages.

and ADME screens including drug likeness calculations have also been incorporated into compound bank library design, combinatorial synthesis design, hit selection, hit-to-lead and lead optimization processes [16–18]. Today's biopharmaceutical profiling tools in drug discovery are shown in *Fig. 1*.

The physiological complexity of the assays at each level differs. For example, the parallel artificial membrane permeation assay (PAMPA) is a single-process *in vitro* assay for transcellular pathway permeation [19–23], and Caco-2 is a multiple-process *in vitro* assay consisting in the transcellular pathway, the paracellular pathway, and active transport systems [11]. *In silico* prediction and single-process *in vitro* assays are frequently employed in the early stages of drug discovery, whereas complicated assays, especially *in vivo* PK studies, tend to be employed at later stages. Single-process *in vitro* assays are usually high-throughput and their data are suitable for structure–property relationship (SPR) studies.

Recent evolution of laboratory automation has increased the throughput of assays. Today, a large number of physicochemical/ADME data are generated daily in drug discovery, but the number of scientists involved is limited. Consequently, the bottleneck of physicochemical/ADME optimization is currently shifting from data generation to data processing.

2. Data Processing

Data processing can be taken as a whole process from purpose determination to knowledge sharing as shown in *Fig. 2* [24]. In other words, drug discovery can be reviewed from the data processing point of view. In this section, the whole process shown in *Fig. 2* is called 'data processing' and distinguished from 'data analysis'. The first step in data processing is to determine its objectives. Then, an assay is developed and the required data are collected. Sometimes, the first data acquired do not contain sufficient information for later steps, and iteration between data acquisition and later steps must occur. Databases are a necessary tool for data processing. Relational databases are currently employed in most cases. Data cleaning is a very important process to reach a right and appropriate knowledge. Various statistical methods have been developed for analysis. To share knowledge with colleagues, it must be presented in an easily available manner, especially for nonspecialists.

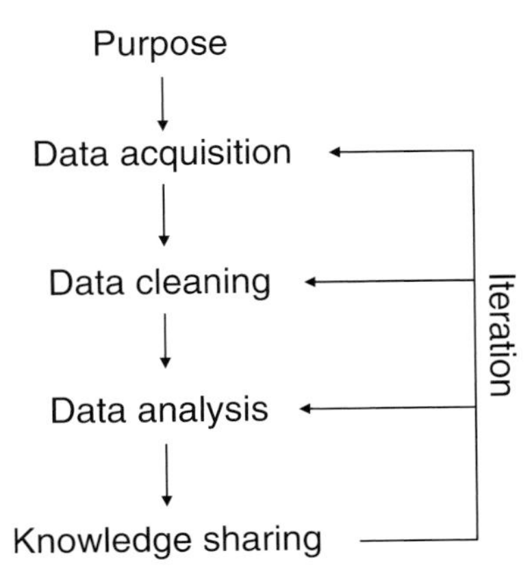

Fig. 2. *Data processing.* Data acquisition is a part of data processing. Today, data acquisition is accelerated by laboratory automation. The total performance of this scheme should be optimized, rather than the partial optimization of data acquisition. Sometimes, the first data acquired do not contain sufficient information for later steps, and iteration between data acquisition and later steps must occur.

2.1. *Objectives of Biopharmaceutical Data Processing in Drug Discovery*

Several objectives exist in ADME data processing, namely:

- structure–property relationships (SPRs);
- *in silico–in vitro–in vivo* correlations;
- mechanistic investigations;
- compound selection;
- project management;
- unconscious objective.

Depending on the objective, the required assay performance differs. Structure–activity relationships require a wider measurement range than compound selection. In the case of solubility, a range higher than 10 μg ml^{-1} is sufficient for compound selection. However, a range larger than 10 μg ml^{-1} is required for SPRs of poorly soluble compounds. The scientists who develop the assays are often conscious about a few of the above objectives. It is preferable that the assays are broad enough to cover most objectives. In addition, unexpected knowledge which is out of our recognition today can be obtained from data processing.

2.2. *Data Acquisition*

As experimental scientists know, experiments are not always performed perfectly. Therefore, in addition to the main data to be used for analysis, information about the assay validity becomes crucial for data analysis. Sample integrity is important for data cleaning, even though this information is not directly used for data analysis. Sample integrity can be ascertained by thin layer chromatography, HPLC, LC/MS, capillary electrophoresis, *etc.* [25][26]. Sample integrity is especially important for assays in which the concentration is measured by UV spectroscopy, *e.g.*, PAMPA [19] and direct UV solubility assay [9][10].

2.2.1. *Solubility*

In the direct UV solubility assay [9][10], the following information should be obtained for data cleaning:

- UV spectral changes after incubation (alert for decomposition);
- standard deviations;
- interplate standard values;

- detection limits;
- date, operator, *etc.*;
- birefringence observations.

Aqueous solubility depends on the solid form. Birefringence observation may reveal whether the measured solubility is from crystalline or amorphous material.

2.2.2. *PAMPA*

In the PAMPA assay, the following information may help data cleaning:

- UV spectral changes after incubation (alert for decomposition);
- standard deviations;
- interplate standard values;
- turbidity of donor solutions (absorption at 650 nm);
- detection limits;
- membrane retentions;
- date, operator, *etc.*

Prefiltration is often employed to remove precipitated material before applying sample solutions to donor compartments. Even though PAMPA is an assay of passive transport, concentration dependence is reported for basic compounds when negatively charged membranes are employed [27][28]. Therefore, the concentration in donor solutions should be unified for every compound. Because PAMPA membranes consist of phospholipids and organic solvents, highly lipophilic compounds are largely retained in the membrane, leading to permeability reduction [29]. At present, it is not clear whether this phenomenon has *in vivo* relevance.

Usually, solubility assays and PAMPA are performed in a 96-well plate format. It is difficult to check the assay manually. Therefore, the above information should be provided automatically.

2.2.3. In vivo *PK Studies*

In vivo PK studies must be carefully designed. The dosage form, the administration route, and the blood collection time course differ depending on the objective. In addition to the usual record of *in vivo* experiments, the following information will be of great help for later analysis:

- dosage form (excipients);
- physicochemical characteristics (in the case of solid or suspension administration: microscope photograph (birefringence), particle size, powder X-ray, calorimetry, *etc.*);
- fasted or fed animals.

PK Scientists, and more-often pharmacologists, tend to pay little attention to the dosage form. Formulation scientists can help them and should therefore take part in drug discovery.

2.3. *Data Cleaning*

It is often forgotten that experimental data always contain errors which can lead to wrong conclusions. Therefore, data cleaning is a very important process. Information about assay validation is of great help for data cleaning. For example, when we perform analysis of PAMPA data, data with alert about, *e.g.*, precipitation, detection limits, or high membrane retention should be excluded. Once data is prepared for cleaning, cleaning itself is easily performed with a sorting function by a computer.

2.4. *Data Analysis* (Data Processing in a Narrow Sense)

Today, various data analysis programs are commercially available. Recent progresses in computational technology have enabled the 2D or 3D visualization of large numbers of data, as well as advanced statistical analysis. Typical data analyses are reviewed below.

2.4.1. *Structure–Property Relationships* (SPRs, *in silico*)

The most frequently employed approach to obtain SPRs is to perform multiple regressions using molecular descriptors [30]. Quantum-chemical descriptors, chemical fragments, physicochemical properties, steric parameters, *etc.* have been employed as molecular descriptors. Artificial neural networks, genetic algorithms, and various other statistical approaches are used to relate the dependent variables to molecular descriptors [16].

Drawbacks of today's *in silico* methods are: *1)* The dependent variables used consist of several physiological processes. For example, bioavailability is influenced by dissolution in the gastrointestinal tract, transcellular permeation, paracellular permeation, active transport, intestinal and hepatic metabolism, *etc.* (*Fig. 1*). *2*) Molecular descriptors may be difficult to interpret. *3*) The relating function may be a 'Black Box'. And *4*) previous findings were sometimes neglected, *e.g.*, the pH-partition hypothesis. These drawbacks may lead to a loss of accountability for users. Good predictability is not a final goal in the pharmaceutical industry, but must motivate synthetic scientists to create adequate candidates. Statistical analysis methods, which yield interpretation (knowledge), have been developed [31].

Mechanism-based approaches have been predicted to be the next *in silico* systems [16]. Thus, *Camenish et al.* developed a theoretical passive absorption model (TPAM) based on a physiological permeation mechanism and simple physicochemical parameters [32][33]. We have extended TPAM based on recent findings [34]. TPAM contains partial models, *i.e.*, a transcellular pathway, a paracellular pathway, and an unstirred water layer permeation (*Fig. 3*). Total intestinal membrane permeability (P_{tot}) is expressed in *Eqn. 1*:

$$\frac{1}{P_{tot}} = \frac{1}{P_{trans} + P_{para}} + \frac{1}{P_{UWL}} \tag{1}$$

where P_{trans}, P_{para}, and P_{UWL} are the transcellular permeability, paracellular permeability, and unstirred water layer permeability, respectively.

When we employ a plug-flow model as an absorption model from the intestinal tube [35], the fraction of a dose absorbed in humans (*Fa*) is expressed in *Eqn. 2*:

$$Fa = (1 - \exp(-Gz \cdot P_{tot})) \cdot 100 \tag{2}$$

where Gz is the lump constant of available intestinal surface area and transit time. In the present study, the value $Gz = 1.39 \times 10^4$ was employed to obtain an identical scale between P_{tot} and the effective intestinal membrane permeability in humans (P_{eff} in cm s^{-1}) [35][36].

Transcellular permeation is basically described by the pH-partition hypothesis, and the transport of cationic species is corrected as an extension to the previous TPAM (*Eqn. 3*) [21][27][37][38]. Recently, it was suggested that cationic species of basic compounds can permeate the negatively charged membrane with the aid of anionic lipids in the membrane, depending on the lipophilicity of the cationic species [21][27][37–39]. The intestinal epithelial membrane contains anionic lipids, *e.g.*, phosphatidylserine and phosphatidylinositol [40][41]. The

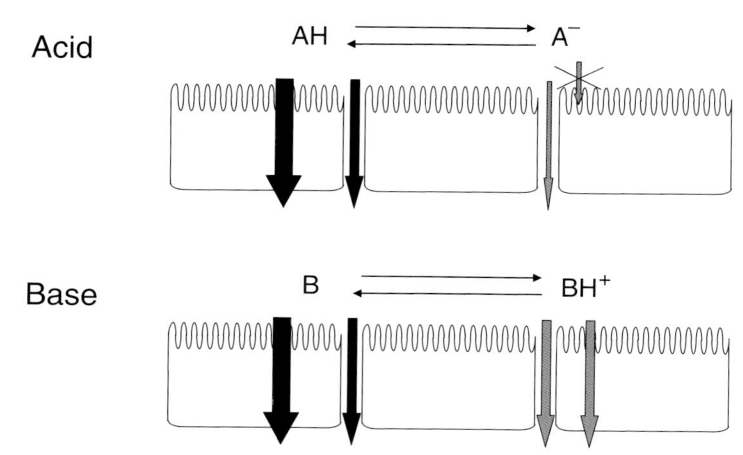

Fig. 3. *Schematic presentation of passive permeation across the intestinal epithelial membrane.* A and B represent an acid and a base, resp. An unstirred water layer covers the membrane (not shown). The intestinal epithelial membrane is negatively charged by anionic lipids, *e.g.*, phosphatidylinositol and phosphatidylserine. Transcellular permeation is basically described by the pH-partition hypothesis. Recently, it was suggested that cationic species of basic compounds can permeate the negatively charged membrane with the aid of anionic lipids in the membrane. Paracellular permeation is diffusion through the negatively charged tight junction between epithelial cells. Small and cationic species can easily permeate the paracellular pathway, whereas large and anionic species permeate little.

lipophilicity of the cell membrane can be modeled by the octanol/water partition coefficient (log P_{oct}) with the help of *Collander*'s equation. The lipophilicity of the cationic species may be scaled by the partition coefficient of neutral species:

$$P_{\mathrm{trans}} = aD_{\mathrm{oct}}^{\alpha} + b \cdot f_{+1} \cdot P_{\mathrm{oct}}^{\beta} \tag{3}$$

where f_{+1} is the fraction of monocationic species, and D_{oct} is the octanol/buffer distribution coefficient at pH 6.0. The value of f_{+1} was calculated from the pK_{a}.

Paracellular permeation is described as a size-restricted diffusion within a negative electrostatic force field (*Eqn. 4*) [42][43]. *Renkin*'s function $F(B)$ (*Eqn. 5*) was employed as the molecular size restrictor. In addition, an electric field of force function $E(Z)$ (*Eqn. 7*) was employed as an extension to the previous TPAM.

$$P_{\mathrm{para}} = A \cdot \frac{1}{M_{\mathrm{r}}^{1/3}} \cdot F(B)\left(f_0 + \sum^{z(z\neq 0)} f_z \cdot E(Z)\right) \tag{4}$$

$$F(B) = (1 - B)^2 (1 - 2.104 \cdot B + 2.09 \cdot B^3 - 0.95 \cdot B^5) \quad (5)$$

where

$$B = \frac{M_r^{1/3}}{R_{M_r}} \quad (6)$$

$$E(Z) = \frac{C \cdot z}{1 - e^{-C \cdot z}} \quad (7)$$

The term f_z is the fraction of each charged species (z: charge number), calculated from the pK_a. The term R_{M_r} (value = 8.46) is the apparent pore size of the paracellular pathway based on a molecular weight (M_r) scale. Previously, the molecular volume was employed as a parameter of the molecular size [43]. In the present work, we used M_r since it is more public in the drug discovery process and easier to calculate. The replacement of molecular volume by M_r did not affect the predictability of the P_{para} model (data not shown). The term A (value = 2.41×10^{-2}) is the lump constant of diffusion coefficient, porosity, and viscosity of water in the paracellular pathway. The term C (value = 2.39) corresponds to the electric potential of the paracellular pathway. A and C were obtained from the literature [43].

The unstirred water layer permeability (P_{UWL}) was modeled as a simple diffusion process in a water layer (*Eqn. 8*). This parameter is the reciprocal of $M_r^{1/3}$ [44]. The effective intestinal membrane permeability in humans (P_{eff}) of glucose, which is rate limited by the unstirred water layer (UWL), was reported to be 10×10^{-4} cm s^{-1} [36]. Since the molecular weight of glucose is 180:

$$P_{UWL} = 10 \times 10^{-4} \cdot \left(\frac{180}{M_r}\right)^{1/3} \quad (8)$$

The parameters P_{oct}, D_{oct}, and pK_a used as chemical descriptors were calculated by the Pallas 3.1 algorithm. The coefficients in the transcellular pathway model (*Eqn. 3*), *i.e.*, $a = 1.4 \times 10^{-4}$, $b = 0.23 \times 10^{-4}$, $\alpha = 0.32$, and $\beta = 0.19$ were obtained by fitting *Eqn. 2* with 258 *Fa* data (fraction absorbed in humans) obtained from the literature. The predictability of this theoretical passive absorption model (TPAM) is shown in *Fig. 4*. The model apparently corresponds to real physiological processes, and its molecular descriptors are simple. The number of fitting parameters is 4, in

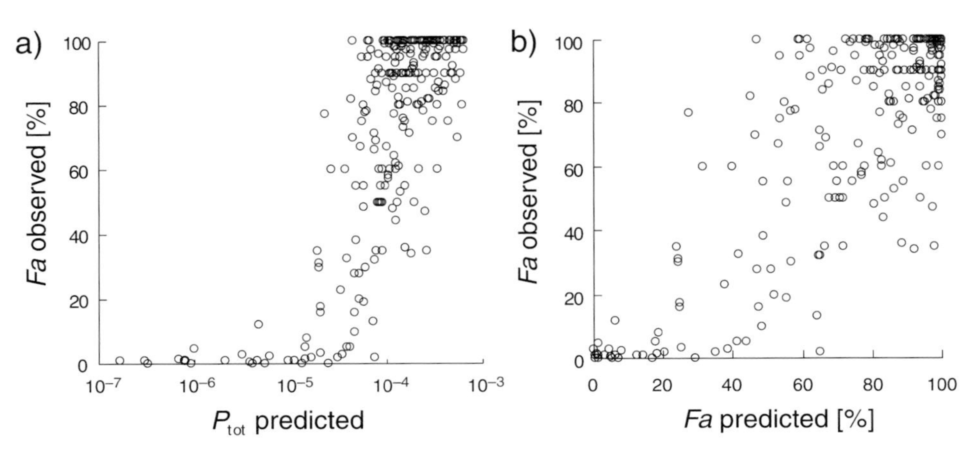

Fig. 4. *Prediction of* Fa *by TPAM. a)* P_{tot} *vs. Fa* observed. *b) Fa* predicted by *Eqn. 2 vs. Fa* observed. Solubility-limited absorption drugs and active transport substrates were excluded.

accordance with *Occam*'s razor rule. TPAM is very beneficial to comprehend membrane permeation characteristics from the point of view of both intestinal physiology and the chemical structure of drugs. By using TPAM, multiple dependent variables with different complexity (*Fa*, P_{eff}, Caco-2 permeability (P_{app}), PAMPA permeability (P_{am}), *etc.*), can be used for training and validation, but their weighting remains an issue. Recent progress of *in vitro* assays of single permeation processes may enable the fine tuning of each partial permeation model. For example, the PAMPA may improve the prediction of P_{trans} [45].

2.4.2. In vitro–in vivo *Correlations*

In vitro–in vivo correlations (IVIVC) often differ among chemical classes. They should, therefore, be checked as soon as possible for each project. Usually, the adequateness of an assay is checked by comparison with upper complexity level assays (*Fig. 1*). For example, PAMPA permeability can be checked by comparison with Caco-2 permeability, *in situ* intestinal membrane permeability, and fraction of dose absorbed *in vivo* (*Figs. 5,a*, *b*, and *c*, resp.) [43][46–48]. To compare PAMPA permeability (P_{am}) with these assays of higher complexity level, it was corrected for paracellular and UWL permeation based on TPAM. Multiple validation of an assay may increase its robustness.

Gastrointestinal absorption from a solid dosage form is a dual process of dissolution and permeation. TPAM does not consider the dissolution

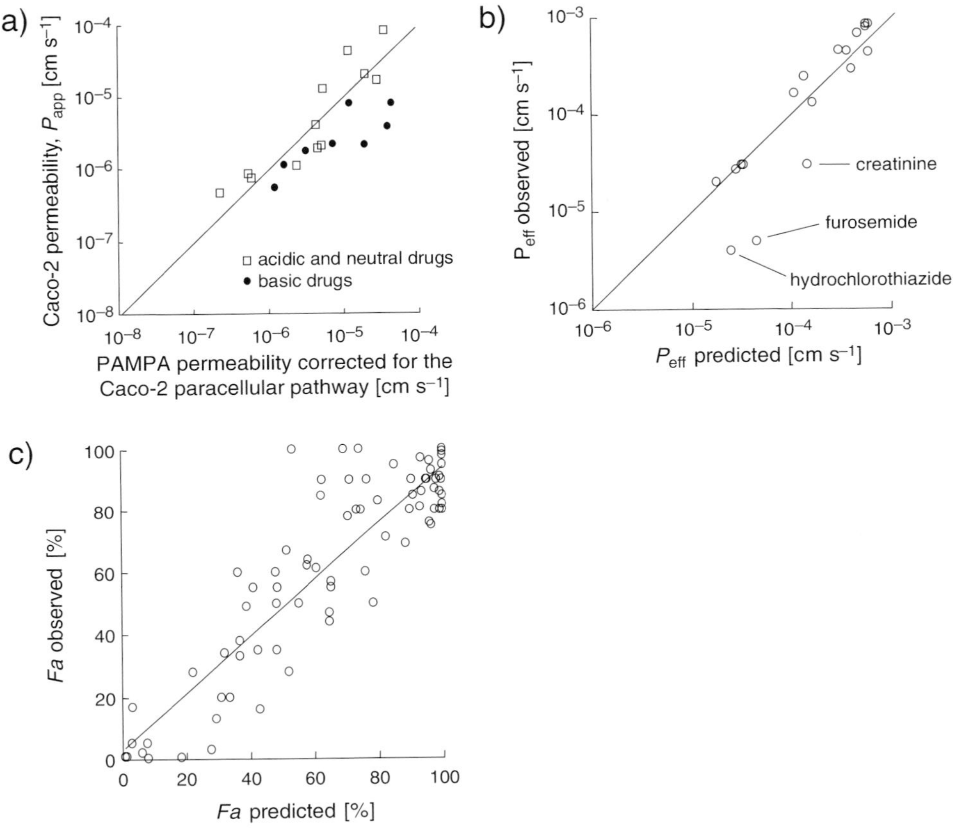

Fig. 5. *Validation of PAMPA by comparison with* a) *Caco-2 permeability* (P_{app}), b) *effective intestinal membrane permeability in humans* (P_{eff}), *and* c) Fa. Solubility-limited absorption drugs and active transport substrates were excluded. PAMPA permeability was corrected for paracellular and unstirred water layer (UWL) permeation based on TPAM (*Eqns. 1–8*). Transcellular permeability was assumed to be proportional to PAMPA permeability. To compare with Caco-2 permeability, paracellular pore radius and electric potential for Caco-2 cell were employed instead of those for humans. To compare with P_{eff}, the effect of UWL permeability was incorporated (*Eqs. 1* and *8*). To compare with *Fa*, corrected PAMPA permeability was converted to *Fa* by a plug-flow model (*Eqn. 2*). See the references for details.

process in the gastrointestinal tract. To simulate gastrointestinal absorption from a solid dosage form, a compartmental absorption transit (CAT) model is available [49]. CAT is also a physiologically based model. CAT enables IVIVC of solubility/dissolution properties.

2.4.3. *Mechanistic Investigations*

To investigate ADME mechanisms, it is important to combine assay data at different levels of physiological complexity. For example, a quantitative comparison between Caco-2 and PAMPA data may suggest the participation of transporters [45] [50] [51]. However, it is important to confirm such mechanisms by enzyme level assays. Mechanisms diagnosed by two assays are summarized in *Table 1*. In addition, a TPAM approach enables the prediction of paracellular pathway contributions (*Table 2*) [48] [52].

Table 1. *Mechanism Investigation by the Quantitative Comparison of Two Assays at Different Levels of Complexity*

Assays	Information
PAMPA *vs.* Caco-2	Influx transport, efflux transport, paracellular pathway
Caco-2 *vs. in vivo* (solution dose)	Metabolism
In vivo (solution dose) *vs. in vivo* (solid dose)	Solubility/dissolution

Table 2. *Predicted Contribution of Paracellular Pathway*

Drugs	Contribution of paracellular pathway[a])		
	Theoretical passive absorption model (log P_{oct}, log D_{oct}, pK_a)[b]) [%]	Theoretical passive absorption model (PAMPA)[c]) [%]	Caco-2[d]) [%]
Chlorothiazide	40	65	69
Cimetizine	36	75	31
Furosemide	1	3	1
Naproxen	2	1	0
Propranolol	18	16	3

[a]) Calculated as $P_{para}/(P_{trans}+P_{para})$. [b]) P_{trans} calculated by *Eqn. 3* using log P_{oct}, log D_{oct}, and pK_a. [c]) P_{trans} calculated from PAMPA permeability [48]. [d]) Ref. [52].

2.4.4. *Criteria for Compound Selection*

To select adequate compounds, it is important to set up adequate criteria. Criteria differ among each target disease. For example, drugs for an acute disease should be absorbed immediately after dosing, while drugs for a chronic disease do not have to.

A database of marketed drugs is of great help to derive adequate criteria for each assay. Today, *ca.* 2500 active ingredients are approved for clinical medication all over the world. The following characteristics of these drugs may help to set adequate criteria: structure, physicochemical properties, therapeutic category, pharmacological target, clinical PK data, clinical toxicology data, dosage form, dose strength, dosing regimen, sales, *etc.* To be of use, these data must be collected as digital numeric parameters and stored in a database. In addition, in-house assay data of these marketed drugs are required to obtain an IVIVC and to derive adequate criteria for each assay. We have constructed a marketed drug database using *Microsoft Access.*

'Must condition' and 'Enough condition' can be employed as criteria (*Fig. 6*). 'Must condition' means that a product (or a compound) must reach this given value to be launched. 'Enough condition' means that further improvement of the property will not make any contribution for the product value. In the case of solubility/dissolution, compound selection criteria are not the same as product criteria, because salt/crystal form screening and formulation study can improve the solubility/dissolution profile. To increase the success rate, it is preferable to examine experimentally the contribution of salt/crystal screening and formulation study

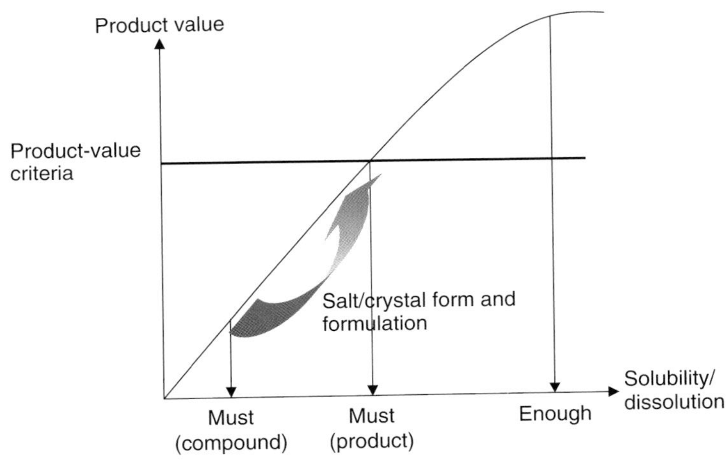

Fig. 6. *Schematic presentation of criteria for the solubility/dissolution property.* 'Must condition' means that a product (or a compound) must reach this value to be launched. 'Enough condition' means that further improvement of the property will not make any contribution for the product value. In the case of solubility/dissolution, compound-selection criteria are not the same as product criteria, because salt/crystal form screening and formulation study can improve the solubility/dissolution profile. To increase the success rate, it is preferable to experimentally examine the contribution of salt/crystal screening and formulation study as early as possible in the drug discovery/development process.

as early as possible in the drug discovery/development process. High-throughput formulation screening can enable this approach [53][54].

In addition, criteria of each property are often interdependent. Each property can be combined to generate a new parameter, which will be clinically more relevant. For example, dose, solubility, and permeability can be combined to predict *Fa*. The predicted *Fa* may help the project manager to decide on 'Go/No Go'. However, parameter conversion often loses profile information. In addition, numerical data are not suitable for an intuitive understanding. To overcome this problem, 2D- or 3D-visualization techniques may be of great help. We employ the biopharmaceutics classification system (BCS) to represent the absorption profile (*Fig. 7*) [10][55]. Dose number (*Do*, *Eqn. 9*) might work as a dose-corrected solubility criterion:

$$Do = \frac{\text{Dose}}{\text{Solubility} \times 250\,\text{ml}} \tag{9}$$

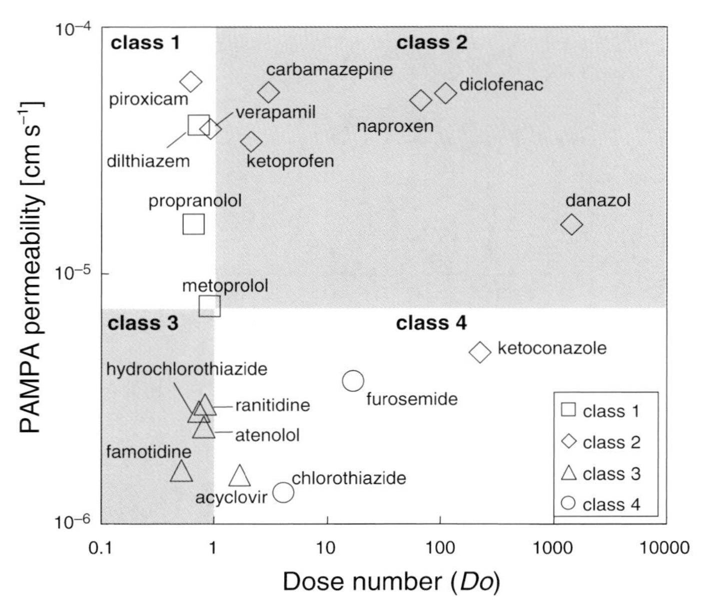

Fig. 7. *Biopharmaceutical classification of marketed drugs by PAMPA and direct UV kinetic solubility assay.* Solubility was converted to dose number (*Do*) by *Eqn. 9*. PAMPA permeability is corrected for the paracellular pathway. Each class indicates a classification previously reported in the literature.

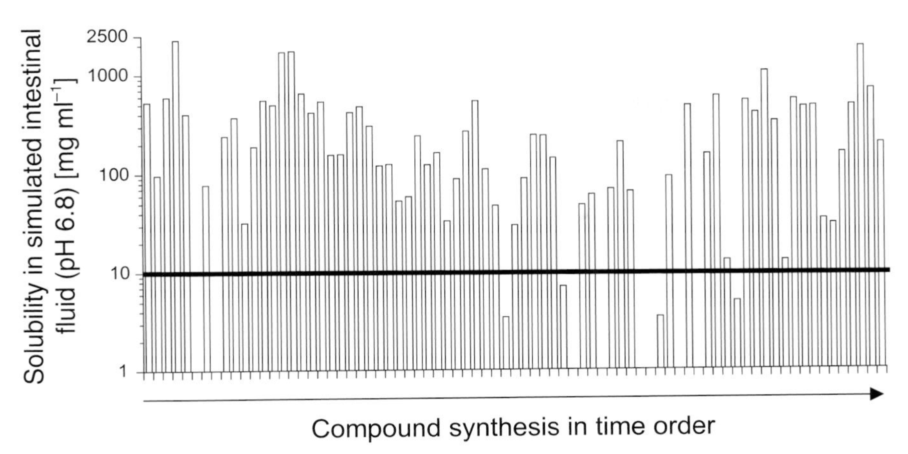

Fig. 8. *A case of solubility change in lead optimization process at* Chugai

2.4.5. *Progress Management*

If you are the director of a project, you may require information about its progress and risks. In this case, 'date' is the important information. With this information, we can draw up the next plan. For example, previously, *Lipinski* indicated that the log P_{oct} of candidate compounds began to increase after the incorporation of high-throughput screening and combinatorial chemistry [56]. This indication is derived from the log P_{oct} date profile. *Fig. 8* shows a case of lead optimization process in our company. In this project, the lead compound had a high solubility. As lead optimization progressed, solubility decreased. It is interesting that when solubility fell too low, it began to increase again. As a result, the final candidate had an adequate solubility.

2.5. *Knowledge Sharing*

After some knowledge is obtained by data analysis, it is important to share it with other researchers. In any drug discovery project, the number of ADME scientists is limited and usually only one or two persons are assigned to it. However, as its acronym indicates, ADME is not a single scientific subject. For the project to progress, the required ADME knowledge must range from physicochemical properties to clinical pharmacokinetics. For example, if the assigned person has a background in drug metabolism, it is seldom anticipated that he/she will give an adequate suggestion on how to improve the intestinal absorption of a poorly soluble compound by formulation. Similarly, if the assigned person has a background in oral absorption, he/she cannot give an adequate suggestion about clinical drug–drug interactions from *in vitro* CYP inhibition assay data. One approach to overcome this problem is to use a computer expert system.

Fig. 9 shows a graphic capture of *Chugai*'s in-house expert system for *Fa* prediction [34]. Input data are M_r, $\log D_{oct}$, $\log P_{oct}$, pK_a, PAMPA permeability, solubility, and Caco-2 data. Output data are predicted *Fa* with its prediction probability, the contribution of transcellular and paracellular pathways, suggestion for participation of active transport systems, and BCS class. This system has both single-run and batch-run modes. In addition, for further processing, the data can be exported for other visual data processing. It is important that the system be user friendly and also friendly for developers. Indeed, it is constructed using *Microsoft Access Visual Basic*, making it is easy to expand and maintain. Currently, our system focuses on intestinal membrane permeation. However, this system can be extended to oral absorption, drug metabolism, distribution, and excretion.

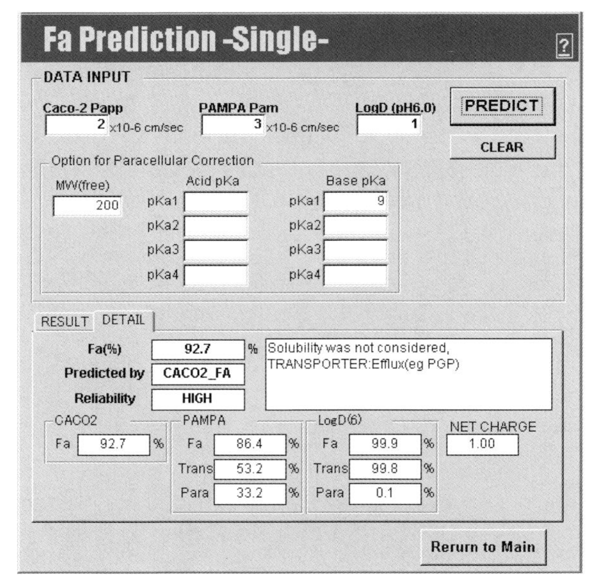

Fig. 9. *A partial graphic capture of* Chugai*'s in-house expert system for* Fa *prediction.* Input data are M_r, $\log D_{oct}$, $\log P_{oct}$, pK_a, PAMPA, solubility, and Caco-2 permeability. Outputs are predicted *Fa* with its prediction probability, contribution of transcellular and paracellular pathways, suggestion for participation of active transport system, and BCS class.

3. Conclusions

It is important to optimize the total ADME screening system to increase the discovery of clinical candidates. Mechanism-based/physiologically based models may offer a basic scaffold *to integrate data throughout from* in silico *to* in vivo. A computer expert system can be used for *knowledge sharing*. The cooperation among discovery ADME scientists, clinical ADME scientists, computer scientists, medicinal chemists, formulation scientists, and other discipline scientist is necessary. In this chapter, data processing of oral absorption data is discussed in detail. The concept of data processing is adaptable to other ADME processes.

Valuable discussions with Dr. *Yoshiaki Nabuchi*, Dr. *Minoru Machida*, and Dr. *Yoshinori Aso* are gratefully acknowledged.

REFERENCES

[1] T. Kennedy, *Drug Disc. Today* **1997**, *2*, 436.
[2] P. J. Eddershaw, A. P. Beresford, M. K. Bayliss, *Drug Disc. Today* **2000**, *5*, 409.
[3] H. Yu, A. Adedoyin, *Drug Disc. Today* **2003**, *8*, 852.
[4] J. F. Pritchard, M. Jurima-Romet, M. L. J. Reimer, E. Mortimer, B. Rolfe, M. N. Cayen, *Nat. Rev. Drug Disc.* **2003**, *2*, 542.
[5] A. Avdeef, *Curr. Topics Med. Chem.* **2001**, *1*, 277.
[6] A. Avdeef, 'Absorption and Drug Development. Solubility, Permeability and Charge State', Wiley-Interscience, Hoboken, NJ, USA, 2003.
[7] K. Box, C. Bevan, J. Comer, A. Hill, R. Allen, D. Reynolds, *Anal. Chem.* **2003**, *75*, 883.
[8] C. D. Bevan, R. S. Lloyd, *Anal. Chem.* **2000**, *72*, 1781.
[9] T. M. Chen, H. Shen, C. Zhu, *Comb. Chem. High-Throughput Screen.* **2002**, *5*, 575.
[10] K. Obata, K. Sugano, M. Machida, Y. Aso, *Drug Dev. Ind. Pharm.* **2004**, *30*, 181.
[11] I. J. Hidalgo, *Curr. Topics Med. Chem.* **2001**, *1*, 385.
[12] S. A. Roberts, *Xenobiotica* **2001**, *31*, 557.
[13] V. P. Miller, D. M. Stresser, A. P. Blanchard, S. Turner, C. L. Crespi, *Ann. N.Y. Acad. Sci.* **2000**, *919*, 26.
[14] I. Kariv, H. Cao, K. R. Oldenburg, *J. Pharm. Sci.* **2001**, *90*, 580.
[15] L. W. Frick, K. K. Adkison, K. J. Wells-Knecht, P. Woollard, D. M. Higton, *Pharm. Sci. Technol. Today* **1998**, *1*, 12.
[16] H. van de Waterbeemd, E. Gifford, *Nat. Rev. Drug Disc.* **2003**, *2*, 192.
[17] S. Modi, *Drug Discov. Today* **2003**, *8*, 621.
[18] C. A. Lipinski, F. Lombardo, B. W. Dominy, P. J. Feeney, *Adv. Drug Delivery Rev.* **1997**, *23*, 3.
[19] M. Kansy, F. Senner, K. Gubernator, *J. Med. Chem.* **1998**, *41*, 1007.
[20] F. Wohnsland, B. Faller, *J. Med. Chem.* **2001**, *44*, 923.
[21] K. Sugano, H. Hamada, M. Machida, H. Ushio, *J. Biomol. Screen.* **2001**, *6*, 189.
[22] C. Zhu, L. Jiang, T. M. Chen, K. K. Hwang, *Eur. J. Med. Chem.* **2002**, *37*, 399.
[23] A. Avdeef, M. Strafford, E. Block, M. P. Balogh, W. Chambliss, I. Khan, *Eur. J. Pharm. Sci.* **2001**, *14*, 271.
[24] P. Adriaans, D. Zantinge, 'Data Mining', Addison Wesley Longman, London, 1996.
[25] E. H. Kerns, *J. Pharm. Sci.* **2001**, *90*, 1838.
[26] C. E. Kibbey, S. K. Poole, B. Robinson, J. D. Jackson, D. Durham, *J. Pharm. Sci.* **2001**, *90*, 1164.

[27] K. Sugano, Y. Nabuchi, M. Machida, K. Saitoh, *J. Pharm. Sci. Technol. Jpn.* **2003**, *63*, S225.
[28] K. Sugano, Y. Nabuchi, M. Machida, Y. Aso, *Int. J. Pharm.* **2004**, *275*, 271.
[29] M. Kansy, H. Fischer, K. Kratzat, F. Senner, B. Wanger, I. Parrilla, in 'Pharmacokinetic Optimization in Drug Research – Biological, Physicochemical, and Computational Strategies', Eds. B. Testa, H. van de Waterbeemd, G. Folkers, R. Guy, Verlag Helvetica Chimica Acta, Zürich, 2001, p. 447–464.
[30] D. Butina, M. D. Segall, K. Frankcombe, *Drug Disc. Today* **2002**, *7*, S83.
[31] D. Zmuidinavicius, R. Didziapetris, P. Japertas, A. Avdeef, A. Petrauskas, *J. Pharm. Sci.* **2003**, *92*, 621.
[32] G. Camenisch, G. Folkers, H. van de Waterbeemd, *Eur. J. Pharm. Sci.* **1998**, *6*, 321.
[33] G. Camenisch, G. Folkers, H. van de Waterbeemd, *Pharm. Acta Helv.* **1996**, *71*, 309.
[34] K. Obata, K. Sugano, R. Saitoh, A. Higashida, Y. Nabuchi, M. Machida, Y. Aso, '124th Annual Meeting of the Pharmaceutical Society of Japan, Osaka, 2004', Abstracts, Vol. 3, p. 66.
[35] L. X. Yu, G. L. Amidon, *Int. J. Pharm.* **1999**, *186*, 119.
[36] H. Lennernäs, *J. Pharm. Sci.* **1998**, *87*, 403.
[37] K. Takacs-Novak, G. Szasz, *Pharm. Res.* **1999**, *16*, 1633.
[38] H. Saitoh, A. Noujoh, Y. Chiba, K. Iseki, K. Miyazaki, T. Arita, *J. Pharm. Pharmacol.* **1990**, *42*, 308.
[39] R. Neubert, *Pharm. Res.* **1989**, *6*, 743.
[40] P. Proulx, *Biochem. Biophys. Acta* **1991**, *1071*, 255.
[41] G. Lipka, J. A. Op den Kamp, H. Hauser, *Biochemistry* **1991**, *30*, 11828.
[42] A. Adson, T. J. Ruab, P. S. Burton, C. L. Barsuhn, A. R. Hilgers, K. L. Audus, N. F. H. Ho, *J. Pharm. Sci.* **1994**, *83*, 1529.
[43] K. Sugano, N. Takata, M. Machida, K. Saitoh, K. Terada, *Int. J. Pharm.* **2002**, *241*, 241.
[44] A. W. Larhed, P. Artursson, J. Grasjo, E. Bjork, *J. Pharm. Sci.* **1997**, *86*, 660.
[45] R. Ano, Y. Kimura, M. Shima, R. Matsuno, T. Ueno, M. Akamatsu, *Bioorg. Med. Chem.* **2004**, *12*, 257.
[46] R. Saitoh, K. Sugano, N. Takata, T. Tachibana, Y, Nabuchi, K. Saito, '17th Annual Meeting of JSSX, Tokyo, 2002', Abstracts, p. 255.
[47] R. Saitoh, K. Sugano, N. Takata, T. Tachibana, A. Higashida, Y, Nabuchi, Y. Aso, *Pharm. Res.* **2004**, *21*, 749.
[48] K. Sugano, Y. Nabuchi, M. Machida, Y. Aso, *Int. J. Pharm.* **2003**, *257*, 245.
[49] L. X. Yu, E. Lipka, J. R. Crison, G. L. Amidon, *Adv. Drug Delivery Rev.* **1996**, *19*, 359.
[50] L. Di, K. Y. Fan, S. L. Petusky, E. H. Kerns, O. J. McConnell, G. T. Carter, M. Farris, L. Rob, P. Jupp, *AAPS Pharm. Sci.* **2002**, *4*, Abstr. M1252.
[51] A. Higashida, R. Saitoh, K. Sugano, K. Obata, Y. Nabuchi, M. Machida, T. Mitsui, Y. Aso, '124th Annual Meeting of the Pharmaceutical Society of Japan, Osaka, 2004', Abstracts, Vol. 3, p. 66.
[52] V. Pade, V. Stavchansky, *Pharm. Res.* **1997**, *14*, 1210.
[53] M. L. Peterson, S. L. Morissette, C. McNulty, A. Goldsweig, P. Shaw, M. LeQuesne, J. Monagle, N. Encina, J. Marchionna, A. Johnson, *J. Am. Chem. Soc.* **2002**, *124*, 10958.
[54] H. Chen, Z. Zhang, C. McNulty, C. Olbert, H. J. Yoon, J. W. Lee, S. C. Kim, M. H. Seo, H. S. Oh, A. Lemmo, *Pharm. Res.* **2003**, *20*, 1302.
[55] G. L. Amidon, H. Lennernäs, V. P. Shah, J. R. Crison, *Pharm. Res.* **1995**, *12*, 413.
[56] C. A. Lipinski, *J. Pharm. Toxicol. Methods* **2000**, *44*, 235.

Part V. Concluding Chapters

Educational and Communication Issues Related to Profiling Compounds for Their Drug-Like Properties

by **Ronald T. Borchardt**

Department of Pharmaceutical Chemistry, 2095 Constant Avenue, 104 McCollum Research Laboratories, The University of Kansas, Lawrence, KS 66047, USA
(phone: 001785-864-3427; fax: 001785-864-5736; e-mail: rborchardt@ku.edu)

Drug discovery and development is a very complex, costly, and time-consuming process. Because of the uncertainties associated with predicting the pharmacological effects and the toxicity characteristics of new chemical entities (NCEs) in man, their clinical development is quite prone to failure. In recent years, pharmaceutical companies have come under increasing pressure to introduce new blockbuster drugs into the marketplace more rapidly. Companies have responded to these pressures by introducing new technologies and new strategies to expedite the drug discovery and development processes. This chapter will focus on two aspects of the new drug discovery/development paradigm, *i.e.*, educational and communication issues associated with the integration of 'drug-like' profiling data early in drug design. Since this author recently wrote a chapter on this same subject [1], he has provided here only a brief summary of his thoughts. For more details about the author's views on this subject, readers are referred to the book chapter referenced above and the PowerPoint slides which the author presented during his lecture at the *LogP2004 Symposium* (see enclosed CD).

Drug discovery and development have traditionally been divided into three separate processes (*i.e.*, discovery research, preclinical development, and clinical development) that should be integrated both organizationally and functionally. To be successful, each of these processes needs highly educated and experienced scientists, clinicians, and engineers who have great depth in their respective areas of expertise. While these individuals all have significant depth in their areas of expertise, they often lack the scientific breadth and experience necessary to communicate effectively across disciplines. The inability of these people to communicate across discipline lines, as well as the increasing size and complexity of pharmaceutical and biotechnology companies, fostered the establishment of separate and distinct discovery research, preclinical development, and clinical development 'silos' within these companies. Because of their isolation, scientists in the discovery research 'silos' in many companies were advancing an increasing number of 'marginal' drug candidates into preclinical development in the 1980s and 1990s. These 'marginal' drug candidates often lacked the characteristics needed to succeed in preclinical and clinical development. The increase in the number of 'marginal' drug candidates being advanced into development arose in part because: *i*) discovery scientists did not fully appreciate the 'complete portfolio' of characteristics that a drug candidate must have to succeed in preclinical and clinical development; and *ii*) a paradigm shift occurred in discovery research in the late 1980s and early 1990s that resulted in the generation of many drug candidates lacking 'drug-like' properties. This drug discovery paradigm shift involved a transition from whole animal disease-based screens to biochemical-based screens, which use isolated and purified macromolecules (*e.g.*, receptors, enzymes) assayed *in vitro*. The whole animal disease-based screens had the advantage that they afforded information about a molecule's 'pharmacological' as well as 'drug-like' properties. In contrast, the biochemical-based screens afforded only information about a molecule's potential 'pharmacological' activity.

Unfortunately, this drug design paradigm that developed in the late 1980s and early 1990s often involved only the interactions of medicinal chemists with biologists (biochemists, cell biologists, molecular biologists, and pharmacologists) in discovery research. Input from preclinical development scientists, who have the knowledge and expertise in areas such as pharmaceutics, biopharmaceutics, pharmacokinetics, drug metabolism, toxicology, and process chemistry, was not sought by the medicinal chemists during this phase of drug design. The result was that scientists in discovery research were advancing what they thought were 'high-quality' drug candidates into preclinical development. However, in reality, these compounds were 'marginal' drug candidates that might better be described as 'high-affinity ligands'. By definition, a drug candidate is a molecule that has high binding affinity and specificity for a validated therapeutic target, as well as 'drug-like' properties (*i.e.*, solubility, permeability, chemical/enzymatic stability, *etc.*). In contrast, a 'high-affinity ligand' is a molecule that has high binding affinity and specificity for a validated therapeutic target but lacks certain 'drug-like' properties. The absence of these 'drug-like' properties could ultimately lead to the failure of this 'high-affinity' ligand in preclinical or clinical development or at a minimum make the development process more time-consuming and expensive.

Retrospective analysis data from the 1980s and 1990s has shown that compounds failed in preclinical and clinical development for various scientific reasons including undesirable toxicity, lack of efficacy, and lack of optimal 'drug-like' properties. The failures arising from the lack of 'drug-like' properties could be due to the paradigm shift in drug discovery described above and/or the availability of new knowledge and more-precise and selective assays for characterizing a molecule's 'drug-like' properties. Even if these 'marginal' drug candidates succeeded in preclinical and clinical development, they would cost more in time and money to develop into commercial products.

Traditionally, incorporating optimal 'drug-like' properties into a structural 'lead' was not considered by medicinal chemists to be their responsibility. Instead, medicinal chemists felt that the undesirable 'drug-like' characteristics in their structural 'leads' and drug candidates would be fixed by preclinical development scientists. However, that view has changed in the past 5–7 years, resulting in another significant paradigm shift in drug discovery. The most-significant aspect of this latest paradigm shift is the recognition by medicinal chemists that the 'drug-like' properties of structural 'hits', structural 'leads', and drug candidates are intrinsic properties of the molecules and that it is the responsibility of the medicinal chemist to optimize not only the 'pharmacological' properties but also the 'drug-like' properties of these molecules. Therefore, assessment of these 'drug-like' properties is now done early in the drug discovery process on structural 'hits' and structural 'leads'. Optimization of these 'drug-like' properties is done through an iterative process in close collaboration with preclinical development scientists. This process is analogous to that used by the medicinal chemist to optimize the pharmacological activity of a molecule. The implementation of this paradigm shift in drug discovery has created some 'people-related' problems (*e.g.*, educational and communicational), which are briefly discussed below.

Scientists themselves are a crucial ingredient for the successful implementation of this new drug discovery paradigm. Because scientists working at this interface between discovery research and preclinical development will be expected to work closely with their colleagues in other disciplines, they will need to have not only depth in their respective scientific disciplines but also the following skills: *i*) scientific breadth in related disciplines; *ii*) communication skills; *iii*) interpersonal skills; and *iv*) mutual respect for colleagues in other disciplines. In other words, these individuals will need to be very well-rounded scientists who are also team players. In many pharmaceutical and biotechnology companies, scientists working at the interface between discovery research and preclinical development often lack the desired level of these skills. Therefore, companies need to encourage these individuals to expand their scientific knowledge and further refine their communication and interpersonal skills so that they can contribute more effectively to the implementation of this new drug discovery paradigm.

Because of a lack of scientific breadth, scientists in discovery research and preclinical development often have different scientific viewpoints. These differences in viewpoints are due in part to a lack of appreciation by discovery scientists for preclinical and clinical drug development. Similarly, preclinical development scientists often lack an appreciation for the process of drug discovery. These differences in scientific perspective can be changed through the increased exposure of all the scientists, clinicians, and engineers in a company to the total process of drug discovery and development. Discovery scientists need to better understand and appreciate the highly regulated nature of preclinical and clinical development. Preclinical and clinical development scientists, clinicians, and engineers need to better understand and appreciate the qualitative and semi-quantitative nature of discovery research, the need for speed in the generation of 'pharmacological' and 'drug-like' data during the 'hit-to-lead' and 'lead' optimization stages of drug discovery, and the difficulties and complexities of balancing the 'pharmacological' and 'drug-like' properties of drug candidates.

If pharmaceutical and biotechnology companies intend to successfully implement this new drug discovery paradigm, they will need to encourage their scientists to attend external or internal short courses and workshops that would expand the scope of their scientific knowledge. These scientists also need to be encouraged (and rewarded) to attend scientific meetings outside of their own disciplines to further expand and refine their scientific breadth. Reward systems need to be developed within companies to recognize individuals who have adapted to this new drug discovery paradigm and are making significant contributions to the discovery of 'high-quality' drug candidates.

Other mechanisms that companies can use to increase the scientific breadth of their employees include: *i*) offering cross-discipline rotational assignments to scientists, clinicians, and engineers; *ii*) creating forums that would encourage close interactions among scientists, clinicians, and engineers in traditionally isolated disciplines; and *iii*) physically structuring organizations in a way that embeds or co-locates scientists, who need to work together and appreciate each other's perspective on a problem.

In addition to expanding the scientific breadth of scientists who work at the interface between discovery research and preclinical development, the communication and interpersonal skills of many of these individuals also need to be refined. To overcome the gap that has existed for decades between scientists in drug discovery and preclinical development, pharmaceutical and biotechnology companies need to create an atmosphere of open communication and mutual respect among all scientists. Medicinal chemists need to be willing to defer to the experts (preclinical development scientists) in helping them optimize the 'drug-like' properties of their structural 'hits', 'leads', and drug candidates. Preclinical development scientists need to embrace this responsibility with enthusiasm, but also must have respect for the delicate balance that medicinal chemists are trying to achieve in designing a drug candidate. While preclinical development scientists are passionate about a molecule's 'drug-like' properties, they need to respect the passion that discovery scientists have for the 'pharmacological' activity of their structural 'hit', 'lead', and drug candidates. All parties involved in drug discovery need to accept the thesis that a balance between the 'pharmacological' activity and the 'drug-like' properties of a drug candidate must be attained. This environment can only occur if discovery scientists and preclinical development scientists respect each other and grant each other equal rights at the 'drug design table'. In addition, a company's incentives and rewards to its employees need to be aligned with reinforcing this culture and this type of organizational behavior.

If the 'people' problems associated with the implementation of this new drug discovery paradigm are appropriately addressed, pharmaceutical and biotechnology companies are likely to see more high-quality drug candidates emerge from their drug discovery groups. The net result is likely to be less attrition of their drug candidates in preclinical development and their NCEs in clinical development.

REFERENCE

[1] 'Pharmaceutical Profiling in Drug Discovery for Lead Selection', Eds. R. T. Borchardt, E. H. Kerns, C. A. Lipinski, D. R. Thakker, B. Wang, AAPS Press, Arlington, VA, 2004.

Present and Future Significance of ADMET Profiling in Industrial Drug Research

by **Werner Cautreels***[a]), **Michiel de Vries**[a]), **Constance Höfer**[b]), **Henk Koster**[a]), and **Lechoslaw Turski**[b])

Solvay Pharmaceuticals Research Laboratories
[a]) C. J. van Houtenlaan 36, NL-1381 CP Weesp
(tel: +31294479624; fax: +31294477173; e-mail: werner.cautreels@solvay.com)
[b]) Hans Böckler Allee 20, D-30173 Hannover

Abbrevations
ADMET: Absorption, distribution, metabolism, excretion, toxicity; CC: clinical candidate; CYP: cytochrome P450; fMRI: functional magnetic resonance imaging; HT: high throughput; NCE: new chemical entity; P-gp: P-glycoprotein; PK/PD: pharmacokinetics/pharmacodynamics; PoP: proof-of-principle; PPB: plasma protein binding; Ro5: rule-of-five; R&D: research and development.

1. Introduction

This chapter provides insight into some of the major challenges pharmaceutical companies now face in the discovery and development of new products. Mastering the decision and selection process of potential new compounds throughout the entire discovery and development process is becoming an art in itself and requires continuous adjustment of the corresponding data review and data integration processes. Making this process transparent and optimal will often require changes in organizational structures so that multidisciplinary approaches and interactions are encouraged. ADMET Profiling has become a major part of the decision algorithms leading to efficient and sustainable portfolio management. It is our strong conviction that the ADMETexpert needs to better understand the context and the environment in which her/his science is applied to optimize its contribution to the successful development of new and innovative medication.

2. The Regulatory and Economical Environment

The landscape of drug development is changing rapidly [1]. In a climate of increased public scrutiny and criticism of industry profits, regulatory requirements are becoming significantly more stringent and costly to fulfill [2–4]. On the other hand, if a favorable benefit/risk ratio is demonstrated together with sound pharmaco-economic profiles, there is willingness to accept premium costs for innovative products.

Improved insight into disease processes, including potential pharmacogenomic parameters, entails market fragmentation through individualized therapy [5]. Ideally, optimal medication should be codeveloped with sensitive and specific diagnostics for patient selection and follow-up to increase product safety, success, and profitability.

New technologies and know-how disseminate more quickly in this information age, and are less-significant barriers for market entry. As a result, product life cycles are shortened, and products with the same mechanisms of action are frequently introduced only a short time after the first-to-market innovative product.

The important factors listed in *Table 1* have changed the economic landscape in which new pharmaceutical compounds have to be developed. On purpose, these factors and their contributions are, at least, controversial. Not so long ago, the process of harvesting new compounds was constrained at the discovery stage, where processes were frequently based on trial and error. Because of the decline in its interest towards the advantage of biochemistry, harvesting new compounds is often limited by the lack of solid and creative medicinal chemistry. Today however, rational drug design, integrated knowledge, and new technologies from all scientific disciplines (combinatorial chemistry, target discovery and functional analysis, genomics and proteomics, predictive ADMET, to name only a few) produce an unlimited stream of new and highly optimized compounds [5–7]. Unfortunately, these compounds are then often developed in an old regulatory paradigm where the clinical development is hampered by patient scarcity. Because of better diagnostic techniques and insights in the disease, blockbuster drugs will increasingly be replaced by products tailored for subgroups of patients. This provides a new challenge for the pharmaceutical industry which was traditionally focused to provide a product rather than a service of therapeutic solutions and integrated diagnostic, sometimes called theranostics [8]. The willingness to pay for such a new service, not only by the patients who should markedly benefit from such approach, but also by public or private insurance systems on the basis of improved pharmaco-economics, should become a major driving force towards such an integrated approach. The last and

Table 1. *Changing Factors in the Economic Landscape*

Traditional factors	New factors
Discovery constraints • rational drug design • new technologies	Development constraints • old regulatory paradigm • patient scarcity
Blockbuster-driven • large patient groups • product focused	Fragmentation opportunities • patient-tailored products • service focused > theranostics
Low-price competition • long product life cycles	Intense-price competition • shorter product life cycles • widely diffused technologies • better therapeutic substitute reached faster

perhaps most-significant factor changing the economic landscape is the ever accelerating diffusion of knowledge and information, leading to much shorter product life cycles [9]. Short patent life is only one factor curtailing return on investment. Indeed, second- and third-in-class products will reach the market very rapidly after the pioneer first-in-class compound. Examples of this new situation are the COX-2 inhibitors, and the – perhaps more-exiting – inhibitors of cGMP specific PDE5 [10]. In the case of COX-2 inhibitors, the second-in-class product was introduced less than one year after the first.

What are we doing wrong (*Table 2*)? With all the new technologies, with all the new insights in biology and underlying disease mechanisms, and perhaps most of all with all the money we spend, where do we derail in our efforts to generate better medication? If we believe the statistics, and that belief is also somewhat controversial, fewer NCEs are currently approved and launched. Or perhaps have we already discovered the easier products, and the challenge is becoming more difficult? Perhaps did we not do a good enough job in public relations? Indeed, pharmaceuticals have significantly contributed to the increased life expectancy of those that have access to them. The combination of modern antibiotics and vaccines, as well as the development of anticancer and cardiovascular compounds, has significantly contributed to a multidecennial increase in

Table 2. *The Paradox in Pharmaceutical R&D*

- Never did we have so much technology,
- Never did we have so much insights into biology and biomechanisms,
- Never have we spent so much money in R&D ....
- ... but never was output so low.

life expectancy [11–13]. But where do we go next? We have perhaps reached an asymptotic level of life expectancy, the next challenge being to make a better impact on the quality of life [14], especially in the later stages of life? So, if this is all true, why do we deserve the current public scrutiny and criticism of our industry? Perhaps the challenge is poorly understood? Perhaps we need to change our public relation policy and re-emphasize our contribution to life expectancy, health, and quality of life, rather than the short-term gains in speculative financial investments? Finally, perhaps the financial pressure is still not high enough to bring dissatisfaction to the level required for further material changes in the way we do business in our industry [15].

3. The R&D Conundrum

Table 3 summarizes some of the key challenges currently faced by the pharmaceutical industry. These challenges can be summarized in one word: productivity.

Table 3. *What Is the Challenge in R&D?*

New technologies are still extensively used in old discovery paradigms:
- Inertia in conventional discovery organizations;
- Filters in discovery process not yet adapted.

New technologies are under-utilized in clinical development:
- New diagnostic techniques not yet applied during early stages;
- Do surrogate endpoints receive regulatory acceptance?

Portofolio decisions and resources allocation should be modified:
- Blockbusters *vs.* patient-tailored products.

With our access to new technologies, and the R&D process no longer being discovery-constrained, where do we fall short? We believe that not enough changes were made in the discovery organizations of the larger pharmaceutical companies [16]. New technologies and insights are still largely applied in the old paradigm of decision making and power structures [17]. Different algorithms are required to integrate the masses of data generated by these new technologies so that new filters can be put in place for decision making [18–20]. The smaller biotech companies may provide good inspiration, with smaller, multidisciplinary teams, and quick decision-making processes. Several large companies have tried to mimic this model by installing small decentralized teams focused on specific targets or disease areas. Unfortunately, those teams are often still restrained in their efforts by the heavy infrastructure they have to work in. Perhaps even worse is the situation in the clinical arena. Often imposed by a restrictive regulatory environment, new technologies and approaches only slowly penetrate into clinical studies. Few compounds enter clinical development with tailor-made tools to measure their pharmacological activity, although this should be possible, even if it would be restricted to techniques such as proteomic fingerprinting [21]. This however will require much closer links between the scientists and the clinicians, and will require additional investments during the discovery phase to develop such surrogate measures or endpoint biomarkers. But how much would be gained if we could better select clinical candidates in the very early phases of clinical studies? Clinical development programs are becoming increasingly larger. Patient recruitment is becoming more and more difficult and often represents the single most-limiting factor in the development phase. As a consequence, we are probably diluting the power of the clinical program by including patients who cannot benefit from the treatment, thereby decreasing the benefit/risk profile of the new treatment.

4. Opportunities to Change the Cost Paradigm

The *Tufts Center for the Study of Drug Development* is a leader in the study of drug development and publishes regular updates on new developments and statistics in this area. In their latest reports, the average cost for a new approved product is *ca.* $US 800 million, including costs for failed compounds [22]. According to the same report, the average probability for a phase-I compound to be approved is *ca.* 21%. Using these data, the average costs for a single compound can be estimated at *ca.* $US 200 million, although with some major differences depending on the therapeutic area and/or disease. Because of increasing regulatory requirements, often related to the need of producing larger patient data base for safety, it is unlikely that these costs will decrease. Therefore, cost optimization will need to come from improved selection and attrition processes.

Table 4 provides a list of critical areas for optimization. Most critical will be to link better discovery with clinical development by developing better tools to avoid failed clinical studies and programs, the biggest drain of resources in the development process.

Table 4. *Costs of Drug Development: Opportunities to Change the Paradigm*

Traditional Costs per Product: $US 800 million Causes	New Costs per Product: $US 400 million? Solutions
Inefficient drug design by trials & errors	Rational drug design, *in silico* technologies
Erratic clinical endpoints	Implementation of clinical endpoints as part of discovery
Many patients → dilution in phase III	Better clinical endpoints → fewer patients and fewer failed studies
Same paradigm for different products	Adaptation of organization and processes to specific situations
Focus on delivery of drug	Focus on disease management (how to match patient to drug?)

5. The Contributions of ADMET Profiling

ADMET Profiling, when applied in creative and proactive decision algorithms, will have the power to significantly contribute to the selection process of the most-optimized compounds during the different phases of the discovery and development process. To achieve this objective, ADMET profiling needs to be completely integrated into all phases of the process, with clear definition of the selection process and predefined selection criteria. Whereas this sounds obvious, it cannot always be easily achieved because of organizational barriers [23]. Within *Solvay Pharmaceuticals*, these barriers were largely overcome by a complete overhaul of the traditional R&D organization (*Fig. 1*).

The total process is broken down into only two systems: programs from early discovery (D) to Proof-of-Principle (PoP) in the clinic, and projects from PoP to registration. A program team will focus on a therapeutic or disease area, and its leadership is shared by a scientist and a clinician. The program team has members from the different scientific disciplines. The program team's mission is to bring compounds to successful completion of clinical PoP, typically in phase-II clinical studies. Alternatively, PoP can be achieved in the preclinical stage for an optimized me-too product, provided that an adequate reference point is available. The project team's mission is to develop this new PoP compound on a global basis. Its leadership is shared by a project director and a clinician.

Working in the program team concept has resulted in more-selective compound selection throughout the program process within better timelines. To achieve this result, the other side of the matrix (the line function area) was also changed. The structure of traditional departments, such as pharmacology, chemistry, and the like was abandoned and new multi-

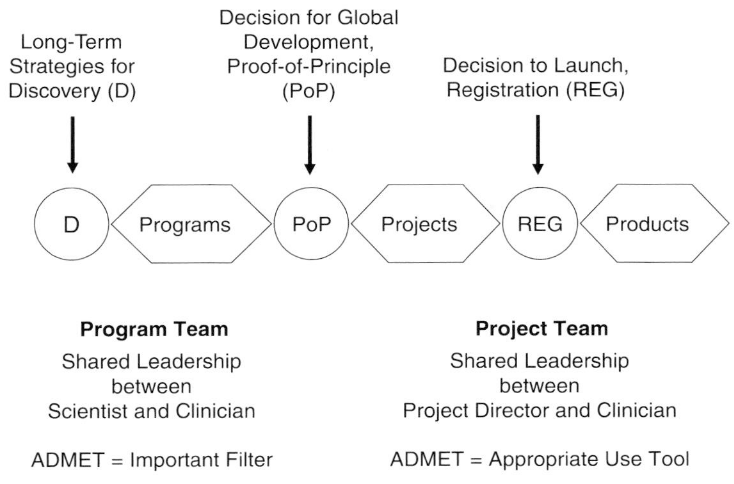

Fig. 1. Solvay Pharmaceuticals' *R&D organizations*

disciplinary units were established with a specific mission in the assembly line of new PoP compounds. *Fig. 2* summarizes this new concept as implemented in *Solvay Pharmaceuticals*. ADMET Research is present in all of these units. Clinical pharmacologists and clinicians are involved in the decision processes for each of the units, and vice versa, the scientists

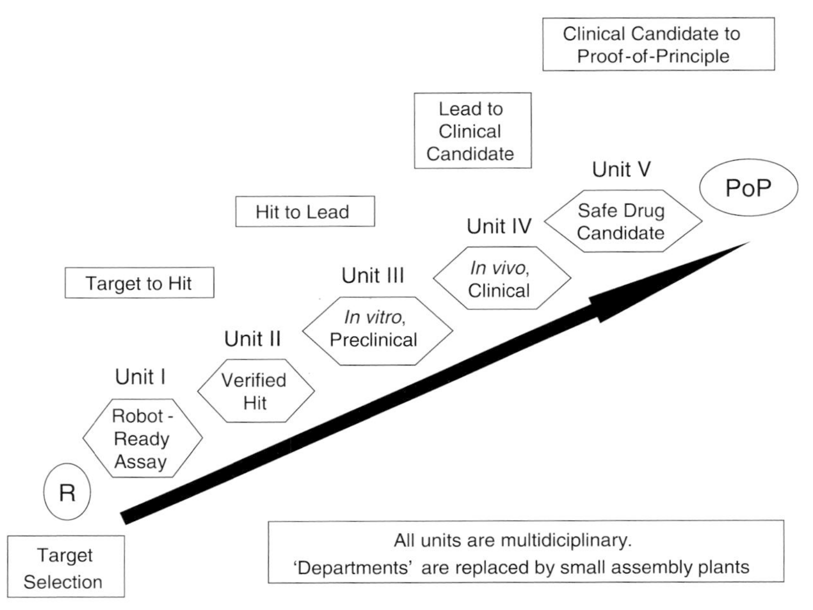

Fig. 2. *Line functions in the new organization*

are closely involved in the design of phase-I and PoP studies. During the program phase (the phase from discovery to PoP), ADMET profiling plays an important role as a filter for the compound selection process, but obviously also generates data for regulatory purposes. In the later project phase, ADMET will play an important role for appropriate use, together with its traditional role to collect data for regulatory purposes.

Within these units, the processes leading from target discovery to hit, to lead, to clinical candidate, to PoP are conducted on the basis of predefined criteria, and ADMET filters are included. Unfortunately, the predictive value of these filters is not always of binary order (yes/no) and, most frequently, is based on exclusion criteria. It is not too difficult to define negative decision criteria, *e.g.*, poor solubility, poor membrane permeability, unwished CYP interactions, and many others. It is however more difficult to define positive and predictive decision criteria, *e.g.*, the absence of *in vitro* or animal toxicity will not predict the absence of an idiosyncratic drug reaction in patients. *Table 5* lists a number of negative and positive decision points during the program phase, *i.e.*, from hit identification to clinical candidate. Other factors should be included based on the specific requirements of the expected therapeutic indications. And obviously common sense is a must when applying these rules to real life cases. Furthermore, it is easier to find reasons to stop a project, and more difficult to continue it and face a possible failure at a later stage.

Fig. 3 provides an example of a decision algorithm for the program phase. Most important in such a paradigm are the feedback loops in the process. To achieve optimal feedback loops, program organization and communication are of utmost importance so that the right information is used by the right team players at the right time. Hence, the need to structure or restructure the organization accordingly, as discussed earlier in this paper.

Clearly, the most-critical and important feedback loop follows entry of the compound in human phase-I trials, when information is at last gathered in the species of ultimate interest, namely humans. It is therefore more important to take a first compound from a specific program or compound class as soon as possible into humans than to continue the preclinical optimization process for too long. Obviously, no compromise can be made regarding the safety of phase-I trials, and adequate ethics must be followed as this compound may not (yet) be drug-able for a variety of reasons. Before entering phase-I trials, clear objectives and clinical endpoints need to be defined, and this should be done not in isolation but in view of future requirements for phase-II PoP studies. And again, one needs to define prospectively a number of negative and positive decision points.

Table 5. *Positive and Negative ADMET Decision Points During the Program Phase from Hits to Clinical Candidates* (CC)

Program phase	Negative criteria	Positive criteria
Hit identification	• Low stability and solubility • Low membrane passage • High synthetic effort	• High solubility and stability • Drug-like properties (Ro5) [24] • Uncomplicated synthesis
Hit to lead	• Low metabolic stability, cell permeability, bioavailability (early animal PK) • High inhibition potential (*in vitro*) • High interaction potential (*in vitro*) • High induction potential (*in vitro*)	• Balanced metabolism and stability • Permeable into biomembranes • Early animal bioavailability >15% • Low risks of inhibition or induction • Metabolic profile in laboratory species same as in humans
Lead to clinical candidate	• Low metabolic redundancy (*e.g.*, >95% metabolism *via* single CYP isozyme) • Low similarity of metabolic profile in animals and humans • High number of active metabolites • Loss of compound by unknown pathways of metabolism and excretion • Tissue retention and accumulation	• Multiple known pathways of metabolic and renal elimination reduce risks of PK/PD problems • Availability of metabolic map for risk assessment in clinical ADMET • Biomarker available

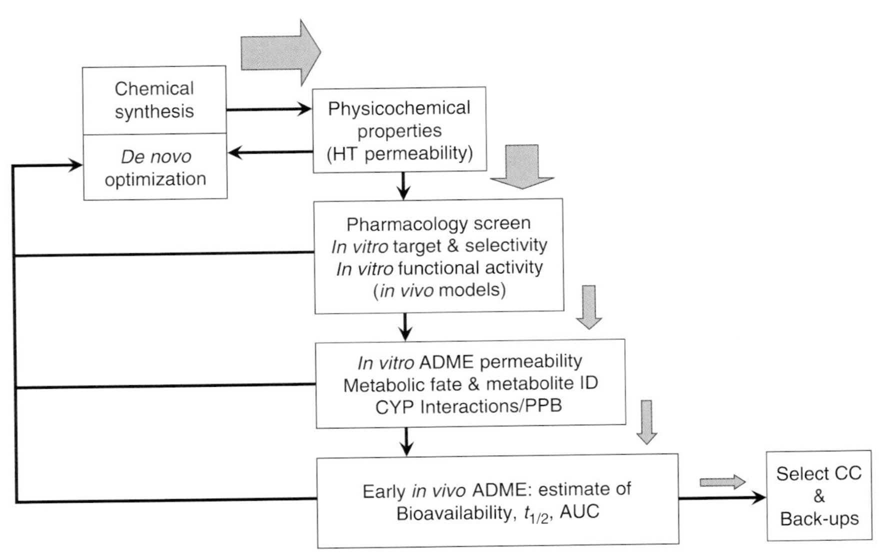

Fig. 3. *Example of a decision algorithm during the program phase*

PoP has become a common term in drug development. It is supposed to generate data in early clinical studies that confirm the relevance of the mechanism of action through the expression of certain biological or therapeutic activities in patients thus warranting further, full development of the compound. This is a very important decision point in the drug development process as the consequence will be an investment of several hundred million Euros for a typical development program. Betting on the wrong compound can have dramatic results as it will also take away resources from other perhaps more-promising compounds. This situation is further complicated by the fact that the ultimate outcome of the PoP decision will not be known before several years, often at the end of phase III. The very long timelines in our industry certainly complicates the decision process when compared to other fields of technology such as electronics.

It is therefore of critical importance to correctly define the objectives of the PoP studies [25], to integrate these objectives not only in the scientific and medical aspects of the compound, but also into commercial reality. Because this exercise has to be done in anticipating the market situation more than five years in advance, long debates around the definition of PoP requirements are routine (*Table 6*).

Technologies are developing rapidly in the area of ADMET profiling, allowing a significant acceleration in data gathering, with much larger numbers of experiments and very large numbers of compounds. High-

Table 6. *Positive and Negative Decision Points During the Phase-I–Proof-of-Principle Phase*

Phase-I–PoP phase	Negative decisions	Positive decisions
Phase I	• Absence of PD activity within safe dose levels • Inappropriate PD duration • Low or erratic bioavailability • Indication of QTc prolongation • Unexpected or unavailable adverse events	• Easy PK profiling • Established PK/PD relation • Availability of biomarkers
Proof-of-Principle	• Absence of PoP definition • Nonlinear kinetics in expected therapeutic dose range	• Validation of biomarkers • Established relation between biomarker and future clinical endpoints • Clear definition of expected dose range for full phase II–III • Compound meets prospective PoP requirements

Table 7. *Approaches in ADMET Screening*

Screens	Past & current approaches	Current & future approaches
Metabolic stability	Microsomes	Intact cells
Routes of metabolism and excretion	*In vitro* (liver), intact animals	Intact cells, *in silico*
Absorption and transport	log *P*, Caco-2 cells, P-gp expressing cells, intact animals	Artificial membranes, human preparations
Urinary excretion	Intact animals	Kidney tubular preparations
Interaction potential	Human microsomes, heterologous expressed CYPs	*In silico*
Reactive intermediates	Time-dependent incubations and enzyme activity	*In silico*
Active metabolites	Biological production followed by purification and structure elucidation	Cultured cells and mini-technologies
Toxicity	*In vitro, in silico*	Toxicogenomics, proteomics
Bioanalytics	Microliter sampling	Nanoliter sampling
Phase I	Biomarkers	Surrogate clinical endpoints, microdosing studies
	PET spectroscopy	fMRI
	Phenotyping–genotyping	Genotyping–proteotyping

throughput techniques are now common to assess ADMET properties [26]. Further development of these techniques will also allow the generation of more-relevant data, with better extrapolability to the (human) *in vivo* situation. *Table 7* provides a summary of the evolution of some of these techniques as applied during the program process.

6. Concluding Remarks

The pharmaceutical industry has made a significant contribution to our longer life expectancy and to the quality of our life in general. This is a remarkable achievement if one considers the relatively primitive tools and limited knowledge that were available to do this. Our industry could have done a better job explaining this to the public.

With this achievement in mind, we can better understand why the development of new – and better – pharmaceuticals is becoming more difficult. Medical needs remain very high and many diseases still receive only symptomatic treatment. The development of new anti-infective agents and treatments is rapidly becoming a top priority. Meanwhile mortality and morbidity treatments are more frequently complemented by Quality-of-Life treatments. The latter require new benefit/risk standards with little tolerance for adverse events. Such products will therefore require modified development approaches.

With new insights in biological processes and development technologies, one should be convinced that this new challenge will eventually be met. This will require that all individuals involved in the chain of events, be they academics, basic scientists, educators, or industrialists, will find new ways to work together. Indeed, few will be able to have all knowledge and capabilities available in one platform. True networking is already today an absolute requirement for success. A further important partner in this new platform will be the regulators. Regulatory systems and requirements have been harmonized across most of the world in the last years, thanks to the efforts of ICH (International Conference on Harmonization) [27], but the basic content and development requirements of new pharmaceuticals have remained largely unchanged, or have increased significantly.

This whole process will need rethinking. The conventional phase I–II–III approach is no longer optimal and new, parallel processes are already required to meet the challenge of delivering better products to the patient. This will come also with a need for additional clinical studies following first approval to define appropriate use of the product in different patient segments. ADME Characterization, together with biomarker and diagnostic techniques, will make significant contributions also to this appropriate use in the phase of development.

ADMET Profiling is used now not only as the descriptive tool supporting safe human dosing, but also as a filter tool to select favorable compounds. This has moved ADMET research to the multidisciplinary area of drug discovery. When the drug selection process improves, 'better' candidates will enter clinical research, which in turn will become more successful and hence less expensive.

REFERENCES

[1] H. Kubinyi, *Nat. Rev. Drug Disc.* **2003**, *2*, 665.
[2] C. S. Goodman, A. C. Gelijns, *Baxter Health Policy Rev.* **1996**, *2*, 267.
[3] M. Sumner-Smith, *Curr. Opin. Drug Disc. Dev.* **2001**, *4*, 319.
[4] J. Drews, *Drug Disc. Today* **2003**, *8*, 411.
[5] J. A. Johnson, *Trends Genet.* **2003**, *19*, 660.
[6] S. Dry, S. McCarthy, T. Harris, *Nat. Struct. Mol. Biol.* **2000**, *7*, 946.
[7] Y. Avidor, N. J. Mabjeesh, H. Matzkin, *South Med. J.* **2003**, *96*, 1174.
[8] M. C. Anon, *Med. Device Technol.* **2000**, *11*, 33.
[9] P. M. Webber, *Nat. Rev. Drug Disc.* **2003**, *2*, 823.
[10] R. C. Renaud, H. Xuereb, *Nat. Rev. Drug Disc.* **2002**, *1*, 663.
[11] R. D. Miller, H. E. Frech, *Pharmacoeconomics* **2000**, *18 (Suppl. 1)*, 33.
[12] R. Levy, *Care Manag. J.* **2002**, *3*, 135.
[13] R. A. Ingram, *Oncologist* **2003**, *8*, 2.
[14] F. R. Lichtenberg, *J. Clin. Psychiat.* **2003**, *64 (Suppl. 17)*, 15.
[15] B. Booth, R. Zemmel, *Nat. Rev. Drug Disc.* **2003**, *2*, 838.

[16] T. Peakman, S. Franks, C. White, M. Beggs, *Drug Disc. Today* **2003**, *8*, 203.
[17] M. Pandey, *Drug Disc. Today* **2003**, *8*, 968.
[18] S. Mullner, *Adv. Biochem. Eng. Biotechnol.* **2003**, *83*, 1.
[19] H. Pearson, *Nature* **2003**, *424*, 990.
[20] D. Cavalla, *Drug Disc. Today* **2003**, *8*, 973.
[21] E. F. Petricoin, K. C. Zoon, E. C. Kohn, J. C. Barrett, L. A. Liotta, *Nat. Rev. Drug Disc.* **2002**, *1*, 683.
[22] J. A. DiMasi, R. W. Hansen, H. G. Grabowski, *J. Health Econ.* **2003**, *22*, 151.
[23] C. M. Cohen, *Nat. Rev. Drug Disc.* **2003**, *2*, 751.
[24] C. A. Lipinski, F. Lombardo, B. W. Dominy, P. J. Feeney, *Adv. Drug Delivery Rev.* **1997**, *23*, 3.
[25] W. A. Colburn, J. W. Lee, *Clin. Pharmacokinet.* **2003**, *42*, 997.
[26] H. van de Waterbeemd, E. Gifford, *Nat. Rev. Drug Disc.* **2003**, *2*, 192.
[27] www.ich.org.

Subject Index

(Compiled by *Stefanie D. Krämer* and *Maja Günthert*)

C

D

E

T

U

V

W